Southeast Asian Water Environment 2

Southeast Asian Water Environment 2

Selected papers from The Second International Symposium on Southeast Asian Water Environment, Hanoi, Vietnam, December 2004 and The Third International Symposium on Southeast Asian Water Environment, Bangkok, Thailand, December 2005

Edited by H Furumai, F Kurisu, H Katayama, H Satoh, S Ohgaki and NC Thanh

IWA Publishing

Published by IWA Publishing, Alliance House, 12 Caxton Street, London SW1H 0QS, UK

Telephone: +44 (0) 20 7654 5500; Fax: +44 (0) 20 7654 5555; Email: publications@iwap.co.uk
Web: www.iwapublishing.com

First published 2007

Printed by TJ International Ltd, Padstow, Cornwall, UK

British Library Cataloguing in Publication Data
A CIP catalogue record for this book is available from the British Library

Library of Congress Cataloging- in-Publication Data
A catalog record for this book is available from the Library of Congress

ISBN: 1-84339-124-4

ISBN13: 9781-84339-124-1

Contents

Preface

In Southeast Asian countries, rapid economic growth in recent years has brought about better living conditions and greater opportunities for global market integration. However, diverse activities of this growth trend have also caused damages of the natural resources and degradation of environmental quality. Evolving demographic conditions, industrialization and urbanization are among the important actors of these impacts. These forces in Southeast Asia are more prominent than in any other areas of the globe.

The water sector is an integral component of the global system, and thus change in the global environmental system will affect the water sector greatly. To a certain extent, the global system in turn, is likely to be affected by how the water resources are planned and managed. These days degradation of water environment has proportionately impacted the health and livelihood of the inhabitants in the region.

In order to meet the challenges imposed by these rapid global changes, existing water management systems must also be addressed intelligently. The water profession has seriously examined the implications of accelerating developments in society and industry which are likely to revolutionize future water availability and use patterns, and rethink about waste water management practices and current approach to water management.

Over the last few years, a number of research projects have been carried out in Southeast Asia to provide applied environmental technology, improve public awareness and help policy makers come up with sound environmental measures. To provide opportunity for the researchers and policy makers from both developed countries as Japan, Europe, the United States and from countries in the region, to exchange experience and research work related to water environmental protection and restoration, a symposium series was originated. The first International Symposium on Southeast Asian Water Environment was successfully organized in 2003 in Bangkok, Thailand and the second symposium held in Hanoi in 2004 also witnessed noticeable results.

Asian Institute of Technology Center in Vietnam (AITCV) was selected to host the second symposium jointly with the University of Tokyo (UT) and in cooperation with the Southeast Asian Center on Water Environment Technology (SACWET). The symposium attracted more than 200 researchers, policy makers, NGO workers, and graduate students in the fields of groundwater, water and wastewater management, water environment monitoring and management from all over the world.

The second symposium created great opportunities for North – South and South – South cooperation in water environment. It was an arena assembling many experts and scientists worldwide to share experiences and knowledge to one another in order to make our society better place to live, enjoy and work.

Following the success of the second symposium, the next symposium was planned and held again at the AIT conference center, Bangkok in 2005. For the third symposium, the following subjects were focused and highlighted; water environment conservation, groundwater issues, and pathogenic contamination and health-related microbiology. In addition, two special sessions were included with assistance by the two 21st Century COE Programs of the University of Tokyo (Center for Sustainable Urban Regeneration) and Yamanashi University (Research & Education on Integrated River Basin Management in Asian Monsoon Region). This cooperation made participants stimulate intensive and fruitful discussion.

The series of the symposium which attracts such a big number of participants is the testimony of the increasing concerns and interest in water environment not only in the region but also worldwide. As the host of the event we feel gratified to contribute our efforts in the organization of the symposium and to be associated with so many colleagues in the field. We appreciate the contribution of the researchers, policy makers and stakeholders from Japan, Europe and other developed countries who had joined us to share the impressive and invaluable pile of knowledge related to water issues in Southeast Asia. The most important articles presented at the first symposium on Southeast Asian Water Environment (Bangkok 2003) have been selected for the first volume of "Southeast Asian Water Environment", which was published from the IWA in December 2005.

Since the last two symposia were also informative and helpful, we have taken the initiative to record the selected papers collected during the deliberation in this book for the sake of information dissemination. Topics discussed in the symposia were technical and management related issues ranging from risk and vulnerability mapping, water quality audit and accounting, treatment efficiency, community engagement, watershed management to conservation policies.

We hope that this book, "Southeast Asian Water Environment 2", is helpful in assisting researchers, policy makers and environmental practitioners in NGOs and IOs of the different parts of the world to deal with water environment issues successfully.

古米弘明

Nguyen Cong Thanh
Chairman of the 2nd Symposium
Director,
AIT Center in Vietnam (AITCV)

Hiroaki Furumai
Co-chairman of the 3rd Symposium
Professor,
The University of Tokyo

Section 1

Sustainable Water Environment and Aquatic Resource Management

Least-Cost Policy Options for Nutrient Pollution Reduction in Taal Lake, Philippines

Arvin B. Vista[1)]

[1)] Department of Agricultural Economics, University of the Philippines Los Baños
College, Laguna 4031, Philippines
arvinvista@yahoo.com

Abstract

Stakeholders of the Taal Lake fish cage industry face the challenge of producing fish in a situation of frequent fish kill occurrences, which have been attributed to the deteriorating water quality due to nutrient pollution from fish cages. This study explores the least-cost policy options for nutrient pollution reduction in the lake. Fifty operators were surveyed and the production data elicited were used in the estimation of a Cobb-Douglas production function. Results show that the types of ownership arrangements in the municipalities of Agoncillo, Laurel, San Nicholas, and Talisay affect the production efficiency and nutrient loadings in the four cage locations. Marginal analysis reveals that operators are over-utilizing stocking density relative to feeding ration. At different percentage reductions in nutrient loadings, the cost-effectiveness of nutrient pollution reduction options differs among municipalities. With the current fish production technology and institutional set-up, nutrient pollution is inevitable.

Keyword Externalities, fish cage, least-cost policy, nutrient pollution, Taal Lake

1. Introduction

Recent fish kill occurrences in Taal Lake in the Province of Batangas, Philippines suggest a deteriorating water quality in cage areas. An excessive amount of nutrients accumulating from the cages led to nutrient enrichment, endangering the industry itself and the general lake ecosystem. The current management options for reducing the loss of nutrients into the surface water are largely limited to controlling the intensity of fish cage production (Naylor *et al.*, 2003 and Marte *et al.*, 1999).

The rapidly growing fisheries trade has had its impacts, both positive and negative. In the Philippines, Batangas Province was the top freshwater fish cage producer in 2000. It contributed a total of 15,268 *mt* of fish equivalent to 52 % of the total freshwater fish cage production in the Philippines (BAS, 2003). While aquaculture accounts for 30 percent of the world's food fish (Delgado *et al.*, 2003), it has significant environmental tradeoffs (Naylor *et al.* 2003). Between 1999 and 2001, six massive fish kill occurrences in Taal Lake were recorded.

Examining the past and present status of the lake is imperative for finding a sustainable solution to the problem. Many scientific studies have been conducted to describe the limnological, ecological, and biophysical characteristics of the lake and its watershed, and the socio-economic and legal considerations of cage culture (Herre, 1927; Hargrove, 1991; Zafaralla *et al.*, 1999; Rosana and Salisi, 2001). Various researchers have suggested different management techniques (Coche, 1980), yet the majority of the fish cage operators are not very receptive to changing their current practices.

This paper explored the tilapia cage production efficiency and the effects of ownership arrangement and cage location in minimizing inefficiencies, which is suspected to affect nutrient loading in Taal Lake. It also presented ranking of three 'waste reduction' and two 'waste management' options that can be explored in Taal Lake.

2. Research Methodology

A fish cage operator survey was conducted in the municipalities of Agoncillo, Laurel, San Nicholas and Talisay since the cages are limited to the barangays of these municipalities. The survey process involved two stages. The first stage, community familiarization and field reconnaissance, acquainted the researcher with the current practices of fish cage farming in the area. The second stage, the actual survey, involved interviews of 50 randomly selected fish cage operators with a varying number of fish cages and types of ownership. A total of 316 cages comprised the survey. Aside from the data gathered through the survey questionnaire, secondary data were obtained from various sources. The Statistical Package for the Social Sciences software was used in the analysis of the data gathered.

The physical relationship between inputs and output in the fish cage production was estimated following the Cobb-Douglas (C-D) production function[1], defined as:

$Y = f(X_F, X_{SD}, X_{D1}, X_{D2}, X_{D3}, X_{D4})$

where Y = yield, kg per cage per cropping cycle;

X_F = total feeding ration, kg per cage per cropping cycle;

X_{SD} = stocking density, number of fingerlings per cage per cropping cycle;

X_{D1} = dummy variable for ownership arrangement, = 1 if owner is the operator or = 0 if owner hire a caretaker.

X_{D2} = dummy variable for location, = 1, if cage is located in Talisay or = 0, otherwise;

X_{D3} = dummy variable for location, = 1, if cage is located in Agoncillo or = 0, otherwise; and

X_{D4} = dummy variable for location, = 1, if cage is located in Laurel or = 0, otherwise.

Data were transformed into their logarithmic values. The production function was estimated employing the ordinary least square method using *Stata* version 8 software. The *C-D* log linear model[2] is written as:

$$Y = AX_F^{\beta_1} X_{SD}^{\beta_2} X_{D_1}^{\beta_3} X_{D_2}^{\beta_4} X_{D_3}^{\beta_5} X_{D_4}^{\beta_6} \quad \text{or}$$

$$\log Y = \log A + \beta_1 \log X_F + \beta_2 \log X_{SD} + \beta_3 X_{D_1} + \beta_4 X_{D_2} + \beta_5 X_{D_3} + \beta_6 X_{D_4}$$

To test whether multicollinearity presented problems, simple correlation among the independent variables was examined. To test homoskedascity, a plot of the residuals (the

---

[1] One limitation of the *C-D* is that it does not have stage III of the production function.

[2] Y=output; X_i=inputs; β_i = factor productivities; A = constant. Error terms are omitted.

difference between the observed Y and the estimated Y) against the independent variables was also made to look for systematic distribution of the deviations around the regression line. The results show no evidence of multicollinearity and heteroscedasticity. To determine the profit maximizing level of production, the marginal product of X_i was calculated:

$$\left[MP_{X_i} = \frac{\partial Y}{\partial X_i} = \frac{\beta_i}{X_i}(Y) \right]$$

The optimum value of X_i was derived by equating MP_{Xi} to the X_i input-output price ratio. The optimal yield was derived from the optimal input levels obtained. Numerical solutions were derived through optimization of the variables X_F and X_{SD} using Microsoft Excel Solver.

A least-cost solution to nutrient enrichment in Taal Lake was based on the comparison of the baseline cost with pollution reduction costs associated with the option(s) selected. The decision variables include nitrogen *(N)* and phosphorus *(P)* level or loading, and the rule used was to choose the option with the lowest cost of attaining the objective and the one with the highest net revenue. First, baseline waste generation was measured. Survey data served as input into the estimation of a production function and the optimal values of inputs and output. Second, pollution reduction options were identified and the costs and benefits of each option were estimated. Third, the costs of each option were compared with the baseline data. Finally, sensitivity analyses were conducted, which indicates how measures of each option's worth might change under different assumptions of percent reduction in nutrient pollution per municipality.

3. The Taal Lake Environment and the Fish Cage Industry

Taal Lake is located in the province of Batangas and lies approximately 60 km south of Manila, Philippines. It is a major tourist attraction and is one of the 10 priority freshwater wetlands in the Philippines. Its watershed has 38 tributary rivers draining into the lake. The only outlet, to Balayan Bay, is the Pansipit River. The lake covers an aggregate area of 24,236 ha, excluding the islands, and has an average depth of 60.1 m. It has a circumference of 120 km and is bounded by nine municipalities and two cities. Cages are mostly concentrated in the fish sanctuary area between Barangay Manalao, Agoncillo and Barangay Gulod, Laurel, although regulation restricts construction of cages in this area. The municipal regulations for cage culture and lake utilization differ among the four municipalities (e.g. different rate of fees when securing an operator's permit).

Two ownership arrangements exist, namely 'owner as operator' and 'owner hires a caretaker. Cage operators (classified either as owner/financiers or caretakers), open water fishermen, middlemen, and the different feed producers/ suppliers/ retailers are the key actors in the industry. Most of the financiers are non-residents, who hire labor to work as caretakers of their cages and significantly influence decision-making. Due to the lack of financial capital, most local residents end up as caretakers for an absentee owner. The legal right to own at most five cages is vested in the local residents only.

Fish cage culture has been practiced for the past two decades, and the present trend is towards increased stocking density and intensive feeding. Each fish cage measures an average of 10 x 10 x 6 m with a stocking of 65 -100 tilapia fingerlings per m^3. In 2001, Agoncillo had a total number of 2,362 registered cages, Laurel had 2,729, San Nicholas had 481, and Talisay had 1,861. The fish are heavily fed with different types of commercial feeds, which comprise the highest proportion of production expenses (76 %). The culture period spans from four to seven months, which depends on the intensity of feeding and the quality of fingerlings. Many operators observed extended periods of intensive tilapia culture with an additional 1-3 months of culture period compared to 1990s situation. At present, the species cultured in cages are tilapia (*Oreochromis niloticus*), bangus (*Chanos chanos*), and maliputo (*Caranx ignobilis*). Among the three species, tilapia is the most commonly cultured.

Production from the open water fisheries dwindled after fish cage culture flourished in the lake. The lower productivity of endemic fish species may be attributed to the disturbance and displacement of their spawning ground brought about by tilapia dominance and fish cage wastes. Tilapia is an exotic and prolific breeder, and hence, a competitor of native fish species. Fish cage production plunged to a lower level starting 1998 due to massive fish kill occurrences, which were attributed to oxygen depletion during overturn and toxic poisoning from the suspected pollutants (NH_3, NO_2 and H_2S) (Rosana and Salisi, 2001).

In Taal Lake watershed, excess nutrients are supplied to the system through point and non-point sources. Table 1 presents the estimated nutrient pollution loading from various sources in 2001. The fish cages mainly contribute to the point source pollution, which are largely supplied by the discharges of nitrogen and phosphorus derived from uneaten feed, feces, and excretion via the gills and urine (Beveridge, 1993; Kibria, 1996; 1998). Beveridge (1993) estimated about 76.56 % of N put into use is being lost into the cage and its surrounding waters, while Phillips *et al.* (1994) estimated about 85 % for P. Consequently, for every 1000 kg tilapia harvested in 2001, an estimated 47 kg N and 9 kg P are being lost. This situation bears witness to high nutrient loading in the lake. In Taal Lake, *N* more than *P* seems to limit primary production (Zafaralla, 1999). This observation supports the findings in other freshwater lakes, where *P* has been demonstrated to be more critical in regulating water quality (Howarth, 2000; Zafaralla, 1999).

Table 1. Estimated nutrient input loading rates from various sources, 2001.

Source of Nutrients	Total N		Total P	
	(mt)	%	(mt)	%
Fish cages	3,758	76	816	82
Household sewage[3]	969	20	146	15
Agriculture/watershed[4]	239	5	35	4
Total	4,966		997	

4. Key Results and Discussion: Comparison of Least Cost Policy Options for Nutrient Pollution Reduction

The average feed conversion ratio (FCR) was 2.78 +/- 0.34 and ranges from 2.32–3.37. An FCR of 2.78 suggests a requirement of 2.78 kg of feed to produce a gain of 1 kg fish weight.

[3] Adopted from Jacinto *et al.* (1998), assuming a 1.96 kg N per person per yr and 0.29 kg P per person per yr loss.

[4] Adopted from Clemente and Wilson (2002), assuming a 6.48 kg N per ha per yr and 0.96 kg P per ha per yr loss and a 10% filtration of nutrients by the watershed vegetations.

Providing more feed to fish beyond their growth requirements results in a higher FCR, increased costs, and nutrient losses. Net revenue of cage operators among the four municipalities and between the two ownership arrangements was not statistically different from each other. However, it should be noted that the feeding ration and practices, and so feed costs, was statistically different among the operators in the four municipalities and under the two ownership arrangements. This observation has significant implications to the amount of nutrient loadings, and so, with the existence of negative existence loop in the fish cage production.

The estimated production coefficients and related statistics are shown in Table 2. The net stocking density, ownership arrangement and location with reference to Laurel were found to be statistically significant at 5 % level. The production coefficients for feeding ration (B_1) and net stocking density (B_2) have the expected positive sign while the ownership arrangement (B_3) and location (B_6) *were* negative. The negative coefficient of ownership arrangement (-0.33) and location (-0.29) means an inverse relationship with tilapia yield amounting to, all else equal, about a 7-8% reduction in yield. In the case of ownership arrangement, tilapia yield will decrease in 'owner-operated' fish cages. In the case of location, Laurel operators are producing significantly less tilapia per cage per cropping cycle than San Nicholas operators. The coefficients of the explanatory variables X_{SDa} and X_D were significantly different from zero at 10 % level based on t-test[5]. The coefficient of the explanatory variable (X_F) was only significantly different from zero at 12 % level, so care is needed in interpreting this result. With regard to the overall regression, the computed F value (6.60) was greater than the critical value (2.09) at the 10 % level of significance. It can be said that the overall regression was statistically significant. The production elasticities were 0.20 and 0.22, respectively. An elasticity of production that is less than one indicates diminishing marginal product. A 1% increase in feeding ration will produce a 0.20 % increase in yield (significant only at the 12% level), while a 1% increase in the net stocking density would produce a 0.22 % increase in yield. Since the sum of the production coefficients was < 1, a decreasing return to scale existed; a doubling of X_F and X_{SD} inputs would less than double yield.

Table 2. Estimated Cobb-Douglas production function coefficients and related statistics for a sample of 50 operators in Taal Lake, Philippines, 2002.

Dependent variable: Log Y [Yield (kg) per cage per cropping]

Variables (Log)	Parameter	Coefficient	Standard Error	P value
Total feeding ration (X_F)	β_1	0.20	0.12	0.12 [a]
Net stocking density (X_{SDa}) [6]	β_2	0.22	0.09	0.03 [b]
Ownership arrangement (X_{D1})	β_3	-0.33	0.13	0.01 [b]
Location, i.e. Laurel (X_{D4})	β_6	-0.29	0.13	0.04 [b]
Intercept	β_0	3.99	1.33	0.00 [b]

R^2: 0.37; F statistics: 6.6

[a] Significant only at 12 %; [b] Significant at 5 % level. H_o: $B_i = 0$.

[5] H_o: $B_i = 0$. T-statistics is computed as B_i/se.

[6] The net stocking density (SDa), derived as stocking density at stocking (SD) less fingerling mortality at stocking (SD_m) was used instead of the actual SD to account for the uncertainty and risk of getting zero output.

The *MP* of stocking density and feeding ration were less than the price ratios (0.08 < 0.30; 0.01 < 0.13); hence, increasing these inputs is discouraged. To be efficient, the MP of stocking density and feeding ration should be equal to the price ratios of fingerlings stocked and feeds used. The estimated optimum-net stocking density was 5,495 fingerlings per cage per cropping cycle (10 fish per m^3) while the estimated optimum-feeding ration was 2,606 kg per cage per cropping cycle (equivalent to and FCR of 1.54). The estimated optimal yield was 1,689 kg per cage per cropping cycle. It should be noted that even at the optimal level, nutrient loss is inevitable given the current production technology.

Waste reduction or minimization reduces the release of nutrients into the surface water. For operators and LGU officials interviewed, the reduction in the total number of cages (*Option* Z_1) is only a short-term solution; hence, this option may not be fully supported by the operators in the long run. Many local residents are dependent upon the cage culture business as their primary source of family income. This strategy has more appeal with the more informed local government officials rather than the operators. The reduction in feeding ration and stocking density (*Options* Z_2 *and* Z_3) are cost-saving strategies, which are assumed to improve yield and increase net revenue. Reducing the amount of feed employed results in less nutrient loss, better water quality, and a possible increase in output. Based on ranking of net revenue per cage per year (Table 3), the reduction in feeding ration (*Option* Z_2) is suggested for the municipalities of Laurel, San Nicholas, and Talisay (except at 15 % reduction in nutrient loading). For the municipality of Agoncillo, reduction in stocking density (i.e. with proportional reduction in feeding ration, *Option* Z_3) is recommended. *Option* Z_2 is advised for Talisay at 5 % and 10 % reduction in nutrient loading and *Option* $Z_{3\ at}$ 15 % reduction in nutrient loading. Based on cost or incentive per year (Table 4), *Option* Z_2 is recommended in all municipalities, expect for Agoncillo at 5 % reduction in nutrient loading and for Talisay at 15 % reduction in nutrient loading. It should be noted that these results implies inefficiency, so caution is warranted for *Options* Z_2 and Z_3.

Table 3. Ranking of *Options* Z_1, Z_2 and Z_3 at 5 %, 10 %, and 15 % reduction in nutrient loadings in Agoncillo, Laurel, San Nicholas, and Talisay, Philippines based on net revenue per cage per year.

Municipality	5 % reduction in nutrient loading			10 % reduction in nutrient loading			15 % reduction in nutrient loading		
	Z_1	Z_2	Z_3	Z_1	Z_2	Z_3	Z_1	Z_2	Z_3
Agoncillo	2.5	2.5	1	2.5	2.5	1	2.5	2.5	1
Laurel	2	1	3	2	1	3	2	1	3
San Nicholas	2	1	3	2	1	3	2	1	3
Talisay	2	1	3	3	1	2	3	2	1

Note: Z_1 - Reduction in the number of cages; Z_2 - Reduction in feeding ration; Z_3 - Reduction in stocking density

Improved waste management entails additional costs to reduce the impacts of nutrient pollution in Taal Lake fish cage industry. Single-line positioning of cages and the use of aerator (Options Z_4 and Z_5) do not reduce the nutrient loadings but rather mitigate harm to fish production. In the implementation of single-line positioning of cages (*Option* Z_4), individual operators will incur additional cost of at least PhP 1,430 per cage per year. This option could result in a better water exchange inside and outside the surface area in cages. Properly spaced

Table 4. Ranking of *Options* Z_1, Z_2 and Z_3 at 5 %, 10 %, and 15 % reduction in nutrient loadings in Agoncillo, Laurel, San Nicholas, and Talisay, Philippines based on cost or incentive per year.

Municipality	5 % reduction in nutrient loading			10 % reduction in nutrient loading			15 % reduction in nutrient loading		
	Z_1	Z_2	Z_3	Z_1	Z_2	Z_3	Z_1	Z_2	Z_3
Agoncillo	1	3	2	3	1	2	3	1	2
Laurel	3	1	2	3	1	2	3	1	2
San Nicholas	3	1	2	3	1	2	3	1	2
Talisay	3	1	2	3	1	2	3	2	1

Note: Z_1 - Reduction in the number of cages; Z_2 - Reduction in feeding ration; Z_3 - Reduction in stocking density

cages could make nutrient concentration more diluted. On the other hand, an estimate of PhP 176,256 per year for each municipal government is needed in implementing *Option* Z_4. Thesuccess of this strategy lies with the participation of all operators in Taal Lake and strong 'political will' of the LGU officials. An aerator (*Option* Z_5) can be employed during critical condition when the dissolved oxygen level dip below the acceptable level (5 mg/L). An aerator costs about PhP 30,000 per unit and is replaced after nine years with a salvage value of 20 % of the initial cost. An estimate of PhP 6,000 per year on operating and maintenance costs is needed. Given the high cost of this strategy, many operators are hesitant in employing this strategy. They believe that it is only applicable in ponds and not in a deep lake. Agoncillo and Laurel are expected to have the most localized effects of nutrient pollution. The use of fish medicine or anti-stress medication by operators in these two municipalities supports such claim. In this regard, timing of stocking is also critical for the survival of fingerlings in cages. Since the transition of monsoon winds from northeast to southwest (May and October) the temperature is higher than normal range and fish kill is most likely to occur, a cropping cycle between September and April maybe the best period to culture fish in Taal Lake.

5. Conclusion

Indeed, the current structure of right ownership and the inherent nature of the fish cage production technology influences the water quality in Taal Lake. There is incompatibility in the lake's resource uses since operators rely on water both as an input into the production and as a place to put waste. The current intensity and nature of cage production is not coherent with the standard set for the allowable cage areas. Changes in the intensity of production activities are central to the suggestions elicited through the survey. A change in the municipal regulations may change the unreceptive behavior of many operators towards changing to sustainable production practices. With the congestion of cages in San Nicholas, Laurel, and Agoncillo areas, it may be possible that more owners will exit or transfer production in the Talisay area. Moving away from the highly concentrated cage areas is promising for expanding the cage production. The role of caretakers is essential in reducing the inefficiencies in the use of inputs, specifically feed.

The open-access problem is predictable with the inherent characteristics of the Taal Lake ecosystem. Externalities, such as nutrient pollution, are inevitable with the current fish production technology. The choice left is to accept a certain level of pollution as a trade-off for the benefits gained from use of the lake's resources and services. Even at the optimal level of inputs and output, nutrient losses are certain. The enforcement of property rights is not sufficient to attain cage production efficiencies. As a result, exclusion through prices and transaction costs may be necessary. Finding a better cage culture technology that minimizes the production and transaction costs is promising and practical. However, a cheaper production process may not be that appealing to the current dominant operators who will lose their incentives once the least-cost technology succeeds. Of great importance also is the need to interpret and modify existing laws and regulations, determining who pays for fixed costs and whose preferences count? Developments in the institutional structures warrant future technological changes in the fish cage production system.

6. References

BAS, 2003. *Fisheries Statistics of the Philippines, 1997-2002,* Bureau of Agricultural Statistics, Department of Agriculture, Available: http://bas.gov.ph, Retrieved March 12, 2003.

Beveridge, M. M. and M. J. Phillips. 1993. *Environmental Impact of Tropical Inland Aquaculture,* Paper presented at the Environment and Aquaculture in Developing Countries: ICLARM Conference Proceedings 31.

Clemente, R. S. and E. N. Wilson. 2000. *Hydrologic and Physico-chemical Modelling of the watersheds draining into Lingayen Gulf,* Quezon City: Environmental Science Program, College of Science, University of the Philippines Diliman.

Coche, A. G. 1980. *Cage Culture of Tilapias.* Paper presented at the International Conference on the Biology and Culture of Tilapias, Study and Conference Center of the Rockefeller Foundation, Bellagio, Italy.

Delgado, C.L., N. Wada, M.W. Rosegrant, S. Meijer, and M. Ahmed. 2003. *Fish to 2020: Supply and Demand in Changing Global Markets*: International Food Policy Research Institute and WorldFish Center.

Hargrove, T.R. 1991. *The Mysteries of Taal: A Philippine volcano and lake, her sea life and lost towns,* Makati, Metro Manila: Bookmark.

Herre, A.W.C. 1927. The Fisheries of Lake Taal (Bombon), Luzon, *The Philippine Journal of Science, 34*(3).

Howarth, R., *et al.* 2000. Nutrient Pollution of Coastal Rivers, Bays, and Seas, *Issues in Ecology (7).*

Jacinto, G.S., *et al.* 1998. *N and P Budget of Manila Bay, Philippines,* Quezon City: Marine Science Institute, University of the Philippines.

Kibria, G., *et al.* 1998. Can nitrogen pollution from aquaculture be reduced? *NAGA, The ICLARM Quarterly, 21,* pp. 17-25.

Kibria, G., *et al.* 1996. Aspects of Phosphorus Pollution from Aquaculture. *NAGA, The ICLARM Quarterly, 19,* pp. 20-24.

Marte, C.L. *et al.* 1999. *Recent Developments in Freshwater and Marine Cage Aquaculture in the Philippines.* Paper presented at the First International Symposium on Cage Aquaculture in Asia, Marine Laboratory, Fisheries Research Institute, Tungkang, Thailand.

Mutia, M.T.M. and L.S. Magistrado. 1999, *Status of Open Fisheries and Aquaculture Productivity in Taal Lake, Unpublished Report.* Taal, Batangas, Bureau of Fisheries and Aquatic Resources.

Naylor, R.L., *et al.* 2003. Effects of Aquaculture on World Fish Supplies. *Issues in Ecology* (8).

Phillips, G.R. *et al.* 1994. The importance of sediment phosphorus release in the restoration of very shallow lakes (The Norfolk Broads, England) and implications for biomanipulation. *Hydrobiologia, 275/276*, 445-456.

Rosana, M.R. and N.C. Salisi. 2001. *Fish Kill Investigation in Taal Lake,*Tanuan, Batangas: Unpublished Report. BFAR-Inland Fisheries and Aquatic Resources.

Zafaralla, Macrina. T., *et al.* 1999. *Taal Lake: Limnological Characterization and Sources of Ecological Perturbations. Unpublished Report,* IBS, CAS, University of the Philippines Los Baños.

Landfill Development Causes Changes in Distribution of Flood-prone Areas: a Case Study in Laguna Plain, Metro Manila, The Philippines

Yuji Hara[1)], Shigeko Haruyama[2)], Satoru Okubo[3)] and Kazuhiko Takeuchi[4)]

1) Department of Ecosystem Studies, The University of Tokyo
1-1-1 Yayoi, Bunkyo-ku, Tokyo 113-8657, Japan
e-mail: antares@mail.ecc.u-tokyo.ac.jp
2) Department of Environmental Studies, The University of Tokyo
7-3-1 Hongo, Bunkyo-ku, Tokyo 113-0033, Japan
e-mail: haruyama@k.u-tokyo.ac.jp
3) Department of Ecosystem Studies, The University of Tokyo
1-1-1 Yayoi, Bunkyo-ku, Tokyo 113-8657, Japan
e-mail: aokubos@mail.ecc.u-tokyo.ac.jp
4) Department of Ecosystem Studies, The University of Tokyo
1-1-1 Yayoi, Bunkyo-ku, Tokyo 113-8657, Japan
e-mail: atake@mail.ecc.u-tokyo.ac.jp

Abstract

Urban development is rapidly expanding into coastal lowlands in Southeast Asian cities, where flooding has become a severe problem. Urban areas are typically built on landfill, displacing historical floodwater-retention areas. To effectively manage lowlands, we must understand the effect of landfilling on flooding distribution. We examined natural landforms, land-use changes derived from landfilling, and changes in flooding distribution in Laguna Plain, Metro Manila, the Philippines. Comparisons among these yielded the following findings. Before the 1960s, human settlements were situated mainly on natural levees and upper lacustrine terraces, and flooding also followed natural landforms. In the 1980s, built-up areas using landfill expanded into lower lacustrine terraces with lower natural elevations than the natural levees and upper lacustrine terraces. Since the 1990s, urban development using landfill has expanded into low-lying lacustrine plain and lacustrine delta landforms. Because the lower natural landform has been changed by landfilling, flooding distribution is shifting to higher natural landforms. The estimated total amount of landfill in our study area to date is 4.4×10^6 m^3, equivalent to 8.8 km^2 of new inundation area with an average depth of 50 cm. We recommend that landfilling for development should consider natural landforms to minimize the risk of flooding.

Keywords: floods, GIS, landfill, landform classification, land-use change, Manila

1. Introduction

In Southeast Asian countries, human populations are primarily concentrated in cities, and several mega-cities are emerging (McGee, 1991). Most of these mega-cities are located on alluvial plain or river delta lowlands. In these lowlands, human settlements were originally situated on relatively high-level micro-landforms, which acted as natural levees, and back marsh was planted with broad rice fields. Rainy season floodwater was stored in paddy fields, and people used waterborne modes of transportation (Tanabe, 1994). Naturally occurring floods were not considered hazardous. These conditions continued until the middle of the 20th century (Takaya, 1987), when many such areas were urbanized.

After urbanization of these lowlands, natural flooding became a severe problem (Yeung, 2001). Urban development in back marsh areas is risky, because these are natural flood zones. Further, the risk of flooding is increased by the development practice of landfilling, which removes natural floodwater-retention areas. Whenever lowland development is planned, it is critical to understand the natural topography and to identify flood-prone areas by mapping (Skilodimou et al., 2003). Basic information for lowlands management, such as micro-landform and historical flood records, is insufficient in Southeast Asia; therefore, flood-prone areas are not well identified. Moreover, the effect of landfilling on flooding has not yet been examined. Accurate maps of flood-prone areas are needed to reduce the risks inherent in developing these areas.

Our study examined natural landforms, land-use changes caused by landfilling, and changes in flooding distribution, and the correlations among them, in Laguna Plain, Metro Manila, the Philippines, which contains flood-prone lowlands. Our study was based upon a field survey and GIS analysis.

2. Study area

Metro Manila is located on the west coast of the middle part of Luzon Island; its tropical monsoon climate has distinct dry and rainy seasons. Geomorphologically, Metro Manila is divided into three parts: the West Coastal Plain, the Central Upland, and the Laguna Plain (Figure 1). The Laguna Plain (our study area) is bordered by the Central Upland to the west, by the Marikina Valley to the north, and by Laguna de Bay to the south. The elevation of most areas is less than a few meters above sea level.

Several large floods have occurred within the last 32 years, notably in 1972, 1976, 1986, and 2000. The largest flood was in 1972, when the water level of Laguna de Bay in August rose 2.3 m above the average level (Figure 2).

Figure 1. Study area (inside gray line)

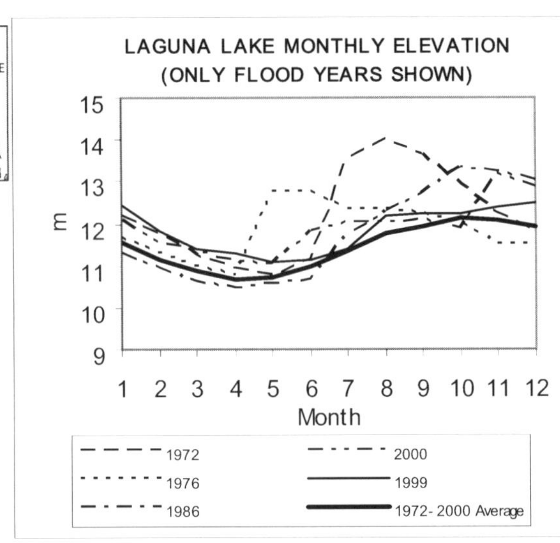

Figure 2. Lake level of Laguna de Bay (modification of *Laguna Lake Development Authority*)

Urbanization of Metro Manila began in the 1950s, and its population has increased rapidly since the 1960s (Murakami et al., 2004). The city core area on the west coast is now extremely populous, and urbanization is increasing on the other side of the Central Upland and particularly on the Laguna Plain, where agricultural land uses are becoming mixed with urban land uses (Sierra et al., 1990).

3. Methods

After conducting the field survey in November 2001, we constructed digital maps for landform classifications, flooding distribution, and land use, using the GIS software TNTmips version 6.5 (MicroImages Inc., Lincoln, NE, USA) and ArcView version 3.2 (ESRI, Redlands, CA, USA). Data sources are shown in Table 1.

We constructed the landform classification map at 1:10 000 scale using field measurements mentioned below and aerial photographs.

Table 1. Data sources

Landform	Field measurement
	Aerial photographs taken in 1982 at 1:32 000
	Topographical maps in 1966 at 1:10 000
	Land condition maps in 1989 at 1:10 000
	Geologic columns (CTI Engineering International Co., Ltd., and Kanto Gakuin University)
	Subsurface geological data (Matsuda et al., 1998)
Flooding	Field survey
	Satellite image; Landsat TM 1989/01/25
	Satellite image; SPOT 1987/11/18
	Satellite image; SPOT 1997/10/26
Land use	Field survey
	Land-use map in 1966 (paper; obtained in mapping museum)
	Land-use map in 1986 (digitized by University of the Philippines)
	Aerial photographs taken in 1996 at 1:12 500

We constructed the flooding distribution maps using satellite images, which are useful when historical records of flooding distribution are unavailable (Yamagata et al., 1988). We identified flooded areas (water surfaces with depths more than approximately 1 m) by using Landsat TM and SPOT images. A Landsat TM image, rather than the landform classification map, was used to study flooding distribution in the entire study area, because the image was taken when the lake level was 0.6 m higher than normal. Two SPOT images, rather than the land-use maps, were used to study the changes in flooding distribution in the southern part of our study area because these two images showed similar lake levels and had similar rainy season dates, which corresponded to the periods of the two land-use maps. Therefore, it was appropriate to compare these images and land-use maps to study changes in flooding distribution resulting from landfilling.

We constructed three digital land-use maps, representing different periods, for the southern part of our study area, which has the lowest elevation and has recently undergone rapid development using landfill. We reclassified land uses into four categories: built-up area, water surface, open space, and paddy field. Although we were not able to distinguish open spaces from paddy fields in the 1966 land-use map, according to local residents we assume that most open spaces at that time were paddy fields.

We conducted ten transect measurements in the field survey to evaluate landfill heights, and interviewed local residents to verify the high-water levels of large floods.

Finally, we used overlay analysis to compare these maps and clarify their relationships. We also evaluated the volume of landfill in the study area by incorporating the field measurements of landfill heights into our GIS analysis.

4. Results and discussion

The landform classification map is shown in Figure 3. The Laguna Plain is partially bordered by upland and hilly areas. Several alluvial cones are evident on the edge of the western upland, and a series of alluvial fans is seen on the edge of the eastern hill. Three geomorphological landforms are evident in the lowlands: lacustrine plain, with elevations less than 2 m above sea level; lower lacustrine terrace, with elevations between 2 m and 4 m above sea level; and upper lacustrine plain, with elevations between 4 m and 7 m above sea level. The lacustrine plain now floods frequently during the rainy season, and has developed an emergent plant zone along the shoreline. The lower lacustrine terraces had already emerged by the Pleistocene; however, the areas along present and former river courses were damaged by large floods. The upper lacustrine terraces are approximately bounded by the Marikina River floodplain; except for small hollows with poor drainage, submergence has not occurred because of its earlier emergent period. The Mangahan Floodway runs diagonally across the center of the map. West of the floodway, the Pasig and Taguig Rivers meander on the lacustrine plain with natural levees along their water courses. The northern part of the map shows the Marikina River valley, where river terraces have formed and sedimentation is dominant toward the downstream area.

Figure 4 shows the flood distribution map derived from the Landsat TM image. When this map is superimposed on the landform classification map (Figure 3), flooded areas overlap lacustrine delta, lacustrine plain, and lower lacustrine terrace landforms. Overflows from the Mangahan Floodway are visible along former river courses.

Table 2. Flooding ratio by each landform unit

Landform	*Flooding ratio*
ALLUVIAL CONE	0%
ALLUVIAL FAN	0.3%
LACUSTRINE DELTA	49.2%
LACUSTRINE PLAIN	19.7%
LOWER LACUSTRINE TERRACE	14.2%
UPPER LACUSTRINE TERRACE	3.3%
EMERGENT PLANT ZONE	36.1%
SAND BANK	56.9%
LOWER RIVER TERRACE	0.7%
UPPER RIVER TERRACE	0.6%
FORMER RIVER COURSE	15.4%
NATURAL LEVEE	12.6%
BACK MARSH	1.8%
VALLEY PLAIN	11.7%
HILL	0.9%

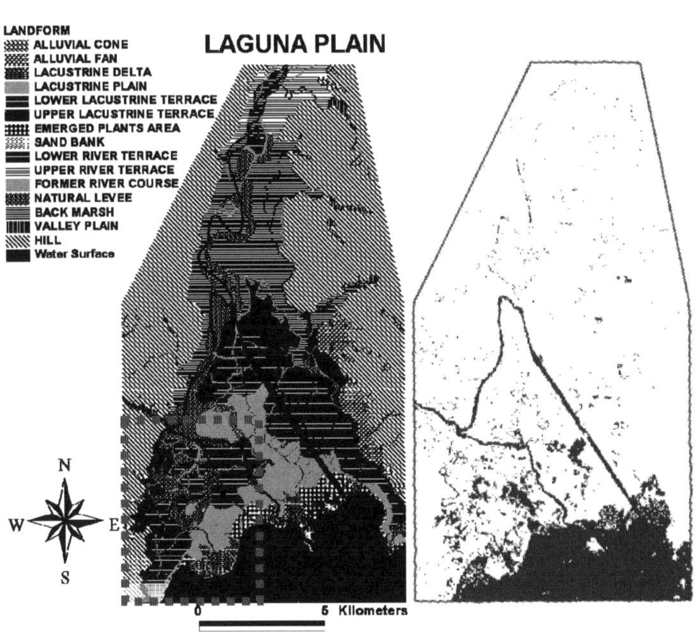

Figure 3. Landform classification map (left)
Figure 4. Flood estimation map by Landsat TM (right)

Flood-prone areas correspond well to natural landforms in the whole study area, as shown in Table 2, which combines data from Figures 3 and 4.

Land-use maps are shown in Figure 5, and detailed flooding distribution maps derived from SPOT images are shown in Figure 6. The landform classification map outlined in Figure 3 has been incorporated into Figure 6 to facilitate comparison. As shown in Figure 5, in 1966, human settlements were situated only on natural levees and upper lacustrine terraces. By 1986, lower lacustrine terraces had been developed, and by 1996, lacustrine plain and lacustrine delta landforms were also developed. Built-up areas increased from 9% of this area in 1966 to 17% in 1986 and 33% in 1996, showing rapid transition from paddy fields to urban

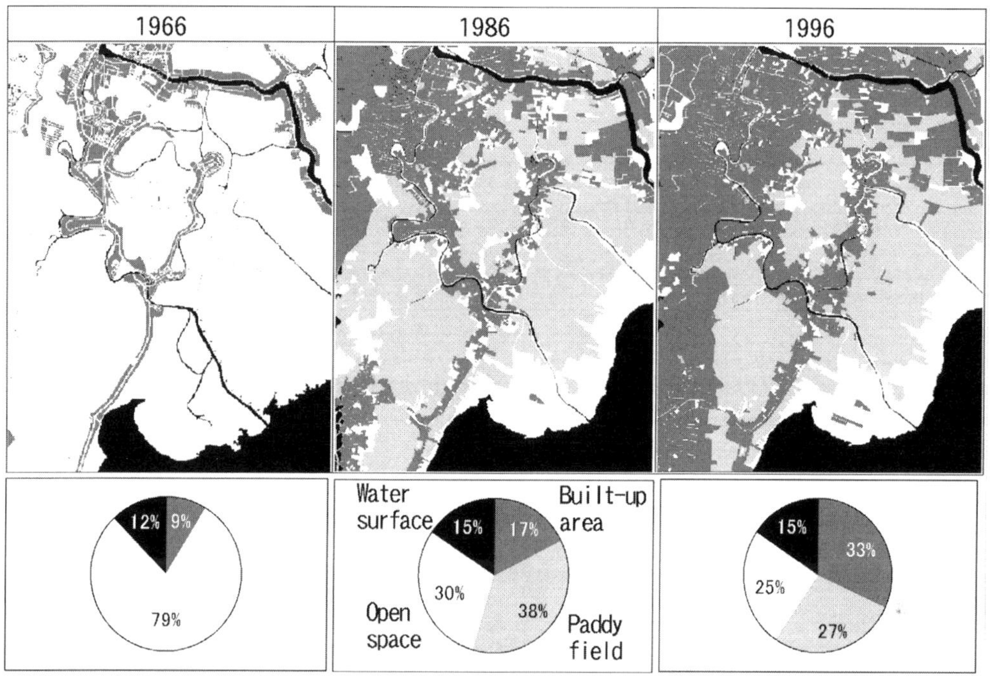

Figure 5. Land-use maps

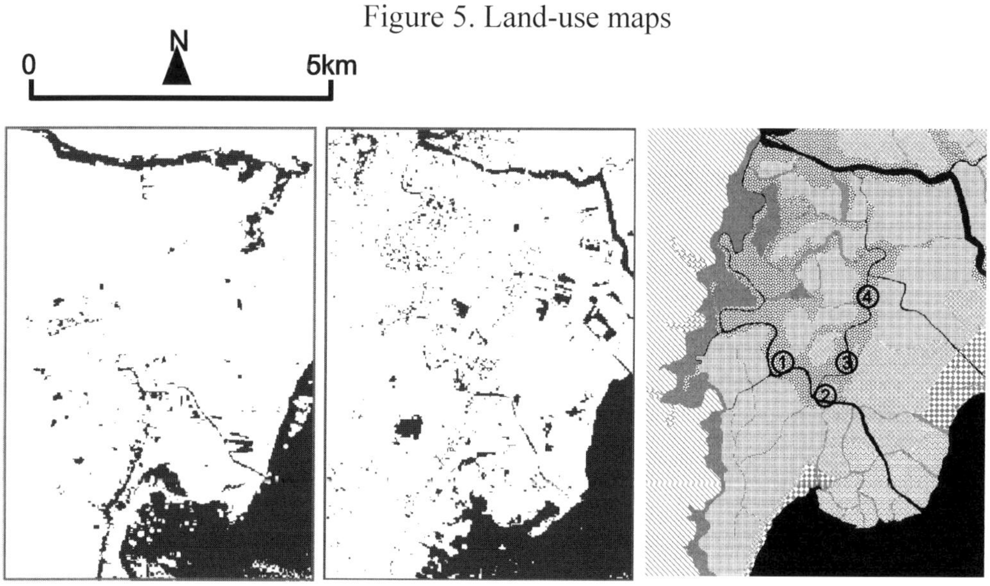

Figure 6. Detailed flood estimation maps by SPOT (left: 1987; middle: 1997; right: landform classification map)

land uses. Figure 7 shows selected field transect measurements of landfill heights. The figure compares original land surfaces with present land surfaces, which were created by landfilling. The landfill heights depended on the natural micro-landforms.

Table 3. Landfill volume

Landform	*Landfill height (m)*	*Development area (m^2)*		*Landfill volume (m^3)*	
		1966-86	*1986-96*	*1966-86*	*1986-96*
ALLUVIAL CONE	0	0	153,641	0	0
LACUSTRINE DELTA	2.0	14,028	29,596	28,057	59,192
LACUSTRINE PLAIN	1.5	126,200	446,436	189,299	669,654
LOWER LACUSTRINE TERRACE	1.0	505,319	1,227,556	505,319	1,227,556
UPPER LACUSTRINE TERRACE	0.5	355,729	1,115,080	177,865	557,540
EMERGENT PLANT ZONE	2.0	66	1,000	133	2,000
FORMER RIVER COURSE	1.0	122,035	54,764	122,035	54,764
NATURAL LEVEE	0.5	1,114,678	518,696	557,339	259,348
HILL	0	878,882	1,024,140	0	0

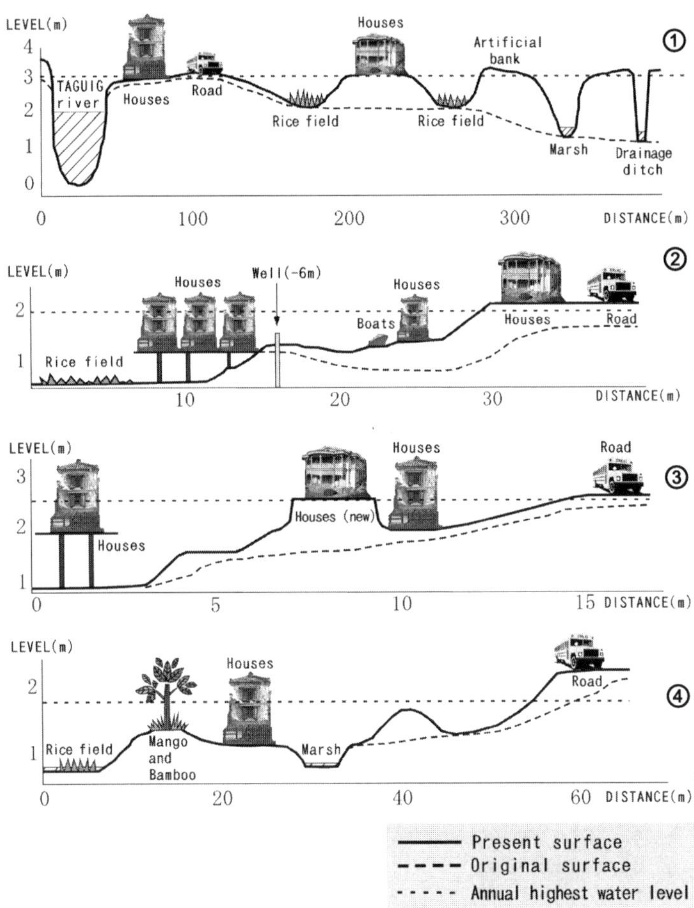

Figure 7. Typical transects (numbers are shown in Figure 6)

Table 4. Flooding distribution changes

	Unit	*Flooding ratio in 1987*	*Flooding ratio in 1997*
Landform	ALLUVIAL CONE	0%	0.4%
	LACUSTRINE DELTA	12.0%	8.2%
	LACUSTRINE PLAIN	7.9%	3.5%
	LOWER LACUSTRINE TERRACE	7.4%	6.7%
	UPPER LACUSTRINE TERRACE	1.4%	3.9%
	EMERGENT PLANT ZONE	7.1%	18.3%
	FORMER RIVER COURSE	12.9%	11.4%
	NATURAL LEVEE	11.3%	10.8%
	HILL	0.1%	3.1%
Land use	BUILT-UP AREA	5.7%	6.2%
	PADDY FIELD	7.2%	7.1%
	OPEN SPACE	10.1%	10.9%

To estimate the volume of landfill within the study area, we first identified new development areas by overlaying land-use maps for three different periods. Next, we classified them into landforms by overlaying these maps with the landform classification map. Finally, we multiplied the calculated area of each landform unit by the height of its landfill, based upon the results of our field transect measurements. The calculated volumes of landfill are shown in Table 3. The estimated total volume of landfill added to the study area during the 30 years from 1966 to 1996 was 4.4×10^6 m^3, equivalent to 8.8 km^2 of new inundation area with an average depth of 50 cm.

We analyzed the relationships among detailed flooding distribution, micro-landforms, and land uses by overlaying Figures 5 and 6, including the landform classification map. Table 4 presents the results of our overlay analysis of Figures 5 and 6, showing that the paddy field and open space landforms retained floodwater. Flooded micro-landform units fluctuated between 1987 and 1997. Based on our comparison of Tables 3 and 4, we inferred that landfilling on the natural low-level micro-landforms caused flood-prone areas to develop in the original, natural high-level areas, which had become lower than the landfilled areas. Figure 7 also supports this observation; the floodwater level on the natural high-level micro-landform is higher than that on the natural low-level micro-landform. Small areas with poor drainage surrounded by landfill development corridors (Figure 5) also became more vulnerable to flooding. These changes in flooding distribution due to landfilling strongly affect the local people's daily lives. In fact, people we interviewed insisted that flood-prone areas were larger and inundations lasted longer than we had found during our field survey.

5. Conclusions and future perspectives

Our results led us to the following conclusions.

1) In the whole Laguna Plain, flooding distribution corresponds well to natural landforms. In the southern part of the Laguna Plain, which is covered with lacustrine sediment, flooding occurs frequently during the rainy season.
2) Landfilling on the natural low-level micro-landforms has caused changes in flooding distribution, particularly in lacustrine plain landforms.
3) Other, smaller areas with poor drainage correspond to a landfilled corridor that was extended from the natural levee.

Based on our study, we recommend the following additions to land-use planning in this area:

a) Master plans should identify flood-prone areas based on landform classification maps. Particularly on the lacustrine plain, where the risk is greatest, open spaces including paddy field should be preserved as far as possible to serve as water-retention areas.
b) District-level plans should avoid expansion of the landfill corridor, which has created areas with poor drainage. Landfill development should be planned based on natural micro-landforms to avoid long-term damage from inundation.

This study implies the effectiveness of land-use planning based on natural landforms on a master-plan scale. At the district plan level, however, the extension of landfilled areas from natural levees, where historical villages lie, should be controlled to minimize flooding risk. The floodwater storage function of paddy fields in the urban fringe areas of low-lying large cities has been investigated and discussed in Japan (Terauchi, 2000). Suitable planning strategies must be developed for growing Southeast Asian cities that are located on coastal lowlands with a tropical monsoon climate. Land-use planning without deterioration of urban green spaces as water-retention areas will be investigated.

6. Acknowledgements

We express our deep appreciation to Ms. Flor Navera, research associate, University of the Philippines, Los Banos; Mr. Ryuichiro Seki, CTI Engineering International Co., Ltd.; and Dr. Norio Abeki, Kanto Gakuin University. The contents of this paper were presented at the annual meeting of the association of rural planning in 2002.

7. References

Laguna Lake Development Authority. Monthly lake level 1972–2001.

Matsuda I., Enomoto T., Banganan E.L. and Narag I.C. (1998). Regional division of Metro Manila on the basis of geological and geomorphological conditions. *Bulletin of Institute of Science and Technology, Kanto Gakuin University*, **25**, 101–112.

McGee T.G. (1991). The emergence of desakota regions in Asia: expanding a hypothesis. In: *The Extended Metropolis: Settlement Transition in Asia*, N. Ginsburg, B. Koppel and T.G. McGee (Eds.), University of Hawaii Press, Honolulu, pp. 3–25.

Murakami A., Zain Medrial A., Takeuchi K., Tsunekawa A. and Yokota S. (2004). Trends in urbanization and patterns of land use in the Asian mega cities Jakarta, Bangkok, and Metro Manila. *Landsc. Urban Plan.* (Article in Press)

Sierra Z.N., Magcale-Macandog D.B., Sotalbo D.D. and Guzman M.E. (1990). Land use inventory of Laguna de Bay watershed area. *U.P. Los Banos Journal*, **1**, 1–15.

Skilodimou H., Livaditis G., Bathrellos G. and Verikiou-Papaspiridakou E. (2003). Investigating the flooding events of the urban regions of Glyfada and Voula, Attica, Greece: a contribution to urban geomorphology. *Geogr. Ann.*, **85**A(2), 197–204.

Takaya Y. (1987). *Agricultural Development of a Tropical Delta: A Study of the Chao Phraya Delta*. University of Hawaii Press, Honolulu.

Tanabe S. (1994). *Ecology and Practical Technology: Peasant Farming Systems in Thailand.* White Lotus, Bangkok.

Terauchi M. (2000). The study of preservation of flood prevention function of paddy-field in urban area: a case study of Fuso Town, Aichi Prefecture. *Journal of Rural Community Studies*, **91**, 60–74. (In Japanese with English abstract)

Yamagata Y., Wiegand C., Akiyama T. and Shibayama M. (1988). Water turbidity and perpendicular vegetation indices for paddy rice flood damage analyses. *Remote Sensing of Environment*, **26**, 241–251.

Yeung Y.-m. (2001). Coastal mega-cities in Asia: transformation, sustainability and management. *Ocean & Coastal Management*, **44**, 319–333.

Assessment of Environmental Health Risks During Monsoon in Nepal

Nawa Raj Khatiwada[1] and Smrity Kamal Rimal[2]

[1] Co-ordinator, Environmental Engineering and Management Program, Nepal Development Research Institute (NDRI), Dhobighat, Lalitpur, Nepal; Tel: +977-1- 5554975, Fax: +977-1-4220161, E-mail: nawa@ndri.org.np; Web: http://www.ndri.org.np

[2]Graduate Student, School of Environmental Management and Sustainable Development (SchEMS), GPO Box 2453, Shanti Nagar, Kathmandu, Nepal, Tel: 977-1-4486803, 4480109, Fax: 977-1-4460317,
E-mail: smritykamalrimal@yahoo.com; Web: http://www.schems.edu.np

Abstract

The empirical evidences indicating the linkages between monsoon or rainy season and water borne diseases are lacking. This study aimed at contributing in assessing the spatial and temporal variations of the waterborne diseases. Emphasis was given on the linkages between the rainfall events, availability of the water and the morbidity and or mortality caused by the waterborne diseases. A rural community and an urban settlement located in the Kathmandu valley were selected in the study. The nationwide data for the last five years indicated that the pattern of diarrhea episodes during the dry and wet seasons has remained unchanged. The site specific research results also showed that most people suffered waterborne diseases during monsoon. The waterborne diseases seem to vary according to seasonal changes, being minimal in winter and peaking in the rainy season. The major risk factor associated with water borne diseases was the contamination of drinking water facilities. The first flush was considered to be the major factor in amplifying the mortality and morbidity figures associated with waterborne diseases in the monsoon period.

Keywords: environmental health risk, Kathmandu valley, monsoon season, waterborne diseases

1. Introduction

Infectious diseases constitute a major fraction of environmental health risks in developing countries like Nepal. Advances in water monitoring technologies are leading to better understanding of certain contaminants and microbes creating acute or chronic health effects in water environment. But as of the distinct seasons in Nepal, the agents causing the diseases and the resulting consequences exhibit seasonal variations adding additional complexity on the understanding of the problems. A factor linked to such variations and the prevalence of diseases is the availability of water in the dry and wet seasons in Nepal.

Nepal is located in South Asia which is in the northern tropics and has dry and wet climates. The monsoon wind brings rain to the regions often of torrential nature with plenty of rainfall. The length of the summer monsoon season varies from year to year, but the precipitation mostly occurs during May to November. The winter monsoon is characterized by dry and cold winds and is not as prevalent as summer monsoon.

Nepal has two rainy seasons. The one falling during June to September is characterized by south-west monsoon and brings about 80% of its total annual rainfall. The other falls on winter and accounts for the rest of the precipitation. The records indicate that an annual precipitation in between 1400mm to 2500mm predominates over most of the country. For example, the average annual total rainfall in the selected study area i. e. Khokana in Lalitpur during 1971-2000 period was 1434mm (Devkota, 2005). The influence of the heavy rain during summer monsoon causes wide variation in river flow, land erosion and floods. With the arrival of monsoon, there is an abrupt change on the availability of water. All water bodies start to receive the rainwater and, often, there are instances of exceeding the carrying capacity, causing flood and plenty of storm-water. The summer and the late spring maximum temperature range from 28°C in mid-part of the country and more than 40°C in Terai with May being the warmest month (HMG, 2001). The effect of monsoon in other environmental sectors is also documented (Hogland et al., 2005).

On the other hand, Nepal is one of the country with highest mortality and morbidity figures related to waterborne diseases. Only diarrhea caused 920 deaths in the year 2002 in Nepal and the morbidity figures related to diarrhea stood at 3.34 percent of the total population. Similarly, epidemics related to water consumption and use such as typhoid and encephalitis is common every year in Nepal. These incidences mostly occur during monsoon season. However, the empirical evidences indicating the linkages between monsoon or rainy season and water borne or water induced diseases are lacking. It is anticipated that such information would lead to the development of a rationale on the relationship between environmental risks and the climatic factors. This would lead to develop strategy in managing such risks and reducing the incidences of mortality and morbidity figures related.

The objective of the research outlined in this paper included the study on spatial and temporal prevalence of waterborne disease and assessment of risk of infection. Further the study also emphasized on the relation of first flush to risk of infection.

Current practices and perceptions of domestic and personal hygiene in the communities, the existing levels of environmental and pre-domestic sanitation and the sudden increase on the health risk with the starting of the monsoon season have also been outlined in this research.

2. Materials and methods

The methodology of the research included literature review and field investigations. Field visits were made to inspect the drinking water facilities, conduct surveys and interviews. In the survey, Khokana Village Development Committee (VDC) of Lalitpur district representing rural communities and Samakhusi area of Kathmandu district representing urban settlements were selected. At Khokana, a total of 50 households were surveyed while at Samakhusi 108 households were surveyed. The analysis of data was conducted and the correlation between various parameters was calculated using routine statistical tools.

The rural settlements around Kathmandu valley are relatively in a better situation in providing drinking water and sanitation facilities compared to the other rural parts of the country. In order to get the empirical evidences regarding monsoon and waterborne diseases, the link between lifestyle, and hygiene and sanitation behavior of people were also considered as the influencing factors. Thus, the Khokana Village Development Committee (VDC) of Lalitpur district, typical in terms of culture, habit, and lifestyle, representing rural communities is considered as the appropriate site. Samakhusi area located in Ward No. 29 of Kathmandu Metropolitan City representing urban settlements is also chosen for the case study and to generate a comparative analysis of environmental health risks.

The Khokana VDC situated in Lalitpur district and 12km south of Kathmandu is well known for oil production. The home-processed mustard oil is one of the most popular products of this village. It is a village on the way to Bungmati and Karya Binayak. The total population is 5566 with 818 households. Out of the total population above 6 years, 1187 cannot read or write, 174 can read only and 280 can read and write. Khokana is a traditional Newari farming community. Most of the households compose of 5 to 6 members in the family. The overall socioeconomic status of the people in the area is not so good. Agriculture is the main occupation or the source of income in the area. Only few are involved in the governmental services and other business.

Samakhusi is located in between Balaju in the west, Lazimpat in the east, Gongabu in the north and Thamel and Lainchour in the South. Samakhusi is highly populated area. People of different areas of Nepal have migrated in this area. It is a newly settled urban area and has a total population of 35259 with 6847 households and is facilitated with various educational, health and official institutions

3. Results and discussion

The findings in the study included a general perspective on the trends of outdoor patients with complains related to diarrhea and visiting the hospitals throughout the country. Similarly the nationwide morbidity figures are also reported. On the other hand, a comparative picture of the environmental health risks in the selected rural and urban areas is also given. A consistent seasonal variation in the total diarrhea visits and a similar trend on the diarrhea episodes during 12 month period were noted during last five years. Figure 1 shows the trend of hospital visits for the five-year period during 1998-2003.

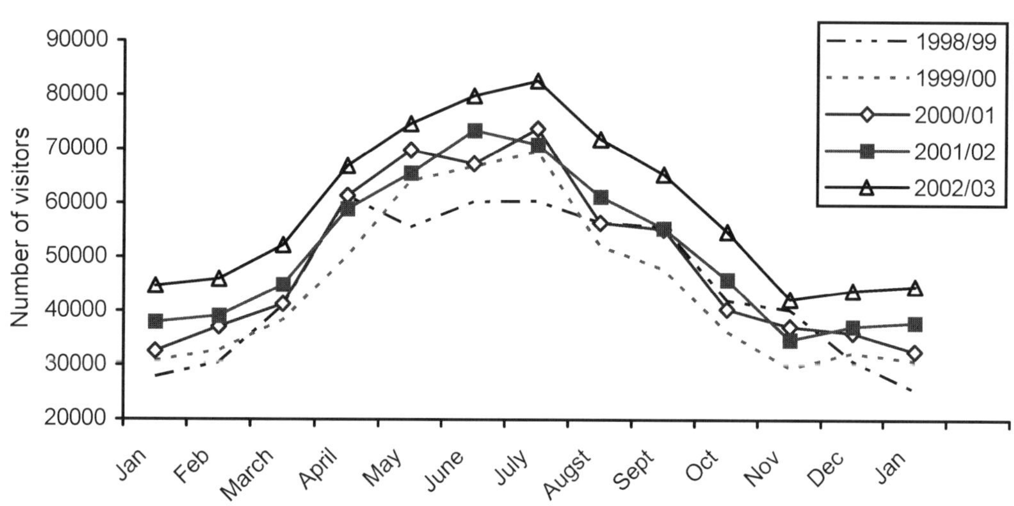

Figure 1. Monthly trends of hospital visitors suffered from diarrhea (DOHS, 2004)

The data indicated that the peak level of diarrhea visits occurs in the month of May, June and July at National Level. The number of patients in January to March period is close to roughly 30,000 but with a slight increase every year. An estimation referring to 23 million population of the whole country, this morbidity figure indicated that an individual per 1000 population is under risk. The figures are even lower during the post monsoon period. Thus the mid-winter season can be categorized as low risk period. But with the arrival of peak dry period and monsoon, the figures gradually increase, reaching an average peak of about 70,000 in June-July period. This shows that the number of individuals suffering from diarrhea increases close to 3 per thousand population. Thus it is reasonable to conclude that the monsoon season magnifies the environmental health risks related to diarrhea by 200%.

Figure 2 shows the percentage distribution of the morbidity rate of the waterborne diseases during 1999 to 2003 period in Nepal. These figures indicated the total new visits only.The population suffered from diarrhea has remained constant during the four years period. However, the worm infection has drastically reduced in 2002/03, while jaundice related infections had a small but sustantial presence during the five years period.

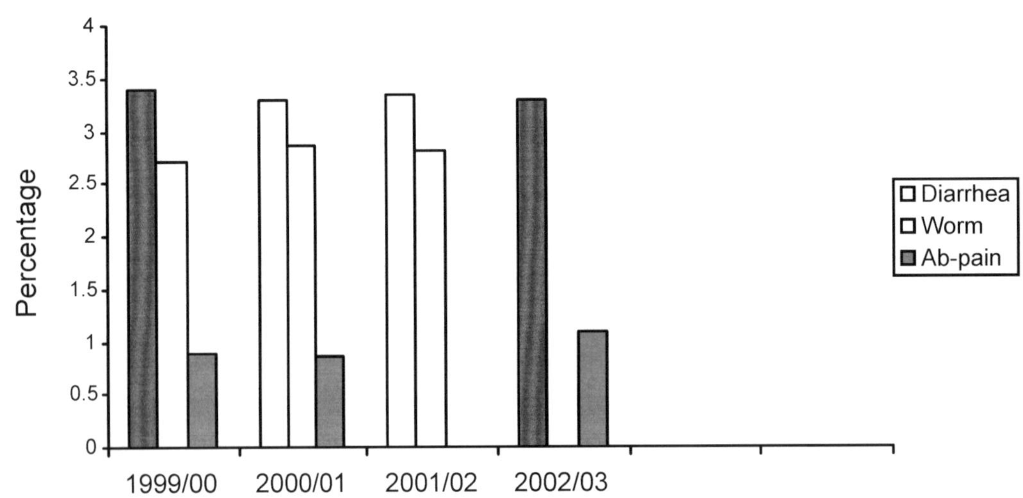

Figure 2. Showing the diseases categorization and the percentage of infection (The percentage figures indicated the population visiting the hospitals out of the total population in the country)

Thus, this national burden created by widespread and heavy rainfall during the monsoon period has been a major concern of the line agencies in Nepal. Ministry of Physical Planning and Works (MPPW) reported that the monsoon risks, waterborne diseases and poor hygiene and sanitary environment lead situation caused the country an enormous cost (MPPW, 2002). This has resulted in the loss of productivity equivalent to more than 10 billions annually. This is estimated to be about 1.6 to 4.1 % of the Gross Domestic Product (GDP) or Rs.1041 to 2646 per household per annum. However, while analyzing the trend in annual basis it was found that there was no expected improvement in the burden of diseases even though a significant investment was made in this sector from the government of Nepal.

The recent census data indicated that the annual population growth was 2.3% in the year 2004. As shown in Figure 1, the annual increase in the figures has remained to be many folds higher than the population growth rate. Thus, it is rather complex to conclude the reasons behind this discrepancy. Nevertheless, general awareness on the hygiene and sanitation, rapid expansion of the health care institutions particularly in urban areas, unplanned migration to urban centers are considered as the contributing factors. Additionally, the epidemics, for example, typhoid incident in Chitwan district in 2002 and the recent Japanese Encephalitis episode (in 2005) in western Terai areas, might have also contributed significantly to the annual increase figures.

The site-specific study at Khokana, showed that most of the people lack the facilities of toilet at their homes and people working in the field often drink water from *Kulo*, a canal system meant for irrigation. The tap water, supplied in the village, is consumed directly by about fifty percent of the people. The record from the health post showed that the patients visiting the health posts were high during three months of monsoon i.e. May, June, and July. Most people were found to not consider waterborne disease as a major source of risk in the village. There are instances when the sick persons are taken to *Guru* for spiritual healing. Health posts and hospitals are visited in serious cases only. The elderly people in the village seldom visit the hospitals, as they are scared with the false notion that people die if they visit the hospital. This showed that waterborne diseases are dependent mainly on the life style of the people, sanitation facilities, water usage and consumption.

Table 2: The results of survey at Khokana VDC (Total respondents = 50)

Items	Details	Respondents
Water Usage	Tap water	28
	wells and spouts	9
	Drinking from irrigation canal *(Kulo)* at field	13
Drinking water consumption	Direct	25
	Filter	7
	Boil	3
	SODIS	15
Waterborne diseases(within last two months, during survey)	Diarrhea	10
	Intestinal worm	5
	Jaundice	1
	Dysentery	4
Treatment	Dhami	9
	Medial Hall	9
	Health Post	12
Use of Toilets		40
No Toilets		10

Table 2 shows the results obtained from the survey at Khokana.

The data from the health post of Khokana VDC showed the top ten diseases in the village. The predominant diseases are acute respiratory diseases, skin infection, injuries, and diarrhea and worm infection (Figure 3).

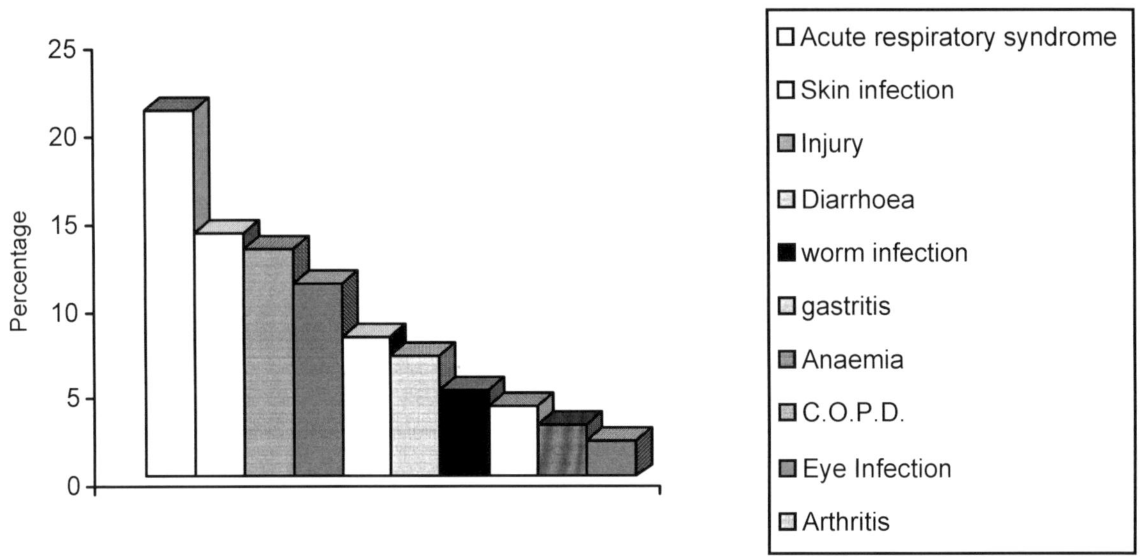

Figure 3. Top ten diseases at Khokana

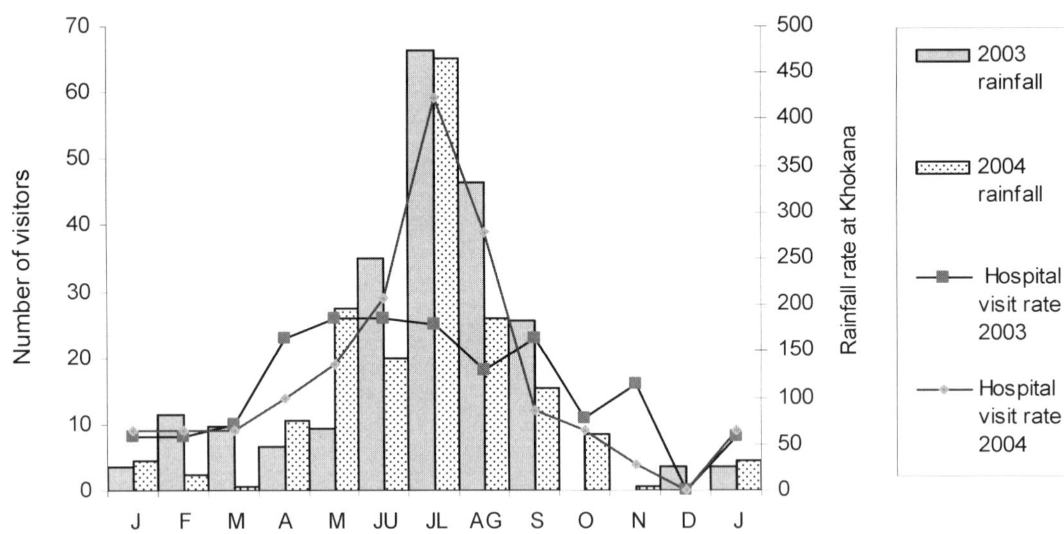

Figure 4. Comparative analysis of the rainfall and waterborne diseases

The health post of Khokana showed that the number of people visiting the health post was high during May to July. The waterborne diseases seem to vary according to seasonal changes, being minimal in winter and peaking in the rainy season (Figure 4). The rainfall data of Khokana station shows the high rainfall rate during May to July and so is the number of patients suffering from diarrhea.

The t-test carried out against rainfall and number of patients suffering from diarrhea showed that there occurs significant correlation between rainfall and occurrence of diseases. The result tested between 2003 rainfall data and 2003 hospital visit records from Khokana gave the result as 2.6. Similarly the data obtained from Samakhusi showed the result as 1.9. The results indicated that the data were more coherent in case of Samakhusi.
The other site Samakhusi representing the urban settlement and close to the facilities, the questionnaire survey showed that most people use the tap water but as the water need is not sufficient people usually use water from wells and spouts. Most of these wells and spouts were found to have coliform contamination.

4. Conclusions

The study highlighted a few empirical evidences indicating the linkages between monsoon and waterborne diseases. The consistent seasonal variation in the total diarrhea visits over last five years show that there was virtually no change in the diarrhea episodes during 12-month period. The increase in diarrhea episodes start in March continues to August.

The study showed that the rainfall is associated with an increased risk of infection.The major risk factor associated with water borne diseases could be the contamination of drinking facilities. The first flush can be considered one of the reasons for occurrence of water borne diseases, which facilitates the pathogens movement in water environment. Rapid growth of population, absence of sewerage system and lack of sanitation faciliites and intensive rainfall in monsoon have combined effect to increase the environmental health risks. However, it is anticipated that the prevalence of waterborne diseases is also influenced by living style of people, cultural and religious factors.

5. Acknowledgement

The authors would like to acknowledge the support provided by School of Environmental Sciences and Sustainable Development (SchEMS) and Nepal Development and Research Institute (NDRI). The participation of the officials in the health posts and general public in the study areas are highly appreciated.

6. References

Devkota L. P., (2005). Rainfall analysis of the Kathmandu Valley. Journal of Hydrology and Meteorology , Society of Hydrologists and Meteorologists –Nepal (SOHAM-Nepal). **2**(1), 34-40.

Department of Health Services (DOHS) (2004). Annual Report, Department of Health Services 2059/60(2002/03), His Majesty's Government of Nepal, Ministry of Health

His Majesty's Government of Nepal (HMG) (2001). Water Resources Strategy Nepal. An official document, Water and Energy Commission Secretariat, Singha Durbar, Kathamndu, Nepal

Hogland, W., Visvanathan C., Marques M., Manandhar D. R., (2005).Landfill in Asia-Coping with Monsoon, Waste Mangement World, July-August 2005.

Ministry of Physical Planning and Works (MPPW) (2002).Thematic workshop on health hygiene and sanitation. His Majesty Government (HMG) and ADB.

Population's Views on the Water Environment: the Case of the Nam Pong Watershed

Mara Regina Mendes[1)1] and Kensuke Fukushi[2)]

1) Department of Urban Engineering, The University of Tokyo
7-3-1 Hongo, Bunkyo, Tokyo, 113-8656 Japan
E-mail: maramendes@yahoo.com
2) Environmental Science Center, The University of Tokyo
7-3-1 Hongo, Bunkyo, Tokyo, 113-8656 Japan
E-mail: fukushi@esc.u-tokyo.ac.jp

Abstract

A survey was carried out among a representative sample of 323 respondents in the Nam Pong Watershed, Northeast Thailand, aiming to identify the population's views on environmental issues, especially issues related to the water environment, as well as their perception on being affected by pollution. The survey consisted of application of a multiple-choice questionnaire by interviewers. Sample involved three groups of respondents randomly selected in three different regions of the watershed. The survey identified health, environment, education, and job opportunity as important issues according population's views. Regarding environmental issues, the population is mainly concerned with the water environment, waste management, and pollution-related health problems. However, air pollution was identified in this survey as the environmental problem that people consider affecting them to the most. The population supports the polluter-pays principle and believes that economic compensation should be provided to those who are affected by pollution. Perception on different problems or issues was in some cases different among the three groups surveyed.

Keywords: population's views, public perception, Nam Pong Watershed, Northeast Thailand, water environment, watershed management

1. Introduction

The management of natural resources and public goods in a given area has to take into account the societal values and public perception, since the local population is the most affected by the decisions made. Defining policies without considering the public views and needs can become a misuse of public's budget and eventually it can lead to negative impacts on society or on the environment. Water is recognized as a public good; however, as broadly reported in the literature, several conflicts over the use of water and the management of river basins have emerged (Dimitrov, 2002; Klug, 2002; Sneddon et al., 2002). To avoid such conflicts, decision-making has to take into consideration population's views and needs. There is a need to integrate collective or societal values in the development of policies related to water use and watershed management.

This paper attempts to discuss the relevant issues on environment, especially the water environment, according to the views of the population in the Nam Pong Watershed, Northeast

[1] Current affiliation: Institute for Global Environmental Strategies, Urban Environmental Management Project. Mailing address: 3-9-30 Asano, Kokurakita-ku, Kitakyushu 802-0001, Japan. E-mail: mendes@iges.or.jp

Thailand. This area was chosen for the research because it is an important development center in Thailand's countryside. Moreover, it faces conflicts related to water use and environmental issues (Sneddon, 2002).

2. The studied area

The Pong River or Nam Pong Watershed, located in Northeast Thailand, serves an agricultural production area that accounts for half of the total area of the watershed. It is also the main source of water in urban areas, including the main city in the region, Khon Kaen, and twelve small towns (Nuntapotidech, 2001). The Nam Pong Watershed is divided into three major sections: the upstream area, middle area, and downstream area. The middle area consists entirely of Ubolratana Reservoir; the upstream watershed consists of the area before the reservoir, while the downstream watershed is the area after the reservoir (KKU, 1995). The land use in the upper and middle part consists of forest, agriculture, and small communities and towns. The land use along the downstream area consists of villages, large communities, irrigation area, and industries including sugar production, distillery, paper, and tapioca mills. Recently there have been intensive fish-farming activities along the lower part of the watershed. According to a water quality monitoring study in the lower part of the watershed, the sources of pollution are industries (point source) and non point sources, such as villages, aquaculture, and agriculture. Mercury contamination was identified as well as presence of mutagens-like substances, organochlorine pesticide, and phenol; other heavy metals were detected at low concentrations (Mungkarndee and Wirojanagud, 2003).

The rapid economic growth of Thailand has affected the rural communities of the Northeast. Forests were depleted by commercial logging and the clearing of land for agriculture, giving rise to environmental problems including soil erosion, drought, periodic flooding, and the loss of habitats for native species. Moreover, poor waste management and sanitation practices also affect the soil and water quality as well as human health.

3. Interactions between people and environment

The interactions between people and the environment in the Nam Pong Watershed are represented in Figure 1. The river is the source of water for farms, fisheries, industry, villages, and cities. It receives the residues from these activities: nutrients and fertilizers from agriculture, organic matter from fisheries, wastewater from industries, and sewage and waste leachate from villages and cities. Cities and villages consume products from farms, fisheries, and industries. Industries also create jobs and generate taxes (Mendes, 2004). The groups that have interests in a watershed include farmers, fishermen, industry, government, NGOs, academia, and civil society in general and they all should be involved in the decisions.

4. Survey methodology

A quantitative population research survey was performed in the following stages: preparation, questionnaire development, sampling design, pre-test, field survey, and data analysis and interpretation. The preparation consisted of literature survey, visits to the watershed, and personal interviews with farmers, fishermen, academicians, and NGO representatives.

Formal survey was conducted through a structured questionnaire containing multiple-choice questions. The questionnaire, besides involving queries regarding socio-economic and educational background, attempted to obtain the views regarding which societal and environmental subjects need more attention as well as the perceptions on sources of pollution and on being affected by it. The questionnaire was pre-tested on 25 respondents aiming to check for clarity and comprehensiveness of the items and then it was revised. The questionnaire was applied to randomly chosen farmers, fishermen, workers, public servants, university students, and academicians. A sample of 323 respondents were interviewed in 23 points (villages, towns, and cities) along the upper, middle, and low part of the watershed. The sample structure was designed to have respondents with different socio-economic and educational backgrounds, which were separated into three groups (Table 1).

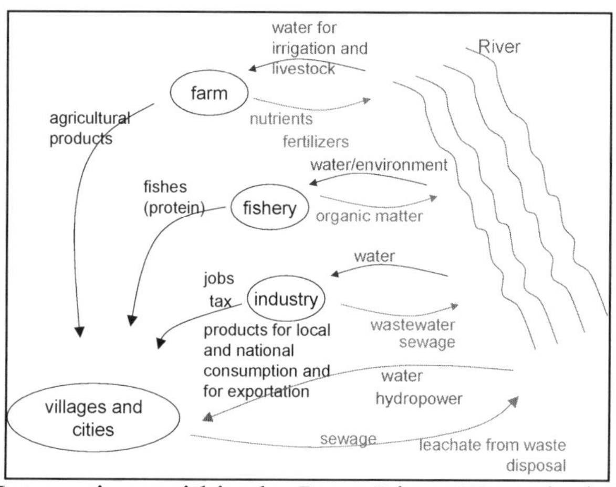

Figure 1. Interactions within the Pong River Watershed (Mendes, 2004)

Table 1. Sampling structure based on social, economical, and educational background

	Group	Interviews	Distribution (%)
Farmers and fisherman	1	112	35
General workers	2	101	31
Businesspersons, government officials, or highly educated individuals	3	110	34

5. Results

5.1 Societal priorities

To provide context to the findings, respondents were asked to consider, from a list, which aspects of public service provision or societal issues were most in need of urgent attention and improvement (Figure 2). They could rate the first three main issues of concern. In average, respondents are most concerned about health services (61%), education (55%), environment (52%), opportunity of jobs (37%), and litter and waste disposal (32%).

The responses by the three groups, farmers (Group 1), general workers (Group 2), and highly educated or high-income workers (Group 3), indicate differences on priorities. Group 3 expressed the highest concern for environmental issues. Second and third concerns are health services and education, which are similar concerns to other groups. The same group expressed also the highest concern for crime prevention. On the other hand, they expressed the lowest concern for water and sewerage services, which might be an indication that they live in areas with good provision of water and face no serious problems with sewerage. Farmers expressed that they consider health services, education, and job opportunities as main issues that need most urgent attention. They had the lowest concern about crime prevention showing that although there are crimes in the countryside, they are not a main issue as they might be in the cities. The chi-square test resulted as statistically significant strengthening that there is a relationship between group categories and views on most urgent societal issues.

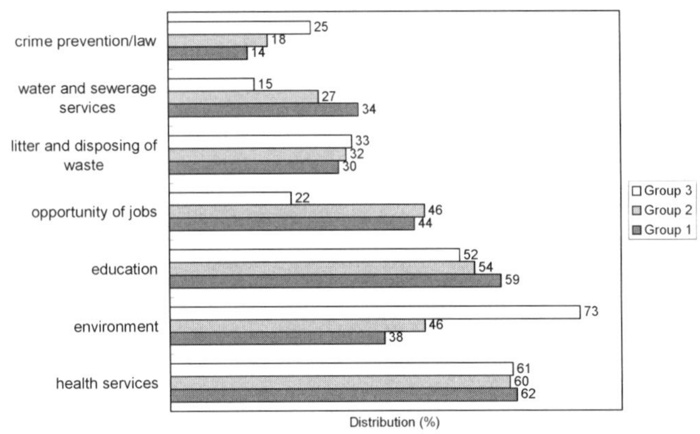

Figure 2. Most urgent societal issues by population group

5.2 Environmental priorities

Respondents were asked to choose which environmental aspects they considered to be in most urgent need of improvement or greater protection from a list of five (Figure 3). In general, the most urgent concern is the issue of water environment (cited by 59% of the respondents) and litter and household waste (cited by 53% of the respondents). The water environment is the issue of main concern for all groups and household waste, the second. The different groups seem to differ on the view about nature conservation and changes in weather pattern. Farmers seem more concerned with nature conservation and weather pattern than other groups, because it might affect them more. Group 3 seems more concerned with air pollution and nature conservation, rather than with changes in weather pattern. However, the chi-square test gave an indicating that there is not a relationship between the group categories and the views on environment.

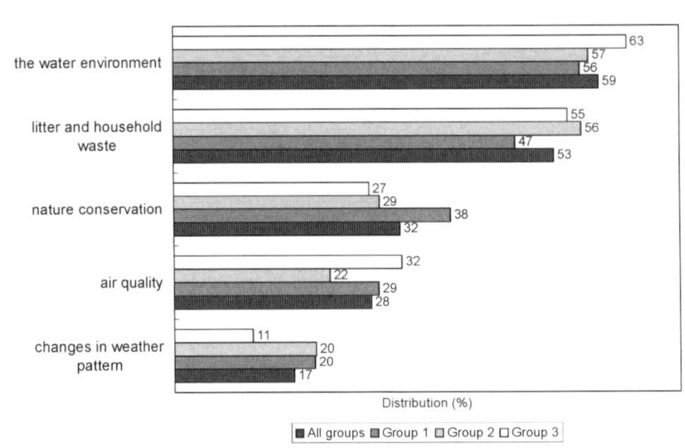

Figure 3. Most urgent environmental issues by population group

5.3 Issues on water environment

Respondents were given a list of four aspects of water environmental issues from which to choose one in most urgent need of attention in terms of either needing improvement or protection. The majority feel that *"avoiding damage by pollution to areas of wildlife and plants, including rivers, water streams, and wetlands"* is the most urgent need of attention – mentioned by 55% of respondents. Although all groups are more concerned with pollution, when looking at the second most cited category, *improving water quality*, it was more cited by farmers (Group 1) than by higher-class workers (Group 3). Farmers also seem more concerned with flooding than other categories. The chi-square test indicated that the difference on response among groups is statistically significant reinforcing that there is a relationship between the group categories and the views on water environment.

5.4 Views on source of water pollution

It was found that 82% of the respondents agree that the rivers in Thailand have become polluted. They consider industrial waste and wastewater (84%) and fertilizer and pesticides from farming (46%) as the main sources of water pollution in the region. Twenty percent of the respondents consider household waste and sewage as sources of water pollution; only 12% consider fisheries as a source of water pollution.

5.5 Attitude towards water environment and pollution

Respondents support the 'polluter pays' principle – 92% agree (43% 'strongly') that *"industries that create water polluting waste or wastewater should pay to protect the environment"*, only 2% mildly disagree (MD). Although the agreement rate is high, the chi-square test indicated that the difference on response among groups is statistically significant, confirming that there is a relationship between the group categories and the support to the 'polluter pays principle'. Strongly agreement (SA) is higher in Group 3 (49%) and lower in Group 2 (37%). Mildly agreement (MA) is higher in Group 2 (52%). Among the people that are unsure (UN) about the issue, the majority belongs to Group 2. It is noteworthy to point out that factory workers, which are the main components of Group 2, may feel reluctant to express their opinion about pollution.

Respondents also believe that economic compensation should be provided to those affected by pollution or environmental degradation: 88% agree (37% 'strongly'), while only 9% are unsure, and 3% mildly disagree. The chi-square test showed that there is no difference on the extension of agreement among the different groups.

Despite the agreement on the right for economic compensation is high, the agreement in protesting in order to obtain such compensation is lower (Figure 4). This might reflect the local culture of not using violence or confrontation. According to the chi-square test, there is no difference on the extension of agreement by the different groups.

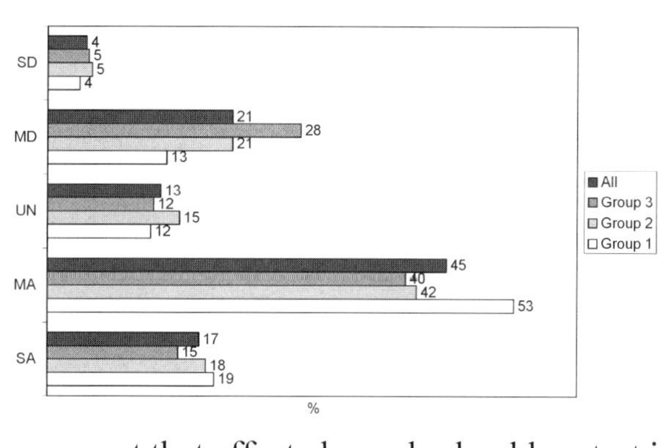

Figure 4. Extension of agreement that affected people should protest in case of being affected by pollution

Additionally, respondents have a very positive attitude towards conservation and the water environment – 97% agree that *"population should use water more wisely, in order to conserve supplies and maintain the water environment"*. Forty-two percent strongly agree with this statement and no one disagree.

5.6 Perceptions regarding being affected by pollution

Respondents were asked if they think that they or their family are affected by pollution. Majority of respondents (88%) believe they are affected by some source of pollution. Respondents were asked about problems that affect their life or of their family in terms of health, business, or well-being (Figure 5). The majority of respondents (61%) feel that there are affected by air pollution. Pollution from farming and industry were also often cited, 35% and 33% respectively. Flooding was also cited as one of the problems that affect people's life. The chi-square test indicated that the difference on response among groups is statistically significant, indicating that there is a relationship between the group categories and the perception of being affected by certain environmental-related problem.

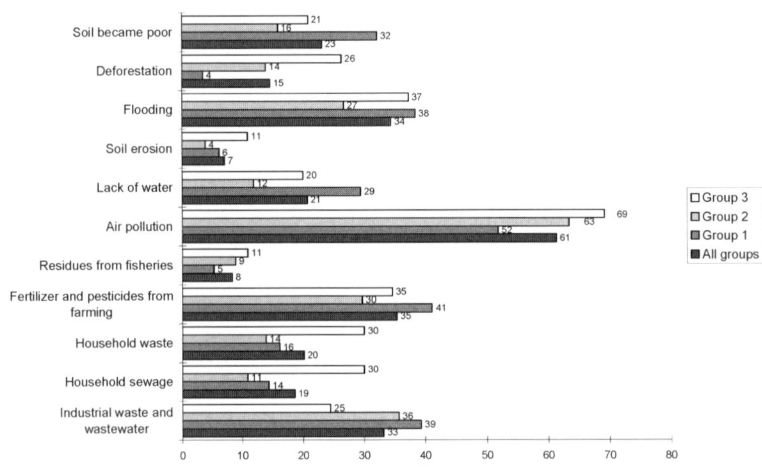

Figure 5. Population's views on environmental issues that affect them (by group)

It is worthy to point out that many respondents when faced with this question had intensely discussed about what they think affect them. Informally, villagers living in villages near-by the pulp and paper mill complained of adverse smell from the factory. They believe that air pollution is the cause of respiratory crisis in those children that suffer this problem.

Figure 6 shows the problems that people consider affecting their lives categorized by geographical region, i.e., upper, middle, and low part of the watershed. Respondents living in the middle part of the watershed, where are located the main industries, believe they are mostly affected by air pollution (30%) and industrial waste and wastewater (26%). Respondents from the lower part of the watershed believe being mainly affected by air pollution (23%). Other environmental problems, such as flooding, pesticides, household waste, and sewage were similarly cited (8-13%) by residents in this area. Regarding the upper part of the watershed, the causes of problems are more scattered: air pollution (17%), pesticides (15%), flooding (13%), and industrial waste and wastewater (11%). In general, the upper part is mainly rural, less prone to flooding, and environmentally less problematic. According to academics and NGOs members, the upper part is mainly affected by deforestation and soil erosion (personal communication).

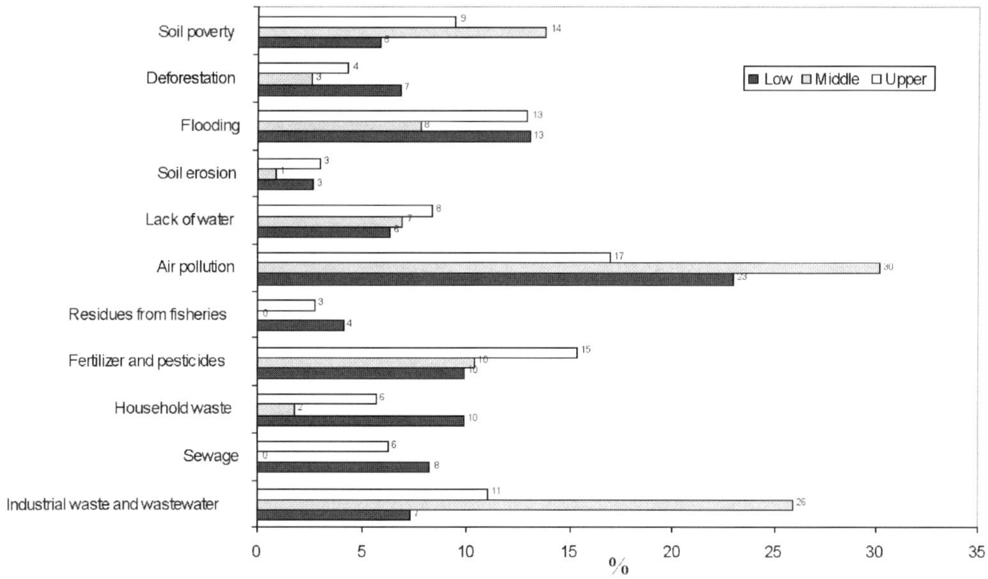

Figure 6. Population's views on environmental issues that affect them (by area)

6. Summary

In this study, population's perception on environmental issues in the Nam Pong Watershed was investigated. Majority of the respondents manifested a positive view towards environment and environmental protection, i.e. they expressed interest in environmental issues. Policy-making needs to take into account population's needs and views; therefore the findings of this survey are expected to be useful for that.

The main findings of this study are listed below.

- The main societal concerns are health, environment, education, and opportunity of jobs. Views on priorities for societal improvements were different for different groups.
- The main environmental concerns are water environment, waste, and pollution-related health problems. The main concern regarding the water environment is its pollution.
- Respondents believe to be affected by environmental problems, such as air pollution (61%), fertilizer and pesticides (35%), flooding (34%), and industrial waste and wastewater (34%).
- Although respondents consider water environment and water pollution as priorities in need of urgent attention, they express more often being affected by air pollution. It might be because of the direct effect of air pollution on health. For instance, several respondents freely expressed their concern regarding respiratory problems and its possible connection with air pollutants released from factories.
- Respondents feel affected by flooding, since it is periodic and causes agricultural losses.
- Respondents support the polluter-pays principle, i.e., they believe that industries that pollute should pay to protect the environment. They also tend to support the need for economical compensation for those affected by environmental pollution, but they do not support the protesting for obtaining such economic compensation to the same extend.
- Perception on different problems or specific issues was different among different groups reflecting the issues that affect each group most.

7. Recommendations

From this study, it become clear that respondents consider industrial wastewater and pesticides as the main sources of water pollution in the region and they feel themselves affected by environmental pollution, both in relation to their health and economic activities (effect in farming and fishery). Therefore, it is recommended that both national and local policies should address these issues and take measures for pollution control and mitigation of its effects. In order to devise mitigation and pollution control measures, further investigation seems necessary to evaluate the extension of pollution in the area and the effects on people's health and on the agricultural production. Research carried out till now consisted mainly in identifying pollutants having limitations in assessing the relationship between sources of pollutants and the effects on local people and local environment.

It is seems necessary to review the strategies for economic development in the region taking into consideration the sustainable management of natural resources including water resources and the local economy, such as agriculture and aquaculture. There is a need for implementing strategies to decrease industrial pollution without decreasing the economy and job availability in the region. Agriculture is the main economic activity of large part of the population in the region and it needs attention. Strategies for decreasing pesticide pollution should be devised as well. Finally, public involvement in the decision making in the watershed is fundamental in order to manage the common resources in a sustainable way.

8. Acknowledgements

The authors would like to thank the "Special Coordination Funds for Promoting Science and Technology, the Ministry of Education, Culture, Sports, Science and Technology, Japan" for providing the financial support and the Environmental Research Center, Khon Kaen University for the logistical support. Special thanks are due to Ms. Aatcharaporn Chaowahem for her assistance as well as to the interviewers. Dr. Ratana Tosakul-Boonmathya, Mr. Akkanit Pongpai, and the Isan Environmental and Resource Network provided valuable information. Finally, the authors want to express their gratitude to the people of the Nam Pong Watershed.

9. References

Dimitrov, R.S. (2002). Water, Conflict, and Security: A Conceptual Minefield. *Society and Natural Resources,* **15**(8), 677-691.

Inmuong, Y. (1998). Thailand Water Pollution Crisis: A case on Massive Fish Deaths in Nam Pong River. *FACT SHEET - Environmental Health,* 1. Ministry of Public Health, Thailand.

KKU - Khon Kaen University. (1995). *Study Project for Preparation of the Action Plan for the Pong River Water Quality Rehabilitation, Khon Kaen Province* (Executive summary). Khon Kaen: Khon Kaen University.

Klug, H. (2002). Straining the Law: Conflicting Legal Premises and the Governance of Aquatic Resources. *Society & Natural Resources,* **15**(8), 693-707.

Mendes, M.R. (2004). *Research into Population's views: Results of the Field Survey in Nam Pong River Watershed.* The University of Tokyo, Tokyo, 88p. (unpublished report)

Mungkarndee, P., Wirojanagud, W. (2003). Impact of Pollution Sources on the Pong River Water Quality. In: *Proceedings of the First Int. Symposium on Southeast Asian Water Environment*, 23-25 Oct., Bangkok, Thailand.

Nuntapotidech, A. (2001). Environmental Disputes in Water Sector: The Case Study of the Pong River Basin, Thailand. In: *Proceedings of DSE-IWMI Workshop on Mediation of the Disputes in Water Sector*, 1-5 Oct., Hambantota, Sri Lanka.

Sneddon, C., Harris, L., Dimitrov, R., & Ozesmi, U. (2002). Contested waters: Conflict, scale, and sustainability in aquatic socioecological systems. *Society & Natural Resources,* **15**(8), 663-675.

Sneddon, C. (2002). Water conflicts and river basins: The contradictions of comanagement and scale in Northeast Thailand. *Society & Natural Resources,* **15**(8), 725-74

Importance of Microhabitats for Benthic Faunal Production in Intertidal Mangrove Forests

Koetsu Kon[1], Hisashi Kurokura[2] and Hisao Ogawa[3]

1) Department of Global Agriculture Science, The University of Tokyo
1-1-1 Yayoi Bunkyo-ku, Tokyo, 113-8657, Japan
aa46273@mail.ecc.u-tokyo.ac.jp
2) Department of Global Agriculture Science, The University of Tokyo
1-1-1 Yayoi Bunkyo-ku, Tokyo, 113-8657, Japan
akrkrh@mail.ecc.u-tokyo.ac.jp
3) Department of Marine Bioscience, Kitasato University
160-4 Uzu, Hokkirai, Ofunato-shi, Iwate, 022-0101, Japan
seaweed@kitasato-u.ac.jp

Abstract

We examined food sources of benthic animals at three microhabitats; a small creek, an inside-forest, and a forest gap. The benthic fauna inhabiting at a small creek and an inside-forest utilized mangrove litters and phytoplankton, but those at a forest gap utilized microphytobenthos and phytoplankton. In addition, biomass of benthic fauna at forest gap was the highest among three microhabitats. These results could suggest that a forest gap has different productive structure in terms of basic producers.

Keywords: mangrove, stable isotope, benthic community

1. Introduction

Mangrove forests are often a prominent feature of the coastline in Southeast Asia, and create unique ecological environment that host rich assemblages of species. Especially, benthic fauna are important primary consumers in mangrove ecosystems (Robertson and Alongi, 1992). Most previous studies have assumed or concluded that detritus from mangroves sustain the benthic fauna (e.g. Robertson, 1986; Camilleri, 1992). However, these detritus are not considered to be good food items, because of low nutritional quality and difficulty for digestion (e.g. Lee, 1999; Kathiresan and Binbham, 2001).
Productivities of microalgaes (phytoplankton or micophytobenthos) are relatively low in mangrove forests where light intensity is restricted (e.g. Alongi, 1994), though they have been found to be important food resource in other intertidal ecosystems such as salt marsh (e.g. Currin et al, 1995; Page, 1997; Riera, 1999), because of the high nutritional quality relative to detritus (e.g. Kathiresan and Binbham, 2001). Recent studies have suggested importance of imported phytoplankton from offshore and in situ microphytobenthos in mangrove forests (e.g.

Bouillon et al, 2002, 2004), and several authors have suggested that their potential role in mangrove ecosystems deserves further study (e.g. Micheli, 1993; Newell et al., 1995; Bouillen et al., 2002).

In mangrove forests, there are some microhabitats in which light restriction dose not exist, such as small creeks or forest gaps. Microalgaes would be high production at these microhabitats. Benthic fauna inhabiting there could utilize microalgaes instead of detritus from mangroves. In this study, we estimated food source of benthic fauna at three microhabitats; a small creek, an inside forest, and a forest gap, and examined the relative importance of those microhabitats for benthic faunal production.

2. Materials and methods

2.1 Study area

The study was conducted in an intertidal mangrove forest located Sikao creek in Trang Province on the west coast of Thailand (7°32'N, 99°21'E) (Figure 1). The forest is dominated by *Rhizophora apiculata*, and neither seagrass nor coral reef occurs in vicinity of the creek mouth. The intertidal forest is subjected to semidiurnal tides with a tidal range of 1.0-2.5m. Sampling were carried out in March (dry season) and July/August (rainy season) 2005 at three stations; a small creek (8m wide) with no mangroves (CR); an inside-forest with mangroves (FO); a forest gap ($150m^2$ area) with no mangroves (GA).

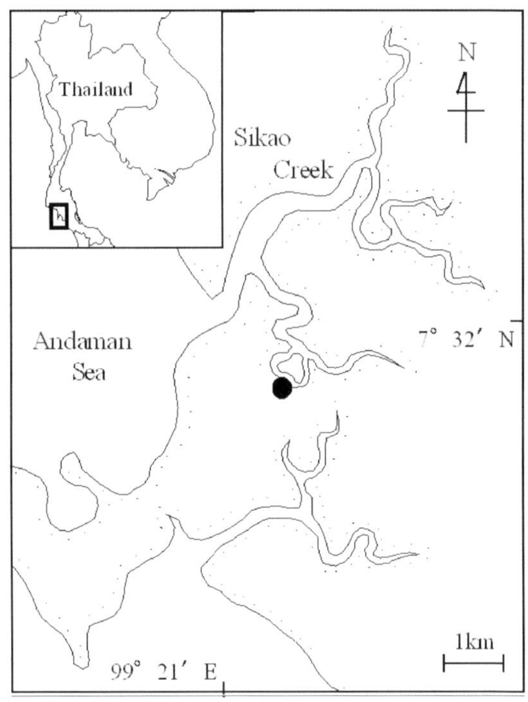

Figure 1. Map of the study area. Closed circle indicates the sampling site

2.2 Ignition loss and biomass

Suspended solids of surface water (100ml) were sampled by filteration on pre-combusted glassfibre filters (Whatman GF/C) at high tide (n=4). Sediments solids were collected using core sampler (10cm depth, 5cm diameter) at low tide (n=4). Ignition losses of solids were analyzed as an index of organic matter contents. Ignition losses of suspended solids and sediments solids were expressed particle organic matter (POM) and sediment organic matter (SOM), respectively. Macrobenthic animals were collected using 30×30 cm quadrat for epifauna. Infauna were collected by excavating below 20cm at same area (n=5). Collected animals were measured wet-weight and preserved in 70% ethanol for species identification.

POM/SOM and biomass were expressed as the mean ignition losses and wet-weights, respectively. A two-way analysis of variance was applied for the comparison in POM, SOM and biomass among stations and seasons. Sheffe tests were used for a posteriori multiple comparisons of means. Before the analyses, POM and SOM data were transformed to arcsin(x), and biomass data were transformed to log(x+1) in order to improve homogeneity of variances (Zar, 1999).

2.3 Stable isotope

POM samples were collected by the same procedure above (n=4). SOM samples were obtain by scrapping the top 1-2mm of sediment from surface mud (n=4). Benthic animals and mangrove reef-litters were collected by hand (n=4). Phytoplankton, defined as material derived from the sea, sampled by filtering 100ml of surface water on pre-combusted glassfibre filters (Whatman GF/F) at offshore of creek mouth during flood tide (n=4). Microphytobenthos were obtained by gently scraping them off the sediment where they formed a conspicuous layer (n=4).

All samples were kept in a cool box, transported to the field laboratory, washed (except for POM, SOM, and phytoplankton samples) with distilled water and dried at 60°C for at least48h. For some of the smaller crab species, the gut and intestinal system were first removed and then muscle tissue of the body was used for the analysis; muscle tissue of larger crab species, was taken directly from the chelae. Whole fleshy tissues of mollusks were analyzed after removal of shells. All samples were ground to a fine powder and treated with dilute HCl to remove possible carbonates and re-dried following to the method by Ute et al. (2005).

Stable isotope ratios were measured on an isotope-ratio mass spectrometer (Thermo/Finnigan Delta plus XP, Kitasato University), and are expressed relative to the conventional standards, i.e. PDB limestone for carbon and atmospheric air for nitrogen, as δ values, defined as:

$$\delta R = \frac{X_{SAMPLE} - X_{STANDARD}}{X_{STANDARD}} \times 10^3 (‰)$$

where $R = {}^{13}C$ or ${}^{15}N$, and $X = {}^{13}C/{}^{12}C$ or ${}^{15}N/{}^{14}N$. Experimental precision (based on standard deviation of replicates of an alanin standard) was lower than 0.15‰ for both $\delta^{13}C$ and $\delta^{15}N$.

3. Results and discussion

3.1 Ignition loss and biomass

The mean ignition loss of POM at GA was higher than those at FO and CR in both dry season and rainy season (Table 1, Figure 2a). In contrast, the mean ignition loss of SOM at GA was lower than those at FO/CR in both seasons (Table 1, Figure 2b). The mean biomass at GA was higher than those at FO in both seasons (Table 1, Figure 2c). The results of comparison of ignition loss and biomass could suggest that a forest gap has different mechanism of organic matter supply.

Table 2. Stable isotope ratio of SOM, POM, and producers.

	Dry season		Rainy season	
	δ ^{13}C± SD (‰)	δ ^{15}N± SD (‰)	δ ^{13}C± SD (‰)	δ ^{15}N± SD (‰)
POM				
Forest	-26.4± 0.7	3.1± 1.0	-27.6± 0.3	1.3± 0.8
Gap	-26.3± 0.8	2.3± 0.7	-27.4± 0.4	2.3± 0.8
Creek	-26.1± 0.5	3.0± 0.5	-27.5± 0.2	2.5± 0.5
SOM				
Forest	-26.6± 0.5	-3.1± 1.5	-27.7± 1.2	1.2± 0.8
Gap	-26.3± 1.0	-0.3± 1.0	-26.0± 0.3	1.5± 0.4
Creek	-25.9± 0.9	-0.7± 1.1	-27.3± 0.1	2.0± 0.2
Producers				
microphytobenthos	-18.2± 0.7	0.8± 0.9	-17.4± 0.3	1.2± 0.3
phytoplankton	-23.1± 0.8	4.7± 0.2	-23.6± 1.8	4.1± 0.1
mangrove	-28.9± 0.7	-1.1± 0.9	-29.2± 0.1	2.0± 0.2

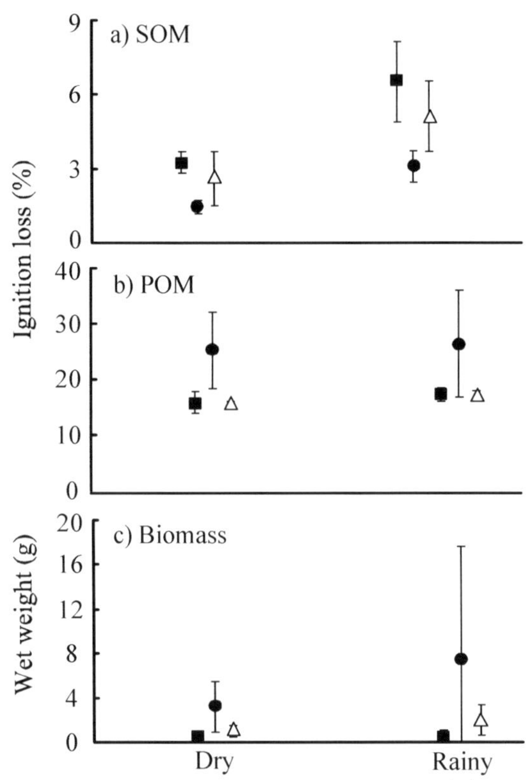

Figure 2. Mean ignition loss of SOM(a), POM(b), and wet-weight of benthic fauna(c). Closed aquare, closed circle, and open triangle indicate forest, gap, and creek, respectively.

3.2 Primary producers, POM, and SOM

In mangrove forests, mangrove trees, phytoplankton, and microphytobenthos are regarded as primary producers (e.g. Bouillon et al., 2002, 2004). Stable isotope ratios of them showed different values; mangrove (dry season: $\delta^{13}C = -28.9 \pm 0.7$, $\delta^{15}N = -1.1 \pm 0.9$, rainy season: $\delta^{13}C = -29.2 \pm 0.1$, $\delta^{15}N = 2.0 \pm 0.2$), phytoplankton (dry season: $\delta^{13}C = -23.1 \pm 0.8$, $\delta^{15}N = 4.7 \pm 0.2$, rainy season: $\delta^{13}C = -23.6 \pm 1.8$, $\delta^{15}N = 4.1 \pm 0.1$), and microphytobenthos (dry season: $\delta^{13}C = -18.2 \pm 0.7$, $\delta^{15}N = 0.8 \pm 0.9$, rainy season: $\delta^{13}C = -17.4 \pm 0.3$, $\delta^{15}N = 1.2 \pm 0.3$) (Table 2). The values of $\delta^{13}C$ and $\delta^{15}N$ of producers fall in ranges which have been reported in other mangrove forests (e.g. Bouillon et al., 2002, 2004). Stable isotope ratios of POM and SOM were not different among stations (Table 2). The values of $\delta^{13}C$ and $\delta^{15}N$ of POM plotted intermediate between mangrove and phytoplankton (Table 2, Figure 3). The values of $\delta^{13}C$ and $\delta^{15}N$ of SOM showed similar to mangrove (Table 2, Figure 3). POM is supposed to be composed of phytoplankton and mangrove, and major component of SOM is mangrove litter.

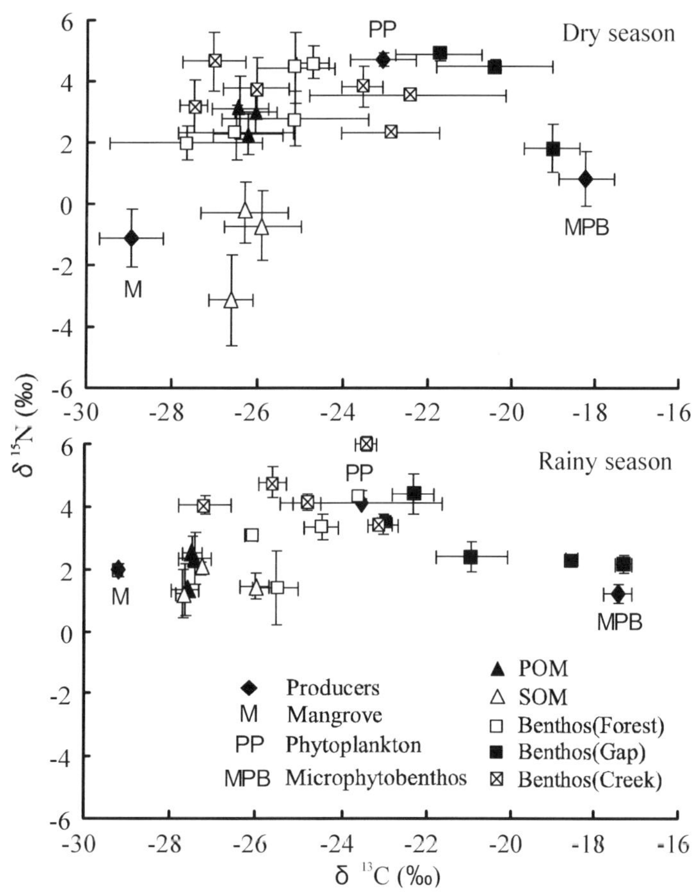

Figure 3. Plot of $\delta^{13}C$ and $\delta^{15}N$ for producers, POM, SOM, and benthic fauna.

3.3 Benthic fauna

Benthic fauna were dominated by deposit feeder, the numbers of deposit feeder individuals were 253 out of 292, and the biomasses of deposit feeder were 12.6g out of 13.5g. The appearance trends of deposit feeder were not different greatly between seasons or among stations (Table 3). For example, *Ovassiminea brevicula* appeared at all stations except for at CR in rainy season (Table 3). The benthic fauna would not be different greatly among stations.

Table 3. Wet weight and stable isotope ratio of main dominant fauna.

	Feeding type	Wet weight (g)	Wet weight (%)	δ ^{13}C± SD	δ ^{15}N± SD
Dry Season					
Forest					
Uca bengali	Deposit feeder	0.09	27.9	-25.1± 0.9	4.4± 1.2
Ovassiminea brevicula	Deposit feeder	0.05	14.8	-26.5± 1.3	2.3± 0.9
Laternula truncata	Suspension feeder	0.04	13.9	-25.1± 1.7	2.8± 0.9
Wolffogebia inermis	Suspension feeder	0.02	6.0	-24.8± 0.4	4.6± 0.5
Sphaeromatidae sp.	Deposit feeder	0.01	4.0	-27.7± 1.8	2.0± 0.5
Gap					
Uca annulipes	Deposit feeder	1.99	61.4	-19.1± 0.7	1.8± 0.8
Cericidea cingulata	Deposit feeder	0.92	28.3	-21.8± 1.0	4.9± 0.2
Ovassiminea brevicula	Deposit feeder	0.18	5.5	-20.4± 1.4	4.5± 0.2
Creek					
Metaplax elegans	Deposit feeder	0.32	31.4	-22.9± 1.1	2.3± 0.1
Wolffogebia inermis	Suspension feeder	0.20	20.1	-27.5± 0.3	3.2± 0.9
Laternula truncata	Suspension feeder	0.18	17.4	-26.0± 0.8	3.8± 1.0
Metaplax distincta	Deposit feeder	0.17	16.4	-22.5± 2.3	3.6± 0.2
Ovassiminea brevicula	Deposit feeder	0.01	0.7	-23.6± 0.5	3.8± 0.7
Cylindrotis quadrasi	Deposit feeder	0.01	0.7	-27.0± 0.7	4.6± 1.0
Rainy season					
Forest					
Uca bengali	Deposit feeder	0.21	37.8	-24.5± 0.4	3.3± 0.4
Perisesarma indiarum	Deposit feeder	0.16	28.7	-26.1± 0.2	3.1± 0.4
Laternula truncata	Suspension feeder	0.06	11.4	-23.5± 0.1	4.4± 0.1
Ovassiminea brevicula	Deposit feeder	0.02	3.0	-25.5± 0.5	1.4± 1.2
Gap					
Cericidea cingulata	Deposit feeder	5.38	73.7	-22.3± 0.5	4.4± 0.7
Uca annulipes	Deposit feeder	1.30	17.8	-17.3± 0.2	2.2± 0.3
Uca forcipata	Deposit feeder	0.41	5.6	-18.5± 0.1	2.3± 0.1
Ovassiminea brevicula	Deposit feeder	0.05	0.1	-20.9± 0.8	2.4± 0.5
Serenella indica	Deposit feeder	0.02	0.1	-23.0± 0.2	3.4± 0.1
Creek					
Metaplax elegans	Deposit feeder	0.87	43.8	-24.8± 0.3	4.1± 0.3
Uca forcipata	Deposit feeder	0.38	19.3	-23.0± 0.4	3.4± 0.3
Cericidea cingulata	Deposit feeder	0.29	14.5	-27.2± 0.6	4.1± 0.3
Laternula truncata	Suspension feeder	0.22	11.1	-25.6± 0.3	4.8± 0.5
Metaplax distincta	Deposit feeder	0.12	6.0	-23.5± 0.3	6.0± 0.2

Stable isotope ratios of benthic fauna were different among stations (Table 3). The values of δ^{13}C and δ^{15}N of benthic fauna at FO/CR were similar to mangrove litter and phytoplankton, while those at GA were similar to microphytobenthos and phytoplankton (Figure. 3).

Even though, same species; *Ovassiminea brevicula*, showed different stable isotope ratios among stations (Table 3), and the values of $\delta^{13}C$ and $\delta^{15}N$ at FO/CR were similar to mangrove and phytoplankton, while those at GA were similar to microphytobenthos and phytoplankton (Figure 3). These results indicate that the benthic fauna inhabiting at a creek and an inside-forest depend their food on mangrove litter and phytoplankton. On the contrary those at a forest gap depends their food on microphytobenthos and phytoplankton. In turn, a forest gap would be a unique environment in terms of basic production.
These results would also explain the contradiction between ignition loss of SOM and biomass at GA; in spite of the lowest SOM, the highest biomass sustained at GA. The benthic fauna inhabiting at forest gap would not utilize SOM derived from mangrove but microphytobenthos.

4. Conclusions

Benthic fauna inhabiting at a creek and an inside-forest depended their food source on mangroves, but those at a forest gap depended their food on micophytobenthos. There are two different systems in terms of basic production, in mangrove forests. Hence, we should recognize the mangrove forests as a mixed system of two different productive structures.

5. Acknowledgement

The Authors are grateful to the staffs of Rajamangala Institute of Technology for assistance in the fieldwork. Constructive comments on the manuscript from Dr. Satoshi Ishikawa were much appreciated.

6. References

Alongi D. M. (1994). Zonation and seasonality of benthic primary production and community respiration in tropical mangrove forest. *Oecologia*, 98, 320-327.

Bouillon S., Koedam N., Raman A. V., and Dehairs F. (2002). Primary producers sustaining macro-invertabrate communites in intertidal mangrove forests. *Oecologia*, 130, 441-448

Bouillon S., Moens T., Overmeer I., Koedam N., and Dehairs F. (2004). Resource utilization patterns of epifauna from mangrove forests with constrasting inputs of local versus imported organic matter. *Mar. Ecol. Prog. Ser.*, 278, 77-88

Camilleri J. C. (1992). Leaf-litter processing by invertebrates in a mangrove forest in Queensland. *Mar. Biol.*, 114, 139-145

Currin C. A., Newell S. Y., and Pearl H. W. (1995). The role of standing dead *Spartina alterniflora* and benthic microalgae in salt marsh food webs: considerations based on multiple stable isotope analysis. *Mar. Ecol. Prog. Ser.*, 121, 99-116

Kathiresan K., and Binbham B. L. (2001). Biology of mangroves and mangrove ecosystems. *Ad. Mar. Biol.*, 40, 81-251

Lee S. Y. (1999). The effect of mangrove leaf litter enrichment on macrobenthic colonization of defaunated sandy sbstrates. *Estuar. Coast. Shelf. Sci.*, 49, 703-712

Micheli F. (1993). Feeding ecology of mangrove crabs in North Eastern Australia: mangrove litter consumption by *Sesarma messa* and *Sesarma smithii*. *J. Exp. Mar. Biol. Ecol.*, 171, 165-186

Newell R., Marshall N., Sasekumar A., and Chong V. C. (1995). Relative importance of benthic microalgae, phytoplankton, and mangroves as sources of nutrition for penaeid prawns and other coastal invertebrates from Malaysia. *Mar. Biol.*, 123, 595-606

Page H. M. (1997). Importance of vascular plant and algal production to macro-invertebrate consumers in a southern California salt marsh. *Estuar. Coast. Shelf. Sci.*, 45, 823-834

Riera P., Stal L. J., Nieuwenhuize J., Richard P., Blanchard G., and Gentil F. (1999). Determination of food sources for benthic invertebrates in a salt marsh (Aiguillon Bay, France) by carbon and nitrogen stable isotope: importance of locally produced sources. *Mar. Ecol. Prog. Ser.*, 187, 301-307

Robertson A. I. (1986). Leaf-burying crabs: their influence on energy flow and export from mixed mangrove forests (Rhizophora spp.) in northeastern Australia. *J. Mar. Exp. Biol. Ecol.*, 102, 237-248

Robertson A. I., Alongi D. M. and Boto K. G. (1992). Food chains and carbon fluxes. In: Robertson A. I., Alongi D. M. (eds) *Tropical mangrove ecosystems*. American Geophysical Union, Washington, D. C., pp. 293-326

Ute J., Katja M., Thomas B., Rainer K., and Kerstin B. (2005). Stable isotope food web studies: a case for standardized sample treatment. *Mar. Ecol. Prog. Ser.*, 287, 251-253

Zar J. H. (1999). *Biostatistical Analysis*, 4th ed. Prentice Hall, New Jersey. 1999

Water Environmental Conservation for Improved Livelihood in the Mekong Delta, Vietnam

Tran Thi Trieu[1)*], Le Anh Tuan[1)**], Mira Kakonen[2)], Marko Keskinen[2)] and Le Duc Toan[1)]

1) Department of Environmental and Water Resources Engineering (DEWRE)
College of Technology (CoT), Can Tho Unversity (CTU)
Campus II, 3/2 street, Can Tho City, Vietnam
Corresponding e-mail: * tttrieu@ctu.edu.vn, ** latuan@ctu.edu.vn
2) Water Resources Laboratory, Helsinky University of Technology
Tietotie 1E, 02150 Espoo, Finland

Abstract

Livelihoods of major population in the Mekong Delta (MD) highly depend on water resources of the Mekong river, especially in Southern Vietnam, lowest part of the Mekong River Basin. While most of Vietnamese farmers in MD use water from the rivers and canals for their daily drinking, irrigation, domestic use, etc., effects of water environment to human life is more crucial. Vise visa, impacts of human activities to water environment are also considered to ensure the sustainable development of the MD.

A case study on water environmental conservation was done in Long An province, Vietnam. Long An, part of the Plain of Reeds where is remarkable vulnerable area with flooding and saline intrusion, has been suffering all water related problems including extra water, shortage of drinking water, saline water, acid water, water pollution by upstream as well as on-stream users. The most water related problems concerned by farmers in this area was access to get clean water.

Water environment conservation is one of the most urgent actions to be done by increasing awareness for local people. Integrated water resources management was considered. "Living with flood" has been trend in the area. Sharing information between provinces is needed to get synthetic data of water resources to adapt flood and orient it become "nice flood" from which local people can utilize its benefit such as alluvia, natural fishes, soil improvement, ...

Key words: livelihood, Mekong Delta, water environmental conservation, water-related problems

1. Introduction

The Mekong River Delta (section of the delta in Vietnam), with its four million hectares of lands for nearly 18 million of Vietnam inhabitants, is great potential for agricultural production. The Delta supplied more than 50 percent of staple food and 60 percent of fish production of Vietnam (Minh, 2000). However, the MD population, especial in the remote areas, is facing to water problems from day to day. Long An province (part of the Plain of Reeds as figure 1 and figure 2) suffer complexity of water related problems including not only flood in the wet season but acid sulfate water, saline intrusion, lack of clean water in the dry season. Water environmental conservation is an urgent action. Better access to water and sanitation will improve the health and productivity of poor rural people and contribute to a sustainable reduction of poverty in the target communities.

This case study aim to understand the most water related problems which link to livelihood of grass root level in the area. The interview and group discussion with officers and farmers at four levels are to comment participatory water environmental conservation for research areas which help to improve living condition of the poor in the MD.

2. Area Description

Long An is located among the nine provinces of the Mekong Delta in the South Vietnam and faces Ho Chi Minh City in the East. The area is 4,492 sq. km, lived with population of 1,392,300 habitants (2003) and Capital Tan An Town. It has 12 districts and as a part of Plain of Reeds. The below map shows the position of Long An in the Mekong Delta.

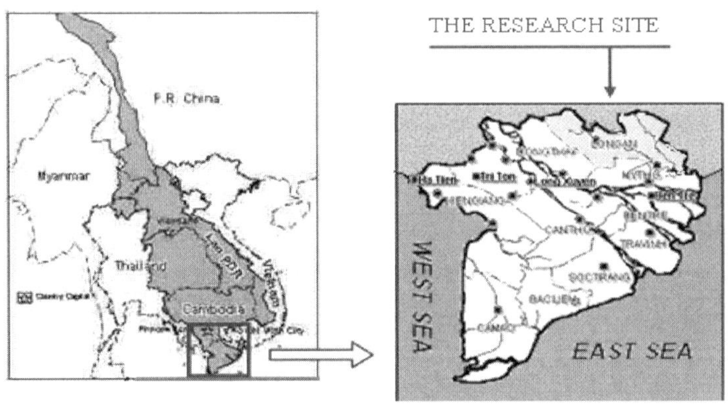

Figure 1: Research site map in the Mekong River Delta

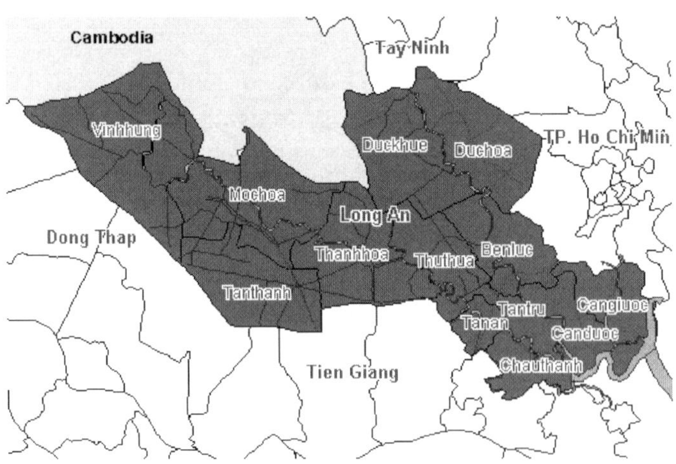

Figure 2: Long An map

3. Method

The survey was carried out at 4 levels including province, district, commune leaders and local farmers. The method based on Participatory Rapid Appraisal (PRA) and Rural Rapid Appraisal (RRA). At the province, district and commune levels, key informants from Provincial Department of Agriculture and Rural Development (DARD), Department of

Natural Resources and Environment (DONRE), Women's Union, Youth's Union were invited for interviewing. At the grass root level, two groups of 15-20 farmers took parts in group discussion. To reach the objectives of the study, mapping, semi-structure checklists, problems ranking were applied.

The result of survey was reported again to get feedback from four levels. The information correctly reflected the situation and was the basis for water conservation. In additional, there were 5-6 private interviews of farmers to confirm and deeply understand farmer 's problems and aspiration on water conservation for their livelihood improvement.

4. Result and discussion

4.1 Water related problems with livelihood

4.1.1 Flood

The MD has 1.2 to 1.9 million hectares (ha) of land affected by annual flooding. In the year of 2000, Vietnam experienced one of largest and most damaged flood in 70 years (Minh, 2000). Long An is affected flood from the Mekong river, it runoffs through Cambodia border, flows down the Tien river and surmerges all main canals in the Plain of Reeds. This province is also affected by flood from Vam Co Tay river. About 24,963 ha of paddy rice, 913 ha of aquaculture, other crops as well as infrastructure were damaged by the flood in the year 2000. The flood killed 78 persons including 39 children and lead 21,600 households to starveling (Report of Long An DARD, 2000). Hamlet number 6, Tan Lap commune lost a lot by flood before the year of 2000 because poor, homeless people lived in the field with unstable, small houses. Children have to leave school in flooding season. Flood occurs in Long An from August to October. In the year 2004, the curve of flood was very strange, it quickly increased to the peak and drained. The officers explain that the flood was depend on rain in the upstream areas. Deforestation for land use change has been listed as a root cause of increased magnitude of flooding (Sitanon, 2004). Damages caused by flooding gradually decreases. Table 1 shows damages caused by flood from 2000 to 2004.

Table 1: Flood damages in Long An

Damaged Items	Unit	2000	2001	2002	2004
Rice (lost)	ha	24 693	539	84	0
Cash crop	ha	1 827	664	25	0
Aquaculture	ha	913	381	-	-
Infrastructure (collapsed bridges)	unit	301	14	-	5
Death	person	78	39	29	7

(Source: Long An DARD, 2005)

However, flood also brings benefit to people. Farmer need flood because it transfers alluvium to the rice field, brings natural fishes, brings plankton which is food for fishes in cases and pools, improves water and soil quality. Flooding flushes toxicity released from acid sulfate soil. An extremely large amount of water is needed to dilute the leachate in order to prevent negative impacts to the environment (Minh, 1998). Every people agreed that "Living with flood" is the best way for the Mekong Delta sustainable development. Sharing information among communes, districts and provinces aim to get synchrony water database and orient flood to become "nice flood" which brings benefit to people.

4.1.2 Acid sulfate soil
Acid sulfate soil (ASS) occupies a large area of Long An. That is an serious problem in the dry season for agriculture and aquaculture. Acid sulfate soil forms when soil containing iron sulfite are exposed to oxygen (Minh et al, 1996). Disturbance of acid sulfate soil will increase the acidity discharging into the water body.

To cultivate crop in the acid sulfate soil, there is necessary to reduce acid sulfate content by a large amount of fresh water for flushing. Lime application is also a way for reduce acidity in the soil but it will increase the production costs to the farmers. The leachate will ultimately discharge into the river which may cause water pollution and kill aquatic life.

4.1.3 Saline intrusion
Saline intrusion affects to Long An through Vam Co Dong and Vam Co Tay Rivers. Vam Co Dong river springs from Cambodia, meets the sea at Xoai Rap estuary, is affected by saline intrusion up to the Cambodia border, including the Plain of Reeds. The most saline intrusion was in 1977 up to Go Dau Ha. In year 1974, saline intrusion reach to Tan Thanh through Vam Co Tay river (Thuan, 2004). The years with high saline intrusion were of years 1992, 1993, 1996, 1997, 1998. Peak of saline intrusion is in March yearly. In year 1996, the concentration of salinity was over 3 mg/L.

Saline intrusion causes low yield of rice crop and damages the boat generators in the rivers and canals. According to an officer in district level, it is very complex to control saline intrusion. The dike system for prevention of saline intrusion will increase the harm of acid sulfate soil to aquaculture and agriculture in the Plain of Reeds. Saline intrusion is one of the principal limiting factors in agricultural production.

Most of the area is under mono-crop rice cultivation, where most of the poor provinces with a high ratio of poor farmers are located. Shortage of drinking water is another constrain for local people. In order to prevent further salinity intrusion, it is necessary to keep flows from dropping too low (WB-ADB, 1996).

As a impact, the more drought leads the more saline intrusion. Saline intrusion period is about 3 months, from March to May yearly. At this period, almost no vegetables, kitchen-gardens and other crops can stand. Many farmers, including women and their children have to move to other places to earn their life. They become stipendiary. Some women do small trading, hired labour. Farmers make use the time to repair their house, canal. In 1996, serious saline intrusion went up to Hung Thanh, Duc Hoa districts. At that time, Vinh Hưng, Moc Hoa districts were in highest shortage of irrigation water. Some of Northern districts had to sow rice and pea nuts twice. Depending on the water supply capacity and land use pattern, there are two major rice cropping systems in the Long An as figure 3:

(i) the single rice crop as *Mua* crop (rainfed rice); and
(ii) double rice crop as *Dong Xuan* crop (Winter-Spring) and *He Thu* crop (Summer - Autumn) or *He Thu* crop and *Mua* crop.

The different in the rice yields in the wet and the dry seasons and the seasonal effects to the livelihood is presented in table 2.

4.1.5 Water quality
Pollution of water resources caused by pesticides, herbicides and fertilizers in the past ten

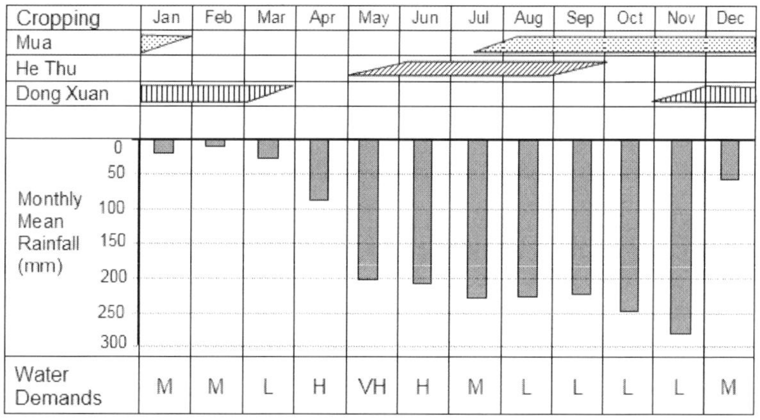

Figure 3: Cropping calendar, monthly rainfall and water demands in Long An
Note: VH: very high; H: High; M: Medium; L: Low

Table 2: Seasonal effects to the rice production and livelihood in Long An

Year	Spring rice yield (ton/ha)	Summer rice yield (ton/ha)	Effect of water regime to paddy field and livelihood
1996	4.80	3.02	• Early and long stayed saline intrusion (4 g/L)
			• Lack of irrigation water
			• Lost 10,497 ha of summer paddy rice
			• Rural women and youth moved to the city
1997	4.70	3.32	• Still suffered effects of the harsh condition of 1996
1998	4.39	2.62	• Complicated climate, long dry season
			• 11,000ha paddy rice of Moc Hoa, Tan Thanh districts was lost by salinity and ASS
1999	4.18	2.97	• The 3rd rice crop was affected by ASS
2000	4.20	2.82	• Big flood
			• 24,693 ha of rice lost
			• 78 person were died by flood
2001	4.40	2.73	• Slowly drainage of flood
			• Spring rice crop was later 1 month
			• 10,675 ha of rice lost
2002	4.50	2.80	• Better climate for rice cultivation
2003	5.02	2.99	• Applying Integrated Pesticides Management (IPM)
			• Improving varieties
			• 80% of population have stable life
2004	5.08	3.25	• Excavating canals
			• irrigated area increased to 40,000 ha
			• Developed semi-dyke system to prevent early flood
			• Improving water resources system management and decentralizing management to district and commune

(Source: Long An DARD, 2005)

years (An, 2000). Water quality issue impacts the people widely. Recent years, development of agriculture adding the expanding population force Long An province to confront water environmental problems. The beginning of the rainy season is the problematic period.

The main concern in the village level was the water quality, especially, the water supply for domestic usage. For the interviewed villagers, water quality cause more worries than the

flood. The main cause of surface water pollution is acid sulfate soil, followed by agro-chemicals and awareness of local people. According to directly interviewed local farmers, some people have had bad habits: dropping dead animals, domestic solid wastes and chemical containers to the water bodies. "Overhung toilets" or "Fish pond toilets" which contributed micro-organism pollution to surface water, is still popular in hamlet No.6
The poor water quality leads to the water supply problems. Immediately need of hamlet No.6, Tan Lap commune is water supply system. They can not use ground water because of very high iron content. Many local people in Hamlet No. 6 and No.4 of Tan lap commune, Moc Hoa district, use rain water in a haft of year, in the rest time they have to buy water from Tan Thanh district with the cost of 2,000 VND /3 liters of water for drinking. For other purposes of domestic usage such as washing, bathing, ... they take polluted water from canals and rivers. Expenditure for high cost drinking water partly affects to their livelihood that may become one of busy and hard loads for women and children in taking care their daily life.

4.2 Water environmental conservation

4.2.1 Behavioral changes for polluted water reduction

Public education is an essential component of a successful water conservation program. Difference organizations at all levels combine to one another in communication of environmental protection. To reduce the effect of intensive agriculture to water environment, experiments of using bio-methodology were carried out. IPM and "3 ups and 3 down" (rising up the yield, the effectiveness and the qualify and decreasing down the seed, the fertilizers and the pesticides) programs have been quite successful and welcomed. Method of row sowing is widely applied to reduce using of fertilizers and seeds.

The villagers issued the regulation that throwing dead animal or pesticide containers into the water body is punished.

Another way to provide public information and education on water environmental conservation is to present the demonstration models, for instant the demonstrative toilets. The "overhung toilets" are replaced by hygiene ones. Biogas systems were introduced to farmer with large animal husbandry farms. The sludge of biogas tanks will be a very good fertilizer for agriculture and horticulture. Mass media were widely used to educate public on water environmental conservation such as announcement in the newspapers, pamphlets, local radio and television system. Public school education is also an important means for instilling water conservation awareness (Grisham and Fleming, 1989).

4.2.2 Integrated water resource management

Water is the main component of the environment and essential element for human life (Igor, 2000). Water management is much more than the management of water. In MD, water is basic of life for millions of people to whom water provides directly or indirectly the source of livelihood. The Mekong immense water resources are one of the key factors for social and economic development in the region (Marko, 2004).

** Identify water resources management*

Provincial authority had official plans on surface water exploitation for each district. Head water resources stations in the provincial level manage lower irrigation systems. In district and commune levels, water management teams are organized. Local people operate and maintain irrigation systems themselves via the control of irrigation teams. Inter-field canals were managed by farmers. In Tan Lap commune, there are also teams spontaneously formed by Youth's Union which were in charge of canals and semi-dike systems. Each team with 4-5 persons managed a part of canal and the team leader was voted by members. He had right to collaborate with other teams and sign contract of irrigation and canal excavation with farmers. The operation and maintenance cost was contributed by farmers on their irrigated area.

* *Socialization of water resources management*
Socialization of water resources management is implemented according to the motto: "Both the Government and People do together". At the same time, the Government advocated rallying all popular efforts and encouraging foreign and domestic investment in water resources exploitation and construction of water structure based on democratic participation and equal benefit (Nghia, 2001). In Moc Hoa district, main canal systems were constructed by province and national 135 program. Lower level canals were invested by district and local people. Almost canal excavation cost was taken from water fee and subsided from national budget.

4.2.3 Legal documents on water conservation in Vietnam
The Law on Water resources which was approved by the National Congress in 1998 and came into effect in 1999. Beside the Law of Water Resources, Vietnam has other laws related to water conservation as:
Forest Protection and Development Law (1991).
Environmental Protection Law (1993).
Rules and Regulations on water usage, water quality, water contamination issued by Ministry of Science, Technology and Environment (MOSTE).
Articles related with environmental protection in Civil Code (1995) and Criminal Code (2000).

There were some important national programs supported and concerned water resources management in the MD as listed below:
Watershed management and reforestation program: The program focus on improvement of environment in general, contributed to improve water resources and flood control in Vietnam.
Food security and rice export (2000): Cultivation of high-yield rice varieties might cause water pollution by agro-chemicals and fertilizers (Ongley, 1996). So, the policy on balancing agriculture development and environment protection is crucial for sustainable development of the Mekong Delta.

4. Conclusion

Sustainable water management (SWM) should not only control the water resources towards the present needs but also consider water-related problems in the future (Tuan, 2004). The research area suffers all problems connected to water: poor water quality, water scarcity, flood, saline water intrusion, acid sulfate soil, which lead local people to poverty. They are difficult to develop the livelihood structure.

The cropping calendar highly depends on weather and river water regime. Lack of clean water is biggest problems which directly impact to local people's daily life. The remote area is the most difficulty area. Water related issues are very complex in the research area and thus challenging for water conservation. The trend of livelihood in Long An is living with flood. The objective should always be the people. The problem and solution should be formulated by citizens, of course, the local people themselves is the most important actor in the water environmental conservation.

5. Acknowledgement

Authors would like to say sincerely thanks the WUP-FIN project for survey funding support and staff members of DEWRE/CoT in CTU for their help on research. We are also not forget that this research cannot finished without the closed collaboration of the Long An province leaders and district farmers.

6. References

An, L.Q. (2000). Bio-diversification and problems in biological Resources Protection. *Journal of Environmental Protection*, 8, 6-10 (in Vietnamese).

Grisham, A. and W. M. Fleming (1989). Long term options for municipal water conservation. *Journal of the American Water Works Association* 81 (3):33

Igor A.Shiklomanov (2000), *Appraisal and Assessment of World Water Resources*, Water International, Volume 25. Number 1, March 2000, pp. 11-32.

Long An Department of Agriculture and Rural Development (2005). *Report of flood damages*. Long An, Vietnam.

Marko Keskinen (2004). The Lake with Floating villages: A socio-economic analysis of the villages around the Tonle Sap Lake, *Proceeding of the Workshop on Integrated Water Resources Management on Tonle Sap Lake*, 1-2 February 2004, Phnom Penh, Cambodia.

Minh L. Q. (2000). *Environmental Governance: A Mekong Delta case study with downstream perspectives*, Can Tho University, Vietnam.

Minh L.Q., Tuong T.P, Xuan V.T. (1996). *Leaching acid sulfate soil and its environmental hazard in the Mekong Delta,* in G.L. Denning and Vo Tong Xuan, Vietnam-IRRI a partnership in rice research, IRRI and MAFI, pp. 99-109, IRRI, Philippines.

Nghia T.T., (2001). Vision of Vietnam's water, life and environment in the 21st century, *From Vision to action-A synthesis of experiences in southeast Asia*, The FAO-ESCAP pilot project on national water vision, pp. 99-103, Bangkok, Thailand.

Onlgley, E.D. (1996). *Control of Water Pollution from Agriculture*, FAO Irrigation and Drainage Pager 55, Rome, Italia.

Sitanon J. (2005). *Managing Transboundary Conflicts in The Lower Mekong Basin, Transboder Issues in the Great Mekong Sub-region*, Ubon Rachatani University, Thailand

Thuan, T.M., N.X. Truong, B.T. Vuong, L.A. Tuan, T.V. Phan (2004). *Monitoring and exploitation of domestic water*, National University of Ho Chi Minh City Publisher, Vietnam (in Vietnamese).

Tuan, L.A., G.C. L. Wyseure , L.H. Viet (2004). *Sustainable water management for rural development in the Mekong River Delta, Vietnam.* The second International Symposium on Southeast Asian Water Environment (Poster presentation), Hanoi, Vietnam.

World Bank/ADB/FAO/UNDP/NGO/GOV. (1996). *Water Resources Sector Review*, Washington, USA.

Strategies for Sustainable Management of Fishery Resources in the Pasak Jolasid Reservoir, Thailand through Physicochemical Assessment

Charumas Meksumpun

Department of Fishery Biology, Faculty of Fisheries, Kasetsart University
Chatuchak, Bangkok 10900, Thailand
e-mail: ffiscmc@ku.ac.th

Abstract

Spatial and temporal variations on physicochemical characteristics of aquatic ecosystem in the Pasak Jolasid Reservoir, Thailand have been critically determined and analyzed under the main objective to provide suitable strategies for sustainable management of fishery resources in the reservoir. Accordingly, field surveys had been carried out in the whole reservoir area at every 2-months interval for 1 year (from February 2002 to February 2003). The research generated considerable data on spatial and temporal variations of water quality that increased our understanding of nutrients, chlorophyll *a*, and other factors in the reservoir. From the data analysis, the Pasak Jolasid has been confirmed as eutrophic or hypertrophic reservoir where can significantly contribute to development of fishery resources. Annual average primary productivity of depth integrated (0-10 m) was found to be *ca* 13,150 g C/m^2/year. Nevertheless, seasonal flooding pattern, inflow load, and water outlet management should be vital factors in controlling its potential on related fishery production. Vertical stratification that usually occurred in the areas with depth of more than *ca* 10 m could also cause deterioration of water condition. Appropriate management scheme for sustainable fishery management should deeply concern categorization pattern of reservoir topography and related ecological habitats. For future effective management of the Pasak Jolasid reservoir, researchers are recommended to monitor suitable stations within 4-5 zones of spatial differences in water environment for 3-4 times a year.

Keywords: water qualities, Pasak Jolasid reservoir, Thailand, spatial and temporal variations, fishery resource management, primary productivity

1. Introduction

Reservoirs have been known to use for fish production in most Asian countries (De Silva and Amarasinghe, 1996), providing an important source of protein for rural communities in the region including Thailand. However, production of fish has been found to decline after impoundment of reservoirs as evident by disappearance of some fish species decrease in overall. Recognizing this problem, research on reservoir ecologoical status was carried out under an integrated research program on "*Strategies for sustainable management of fishery resources in the Pasak Jolasid Reservoir, Thailand through ecological and socio-economical assessment*" finded by the ASEAN Regional Center for Biodiversity Conservation (ARCBC) and the European Commission. The Pasak Jolasid Reservoir was selected based on its characteristics of a newly impounded reservoir established in 1998, including existence of transitional system from riparian to lacustrine. The reservoir was created by construction of

Pasak Jolasid Dam, an initiated by His majesty the King Bhumiphol. The dam was aimed to regulate seasonal discharge from head watersheds in Loei provine and prevents flooding in lower Chao Praya Basin including Bangkok and nearby areas. The total water body area of the reservoir is 18,256 ha or 183 square kilometer at MSL +43.0 m (RID, 1993).

Existing information and researches on reservoir ecosystems in Thailand remain inadequate for policy makers to make effective plans for conservation and rehabilitation of water resources in the country. This scientific study which integrates the reservoir's bio-chemical-physical conditions with related human impact, is essential in providing sound recommendation to policy makers. With more knowledge and understanding on the reservoir, strategies on sustainable management can be developed and later adopted for the management of other reservoirs in the country.

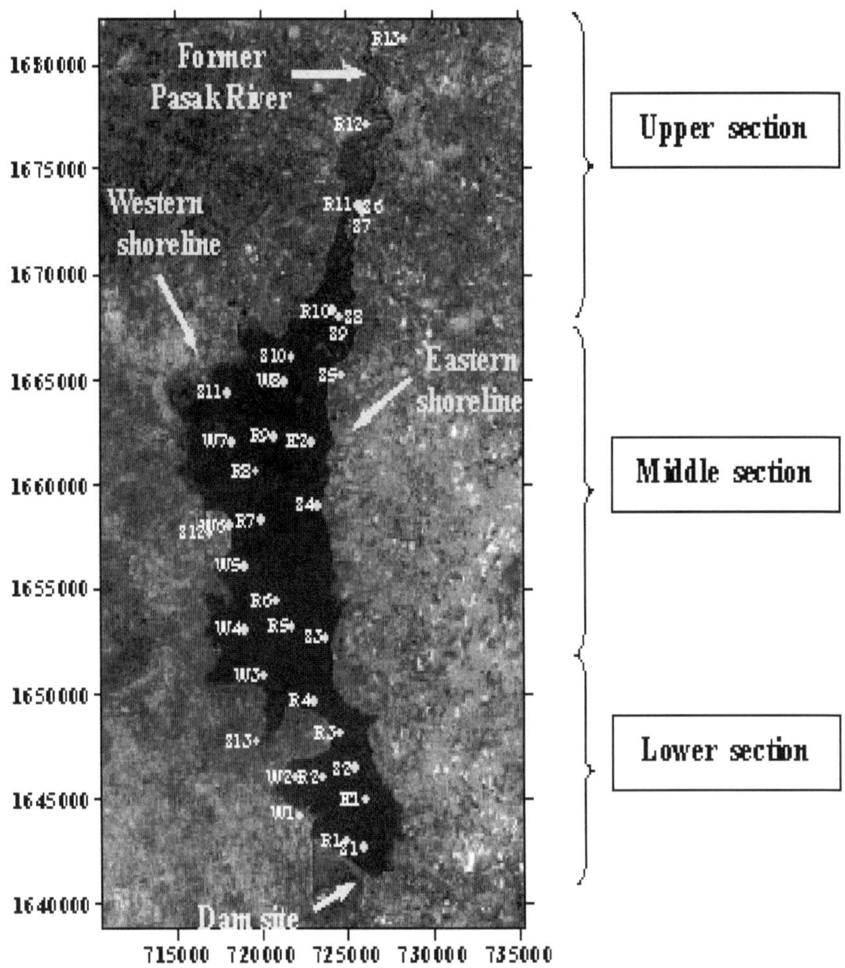

Figure 1. Sampling stations for water and sediment quality surveys in the Pasak Jolasid reservoir (R1-R13: located offshore / along the previously existed Pasak river, E1-E2: located offshore / in the eastern part of previously existed Pasak river, W1-W7: located offshore / in the western part of the previously existed Pasak river, S1-S13: located along the shoreline areas)

2. Materials and methods

Survey stations of the Pasak Jolasid reservoir were illustrated in Figure 1. The survey on physicochemical and biological properties of the reservoir water and sediments had been done for 2-month interval. Water depths and surface redox potential in all marked stations were measured in the field survey. During the survey, investigation on reservoir environment and community activities that may impact on water and sediment qualities was simultaneously conducted. Field examination on vertical gradients (1 m interval) of biological important nutrients (NH_4^+-N, NO_2^-+NO_3^--N, SiO_4-Si, PO_4^{3-}-P), dissolved oxygen concentration (DO), pH, temperature, conductivity, chlorophyll *a*, and total suspended solids (TSS) concentrations in all marked stations were carried out. Bottom sediments in some near shore and former Pasak river line were collected for further analysis of water content, total organic matter content, and acid volatile sulfide contents in laboratory. During the survey in April and October, study on primary productivity of the water column was also performed. Here we collected water from the depths of 0, 1, 2, 3, 4, 5, 6, 8, and 10 m, respectively, and then measured the initial DO on board. All water samples were filled very calmly into glass DO bottles (light/dark bottles) and lowered down again to their former depth for incubation at least 6 hr. After *in situ* incubation, all bottles were retrieved and DO levels were immediately measured for further calculation of depth-dependent primary productivity.

3. Results and discussion

Amount of rainfalls over the Pasak Jolasid reservoir during the year 2002 increased from below 30 mm to more than 170 mm during May. The levels of rainfall remained high until September, peaking at 248 mm in August, before drastically decreased toward the end of the year. Pattern of rainfall during the last decade consistently produced two peaks; the first during May and the second during August to September. Inflow into the Pasak Jolasid reservoir, however, increased at later period after the rainfall. In 2002, the reservoir received more than 2,000 million cubic meters of inflow during September. Water level remained high during October before gradually returned to normal levels during November to December. Air temperature over the reservoir in 2002 ranged between *ca* 26-31 °C with maximum temperature of 30.5 °C during April. There was no significant difference in air temperature throughout the year. Lower temperature was recorded during December and January, a common occurrence laving the seasonal Northeast monsoon. Humidity fluctuated within the ranges of *ca* 60-81 % throughout the year. Higher levels of humidity were found during the period of high rainfall (from May to October), while lower levels were recorded during dry seasons (from November until April). Wind speed was comparatively higher during the period of seasonal changes, particularly from rainy season to dry season at the end of the year.

3.1 Appearance characteristics of aquatic environment

With average depth of only 5 m, water color in Pasak Jolasid does reflect changes in water conditions in water column, geological topography, and littoral and sublittoral vegetation. Normally, the water color ranges from light green to turbid green. Water near shore was more turbid due to greater concentration of organic and inorganic materials from nearby land-based origins. Water in middle section of the reservoir is comparatively clearer, except for some periods of heavy bloom of phytoplankton. Water in upper part of the reservoir frequently turns yellow brown to brown by suspended solids from erosion and river inflow caused by continuous precipitation during rainy season. In 2002, the water levels dropped more than 4 m before the start of rainy season. This period led to heavy bloom of blue green algae occurring throughout the reservoir area during August.

Phytoplankton bloom is common in the middle section of the reservoir. Great accumulations of phytoplankton were also found along the western shoreline of the reservoir due to westward wind. In addition, inspection of the sediment corer obtained in the study revealed that surface sediment here is comprised of accumulation of dead phytoplankton from the water column and inorganic particulates discharged into the water by erosion and inflow.

3.2 Temporal variation of water levels1 and shoreline characteristics

Variation of water levels in the reservoir depends directly on amounts of inflow (inlet) and the outlet. In 2002, they were dramatically decreased (more than *ca* 4 m) during summer. The levels gradually increased at the beginning of rainy season (May) before stabilized at high levels during a period between October and February 2003.

The previously existed Pasak river forms the eastern part of the reservoir, while the western part connects to a large area, spreading out several kilometers. Shorelines with steep slopes can be generally found in narrower section north of the reservoir. Areas with rather flattened slopes can be easily found in the middle part of shorelines. The structure of former Pasak River (dominating in most of the upper part) does provide large amount of inflow during high loading period. Such inflow is partly regulated by an underwater road (located below Station S5, Figure 1) which crosses from east to west of the reservoir.

3.3 Spatial and temporal variation of physicochemical properties

Temperature

Surface temperatures of surface water in the Pasak Jolasid reservoir water recorded during the survey ranged between 24.8-35.1 °C. Comparatively higher temperature was normally found in the narrow upstrem areas (in the upper section of the reservoir) and the shallow area near shoreline, particularly along the western shore. Surface water temperature of the reservoir was found to fluctuated slightly (in the ranges of 2-3 °C) between stations during well-mixing period. However, more fluctuations were observed (in the ranges of 5-8 °C) during the water stagnant periods (in April and October 2002). During those periods, apparent vertical stratification of water temperature can also be found in the stations deeper than *ca* 10 m. Such stratification was found as the like cause for significant separation of upper and lower water mass during April. The stratification are weakened during October, allowing some biological components in the water mass (e.g. phytoplankton chlorophyll) to distribute throughout the depths of water column.

Dissolved oxygen concentration

Dissolved oxygen concentrations (DO) were generally found to be at very high to over-saturated levels. Surface water in the middle section of the reservoir area usually has higher DO levels than those found in the upper and lower sections. In each sampling, horizontal distribution of DO between stations was seemed to be consistent with levels of contribution by primary producers (chlorophyll *a* concentrations). The DO levels were found to be adversely effected by heavy load of particulate materials during late rainy season (in October 2002). The surface DO levels found in the survey period ranged between 4.4-11.1 mg/L. In addition to impacts from primary producers, the DO levels are known to be influenced by complex chemical reactions of substances from inflow or those of bottom origins. The decrease of DO with depth was also investigated at the stations with stagnant water and greater depth (e.g. Stations R5-R13). DO levels was found to drop to nearly zero at the depth of more than *ca* 5-8 m. Comparatively higher DO levels were found in middle section of the reservoir and the shallow water near shorelines (particularly along the western shore). Fluctuations of DO levels found are likely to be depended on quantitative distribution of chlorophyll *a* concentrations.

pH, turbidity, and other related parameters

pH and redox potentials in the Pasak Jolasid reservoir water ranged between *ca* 7.4-8.6 and 260-350 mV, respectively. Such ranges indicate oxidative conditions of the water column. In addition to influence from land-base particulates, relative high pH value could be the result of primary production. Turbidity and conductivity revealed other complex influences to water quality including effects from precipitation, inflow and mixing regime of water column. The conductivity of the reservoir water was at relative high level (with normal ranges of 320-460 microceimens). Slightly higher conductivities were often found in the upper section of the reservoir (e.g. in Stations R10-R13), particularly during inflow periods.

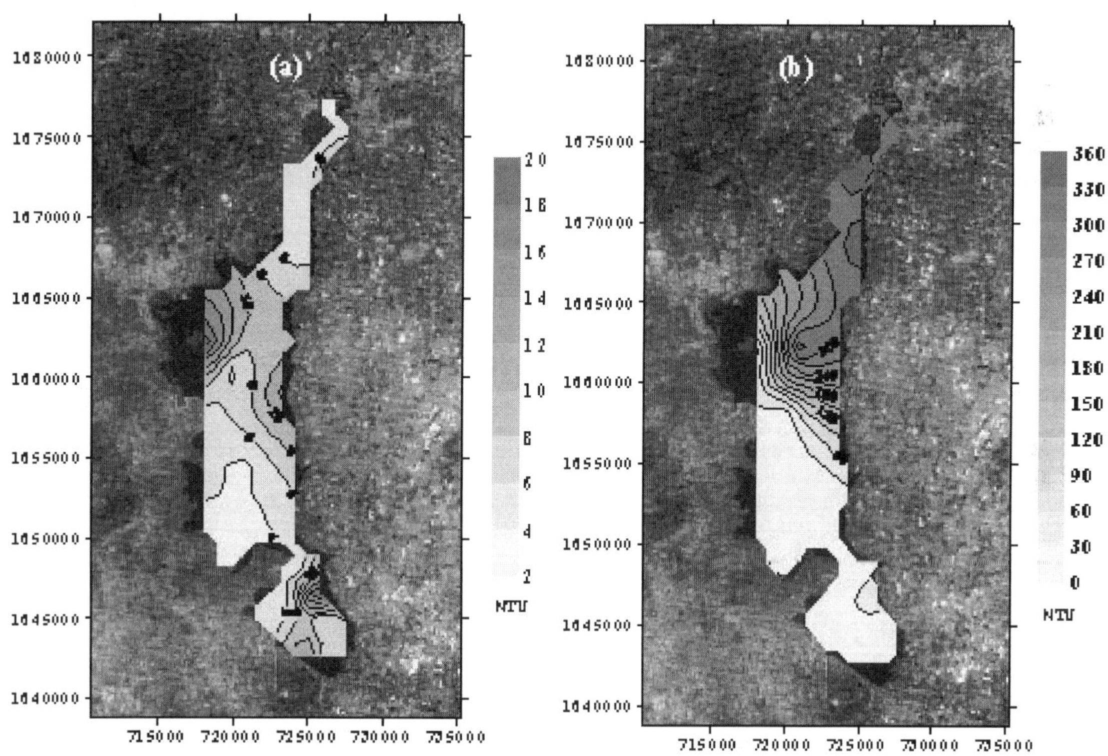

Figure 2. Comparison of turbidity (NTU) contours of surface water of the Pasak Jolasid reservoir during the beginning period of rainy season in June 2002 (a) and during mid-rainy season period in August 2002 (b)

Higher levels of turbidity (reaching over 350 NTU) in surface water were found during heavy precipitation and influx of turbid water in June (Figure 2b). The levels were 15-35 times of the normal conditions (Figure 2a) and can be found around upstream area and in the upper section.

3.4 Spatial and temporal variation of biological important nutrients

Ammonium nitrogen

The levels of ammonium nitrogen concentrations were found to be at moderate level (less than 5 μM). However, the levels did double in value during April when the water level was at the lowest and the water temperature was comparatively high. During this period, high concentrations were particularly high in areas with greater depths.

Nitrite-nitrate nitrogen

The levels of nitrite-nitrate nitrogen concentrations were found to be low (less than 6 μM). However, the levels were gradually increase to over 5 times of the normal level in August (with the maximum of *ca* 40 μM). Such an ocurrence was caused by heavy inflow in rainy season. During this period, the concentrations were particularly high in upstream areas with greater loads. The increases concentrations clearly demonstrated impacts of precipitation and inflow. The concentration should also play an important role in promoting the bloom of phytoplankton in the reservoir.

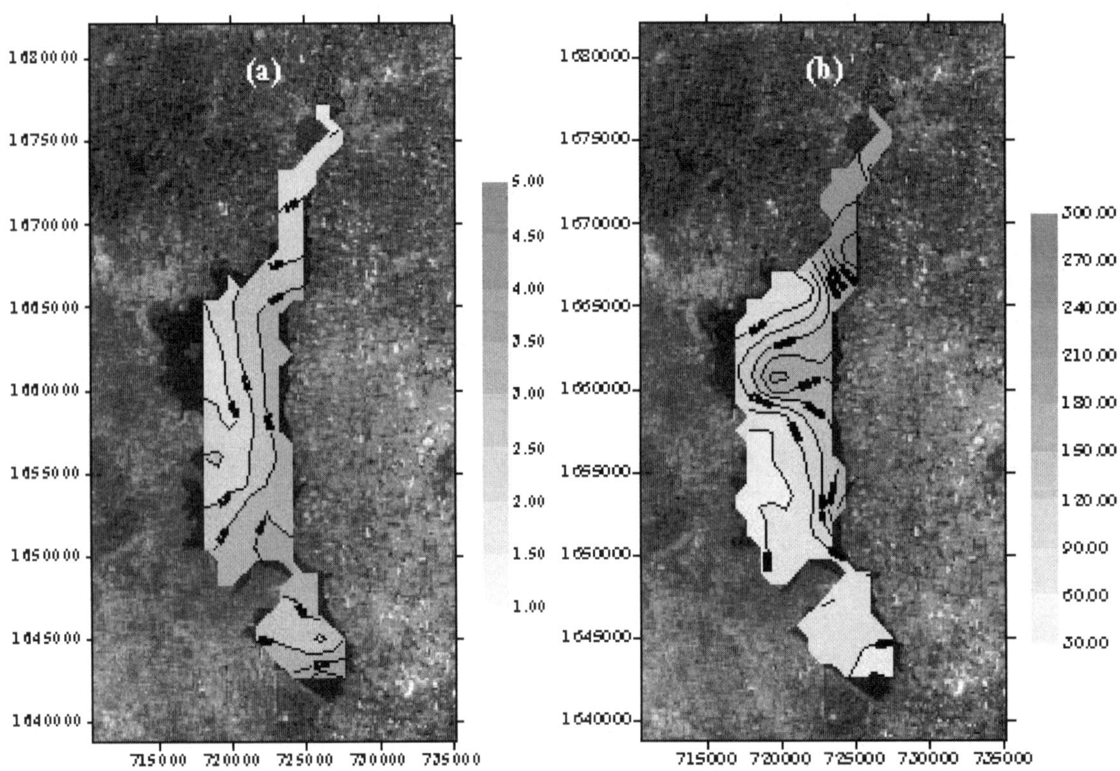

Figure 3. Comparison of silicate silica concentration (μM) contours of surface water of the Pasak Jolasid reservoir during the summer season in April 2002 (a) and during mid-rainy season in August 2002 (b)

Silicate silica

Silicate silica concentrations were found to be at high level (> 1 μM; Figure 3a). The levels rapidly increased from the start of rainy season (in June) to more than 30 times of the normal level in August (with the maximum levels of *ca* 300 μM; Figure 3b). Similar to nitrite-nitrate nitrogen concentrations, the high levels of silicates were due to heavy inflow that can be particularly observed in the upstream areas.

From the data, the reservoir does have excess amount of silicates (Figure 4a). Their movement was found to correspond with physical movement of the water and can thus become useful information managing newly settling particles and in establishing knowledge on movement regimes of related living species in the reservoir.

Orthophosphate phosphorus

Orthophosphate phosphorus were found to be in low to moderate levels (usually lower than *ca* 0.2 μM). The levels decreased during the start of rainy season and increased or became fluctuated toward the end of the year. There was no particular pattern of spatial distribution of the orthophosphate phosphorus. Nevertheless, some temporal and spatial variations could be due to the results of turnover from lower water mass and/or bottom deposits.

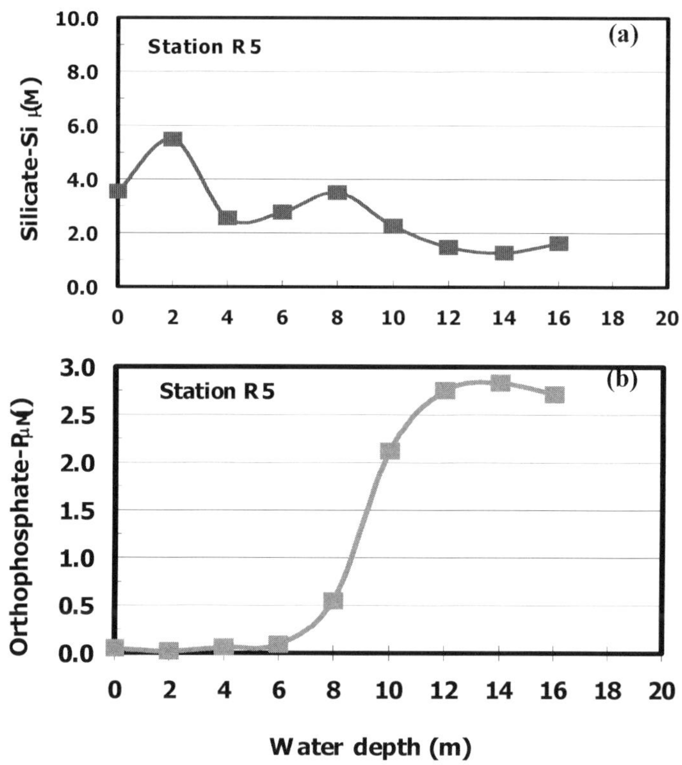

Figure 4. Comparison of vertical profiles of silicate silica (a) and orthophosphate phosphorus (b) concentrations (μM) in the water column at Station R5 of the Pasak Jolasid reservoir during April 2002

3.5 Primary production and productivity of reservoir waters

Chlorophyll a

High chlorophyll *a* concentrations (*ca* 20-80 μg/L) clearly confirmed the reservoir as a very productive water body throughout the year. The levels increased up to 100 μg/L during mid-rainy season. The high levels were frequently observed in the western part of the mid-reservoir area. Certain vertical stratification of chlorophyll *a* levels did reflect insufficient concentration of nutrients and influence of mixing regime in the water column. The stratification was much greater during the summer season period, when the water mass was more stagnant. Very little or none was found when the reservoir water received new inflow loads, particularly during mid- to late- rainy season. With high chlorophyll levels (OECD, 1982), it is possible to categorize Pasak Jolasid as a "eutrophic" to "hypertrophic" reservoir.

The levels directly responded with nutrient concentration/distribution patterns previously discussed and had significantly impacts to DO levels and related factors in the water column.

Primary productivity
Dark and light bottles technique was used in studying vertical distributions of primary productivity during periods with the lowest (April 2002) and highest (October 2002) water levels. In April 2002, the vertical stratification pattern of primary productivity was identified. Maximum primary productivity (*ca* 8 g O_2/m^3/day) was found at the depth of 4-m and the primary productivity was found within the depth of *ca* 8 m. Despite lesser chlorophyll *a* concentrations, primary productivity during October 2002 was found to be comparatively higher. They were found with higher rates (*ca* 15-20 g O_2/m^3/day) throughout the depth of water column. Primary productivity in the study area was assessed with the ECOPATH model. From the calculation, the annual average primary productivity of depth integrated (0-10 m) was found to be *ca* 13,150 g C/m^2/year.

4. Conclusions

The research has confirmed Pasak Jolasid as highly productive reservoir where can contribute significantly to development of fishery resources. It also found seasonal flooding pattern, inflow load, and water outlet management to be vital factors in controlling temporal and spatial fluctuations of water quality in the reservoir. In this reservoir, mixing of water caused by wind seemed to be significant in the shoreline areas with the depth of less than 4-5 m. Nevertheless, the areas with depth of over *ca* 10 m usually illustrated vertical stratification. Long term monitoring on organic sedimentation, decomposition, and regeneration processes would be useful for development of appropriate management strategies.

5. Acknowledgement

This work was financially supported by the ASEAN Regional Center for Biodiversity Conservation (ARCBC) and EU. Sincere gratitude is due to Department of Fisheries and Royal Irrigation Department for providing facilities and kind cooperation on the field survey. I gratefully acknowledge Prof. Ladda Wongrat and Dr. Prajit Wongrat for their invaluable advice throughout the works. Thanks are also due to Chira Ourmaim, Warin Pismoke, Kanokwan Pramong-udomrat, and Wirote Purat for ensuring successful field survey and laboratory analysis.

6. References

De Silva, S.S., Amarasinghe, U.S. (1996). Reservoir fisheries in Asia. In: *Perspective in Asian Fisheries. A Volume to Commemorate the 10th Anniversary of the Asian Fisheries Society*, S. S. De Silva, (Ed), Asian Fisheries Society, Makati, Philippines, pp. 189-216.

Royal Irrigation Department (1993). *Environmental Impact Assessment of the Pasak River Impoundment, Saraburi and Lopburi Provinces*. Final Report prepared by TEAM Consulting

Engineering Co. Ltd., South East Asia Technology Co. Ltd. and ASDECON Cooperation Co. Ltd.

Role and Impact of Small-scale Fishing in Cambodia -Household Survey in Kompong Thom Province-

Mina Hori[1)], Satoshi Ishikawa[2)], Ponley Heng[3)], Somony Thay[3)], Vuthy Ly[3)], Thuok Nao[4)] and Hisashi Kurokura[5)]

1) Department of Global Agricultural Sciences, The University of Tokyo
1-1-1 Yayoi Bunkyo-ku, Tokyo, 113-8657, Japan
dolphinkick37@hotmail.com
2) Japan Science and Technology Agency
Kawaguchi Center Building, 4-1-8, Honcho, Kawaguchi-shi, Saitama, 332-0012, Japan
oounagi@hotmail.com
3) Community Fisheries Development Office, Department of Fisheries
#186 Preah Norodom Blvd, P.O.Box 582, Phnom Penh, Kingdom of Cambodia
cfdo@camnet.com.kh
4) Department of Fisheries, Ministry of Agriculture Forestry and Fisheries
#186 Preah Norodom Blvd, P.O.Box 582, Phnom Penh, Kingdom of Cambodia
naothuok@mobitel.com.kh
5) Department of Global Agricultural Sciences, The University of Tokyo
1-1-1 Yayoi Bunkyo-ku, Tokyo, 113-8657, Japan
akrkrh@mail.ecc.u-tokyo.ac.jp

Abstract

In Cambodia, more than 75 % of animal proteins are supplied from fishery products (Ahmed *et al.*, 1998). Recent report mentioned that small-scale fishing provides approximately 60 % of the total fish catch (van Zalinge and Nao, 1999). In this connection, information of the small-scale fishing in Cambodia is quite important for establishment of rational fishery resource management and making policy on the food security for the Cambodian people. However, the information and data of small-scale fishing are limited, e.g. fishing area, fishing gear, the purpose of fishing. In order to collect detailed information and data of small-scale fishing, household surveys have been conducted in two villages in Kompong Thom province. The data of family structure, protein sources and consumptions, land ownership, agriculture, aquaculture, live stocks, and fishery have been collected from households engaged in small-scale fishing activities. Data suggested there were two kinds of groups of fishermen in both villages, one conduct fishing at "Tonle Sap Lake" and another at "Around village". Results indicated that almost all fish caught by small-scale fishing were commercially sold, although it had been said that fish were consumed at fisherman's home in previous researches. And our data clarified that the income from small-scale fishing possessed main part of the fisherman's income who is fishing at Tonle Sap Lake in both villages.

Keywords: Cambodia, household survey, inland fisheries, small-scale fishing, Tonle Sap Lake

1. Introduction

Most of the animal protein intakes of the Cambodian people are from fishery products. Therefore, fisheries, especially inland fisheries, are very important for Cambodian people, as it is providing main protein source. Inland fisheries are composed from 3 types of fishing, namely large-scale fishing, middle-scale fishing and small-scale fishing. Large-scale fishing is a demarcated fishery which performed in exclusively licensed area called "fishing lot". Large-scale fishing is a commercial fishery and the licenses are sold at auction. Middle-scale fishing is commercial fishery as well and it also needs license to conduct fishing. Small-scale fishing is sometimes called family fishing or rice field fishing and it is conducted in very small-scale in lakes, rivers and rice fields. It is said that these types of fishing are implemented for mainly home consumption by farmers and fishermen.

Total fish catch has been estimated from 1980 in Cambodia, though the catches by small-scale fishing were neglected in those statistical year book at first. On 1995-6, Mekong River Comission (MRC) conducted the research about small-scale fishing (Ahmed *et al.*, 1998). Since then from 1999, estimated fish catch by small-scale fishing has been added to the total fish catch and, the total fish catch has almost doubled (FAO, 2000). It means that small-scale fishing makes up approximately 50% of the total catch. After that, small-scale fishing has been focused on for the establishment of functional resource management and stable supply of protein. However, we cannot obtain information such as people who is carrying on small-scale fishing, fishing area, fishing gear, the purpose of fishing etc. This is the first report about the people who conducts small-scale fishing around Tonle Sap Lake in Cambodia.

2. Materials and methods

2.1 Study areas

The data collection and the household survey had been conducted both in Svay Ear village and Srey Rangit village, Chamnakrown commune, Stoung district, Kompong Thom province (Figure 1). These two villages are neighboring and they are 2 km apart from each other. These villages locate along national road number 6, and approximately 35 km away from the capital, Kompong Thom city.

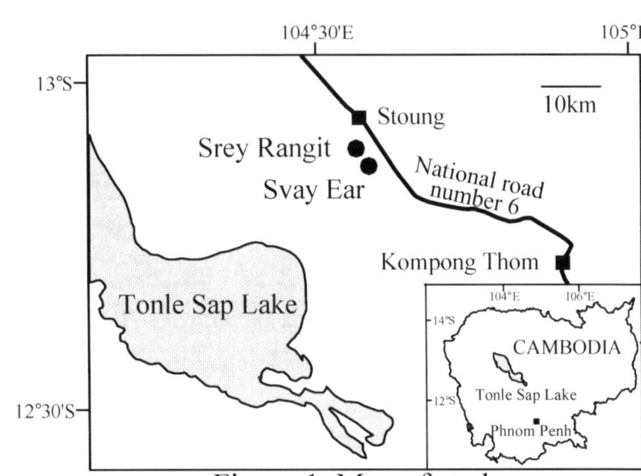

Figure 1. Map of study areas

Data collection

Existing data and information collection had been conducted on 29th January 2004. These data and information were mainly collected through interview to the leaders of the villages.

We prepared questionnaire which was used for the household survey based on the general information and data obtained from the interview.

2.2 Household survey

Household surveys in two villages have been conducted from 9th to 20th March 2004. The family which is living mainly on fishing was selected as target for the household survey. The questionnaire was composed of seven major components; family structure, protein sources and consumptions, land ownership, agriculture, aquaculture, live stocks, and fishery. Regarding family structure, we asked 5 items (name, sex, age, occupation and income). Regarding protein sources and consumptions, we categorize protein source in to 7 groups (fish, beef, pork, chicken, duck, egg and others). Regarding land ownership size of the land was asked. Regarding agriculture 8 items (rice consumption, rice production, varieties of rice, period of raising, kinds of vegetable, purpose and price, kinds of fruits and purpose and price) were asked. Regarding live stocks, 4 items (owned number, purpose of keeping, how to sell, price) were asked about 6 kinds of animals (buffalo, horse, cow, pig, chicken, duck). Regarding aquaculture 3 topics were asked (operated or not, details, future request). Regarding fishery 11 topics were asked (fishery income, fishing ground, period of fishing, maximum catch and month, minimum catch and month, percent of selling, used fishing gears, caught fish species, how to sell, price to sell, amount and price of processed fish). Three groups of investigator collected data and information using the questionnaire prepared by authors. In order to standardize the technique of questioner, the author checked paper of answer every day.

3. Results and discussion

3.1 The general information of the villages

Svay Ear village

Population of the village was 1140 (male: female = 47.2:52.8), 183 households belong to this village (Table1). In this research, the data of 101 households have been collected. The average number of people per household was 6.5 ± 2.0. The villagers of Svay Ear had not operated fishing before year 2000. In 2000, all the rice of this village was slashed by flood. After the flood villagers could not produce enough amount of rice for home consumption. This would be the main reason that the villagers of Svay Ear village stated fishing as an alternative way to live on.

Srey Rangit village

Population of the village was 844 (male: female = 49.4:50.6), 140 households belong to this village. In this research, the data of 103 households have been collected. The average number of people per household was 5.8 ± 2.1. Almost all villagers carry on both rice farming and fishing.

Table 1. The general information of the studied villages

Village	Population			Number of households	Researched number of households	Household size average ± SD
	Total	Male	Female			
Svay Ear	1140	538	602	183	101	6.5 ± 2.0
Srey Rangit	844	417	427	140	103	5.8 ± 2.1

3.2 Kinds of fishing grounds

The fishing grounds of these villagers were divided into two major categories, one is "Tonle Sap Lake", and another is "Around Village" (Table2). "Around Village" includes ponds, streams, flooded forests and rice fields. The people who conduct fishing at "Tonle Sap Lake" stays around the fishing ground, approximately 25km away from home, during fishing season(October to June) . On the other hand, people who do fishing around village go back home every day after fishing. In this context, flooded forest can be categorized as the fishing ground of "Around Village". In this study, flooded forest connected with Tonle Sap Lake is included in "Around Village".

In Svay Ear village, 79 households conduct fishing at Tonle Sap Lake, and 17 households at around village. In Srey Rangit village, 72 households conduct fishing at Tonle Sap Lake, 23 households mainly at around village.

Table 2. Number of households using different types of fishing grounds

Village	Fishing ground					Total
	Tonle Sap Lake	Around village			Others	
		Pond/Stream Rice field	Flooded forest	Total		
Svay Ear	79	15	2	17	5	101
Srey Rangit	72	16	7	23	8	103

3.3 Gross income and fishery income

There were wide differences of average annual gross income per household. In Svay Ear village, the number of annual gross income (1000 riel*) per household varied from 100 to 1745 and the average was 730 ± 400 (n=80), while fishery income contributes 357±324 (n=80) (Table 3). In Srey Rangit village the number of annual gross income varied from 80 to 2700 and the average was 615 ± 429 (n=90), fishery income was 223 ± 175 (n=90) (Table 3). Popular income sources except fishery were selling live stocks, vegetables, fruits or wages from daily labors. The fishery income and annual income of Svay Ear village was significantly higher than that of Srey Rangit village ($p<0.05$ t-test). Difference in annual income might come from how much live stocks they sold. Svay Ear village is known to be pig selling village. Fishery income of Svay Ear village at Tonle Sap Lake was also significantly

higher than that of Srey Rangit village ($p<0.05$ t-test). However, there were no significant differences between fishery incomes of two villages whose fishing ground is around village.
*Cambodian currency; 4000 riel = 1 US dollar (March 2004)

Table 3. Averages of annual gross income and fishery income per household (1000 riel)

Village	Types of fishing ground	n	Gross income average ± SD	Fishery income average ± SD
Svay Ear		80	730 ± 400	357 ± 324
	Tonle Sap Lake	63	760 ± 403	403 ± 306
	Around village	14	677 ± 373	204 ± 368
Srey Rangit		90	615 ± 429	223 ± 175
	Tonle Sap Lake	63	617 ± 456	243 ± 171
	Around village	20	557 ± 346	163 ± 178

3.4 Fish catch and the ratio for selling

In Svay Ear village, the averages ± SD of annual fish catch per household was 285 ± 253 kg (n=88) (Table4). Annual fish catch per household whose fishing ground is Tonle Sap Lake was 333 ± 256 kg (n= 67) and around village was 145 ± 190 kg (n=16). In Srey Rangit village, annual fish catch per household was 207 ± 188 kg (n=94). Annual fish catch per household whose fishing ground is Tonle Sap Lake was 227± 190 kg (n=64) and around village was 114 ± 103 kg (n=22). For annual fish catch, the differences between two village as significant ($p<0.05$ t-test), also there were significant differences between households fishing at Tonle Sap Lake and households fishing around village in both villages. Svay Ear village was higher ($p<0.05$ t-test) and households fishing at Tonle Sap Lake was higher ($p<0.05$ t-test) than households fishing around village.

Table 4. Averages of annual fish catch and the percentage for selling per household

Village	Types of fishing ground	n	Annual fish catch (kg) average ± SD	Percentage for selling average ± SD
Svay Ear		88	285 ± 253	86.3 ± 27.1
	Tonle Sap Lake	67	333 ± 256	96.0 ± 8.2
	Around village	16	145 ± 190	52.2 ± 40.2
Srey Rangit		94	207 ± 188	80.4 ± 29.8
	Tonle Sap Lake	64	227 ± 190	88.6 ± 19.0
	Around village	22	114 ± 103	59.9 ± 40.2

Percentage of fish for selling out of total catch per household in Svay Ear village was 86.3 ± 27.1 % (n=88). Within those data, households fishing at Tonle Sap Lake was 96.0 ± 8.2 % (n=67) and households fishing around village was 52.2 ± 40.2 % (n=16). In Srey Rangit village, percentage of fish for selling out of total catch per household was 80.4 ± 29.8 % (n=94). The value was 88.6 ± 19.0 % (n=64) in households fishing at Tonle Sap Lake and was

59.9 ± 40.2 % (n=22) in households fishing around village in both villages, the ratio was significantly higher ($p<0.05$ t-test) in households fishing at Tonle Sap Lake than households fishing at around village.

3.5 Annual consumption of fishery products

In Svay Ear village, average annual fish consumption per household was 241 ± 161 kg (n=101) (Table5). Fish consumption of a household fishing at Tonle Sap Lake was 239 ± 145 kg (n=79) and the consumption of household fishing around village was 260 ± 226 kg (n=17). In Srey Rangit village, average annual fish consumption was 210 ± 185 kg (n=103). The consumption by household fishing at Tonle Sap Lake was 206 ± 166 kg (n=72) and the consumption by household fishing around village was 251 ± 246 kg (n=23). In both villages, there were no significant differences between households fishing at Tonle Sap Lake and households fishing around village.

About the processed fish consumption, we only got the data of the prahoc (fish paste) and phock (fermented fish). Average annual consumption of prahoc in Svay Ear village was 17.5 ± 13.0 kg (n=70), and the consumption by households fishing at Tonle Sap Lake in the village was 17.8 ± 13.0 kg (n=57) and the consumption by households fishing around village was 17.8 ± 14.2 kg (n=10). In Srey Rangit village, annual consumption of prahoc was 22.0 ± 21.1 kg (n=94), it was 23.0 ± 22.2 kg (n=65) in the households fishing at Tonle Sap Lake and was 21.5 ± 20.9 kg (n=21) in households fishing around village. Annual consumption of phock in Svay Ear village was 20.0 ± 15.9 kg (n=30), and, annual consumption of phock in Srey Rangit village was 10.1 ± 9.3 kg (n=37). For prahoc consumption, there were no significant differences between two villages, also between households fishing at Tonle Sap Lake and households fishing around village. For phock consumption, the difference between two villages were significant ($p<0.05$ t-test), though the difference was not observed between households fishing at Tonle Sap Lake and households fishing around village.

Table5. Averages of annual fishery products consumption per household

Village	Types of fishing ground	Fish consumption (kg/yr)		Processed fish products consumption (kg/yr)			
				Prahoc		Phock	
		n	average ± SD	n	average ± SD	n	average ± SD
Svay Ear		101	241 ± 161	70	17.5 ± 13.0	30	20.0 ± 15.9
	Tonle Sap Lake	79	239 ± 145	57	17.8 ± 13.0	22	21.5 ± 17.2
	Around village	17	260 ± 226	10	17.8 ± 14.2	7	16.9 ± 11.3
Srey Rangit		103	210 ± 185	94	22.0 ± 21.1	37	10.1 ± 9.3
	Tonle Sap Lake	72	206 ± 166	65	23.0 ± 22.2	25	10.4 ± 10.6
	Around village	23	251 ± 246	21	21.5 ± 20.9	7	8.9 ± 5.3

Amount of fish brought back home is not enough for home consumption even though we consider the amount of processed fish which might be included in fish consumption. Amount of fish brought back to home in Svay Ear village was 19.2 ± 38.4 kg/yr, in Srey Rangit village 26.0 ± 37.3 kg/yr (calculated from Table4). On the other hand, amount of fish consumption in Svay Ear village was 241 ± 161 kg/yr, in Srey Rangit village 210 ± 185 kg/yr (Table5).

It is certain that most of the fish which were consumed are not fish brought back home from fishing grounds, but bought from markets. Figure 2 shows how people gets fish for daily consumption. In both villages more than half of households answered that they buy fish for daily consumption even though they conduct small-scale fishing.

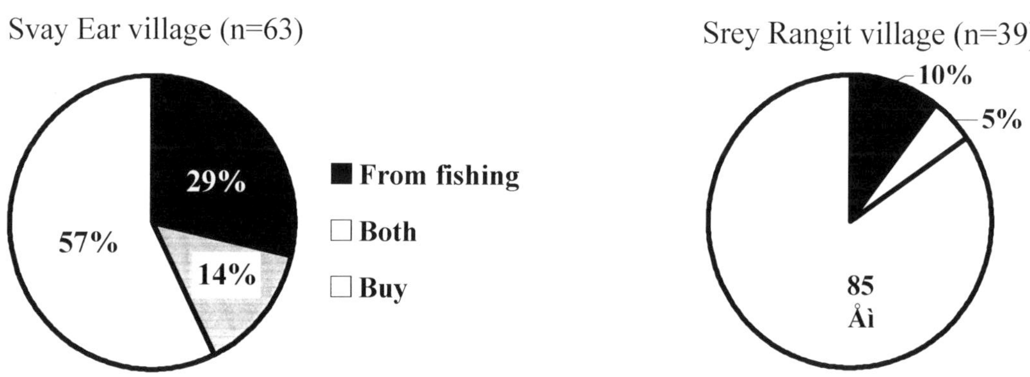

Figure 2. How to get daily consumption fish

3.6 Price and the usage of fish

From the interview, the relationship of price and usage of fish were seen. It seems to be that high price fish, of which price is around 2000 riel/kg, is sold to middleman, middle price fish around 1000 riel/kg is sold or consumed at home or used for processing and low price fish around 500 riel/kg is consumed mainly at home or used for processing.

4. Conclusions

In previous studies, the catch from small-scale fishing is regarded to be consumed at each home. However, the results of our survey showed other important roles of small-scale fishing. And the fish trade system around Tonle Sap Lake is not simple as mentioned before like "the large and middle-scale fishing produce commercial catch and small-scale fishing provides protein for rural people". Practically, lots of fish collected at Tonle Sap Lake by small-scale fishing were sold. And the activities of small-scale fishing occupied the major part of gross income for many households, at least in households fishing at Tonle Sap Lake in two villages.

5. Acknowledgements

Authors are grateful to the staffs of the Community Fisheries Development Office, Department of Fisheries, and the staffs of the Fishery Office of Kompong Thom Province, Cambodia for assistance and coordination of this work. We also thank our supporting staffs for assistance of collecting data. This work was supported in part by grants of Water Use and Management System of the Mekong River, Core Research for Evolutional Science and Technology, Japan Science and Technology Agency, Japan.

6. References

Ahmed M., Hap N., Ly V. and Marites T. (1998). *Socioeconomic Assessment of Freshwater Capture Fisheries of Cambodia – Report On a Household Survey,* Project for Management of Freshwater Capture Fisheries of Cambodia, Department of Fisheries, Cambodia; Danish International Development Assistance (DANIDA); and Mekong River Commission (MRC), Mekong River Commission Secretariat Fisheries Unit, Phnom Penh, Cambodia

FAO yearbook Fishery statistics (2000). vol.76, Food and Agriculture Organization of the United Nations, Rome

Keskinen, M.(2003). *Socio-Economic Survey of the Tonle Sap Lake, Cambodia.* Master thesis, Department of Civil and Environmental Engineering, Helsinki University of Technology.

Rainboth, W. J. (1996). *Fishes of the Cambodian Mekong,* FAO species identification field guide for fishery purposes. Food and Agriculture Organization of the United Nations, Rome

van Zalinge, N. and Nao T. (1999). *Present status of Cambodia's freshwater capture fisheries and management implications.* In: van Zalinge, N., Nao T., Deap L. (eds). Present status of Cambodia's freshwater capture fisheries and management implications, Mekong River Commission and Department of Fisheries, Phnom Penh, Cambodia

Analysis of Reforestation as a Watershed Management Policy for Soil Erosion Control: a Case Study in the Upstream Phong River, Thailand

Suthipong Sthiannopkao[1), Satoshi Takizawa[1) and Wanpen Wirojanagud[2)

[1) Department of Urban Engineering,
Graduate School of Engineering, The University of Tokyo,
7-3-1, Hongo, Bunkyo-ku, Tokyo 113-8656, Japan.
Email: suthisuthi@hotmail.com
2) Faculty of Engineering, Khon Kaen University, Thailand.

Abstract

The impacts of reforestation on soil loss and sugarcane production in the upstream Phong River are simulated for a 30-year period using a system dynamic model. The highland sugarcane plantation areas are converted into forestlands at the mean annual rate of 0, 2, 4, 6, 8 and 10%. The simulation results reveal that the highland sugarcane plantation area gives the highest soil loss rate at 22.25 tons/ha/year, followed by flatland sugarcane plantation (9.31 tons/ha/year), forest (4.63 tons/ha/year) and paddy field (2.06 tons/ha/year), respectively. Transforming the highland plantation areas to forestlands would reduce the amount of total soil loss due to the decrease in highland plantation areas. The topsoil fertility of highland sugarcane plantation substantially depletes after 20 years. This makes it possible for Thai Government to implement a reforestation program by giving farmers a small subsidy. Finally, in order to reach the goal sugar stock set at 1,216,800 tons/year at the same time of converting the highland sugarcane plantation areas into forestlands, the sugarcane yield should be increased from the current yield of 57 tons/ha/year.

Keywords: reforestation, soil loss, system dynamic model, and sugarcane yield

1. Introduction

Forestry development in Thailand changed after devastating floods and land slides wiped out two villages in the Surat Thani province, southern Thailand in 1988. Following this in a general debate, loss of forest cover in the watershed area was identified as the main cause of extreme flooding events. Forest plantations and their environmental impacts have received much attention during the last few decades throughout the tropics including Thailand. Forest plantations were often established on degraded areas to halt the deterioration of the land, and to help restore the environment (Niskanen, 1998).

The upstream Phong River is located in the northeast Thailand and is one of the four main tributaries in the upper region of the Phong watershed. It is facing a problem of soil erosion due to the transformation of forestlands to agricultural lands. According to Khon Khon University (KKU, 1999), one of the pollution sources in this river is agriculture, which was characterized by cash crop plantations in the higher slope areas and by paddy fields in the plain areas. The KKU report (1999) states that the promotion of cash crop plantations resulted in the conversion of 96,000 ha of natural forests into agricultural lands.

It was proposed in the KKU reports that in order to alleviate the problem of soil loss, trees should be planted in the upstream Pong River. The objectives of this paper are (1) to evaluate the effects of reforestation on soil erosion reduction and sugarcane production and (2) to find the reforestation program that has a minimum impact on farmers' income. Sugarcane is assumed as the cash crop promoted in this simulation because it is one of the main crops planted in the area. In addition to optimal climate and soil quality, sugarcane is received the support from both the Thai Government and Thai Sugarcane Association. Furthermore, there are three sugar mills located within the Phong Watershed with the total production capacity of 480,000 tons sugarcane/day (KKU, 2003).

The FAO (1989) reports that since 1960s Thai Government has implemented the policy of using disturbed public forestland to solve problems of rural poverty and landlessness, which can be divided into two broad categories: those planned in relation to reforestation activity; and those designed principally to provide farmland to landless farmers. Prominent among first category programs are those using forest village approach to involve villagers in reforestation while providing them with land to cultivate. Therefore, the 20-year highland sugarcane plantations could be used in order to serve the Thai Government policy of use of disturbed forestlands for poverty alleviation.

2. Materials and methods

The study was first initiated by studying the KKU reports on the soil loss data of different land use and other secondary data related to soil erosion, land use and deforestation in the upstream Phong River. Next, the soil loss (tons/ha/year) of different land use is calculated using the Universal Soil Loss Equation (A=R*K*LS*C*P). Afterward, the system dynamic model was constructed in order to simulate the amount of soil loss caused by reforestation at annual rates of 0, 2, 4, 6, 8 and 10%. The simulation model was composed of three sub-models, namely the soil erosion, the land transformation and the agricultural productivity models. Next, the simulations were run at the soil depths of 30 cm and 90 cm for a 30-year period. According to the Land Development Department (2004), the shallow soil is set at 50 cm below the soil surface. The sugarcane yield was set at 88 tons/ha/year. According to Foodmarketexchange (2004), regarding the 5-year strategic plan (2004/08) for the Thailand sugar industry developed recently to strengthen industrial competitiveness, the production target aims to improve cane yield from the current 57 to 88 tons/ha/year. In this simulation, the increase of forestlands is the result of transforming the highland plantation areas (7,819 ha). The highland area means an area with the slope length and steepness at 0.21 while the flatland area has the slope length and steepness at 0.08.

3. Results and discussion

3.1 Identification of deforestation, soil loss and reforestation in the upstream Phong River

The upstream Phong River covers parts of Loei, Khon Kaen and Nong Bua Lumphu provinces. According to Office of Agricultural Statistics (2004), the agricultural sector has contributed to more than 55% of the provincial economies of Loei and Nong Bua Lumphu, which is significantly higher than other two sectors of manufacturing and wholesale. According to KKU (1999), a natural forest in the upstream Phong River was found only in the area named Phu-Kradueng covering the area of Phu Tumyai and a part of Yod Kao Phu Pueng in the Phu Pa Man natural forest. The remaining was disturbed forests and agricultural lands found also at an elevation of 200 m. From Tak Hong Waterfalls downwards, river flowed in a broad valley through dense human communities. The Donglan forest and other natural forests were converted into agricultural lands. The forest area remained only in the mountainous areas with agricultural lands in between. Crops are the main plant followed by paddy fields and fruit trees. The KKU report (1999) states that an estimated amount of suspended solids carried down in the upstream Phong River is 274,505 tons/year, the highest amount compared to other three main tributaries in the upstream region, namely Prom River (13,702 tons/year), Choen River (91,916 tons/year), and Phaneng River (40,118 tons/year).

According to FAO (1989), the Thai Government has been involved in reforestation for a long period but over a relatively limited area. The RFD (Royal Forest Department) has been the main agency responsible for reforestation but the FIO (Forest Industry Organization) has been involved with the forest village approach since 1967. The fifth development Plan (1981-1986) departed from its predecessors by stressing income distribution and poverty alleviation more than economic growth. Its forestry measures aimed at conservation rather than commercial exploitation and tried to compromise with the population pressure causing encroachment, for example, by fostering the extension of forest villages under the STK program. In 1985, the Thai government established the national forest policy. Its aim was to recover 40% of the total land area as forest, consisting of economic and conservative forest (25% and 15% of the country, respectively). Since the degraded area was so large, however, the government could not adequately implement reforestation solely with its budget. Thus, in order to reforest the country more rapidly, it encouraged the private sector to participate (Shimamoto *et al.,* 2004).

Figure 1 shows that when topsoil is lost caused by the highland sugarcane plantations, the sugarcane productivity is decreased. Consequently, the sugar stock, which is the difference between sugarcane productivity and sugar consumption, is reduced. Moreover, the sugar stock at sugar mills has an effect on plantation areas. When the sugar stock exceeds a goal sugar stock set at 1,216,800 tons/year (calculated from 30% of the total annual sugarcane productions in the provinces of Loei, Nong Bua Lumphu and Khon Kaen), sugar mills will try to reduce the production next year by decreasing sugarcane plantation areas. At some period of time, the highland sugarcane productivity would be decreased to the level of no more being worth plantation. As the result, this piece of land would finally be abandoned. Therefore, the reforestation program promoted and supported by the Thai Government could be implemented to curb the highland soil loss and also to provide this piece of land to landless people. On the other hand, by transforming the abandoned highland sugarcane plantation areas to forestlands would decrease the total amount of soil loss. However, the sugarcane yield should be increased from the current at 57 tons/ha/year in order to compensate the lost from highland sugarcane production due to the reduction of highland sugarcane plantation areas. The increasing of cane yield can be achieved such as by increasing the plant density by reducing row spacing and leaving the dead leaves on the soil surface.

3.2 Soil loss of different land use

Table 1 reveals that the soil loss rate from the highland sugarcane plantation area is the highest at 22.25 tons/ha/year followed by flatland sugarcane plantation area (9.31 tons/ha/year), forestland (4.63 tons/ha/year) and paddy field (2.06 tons/ha/year), respectively. The slop length and steepness (LS), soil erodibility (K) and rainfall erosivity index (R) are the key factors that influence of soil loss in the upstream Phong River.

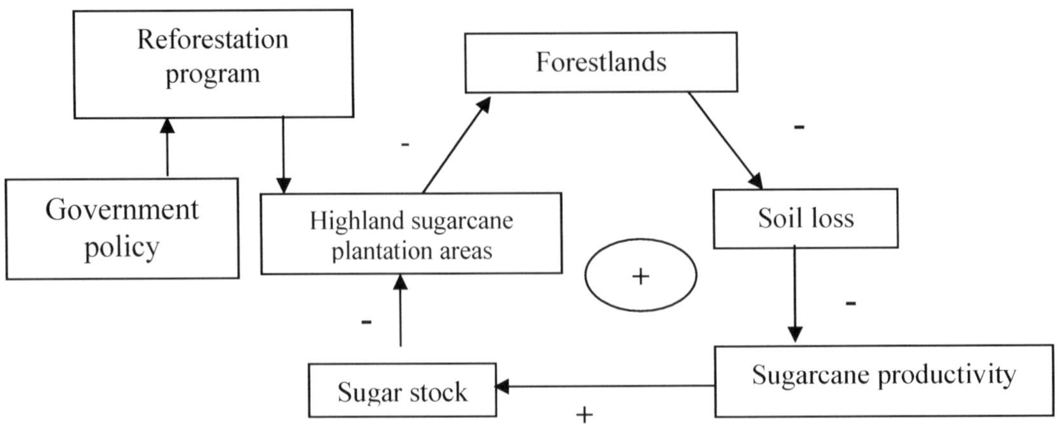

Figure 1. A causal loop of soil loss in highland areas

Table 1. Results of soil loss simulated in the upstream Phong River

Land uses	Soil loss simulated (tons/ha/year)	Soil erosion classification by LDD*
Paddy	2.06	Very slight (0-12.50)
Forest	4.63	Very slight
Highland crops	22.25	Slight (12.56-31.25)
Flatland crops	9.31	Very slight

* LDD= Land Development Department, Thailand (2004)

3.3 Simulation of soil loss and its impact on sugarcane production at topsoil depth 30 cm

According to Prammanee (2004), sugarcane plantations have the most significant impact on the soil quality compared with other crops. The open up of forestlands for cane plantations, the soil preparation, the burning of bagasse at the site after canes harvesting and the transportation of cane harvested out of the plantation sites have caused the deterioration of soil quality, both in its physical and chemical characteristics. In addition, the loss of topsoil as a result of the above activities has finally led to giving up of cane plantations in some countries including the Hawaii. The topsoil loss rate calculated from the simulation results at the starting time is 0.276 mm/year.

The reforestation rates are set at 2, 4, 6, 8 and 10% annually in order to make comparisons of results modeled by KKU. The simulation results reveal that by increasing a forestland at the annual rate of 2, 4, 6, 8 and 10%, the amount of total soil loss is decreased in each decade of 30-year simulations. This is mainly associated with the decrease of highland sugar plantation areas, which gives the highest soil loss rate at 22.25 tons/ha/year. However, the soil loss forest is also increased but at a slower rate of 4.63 tons/ha/year. The soil loss from reforestation at the annual rates of 0% and 10% is shown here in Figures 2 and 3, respectively. Figures 2 and 3 show that the amounts of soil loss from the plain areas like the locations of paddy fields and flatland sugarcane plantations are constant during the three decades of simulations. In case of paddy field, it can be explained that water storage in a field at the water level ranging from 5-10 cm helps to protect soil being transported away from its sources. As a result, the increasing amounts of total soil loss are mainly dependent on the soil loss on highland plantation areas and forestlands located in higher slop areas. Furthermore, Figure 3 indicates that the total soil loss from reforestation is slightly increasing during the second and third decades of simulations compared to the control simulation of 0% land conversion. As the soil loss rate in the highland plantation areas is the highest (22.25 tons/ha/year), the 30 cm depth of topsoil is depleted during the second and third decades of simulation. In order to prove the limitation of topsoil depth impact on the total amount of soil loss, the next section is described the simulation results of topsoil depth set at 90 cm. By increasing the sugarcane yield to 88 tons/ha/year significantly increases in the amount of sugar stock to 1,605,900, 1,124,300 and 1,137,600 tons in each decade of 30-year simulations compared to the same amounts obtained at the yield 57 tons/ha/year (990,900, 579,800, 404,800 tons also in each decade of 30-year simulations). However, Table 2 shows that by transforming the existing highland plantation areas to forestlands would slowly decrease the sugar stock. Therefore the main sugarcane production would come from flatland plantation areas. Furthermore, Table 2 shows that the highland sugarcane production is significantly decreased after the 20-year plantation ranging from 9.20% to 17.99% annually.

3.4 Simulation of soil loss and its impact on sugarcane production at topsoil depth 90 cm

In order to prove the dependence of total soil loss on land use located in the higher slop areas (the soil loss from highland plantation areas and forestlands), the topsoil depth is set at 90 cm. It is found that the total soil loss of annual reforestation at 2, 4, 6, 8 and 10% is decreasing in each decade of 30-year simulations. The total soil loss at reforestation 0% and 10% annually displayed in Figures 4 and 5 are selected to show these phenomenon. It is interesting to see that the graphs of total soil loss and highland soil loss have the same decreasing trend. In addition, when compare the amount of total soil loss in each decade with the control simulation at 0%, they are also less than the amounts of total soil loss obtained from the latter simulation. However, the soil loss forest is almost the same as the simulation results at the topsoil 30 cm depth, which is increasing at a slower rate (Figures 3 and 5).

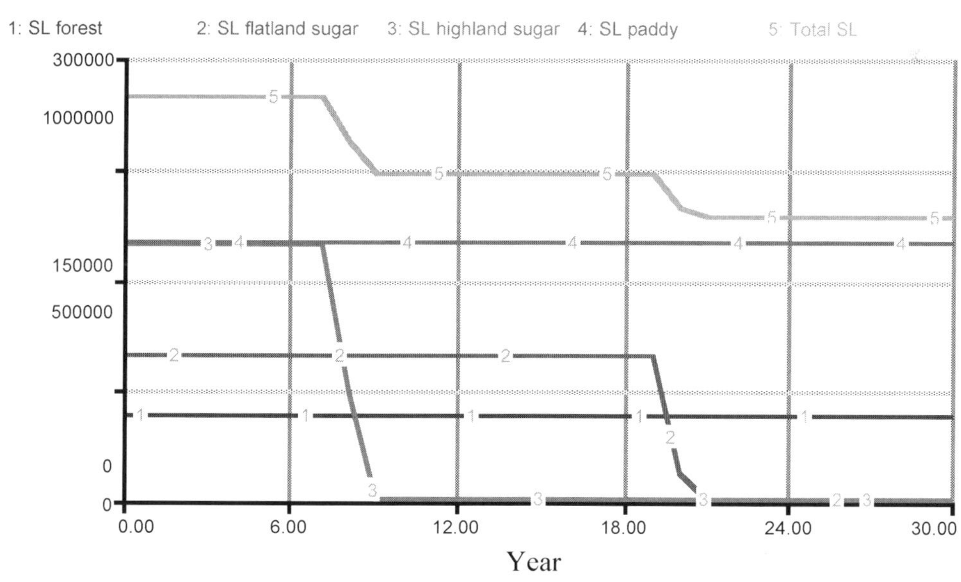

Figure 2. The soil loss of different land use: simulated at topsoil 30 cm, no land conversion and sugar cane yield at 88 tons/ha/year (SL=Soil loss)

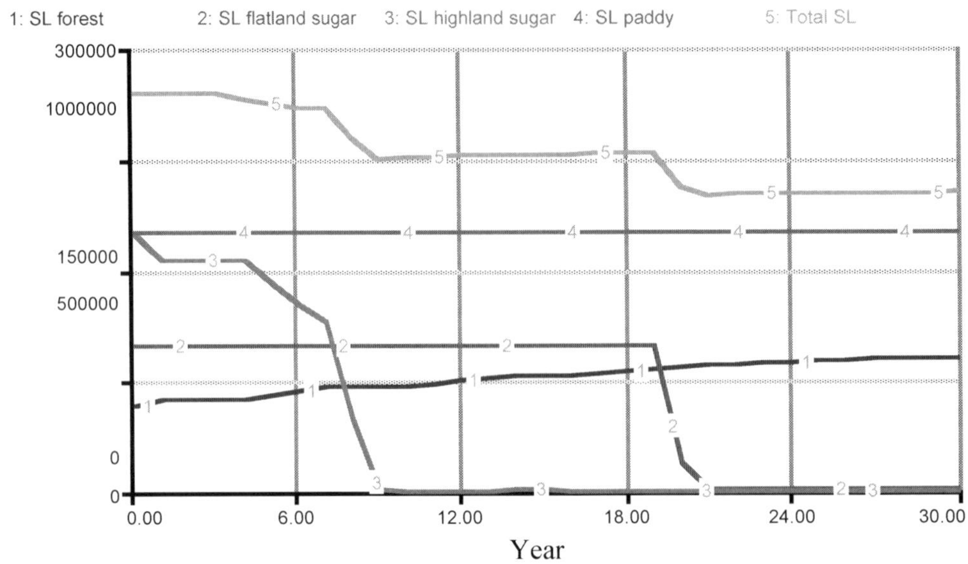

Figure 3. The soil loss of different land use: simulated at topsoil 30 cm with annual land conversion 10% and sugarcane yield at 88 tons/ha/year (SL=Soil loss)

Furthermore, the decrease in highland plantation areas would result in also the decrease in total soil loss. In addition, it can be stated that by transforming highland plantation area to forestlands would increase the forest soil loss (Figure 5). However, the total soil loss would be decreased because the highest rate of soil erosion is from highland plantation areas (22.25 tons/ha/year). Finally, Table 2 shows that by increasing the topsoil depth to 90 cm, the sugar stock is slightly decreased in the first decade but is considerably decreased in the second and third decades. This is because of the decreasing in highland production in the same periods as a result of the highland conversion.

3.5 Comparisons of results modeled

According to Wirojanagud *et.al* (2004), increasing the forest areas at rates of 2, 4, 6, 8, and 10% would decrease the loading of suspended solids in the upstream Phong River at 0.78, 1.33, 1.82, 2.27 and 2.68%, respectively. This is in accordance with the simulation results show that the reforestation would reduce the total soil loss in the upstream Phong River. The differences between the total soil loss at reforestation rates 2, 4, 6, 8, and 10% annually and the control simulation at 0% land conversion are 3,875, 7,617, 11,233, 14,725, 18,098 tons/year, respectively. This indicates that the decrease of soil loss on land as the result of reforestation would lead to the reduction of the concentrations of suspended solids in the upstream Phong River.

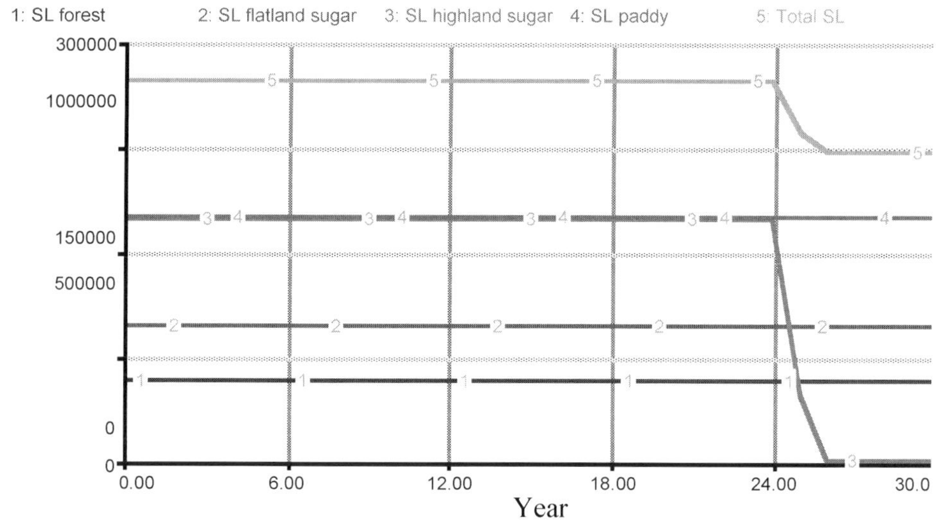

Figure 4. The soil loss of different land use: simulated at topsoil 90 cm with annual land conversion 0% and sugarcane yield at 88 tons/ha/year (SL=Soil loss)

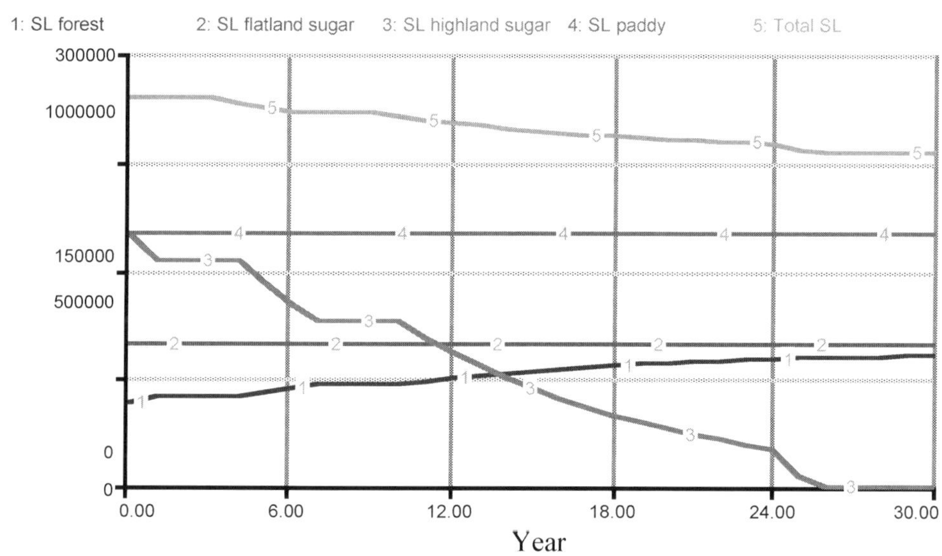

Figure 5. The soil loss of different land use: simulated at topsoil 90 cm with annual land conversion 10% and sugarcane yield at 88 tons/ha/year (SL=Soil loss)

Table 2. Mean annual rates of changes of sugar stock, flatland and highland sugarcane production at the sugarcane yield 88 tons/ha/year (% per year)

Scenarios	Reforestation (%)	Sugar stock (tons)			Flatland sugarcane production (tons/year)			Highland sugarcane production (tons/year)		
		1-10 (year)	11-20 (year)	21-30 (year)	1-10 (year)	11-20 (year)	21-30 (year)	1-10 (year)	11-20 (year)	21-30 (year)
1	0	22.36	25.62	-3.05	35.52	47.28	-9.23	30.15	51.41	-9.20
2	2	21.14	20.99	-3.14	28.14	34.53	-9.97	20.79	35.99	-11.40
3	4	19.98	17.96	-3.83	2820	27.18	-9.51	2882	26.46	-13.10
4	6	19.29	10.49	-4.49	64.99	20.99	-9.26	56.86	17.98	-14.68
5	8	18.64	5.49	-5.32	37.77	16.89	-9.04	27.65	12.00	-16.29
6	10	18.01	1.81	-6.11	27.31	14.03	-8.92	15.69	7.35	-17.99

4. Conclusions

The increase of forestlands would reduce the total amount of soil loss. This is due to the decrease in the highland plantation areas, which gives the highest soil loss rate at 22.25 tons/ha/year. However by increasing a forestland would also reduce highland sugarcane production. Therefore, the flatland sugarcane productivity should be increased from the current yield of 57 tons/ha/year. Consequently, this would assist sugarcane farmers to maintain their income. Finally, the forest village approach should also be implemented to involve sugarcane farmers in reforestation while providing them with land to cultivate.

5. References

FAO, (1989), Forestland for the people: A forest village project in Northeast Thailand

Foodmarketexchange, (2004), Thai sugar, www.marketexchange.com.

Khon Kaen University (KKU) (2003) Environment management of the Phong watershed.

Khon Kaen University (KKU), (1999), Environment management of the upper Phong Watershed

Land Development Department (2004), Soil Erosion Classification of Thailand, www.ldd.go.th

Mihoko Shimamoto, Fumikazu Ubukata and Yoshiki Seki, (2004), Forest sustainability and the free trade of forest products: cases from Southeast Asia, Ecologiocal Economics 50, pp 23-34.

Niskanen Anssi, (1998), Value of external environment impacts of reforestation in Thailand, Ecological Economics 26, p 287-97.

Office of Agricultural Economics, (2004), Agricultural Statistics of Thailand, www.oae.go.th.

Prammanee Precha, (2004), The structure, and physical and chemical characteristics of soil in the sugarcane plantations: The evaluation of soil quality and optimal amount of fertilizer use, the document prepared for the training of technology transferred to sugar cane farmers.

Wirojanagud W, Nantaphotidech A and Jensen G. P, 2004, Spatial planning for integrated watershed management, the paper presented at the 8th International Diffuse/Non-point Pollution, Kyoto, Japan, October 24th-29th, 2004.

Participatory Irrigation Management Practices in Vietnam: A Pathway in Progress towards Effective and Sustainable Use of Agricultural Water

Pham Thi Dung[1] and Ganesh P. Shivakoti[2]

1) Center for Water Resources Development and Environment
Vietnam Institute for Water Resources Research
171 Tay Son, Hanoi, Vietnam
e-mail: phamthidung@hn.vnn.vn

2) Agricultural and Natural Resources Economics
Asian Institute of Technology
P.O. Box 4, Klong Luang, Pathumthani 12120, Thailand
e-mail: ganesh@ait.ac.th

Abstract

Participatory Irrigation Management (PIM) is considered a sound policy approach towards effective and sustainable use of irrgiation water in agricultural countries. In Vietnam, PIM was launched nearly ten years ago. So far, the implementation is still facing obstacles. The overall objective of this study is to explore diversified empirical experiences then determine constraints and opportunities for improvement of PIM in the country. The findings of the study showed that, the practices of PIM implementation again confirm the advantage of PIM as a proper policy alternative to improve performance of irrigation systems. Both successes and limitations in PIM practices imply that the provincial governments can carry out fruitfully the reform of irrigation management. So, the decentralization in irrigation management is a good solution to promote PIM development. Farmers' perceptions of performance of irrigation systems have consolidated the logic of PIM as a good solution to improve irrigation performance. Farmers in all three typical types of irrigation systems studied expressed their willingness to participate in irrigation management. This is a very convenient condition to conduct successfully the transfer of irrigation systems from local authority to farmers. The major obstacles constraining PIM development in Vietnam found here is the insufficiency, inappropriateness, and the lack of harmony of the current policy system.

Keywords: participatory irrigation management, irrigation management transfer, Bac kan, tuyen quang, Vietnam

1. Introduction

Vietnam has experienced through a stagnant socio-economic development due to its long period of the centralized management. Under this framework, water management was controlled by state, provincial and district governments. Farmers were absent from the water distribution process. Water, in this sense, was allocated according to the government's wishes based on supply control, not on users' demand. The backward nature and failure of this centralized mechanism was revealed more clearly than ever when Vietnam opened her door to integrate in the world economy. The comprehensive reform in 1986 has helped the country to improve situation but the irrigation institution reform is still in an incomplete process and somewhere, the implementation has been operating in a vicious cycle.

According to current statistics (MARD, 2004), 91% of the total irrigation systems serving for 80% irrigated area of the country are still under the control of state irrigation management companies or local authorities. The survey by Department of Irrigation-MARD (2004) also showed that most of these irrigation systems are now in a unprosperous condition. On average, only 50-60% of designed area has stable irrigation supply through its existing hydraulic systems. Some systems, mainly the small ones, can ensure irrigation for only 25-30% of the designed area. In inter-communal canals, at the head-end area, water use is normally more wasted by 5-7 folds than the designed level, causing water shortage for around 50-70% area in the immediate communes and almost 90-100% area in the tail-end communes. This is also the reason for a significant increase in service cost for irrigated area in the tail communes. The poor quality of irrigation has resulted in a reduction of about 1.55 – 2 times in productivity compared to that achieved at upstream area of canals. The special consequence is the conflicts between communes and quick deterioration of the hydraulic works (Tiep, 2004).

The underlying causes for these failures are that the state irrigation management agencies, although given the right of balance their own account, still fail in collection of sufficient irrigation fee for O&M. Besides, officials and staff of these agencies don't have strong incentive to improve their irrigation service because they still expect additional subsidy from Government in case of risks or failure in fee collection. Meanwhile, farmers are not willing to pay irrigation fee because they don't receive expected quality of water supply services from government agency. Furthermore, farmers still have the habit of heavy reliance on government subsidy. But the Government can not continue providing subsidy for O&M due to their budget shrinkage. Consequently, irrigation structures are quickly degraded due to lack of budget for proper O&M (Fontenelle, 2000).

In responding to this situation, the Government has tried to involve farmers in irrigation management in order to create the sense of ownership and strengthen their responsibility in O&M. Despite the Government's effort, the expansion of participatory irrigation management in Vietnam is still facing obstacles.

The overall objective of this study is to explore both good and bad empirical experiences in participatory irrigation management in two provinces of Bac Kan and Tuyen Quang then determine constraints/opportunities and policy implications for improvement of participatory irrigation management in Vietnam.

2. Materials and methods

The study was conducted at three levels: The national, provincial and field level. At national level, the study has mainly relied on secondary data, including legal documents related to water resources, especially on irrigation management.

At provincial level, two provinces of Bac Kan and Tuyen Quang were focused. Both of them are located in the mountainous area in North Vietnam. Among them, Tuyen Quang is an example of better performance in term of PIM, while Bac Kan has much limitation in PIM implementation. Legal documents related to PIM implementation of those provinces were reviewed and relevant provincial officials were interviewed to see the current situation as well as their perceptions on PIM.

At field level, the study was conducted in three irrigation systems: Tan An in Tuyen Quang and Phai Lai and Thom Trao in Bac Kan. Those irrigation systems were similar in term of physical characteristics (small scale: 30-60 ha, rehabilitated in 1998-2000) but different in terms of type of governance and level of government intervention. In those systems, besides field observation, semi-structured interviews were conducted to get real situation of PIM and standardized questionnaire was applied to farmers to know their perceptions on PIM. Cross tabulation, frequencies, percentage and means were applied for data description in various indicators of respondents' characteristics as well as their weight given to organizational effectiveness, equity and participation level based on recods of farmers' perception.

3. Results and discussion

3.1 Current Arena of National Policy and Interventions through PIM in Vietnam

The reviews of current national policies and interventions were done together with some interviews of the relevant officials of the Vietnam Network for Participatory Irrigation Management (VNPIM) and Ministry of Agriculture and Rural Development (MARD). The main findings show that, Vietnam has already had a general policy framework for participatory irrigation management. The statement of "encouraging farmers to participate in investment and management of irrigation systems" was scripted in the relevant legal documents such as Resolutions by the Central Communist Party, the Law on Water Resources, the Ordinance on Exploitation and Protection of Hydraulic Works and other by-laws on irrigation management. The definition of water users' organization and its function of irrigation management, the devolution of rights and responsibilities from central government to local governments in decision making in Irrigation Management Transfer (IMT) were first raised in the Ordinance on Exploitation and Protection of Hydraulic Works and a bit concretized in the most recent Decree by the Prime Minister in guiding the implementation of PIM/IMT. The establishment of the VNPIM, more or less is a tangible evidence of effort made by the central government in promoting the development of PIM in the country.

However, practical experiences from many countries showed that, to implement successfully the reform in irrigation management institution with farmers' participation, it requires that the government should have a comprehensive policy system on PIM. This policy system should define clearly the rights and responsibilities of relevant stakeholders, especially water users' association, in the process of decision making in all aspects of investment, management organization, monitoring and evaluation, etc. Hence, the current policy system in Vietnam still reveals much of gap and inappropriateness for strengthening PIM in practice. The weakness of current policies can be listed mainly as follows:

(i) There is no a clearly defined system of rights and responsibilities of various stakeholders, especially water users' association in the process of PIM/IMT.

Farmers' perspective: Although farmers are encouraged to participate in investment and management of irrigation systems, it is not clear that to what extent, which aspects and by which way farmers can participate. So far, the Water Users' Association (WUA), which is the main partner in the process of IMT/PIM, has not been recognized as a legal entity. So, it is not legally protected or supported by the government.

The local authority's perspective: The provincial authority is assigned the responsibility to formulate its own strategy and regulations for implementation of PIM within the province. However, it does not specify to which level of canals (primary, secondary or tertiary) or to which scale of irrigation systems (large, medium or small irrigated area) they should transfer management right to farmers.

The State owned irrigation management company (IMC): There is no personnel policy for redundant officials and staff of the IMC when amount of their work is shrunk, even finished after transfer (if dissolving IMC as an example of Tuyen Quang province). Besides, the government still maintains the subsidy policy for IMC in some cases including the failure in fee collection due to poor crop harvest. This policy may cause the persistence of reliance habit on the government's subsidy by the IMC then diminishing their willingness to transfer irrigation systems to farmers.

(ii) The current policy was also found inappropriate for PIM adoption in some areas when people in this area are exempted from water fee. This is one obstacle constraining the active participation of farmers in irrigation management.

(iii)So far, there is no a comprehensive implementation strategy of PIM at national level.

(iv)The interview with the official of VNPIM also showed that there seems still lacking a strong political will or a proper attention from some high ranking government officials in supporting PIM. The current implementation of PIM is mainly engaged with projects or programme funded by external donors rather than government projects. It may be one among causes of the slackness and insufficiency of current policy on PIM in Vietnam.

3.2 Implementation Strategies and Practices of PIM in Bac Kan and Tuyen Quang Provinces

Both of Bac Kan and Tuyen Quang provinces are located in the North mountainous area of Vietnam. They are poor provinces and similar in natural characteristics of such as topography, climate, and irrigation structures. Agriculture production plays very important role in livelihood of local people. However, two of these provinces have different strategies in irrigation management with farmers' participation and their performance in PIM is also different from each other.

Findings and Lessons Learnt from Tuyen Quang Province

Tuyen Quang started its IMT programme in year 1996. It is the first province in Vietnam conducting the IMT in the whole province. Now, all irrigation systems are under management of ACs.

Before IMT, Tuyen Quang had set up a comprehensive system of policies and mechanisms for implementation of IMT/PIM. The rights and responsibilities of all stakeholders involved (local authorities at all levels, professional agencies, IMC, AC, farmers) were clarified. All aspects related to PIM including training, investment, finance, irrigation fee level and collection method, organizational structure for management of each kind of irrigation, monitoring and evaluation, personnel resolution for IMC, etc were clearly defined. Those policies were also well enforced. This is the most important factor deciding the success of Tuyen Quang in transfer of irrigation systems to farmers.

Impact of the IMT/PIM: The contribution of farmers in investment for irrigation system construction and their payment of irrigation fee increased significantly. The fee level increased from 200 kg/ha/year (before transfer) to 749 kg/ha/year (after transfer). The rate of irrigation fee collection increased from 440-450 ton of paddy in year 1995 (before IMT) to 5,100 ton in recent years (after IMT). Irrigated area was expanded and stabilized remarkably. Crop yield increased from 3.32 ton/ha in year 1995 to 4.78 ton/ha in year 2002.

Practice of PIM at irrigation system level: The farmer managed irrigation system of Tan An (FMIS-T) was studied in term of its performance. This system is under the management of the Agriculture Cooperative of Tan An. It has small scale of serving area. The finding was that this system is operating under a well-defined set of rules including input rule (irrigation fee and labor contribution), boundary rule (irrigators, role and responsibilities of irrigators), distribution rule (rotated water distribution) and monitoring and sanction rule. Those rules were de jure set up. The system also performs well in terms of water supply quality, finance, and agricultural output. The cropping intensity is 235% and crop yield is 6.4 ton/ha of spring rice, 4.2 ton ton/ha of monsoon rice and 3.5 ton of winter maize.

Farmers' perceptions of irrigation performance: The detailed survey of farm household heads were conducted in the FMIS-T to explore farmers' perceptions of performance of their irrigation system. It was found that farmers in this system perceived their system operating effectively and equally. Their level of farmers' participation in irrigation activities was also found active (Weight Average Indices of effectiveness, equity and participation level are 0.76, 0.77 and 0.67 respectively). All farmers interviewed expressed their willingness to participate in irrigation management.

Findings and Lessons Learnt from Bac Kan Province

Currently, most of irrigation systems in Bac Kan province (98% total systems) are still under management of local authorities. In fact, those systems are either lacking a proper management or having no real owners.

The current policy system in Bac Kan is very poor and insufficient for development of PIM. The only legal document found was the temporary regulation on organization for management of irrigation systems within province. However, this regulation seems still trying to consolidate the role of Commune People's Committee in management of irrigation systems while the rights and responsibilities of village and hamlet community in irrigation management are not clearly defined. Besides, Bac Kan does not have policy on irrigation fee. Two of the important areas of policy (management organization and irrigation fee) although play decisive role in effective PIM, but now are insufficient and contrast each other.

Performance of current irrigation institution in aspects of investment, finance, water supply service and irrigated agriculture is very poor. The poor performance in investment is revealed through the low investment efficiency. The structures were quickly degraded after several years of operation. Farmers' contribution was very low, just accounting for 8% total investment budget in irrigation development of the whole province. The lack of a policy on irrigation fee has affected the availability of budget for O&M activities and the sustainability of irrigation systems. Cropping intensity and crop yield were found rather low compared to potential of this area.

Practices of PIM at irrigation system level: Two irrigation systems of Phai Lai (Jointly managed irrigation system, JMIS-B) and Thom Trao (Commune managed irrigation system CMIS-B) were surveyed to see the actual performance at irrigation system level. Both of these systems have similar physical characteristics but are different in institutional arrangement for management. The results show that in overall, the JMIS-B performs better than the CMIS-B. The JMIS-B has a de facto rule on irrigation fee set out by water users although this fee level is very low (2-3 kg/sao/crop) while there is no rule on irrigation fee in CMIS-B. An irrigation institution was also de facto set up in JMIS-B. This institution was established mostly based on the existing social institutions of veteran association and village institution. There is rule on water distribution in this system but it is not written, just orally agreed among village leaders and irrigation management board. The CMIS-B does not have this kind of rule. Overall, the JMIS-B has some de facto rules for irrigation management, meanwhile CMIS-B has no rule and the system is in situation of no real owner. Regarding agricultural performance, the cropping intensities in JMIS-B and CMIS-B are 183% and 122%, respectively. The crop yield in JMIS-B is 4.4 ton/ha of spring rice and 4.0 ton/ha of monsoon rice. In CMIS-B, these figures are 4.0 ton/ha of spring rice and 3.8 ton/ha of monsoon rice.

Farmers' perceptions of irrigation performance: The detailed survey was conducted to explore farmers' perceptions on performance of two irrigation systems of JMIS-B and CMIS-B. It was found that farmers in JMIS-B perceived their system operating more effectively and more equally than farmers in CMIS-B. The level of farmers' participation in irrigation activities was also found more frequently in JMIS-B compared to CMIS-B. Although farmers perceive differently about effectiveness, equity and participation level of the irrigation systems, almost all of them are willing to participate in irrigation management.

Table 1 shows the frequency and percentage of responses of farmers towards their willingness to participate in irrigation management. The result shows that in all three irrigation systems, farmers are willing to involve in irrigation management. In CMIS-B, this figure is even higher compared to JMIS-B. This may be due to that, right now this system is left in very seriously degraded situation. The water supply is also very poor. So, the willingness to participate in system management now is an urgent need of local farmers.

Table 1: Farmers' Willingness to Participate in Irrigation Management

Level of Willingness	Group of Farmers by Irrigation System			Total
	FMIS-T	JMIS-B	CMIS-B	
Very high	4 (13.3)*	3 (10.0)	5 (16.7)	12 (13.3)
High	26 (86.7)	22 (73.3)	24 (80.0)	72 (80.0)
Low	-	2 (6.7)	1 (3.3)	3 (3.3)
Very low	-	3 (10.0)	0	3 (3.3)
Total	30 (100.0)	30 (100.0)	30 (100.0)	90 (100.0)

(*) Figures in the parentheses indicate percentages

<u>Other FindingsThe role of international organizations in IMT/PIM process: It was found that</u> both in Tuyen Quang and Bac Kan provinces, the IMT/PIM was initiated and facilitated by

some international organizations through their funding process. The project of Participatory Natural Resources Development and Management in Tuyen Quang funded by IFAD is an example. This project started in 1992 with various components including component of institutional strengthening in irrigation development and management with farmers' participation. This was considered as a starting point for the successful implementation of IMT four years later in Tuyen Quang.

Awareness of local government officials: The interviews were conducted with some relevant officials in two provinces of Bac Kan and Tuyen Quang. In Tuyen Quang, most of provincial officials interviewed highly appreciate the role and ability of farmers in irrigation management. They also confirmed that the success of PIM/IMT in their province is due to the proper concern by the high ranking government officials together with the willingness of farmers in taking over management of irrigation systems. Meanwhile in Bac Kan province, officials seem lacking a proper concern to irrigation issues in their province. Besides, provincial officials in Bac Kan also lack proper awareness of role of farmers in irrigation management. Some officials hesitate in transferring the right of irrigation management to farmers. The reason may be due to they will lose the chance to continue receiving subsidies from central government unless they try to keep those systems under their control.

4. Conclusions

The practices of PIM/IMT implementation in Tuyen Quang and Bac Kan provinces again confirm the advantage of PIM as the better policy alternative to improve performance of irrigation systems.

Current policy system for PIM in Vietnam is still too general, insufficient and inappropriate for strengthening PIM in the country. Rights and responsibilities of relevant stakeholders are not clearly specified. There is no legal framework supporting or guaranteeing WUAs. The subsidy policy for IMC is still working. These are the most obstacles constraining the development of PIM throughout the country.

The success of Tuyen Quang province in IMT implies that the provincial governments can carry out fruitfully the reform of irrigation management within their province. In addition to an adequate national policy system on PIM, it also requires that the provincial government should formulate an appropriate strategy for implementation of IMT/PIM in the province based on national policy framework.

The limitation in implementation of PIM in Bac Kan shows that the province lacks a strategy for implementation of PIM. However, the underlying cause here is that the provincial officials do not have strong incentives to conduct the reform in irrigation management in their province. This may be due to the inappropriateness of the current national policy system. The subsidy policy together with the exemption of water fee in the poor areas have affected negatively perceptions of not only local government officials but also farmers. The subsidy policy often creates the opportunism and corruption problems among government officials meanwhile the exemption of water fee policy often disable the willingness of water users to share responsibility in irrigation management with the government.

Farmers perceived the farmer-managed irrigation system working more effectively and more equally than the systems managed by the local authority or jointly managed by farmers and local authority. Besides, farmers in all three types of irrigation systems studied expressed their willingness to participate in irrigation management. This is a very convenient condition to conduct successfully the transfer of irrigation systems to farmers.

One set of dominant actors who play very important role in the development of PIM in Vietnam are international donor community. Those donors are acting as facilitators to accelerate the expansion of PIM in Vietnam.

5. Acknowledgement

Acknowledgement is given to the International Water Management Institute, Sri Lanka and the Government of Netherlands for their financial support to conduct this study.

6. References

Bruns, B. (1997). *Participatory management for agricultural water control in Vietnam: Opportunity and challenges,* Background Paper for the National Seminar on Participatory Irrigation Management, April 7-11, 1997, Vinh City, Nghe An Province.

Fontenelle, J. P. (2000). *Water management decentralization in the Red River delta, Vietnam: An uncompleted transition process toward local governance*, Presented at "Constituting the Commons: Crafting Sustainable Commons in the New Millenium", the Eighth Conference of the International Association for the Study of Common Property, Bloomington, Indiana, USA, 31/05- 4/06/2000.

MARD (2004). Draft Framework Strategy on Development of Participatory Irrigation Management in Vietnam. Presented at the Regional Workshop on PIM – Pathways for Progress in Vietnam, 30 March – 2 April, 2004.

Ostrom, E., Burger, J., Field, C. B., Norgaard, R. B., & Policansky, D. (1999). Revisiting the commons: Local lessons, Global challenges. *Science*, 284, 278-282.

Shivakoti, P. G. (2000). Farmer perceptions of the organisational effectiveness of farmer and agency managed irrigation systems in Nepal. *Water Resources Journal* (67-79). S.l. : United Nations, 2000.

Tiep, N. X. (2004). *Participatory Irrigation Management – Roles and Operations of VNPIM.* Paper presented at the Presented at the Regional Workshop on PIM – Pathways for Progress in Vietnam, 30 March – 2 April, 2004.

Adaptation of Rice Farming to the Changes in Water Regime Due to Dykes and Sluice Gates Built against Salinity Intrusion in Mekong Delta

Mayu Aizawa[1], Mai Thi Truc Chi[2], Nguyen Khac Thanh[3], Nguyen Van Hao[4], Nguyen Duy Can[5], Hisashi Kurokura[6]and Kazuhiko Kobayashi[7]

1) The Graduate School of Agricultural and Life Sciences, The University of Tokyo
1-1-1 Yayoi Bunkyo ku, Tokyo, 113-8657, Japan
e-mail: negibouzu6@hotmail.com
2) Research Institute of Aquaculture No.2
116 Nguyen Dinh Chieu St. District 1. Ho Chi Minh City, Vietnam
e-mail: Trucchi9@yahoo.com
3) Research Institute of Aquaculture No.2
116 Nguyen Dinh Chieu St. District 1. Ho Chi Minh City, Vietnam
e-mail: khacthanh0205@yahoo.com
4) Research Institute of Aquaculture No.2
116 Nguyen Dinh Chieu St. District 1. Ho Chi Minh City, Vietnam
e-mail: haoria2@hcm.vnn.vn
5) Mekong Delta Development Research InstituteCan Tho University
University campus 2, 3/2 Street, Can Tho City
e-mail: ndcan@ctu.edu.vn
6) The Graduate School of Agricultural and Life Sciences, The University of Tokyo
1-1-1 Yayoi Bunkyo-ku, Tokyo, 113-8657, Japan
e-mail: akrkrh@mail.ecc.u-tokyo.ac.jp
7) The Graduate School of Agricultural and Life Sciences, The University of Tokyo
1-1-1 Yayoi Bunkyo-ku, Tokyo, 113-8657, Japan
e-mail: aclasman@mail.ecc.u-tokyo.ac.jp

Abstract

Rice cropping in the coastal areas of the Mekong Delta is influenced by salt intrusion because of reduced water discharge of the Mekong River and the tidal fluctuation during the dry season. Dykes and sluice gates have been constructed to prevent the salinity intrusion, but the constructions have changed the water regimes for rice farming, to which the farmers must adapt themselves. We studied the cropping pattern, salinity in the river, gate operation, and water quantity & quality in canals in the island of Con Sau Xa, where dykes and sluice gates were constructed in 2001. We found that the sluice gates were managed in response to salinity in the river and pH in canal in early rainy season, and that the pH is the dominant factor in determining the date of starting water intake from the river. The pH rapidly fell below 3 due to the leaching of acid sulphate after a rain fall at the beginning of rainy season, and farmers had to delay seeding until the pH in canal returned to a non-acidic level. We estimated that the water stored in canals after the gate closure would run out within one month when used for flooding the rice fields in dry season. We then explored an alternative strategy for the management of the sluice gate to expand the period of rice cultivation, but found that such a possibility is limited.

Keywords: rice farming, salt intrusion, Mekong Delta, Acid Sulphate Soil (ASS)

1. Introduction

The Mekong Delta occupies about 3.9 million ha in the South of Vietnam. The delta is located in the tropical monsoon climate, which has two distinct seasons: the rainy season and the dry season. Rainy season runs from May to November, and more than 70 % of the annual precipitation falls during these months. Flooding occurs at the end of August until the end of October. This hydrological environment is suitable for rice cultivation. However, the delta still has several other limitations such as flooding, poor drainage, salinity and acid sulfate soil. Rice cultivation in the coastal areas of the Mekong Delta suffers severely from salt intrusion caused by reduced water discharge of the Mekong River and tidal fluctuation during the dry season. This salt intrusion raised salinity of water in canals and fields leading to a restriction on agricultural practices and crop productivity in this area. The salinity constraint is severest in coastal areas, and the farmers can grow only one crop of rice during the rainy season, when rainfall and fresh water from the Mekong River pushes back the tidal intrusion.

Recently in some areas canal networks have been developed and sluice gates were installed to protect fields from salt intrusion. Salinity damage has been efficiently reduced, and the rice production has drastically increased as shown in the Go Cong Water Control Project (GCP) in Tien Giang Province of the Mekong Delta as an example. The project facilitated triple rice crops over the area of at least 35,000 ha, and the rice production has increased from 293,031 tons in 1995 to 418,853 tons in 2000. However, in many areas in the Mekong Delta, some farmers still suffer from the salt intrusion, yet some others are unsuccessful in increasing the crop production despite the sluice gate constructions.

In Con Sau Xa, an island located in Go Cong District, dykes and sluice gates were constructed in 2001. The land use and cropping pattern depends mostly on the water regimes, and can be classified into triple rice crops, double rice crops and one rice crop. Some farmers' livelihood has improved by the construction of dykes and sluices gates, whereas that of others have deteriorated. A large scale project like this can have both positive and negative impacts on the environment, and the social and economic conditions for the people. A clear understanding of the consequences of the salinity prevention is needed to enhance the benefits and mitigate the negative impacts of the project. We investigated in Con Sau Xa the relationships between salt intrusion and sluice gates management, water quantity and quality in canal and adaptation of rice farming to the changes. Our goal was to propose the optimum water management and agricultural systems to improve the farmers' livelihood in the salt intrusion controlled area.

2. Materials and methods

2.1. Study site

The island of Con Sau Xa has 36 km length and 5 km width, and is located between Tieu River and Dai River in Go Cong District in Tien Giang Province of Vietnam. The climate has a dry season (December-April) and a rainy season (May-November), of which the latter has more than 90% of the annual rainfall (c. 1300 mm). Acid sulfate soils (ASS) cover almost the entire areas of this island. More than 70% of the population is engaged in agriculture and almost all the villagers cultivate rice. The island consists of 5 villages: Tan Thoi, Tan Phu, Phu Thanh, Phu Dong, and Phu Tan, and all villages face the both Tieu and Dai Rivers (Fig.1). Tan Thoi (referred to as site A hereafter) is farthest (24-36 km) from the river mouth. This village has no dykes or sluice gates, yet the farmers can practice triple rice crops. Tan Phu (site B) is located at 18-24 km from the river mouth, and has dykes and 6 small sluices gates built in 2000. The farmers can practice double rice crops in rainy season. Phu Thanh (site C) is located at 11-18 km from the river mouth, and has dykes and 2 large sluice gates: Ba Tai (g1) and Rach Muong (g2) gates built in 2001. Phu Dong (site D) is located at 7-11 km from the river mouth, and has dykes and 3 large sluices gates: Ba Lam (g3), Ly Quan (g4) and Ly Nguon (g5) gates built in 2001. In sites C and D, almost all farmers practice only one rice crop. Phu Tan (site D) is closest to the river mouth, and is outside the dyke, hence all the farmers are engaged in shrimp culture. We studied the 4 villages (sites from A to D), where the farmers grow rice. The canal network data was constructed by field survey with GPS.

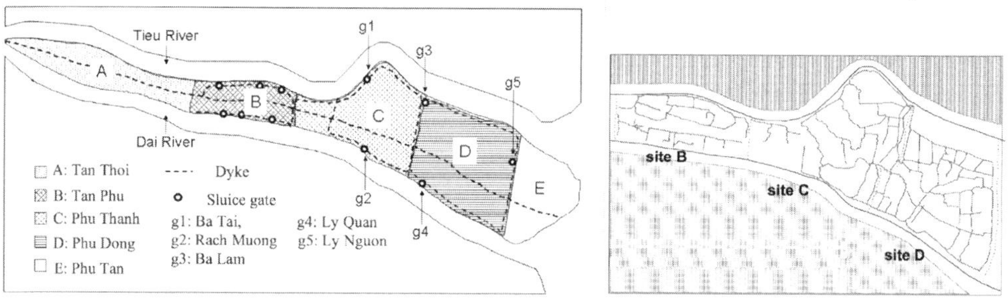

Figure1. Location of the study sites in Con Sau Xa, Go Cong District, Tien Giang Province, Vietnam (left) and the canal network in Sites B, C and D (right).

2.2. Measurements of salinity, pH and water level in Tieu River and canals

In site B, the 6 gates are operated by the village office, and we could not obtain the records of salinity in the river, gates operation, or pH in canals. In sites C and D, the 5 sluice gates are managed by a governmental company, where they have measured salinity, pH and water level outside and inside the gates 5 times a day, twice in the highest tide, twice in the lowest tide, and once in the middle between the high and low tides. We obtained these data from Department Agricultural Rural Development of the government of Tien Giang Province.

2.3. Field survey

Electric conductivity (EC) in Tieu River was monitored by CONPACT-CT sensor with a built-in data logger. EC was measured every 10 minutes from March to August in 2005. The sensors were installed along Tieu River at 4 points, at the distance of 31.2 km(site A), 20.4 km(site B), 15.3 km(site C), and 11.5 km(site D) from the river mouth.

We made a map of the canal network with a survey by a Global Positioning System (GPS). We also measured canal length (m), width (m) and maximum depth (m) in sites B, C and D to estimate the potential volume of canal water storage (PVCWS, m^3) as follows:

$$\text{PVCWS} = \text{length x maximum depth x width} \quad (1)$$

In this calculation, we regarded the canals as a collection of ditches with rectangular cross section. We then divided PVCWS by the area of rice fields (ha) to estimate the potential availability of canal water (m^3 / ha) for irrigating a unit area (ha) of field after the gate closure. Assuming a 5 mm/d water loss by evaporation, we calculated the maximum available period of stored canal water (MAPSCW, d), which farmers can use for rice cultivation after the gate closure as:

$$\text{MAPSCW} = \text{potential availability of canal water} / 50 \quad (2)$$

We assumed that percolation loss of water is negligible at this regional scale: most of the percolated water should be available for use within the same region. We assumed in this calculation that the farmers can easily use the stored canal water even after the gate closure.We also conducted socioeconomic survey to determine the consequences of the salt intrusion prevention in the cropping pattern, farmers' perception of the changes in the water regimes, their economic status, and the farmers' strategies to adapt to the changes in the water regimes. Preliminary interview survey was carried out with 45 households at the 4 sites (A-D) in June, 2005. A larger scale interview survey was carried out with 240 households in August, 2005, but the results are yet to be analyzed.

3. Results

3.1. Salt intrusion in the Mekong River

Fluctuation of daily salinity and water level in Tieu River in 2004 is shown in Fig.2. Water level changed between 2.04 m and 0.70 m corresponding to monthly tidal fluctuation. In dry season from November to May, salinity fluctuation synchronized with water level changes. As rainy season began in May, salinity in the river declined and then rapidly dropped to 1.89 g/l on 28th of July, when the integral rainfall has reached 493.2 mm. Thereafter, salinity was maintained less than 2 g/l until 24th of November. Daily mean salinity is shown in Fig.3 for the 4 points: at 31.2 km (site A), 20.4 km (site B), 15.3 km (site C), and 11.5 km (site D) from the mouth of Tieu River for the period from March to May in 2005. The temporal fluctuation was similar among the 4 points under the influence of the tidal fluctuations. There was, however, a negative correlation between the distance from the river mouth and salinity in the river. Salinity was lowest at site A and highest at site D (Fig. 3). This suggests that duration of fresh water supply is longer at the sites in the upper reaches, and the farmers there can have fresh water for rice cultivation earlier in rainy season.

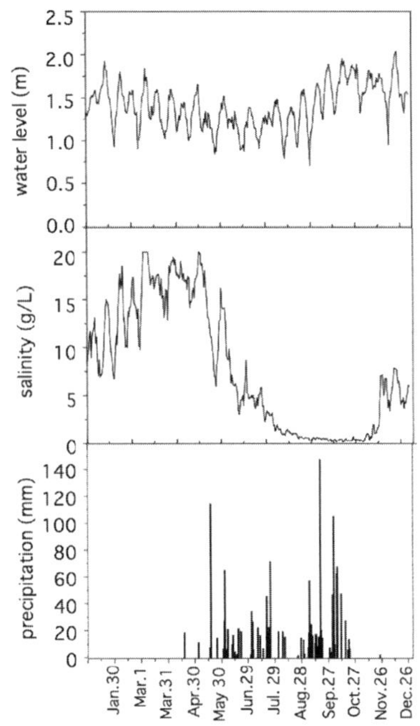

Figure 2. Salinity, water level and precipitation in Tieu River at g5 in 2004.

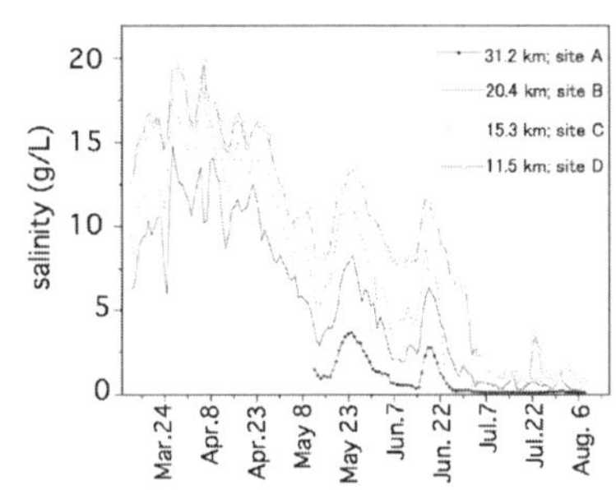

Figure 3. Salinity in Tieu River at the distance of 31.2 km (site A), 20.4 km (site B), 15.3 km (site C), and 11.5 km (site D) from the river mouth in 2005.

3.2. Gate management

In site B, the 6 gates are operated by the village office. Farmers living near the gates check salinity in the river with his tong early in the rainy season, and based on his judgment, the village office decides when start to open the gates at the beginning of rainy season. During rainy season, the village office manages to take in / out water depending on daily salinity, tidal fluctuation and the rice field condition. In 2004, the gates began to open in the middle of June and were completely closed on 25th of January in 2005. Thereafter, the gates were kept closed until beginning of rainy season.

In sites C and D, salinity in the river and pH in canal changed at the timing shown in Table 1, which also shows the dates of opening and closure of the gates g1-g5. The date of pH getting higher than 7 was different among the gates. The pH at g1 and g2 in site C exceeded 7 in mid-June, whereas the pH in g3 and g5 in site D at the lower reach became 7 on 1st of August. On the other hand, the reduction of salinity below 2 g/l occurred earliest at g1 followed by g2. Gate operation was dependent on the salinity in the river and pH value in canal. All the 5 gates started to drain water on the same day, 15th of June, before salinity became less than 2 g/l. This operation flushes away the water with low pH and pollution accumulated during the gate closure in dry season. It was observed that the date of pH exceeding 7 coincided with that of the beginning of water intake at the 4 gates. This shows that the date of freshwater supply is more tightly governed by pH in canal than salinity in the river. All gates were closed before the salinity in the river rapidly increased at the beginning of dry season.

Table.1 Timings of the changes in pH in canal, salinity in river, gate opening for drainage, gate opening for irrigation, and gate closure at gates, g1, g2, g3, g4 and g5.

	Date (month / day)					
	Let out water	Take in water	Close gates	pH in canal < 7	Salinity in river > 2 g/l	< 2 g/l
g1	6/ 15	6/ 20	11/ 20	6/ 20	6/ 19	11/ 24
g2	6/ 15	6/ 16	11/ 19	6/ 16	7/ 3	11/ 23
g3	6/ 15	8/ 1	11/ 15	8/ 1	7/ 28	11/ 24
g4	6/ 15	-	11/ 15	-	7/ 28	11/ 16
g5	6/ 15	8/ 1	11/ 15	8/ 1	7/ 27	11/ 16

3.3. Water quantity in canal

After the complete closure of the gates at the beginning of dry season, the canals can be regarded as a reservoir. The potential volume of canal water storage (PVCWS, m^3) was calculated based on the canal network map data as shown in Figure 1. Maximum period of water availability after the gate closure was calculated as noted before. The result shows that water in canals would run out in only 13 days in site B, 25 days in site C, and 27 days in site D after the gate closure (Table 2). In site B, g1 and g2 were closed in the end of January, which suggests that farmers could use canal water for rice cultivation until the beginning of February. While in sites C and D, the sluice gates were closed in late November and farmers could use canal water for rice cultivation until late January.

Table.2 Potential volume of canal water shortage (PVCWS), the area of rice field, PVCWS by the area of rice fields (ha) and Maximum available period of stored canal water (MAPSCW), water volume per unit in sites B, C and D.

	PVCWS (m^3)	The area of rice field (ha)	PVCWS by area unit (m^3 / ha)	MAPSCW (days)
site B	412574.3	649	635.7	12.7
site C	1211609.4	967	1253.0	25.0
site D	1419977.3	1068	1320.0	26.6

3.4. Water quality in canal

Salinity in canals at the 5 sluice gates in sites C and B in 2004 is shown in Figure 4(left). Line "P" shows the date of the first heavy rain, and line "out" is the date of starting water drainage. The arrow "in" points to the date of starting water intake from the gates, and arrow "close" points to the date of gate closure completely. For all the 5 sluice gates, the salinity increased during dry season through to the gate closing. This probably indicates that the saline water outside the gates leaked inside the gates and the canals. Salinity in canals sharply declined on 17th of May at all the sluices gates because there was a heavy rain event on that day, and the rain water diluted salinity in canal. When gates started water drainage on 15th of June, the salinity in canals decreased rapidly, because salinity water stored near the gates was flushed away to the river. From August on, the salinity stayed less than 2 g/l until the end of 2004. At all 5 gates, pH value in canal rapidly decreased to less than 3 on 18th of May because the heavy rain on 17th of May washed acid sulphate soil and the acid water was released into the canals. The pH value increased and restored the non-acidic level on 15th of June, when the 5 sluices gates were opened to flush out canal water to the river. After sluice gates opened, the pH fluctuated between 4.7 and 7.8 in response to rain events and gate operation (Figure 4 on the right). These suggests that the farmers in sites C and D could get fresh water from June, however, they could not use the canal water for rice cultivation until pH in canal returned to non-acidic level by flushing the acid water via the sluice gates.

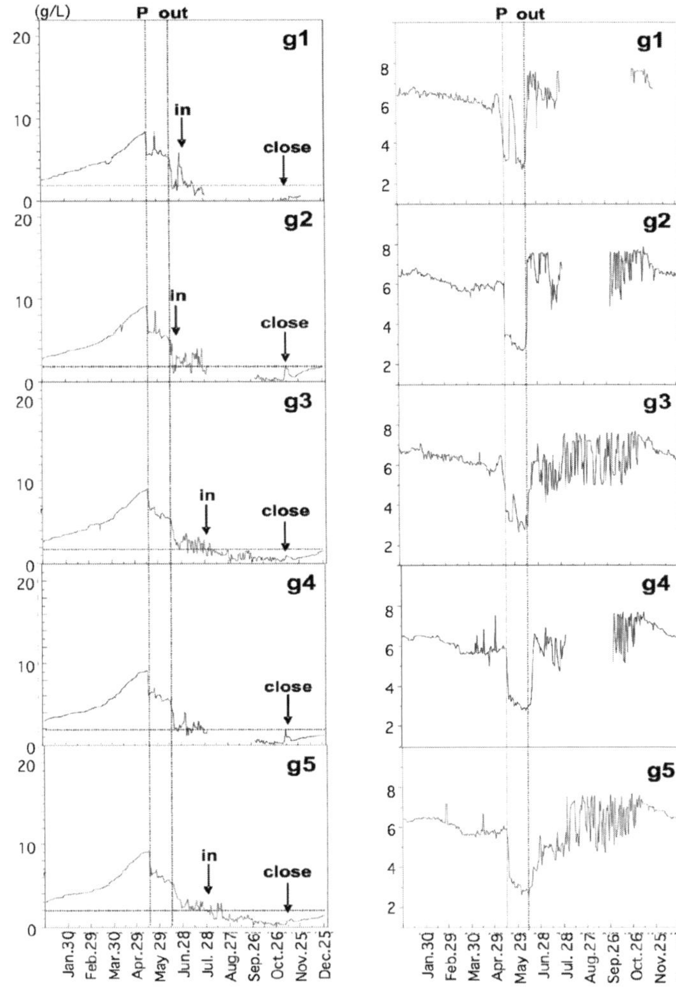

Fig.4 Salinity(left) and pH(right) fluctuations in canal at g1, g2, g3, g4 and g5 in 2004. Arrow "in" points the date of starting water intake and arrow "close" points the date of closing date.Line "P" is the date of first heavy rain and line "out" is the date of starting water drainage.

3.5. Rice cropping calendar

With the preliminary interview survey, we established the rice cropping calendar for each site (Figure 5). In site A, farmers practice triple rice crops in a year with modern short duration varieties. The 1st rice is a summer-autumn crop from mid-May to the end of August, whereas the 2nd rice is autumn-winter crop from the beginning of September to mid-December, and the 3rd rice is winter-spring crop from the end of December to early March.

In site B, almost all farmers practice double rice crops: a summer-autumn crop from mid-June to the beginning of October followed by an autumn-winter crop from mid-October to mid- January. In site C, some farmers living near site B and near the gates practice double rice

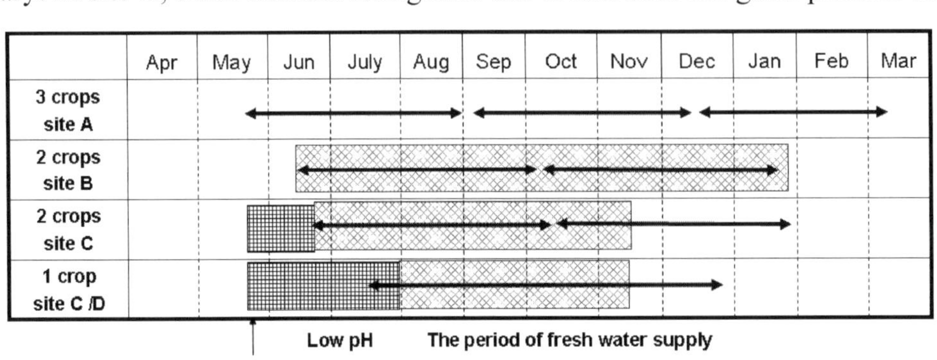

Figure 5. The calendar of rice cultivation in site A, B, C and D. Arrows point the period of rice cultivation. ▦ Period of low pH, ▨ Period of fresh water supply

crops. However, most farmers in site C practice only one crop with traditional long-duration varieties and depend on rain water for the period from mid-July to the end of December. In site D, the farmers practice only one rice crop like those in site C.

4. Discussion

We explored possibilities of extending rice cultivation period to increase the number of crops in sites B to D. To this end, we first tried to simulate the current practices for the gate management. It is assumed that potential period for rice cultivation (PPRC) is the sum of the period of canal water supply (PCWS) and maximum available period of stored canal water (MAPSCW) viz.,

$$PRCP = PCWS + MAPSCW \tag{3}$$

where PCWS is defined as the period from the date of water intake to the date of complete gate closure. MAPSCW has been defined in eq. 2. According to the findings of this study, it is suggested that the gates are operated with the following logic (L1).

At the beginning of rainy season:

Step 1: When salinity at low tide in the river becomes less than 3 g/l, start opening the gate to flush canal water out to the river at low tide, and

Step 2: When pH in canals is restored to 7 or higher, and the salinity at high tide in the river becomes lower than 2 g/l, open the gate at high tide to take water into canals.

At the end of rainy season:

Step 3: When salinity in the river exceeds 2 g/l at high tide, close the gate completely.

PCWS in eq. 3 can be calculated with the above logic using the observed timing of the critical salinity and pH (Figure 4), whereas MAPSCW (eq. 3) has been calculated in Table 2. PRCP, possible rice cultivation period, thus calculated for site B is 7.5 months from mid-June to the beginning of February, that for site C is 6 months from the end of June to the end of December, and that for site D is 4.5 months from the beginning of August to the end of December. Therefore, some extensions of the rice cultivation period may be possible, but they may not be large enough to allow for one more crop in any of the sites investigated.
We then considered three possible changes in the timing of gate operation to increase PCWS. The first is to let out canal water earlier to accelerate recovery of pH to non-acidic level. The second is to start taking in water at low tide in the river, since salinity at low tide drops earlier than that at high tide. The third is to keep taking water into canals at low tide until the salinity at low tide rises at the beginning of dry season. These changes can be incorporated into the logic (L2) of the gate operation, via

At the beginning of rainy season:

Step 1': When pH in canals becomes less than 3 after having more than 500 mm rainfall, open the gate to let the acidic canal water out.

Step 2': When pH in canals is restored to 7 or higher, and salinity at low tide in the river becomes less than 2 g/l, open the gate to take water in.

At the end of rainy season:

Step 3': When salinity in the river at low tide exceeds 2 g/l, close the gate completely.

With this logic, PPRC would be 9 months from the beginning of May to the beginning of February for site B, 7.5 months from the beginning of June to the end of January for site C, and 6 months from the beginning of July to the beginning of January for site D. A period of at least 9.5 -10 months is required for triple rice crops, and that of 6.5-7 months is required for double crops. Hence a transition from double to triple rice crops in site B may not be feasible, while, in site C, they may be able to convert to double rice crops in fields near the gates. In fields far from the gates, however, water exchange is slower and the acidic water may stay in canals longer. Therefore, it may be hard to convert the entire fields to double crops. In site D, double crops would not be possible even with the extension of rice cultivation period.

The above evaluation illustrates the difficulty to increase the number of crops by changing the gate operation. One possibility for intensifying the rice faming would be to introduce fresh water supply from upper reaches of the river to the lower reaches. This option may, however, encounter some difficulties such as salinity intrusion to the upper most reaches, costs for building the water supply system, and, over all, the declining rice price. Another option would be to introduce aquaculture, such as shrimp culture, which is rapidly spreading outside the dykes, e.g. site E and along the coast of site D. A good combination of rice farming and aquaculture should be sought out on the basis of good understandings of the hydrological and agronomic conditions as well as the farmers' requirements in different parts of the area.

5. Conclusions

This study confirmed that rice cultivation in the study area is limited by pH in canals at the seeding time and water quantity in canals in dry season. It might be difficult to increase the number of crops by extending the period of rice cultivation. Constructing dykes and sluice gates and expanding salt-free area may, but may not necessarily be effective for raising the farmers' living status, especially in the area where the fresh water supply is insufficient. A holistic approach including aquaculture may be more promising than only focusing on salinity control for rice production.

6. Acknowledgement

The authors are grateful to the staffs of the Research Institute of Aquaculture No.2 in Ho Chi Minh City for assistance and coordination of this work. We also thank the staffs of Can Tho University for assistance in the interview survey. This work was supported in part by grants of Water Use and Management System of the Mekong River, Core Research for Evolutional Science and Technology, Japan Science and Technology Agency, Japan.

7. References

Kono Y. (2001). Canal Development and intensification of rice cultivation in the Mekong Delta: A case study in Cantho Province, Vietnam. *Southeast Asia studies*, 30(1), 70-85.

Minh Q. L., Tuong P. T., Mensvoort van M.E.F. and Bouma J. (1997).Contamination of surface water as affected by land use in acid sulfate soils in the Mekong River Delta, Vietnam. *Agriculture, Ecosystems and Environment*, 60, 19-27.

Nguyen H. C. (1994). Former and present cropping patterns in the Mekong Delta. *Southeast Asia studies*, 31(4), 345-384.

Tanaka K. (1995). Transformation of rice-based cropping patterns in the Mekong Delta: From intensification to diversification. *Southeast Asia* studies, 33(3), 363-378.

Tuong T. P., Kam S. P., Hoanh C. T., Dung L. C., Khiem N. T., Barr J.and Ben D. C. (2003). Impact of seawater intrusion control on the environment, land use and household incomes in a coastal area. *Paddy Water Environment*, 1, 65-73.

Xuan V. T. and Matsui S. (1998). Development of farming systems in the Mekong Delta of Vietnam. *Ho Chi Minh City publishing house, Saigon times group Vietnam Asia pacific economic center.*

Assessment of Urban and Industrial Pollution on Water Quality: Kelani River Sri Lanka

Gemunu Herath[1)] and Thusitha Amaresekera[2)]

1) Department of Civil Engineering, University of Peradeniya,
Peradeniya 20400, Sri Lanka.
e mail: gemunuh@pdn.ac.lk
2) National Water Supply and Drainage Board, Towns north of Colombo
No. 11, Dippitigoda Road, Kelaniya, Sri Lanka
e mail: mtnc@sltnet.lk

Abstract

A previous study on the water quality of Kelani river has identified that the pollution in the river is a threat to the Greater Colombo water supply intake at Ambatale, estuary ecosystem and socio-cultural uses of the river. Among these, the most important constraint for development is the need to protect the water quality at Ambatale. Hence the intake is currently protected from further large industrial effluent discharges by a government policy not to site such industries above Ambatale. However, this decision has been made without a detailed study of the basin and could well lead to unnecessary constraints on industrial development in the region. Therefore a river water quality monitoring program was instigated in 2002 to model the river water quality so as to understand the limits of future development. The current progress of this on-going study is presented here. The study so far has yielded that the water in the river polluted both with industrial and urban effluents and the present quality is found to be within the standard for drinking water only with conventional treatment. The water quality below Ambatale is poor endangering the aquatic life and severely effecting the ecology of the estuary. Analysis of organic loads show that the contribution from industrial sources very much higher compared to urban load. Nutrient loads measured in the river are somewhat less than expected compared to the applied fertilizer load to the catchment and trapping of nutrient by the hydropower reservoirs may be the main reason for this difference. Fairly high dissolved Cr and Pb concentration obtained necessitates a more extensive investigation of the impact of heavy metals and toxic organic.

Keywords: drinking water, ecosystem, industrial pollution, Kelani river, socio-economical Issues, urban pollution

1. Introduction

Asian developing countries particularly that are industrializing, have most of the present water quality problem of developed countries. With very little technical, scientific and economic resources available, these countries are under tremendous pressure to cope up with the situation. According to the United Nations world water report 2003, all over the globe some two million tons of waste per day are disposed of within receiving waters, including a major portion of an estimated amount of 1,500 km^3 of global wastewater generation. The report further states that Asia shows the highest number of people (estimated over 0.72 billion, 65% of global figure) lack access to improved water supply. Major concern here is that, a large number of developing countries especially in Asia region is at present facing water shortages, a problem exacerbated by severely degraded water quality (Ongley 1998). Further, with UNIDO's projections for water quality situation in the developing world over the next 25 years, public health problems and a possible complete ecological collapse is inevitable unless no quick remedial actions are introduced. Consequently, all these facts points the need for a global policy shift in the direction towards better water quality management protecting the available water resources.

The present situation in Sri Lanka is no exception from the above even located in the tropics. In tropical countries it is commonly expected to prevail relatively constant environmental characteristics such as water quality (Payne, 1986). However most Sri Lankan rivers are frequently polluted by untreated waste discharges from domestic, commercial and industrial sources and uncontrolled solid waste dumping.

The Kelani river which is the focused stream in this study is the third largest watershed in Sri Lanka and plays an important role with respect to the Island's overall economy, since it drains the most fertile land in the wet zone and intercepts the most populated and economically important administrative district of Colombo the capital of Sri Lanka. The principal use of water in the Kelani river is for water supply including to the Greater Colombo area at Ambatale. Further to this the hydropower reservoirs in the upper catchment, the tributaries in the forested areas, the main body of the river and the estuary are the important aquatic ecosystems within this basin. This river is 144 km long and drains an area of 2292 km^2 originating at levels above 1500m on the steep slopes of the western rim of the central highlands. In its descent the main stream and its numerous tributaries travel through deep structurally controlled valleys, generally oriented in many directions at higher and lower elevations. The main river eventually empties into the Indian Ocean on the west coast of Sri Lanka just few kilometres north of Colombo. The entire river catchment receives an average annual precipitation of 3718 mm which corresponds to a volume of about 8,600 million cubic meters (MCM) of which nearly 65% discharges into the Indian Ocean.

A previous study on the Kelani river by the Danish Hydraulic Institute 1999, suggest that the even though the water quality of the mid-catchment was within expectable limits apart from localized pollution close to small towns, the quality in the lower flood plains had indicated a trend of increasing pollutant levels with occasional fish kills due to industrial and domestic effluent discharge. It was also identified that this pollution is a threat to the Greater Colombo water supply intake at Ambatale, estuary ecosystem and socio-cultural uses of the river. Among them, the most important constraint for development is the need to protect the water quality at Ambatale. Though the pollution levels were not critical at that time, this need has lead to a decision not to site any more high-polluting industries above the intake. However, this decision has been made without a detailed study of the basin and could well lead to unnecessary constraints on industrial development in the region. Hence in year 2002 the authorities instigated a comprehensive program to monitor the river basin with the aim of developing a policy for sitting new industries in the region and to review the existing pollution standards. Under this a detailed river quality-monitoring study was commenced in November/December 2002. The current progress of this on-going study is presented here.

2. Materials and methods

There is a common understanding that water quality programs conducted in the developing regions collect wrong parameters, from wrong places at inappropriate sampling frequencies (Ongley, 1998). In this study, two samplings per month at twelve pre-selected locations is done (figure 1). This frequency is well above the required minimum for an effective study (Blind and Aaiderink 1998). As most of the pollution occurs inside the lower 55 km segment of the river, six stations along the main river and six major polluting tributaries within this 55 km stretch upstream of Greater Colombo water intake is investigated (figure 1). The six stations along main river are Seethawaka confluence (S), Glencorse (G), Hanwella (H), Ambatale (A), Victoria Bridge (V) and Paliyagoda at 55, 52, 34, 16, 2.5 and 1 km upstream of ocean outfall respectively. The six tributaries are Eswathu oya (EO), Pugoda ela (PE), Wak oya (WO), Pusseli oya (PO), Maha ela (ME) and Raggahawatte ela (RE) with their confluences at 51, 45, 35, 31, 20 and 18 km upstream ocean outfall respectively. Quality parameters pH, conductivity, turbidity, temperature, dissolved oxygen (DO), biochemical oxidation demand as BOD_5, chemical oxidation demand (COD), Chloride, dissolved Chromium, dissolved Lead, Nitrate, Phosphate and total coliform concentration are examined.

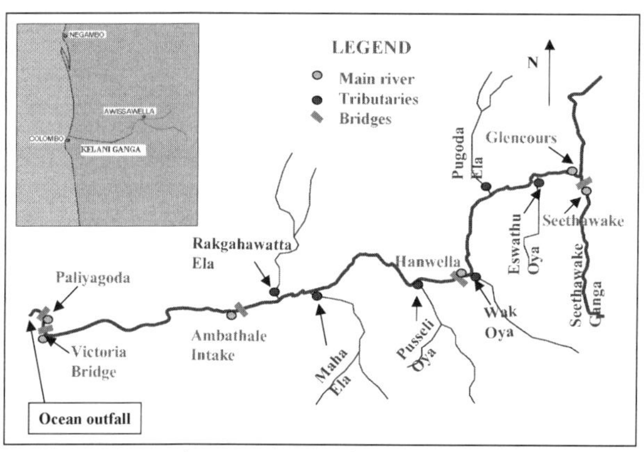

Figure 1. Study Area

The temperature, DO, pH, conductivity and turbidity are measured in the site. BOD_5, COD, NO_3-N, Orthophosphate, Cr, Pb (Spectrophotometer HACH) chloride, and total coliform (plate counting with membrane filtration) are measured in the laboratory. All samples for laboratory investigations are collected in pre-cleaned glass containers following the Standard Methods (Eaton et. al., 1995).

3. Results and Discussion

Referring to the land usage in the basin, it is expected that the run-off of fertilizer, from tea lands in the upper catchment and from rubber and paddy lands in the mid-catchment, is the principal source of nutrients into the river. The current application of Nitrogen and PO_4 as fertilizer is estimated to be about 6900 MT and 1,920 MT per year-of which 4,000 and 845 MT are applied to Tea, 2000 and 370 MT to paddy and 500 and 650 MT to rubber respectively (Danish Hydraulic Institute, 1999). In addition currently urban run-off, solid waste, as well as the effluent from several industries too pollutes the river. The effluent discharge above Ambatale includes the large volume of treated effluent from the industrial parks. Occasional discharges of oils and sludges from these industrial parks have forced the closure of the intake on a few previous occasions.

The obtained average, maximum and minimum water quality parameters at all above mentioned stations are tabulated in table 1 and table 2 below. Table 1 presents the pH, conductivity, turbidity, chloride, temperature, dissolved chromium and dissolved lead contents while the table 2 presents the DO, COD, BOD_5, NO_3-N and PO_4 concentration respectively. Although there is some variation in monthly temperatures, sampling points indicates almost a constant temperature along the river (table 1). Observed temperature variation during the study was in the range of 24-31^{0}C. This variation can mainly be due to seasonal fluctuations in ambient temperatures. pH in the steam was almost constant but, occasional fluctuations persisted due to high industrial discharges.

Table 1, Maximum, Minimum and Average water quality parameters I

Location		pH	Cond. (mS/cm)	Turb. (NTU)	Temp. (^{0}C)	Chloride (mg/l)	Cr (ppb)	Pb (ppb)
Seethawaka	Average	6.84	0.03	24.87	27.2	8.48	0.66	0.13
(S) at 55 km	Range	6-7.4	0.02-0.04	5-68	24-31	0-65	0-4	0-1
Glencorse	Average	6.83	0.03	34.47	26.9	8.47	0.70	0.22
(G) at 52 km	Range	6-7.6	0.02-0.05	10-126	25-30	1-30	0-4	0-1
Hanwella	Average	6.84	0.03	41.90	27.1	8.77	0.59	0.33
(H) at 34 km	Range	6-7.7	0.02-0.05	6-263	25-30	2-40	0-2	0-1
Ambatale	Average	7.02	0.05	38.07	27.5	14.97	0.93	0.27
(A) at 16 km	Range	6-7.9	0.02-0.23	7-237	25-30	2-80	0-8	0-1
Victoria bridge	Average	6.99	1.49	42.23	27.5	529.52	0.87	0.75
(V) at 2.5 km	Range	6.5-7.8	0.02-22.2	10-417	25-30	2-6170	0-4	0-3
Paliyagoda	Average	7.01	1.40	40.10	27.5	592.01	0.69	0.49
(P) at 1 km	Range	6.5-7.8	0.02-21.9	7-347	25-30	2-8500	0-2	0-2
Eswathu Oya	Average	6.59	0.03	15.17	26.6	12.21	0.74	0.33
(EO) at 51 km	Range	6-7.4	0.03-0.06	4-149	24.-29	1-125	0-4	0-2
Pugoda Ela	Average	6.56	0.04	16.80	27.2	9.17	0.57	0.23
(PE) at 45 km	Range	6-7.2	0.04-0.08	6-105	25-29	2-50	0-4	0-2
Wak Oya	Average	6.70	0.03	22.23	27.3	11.40	0.62	0.45
(WO) at 35 km	Range	6-7.4	0.03-0.05	8-139	25-29	1-110	0-3	0-3
Pusseli Oya	Average	6.55	0.04	31.30	27.9	12.92	0.61	0.27
(PO) at 31 km	Range	6-7.2	0.02-0.07	7-119	25-30	1-95	0-4	0-2
Maha Ela	Average	6.55	0.12	16.03	28.6	16.01	0.76	0.69
(ME) at 20 km	Range	6-7.7	0.04-0.83	6-53	26-31	3-100	0-4	0-5
Raggahawatte	Average	6.92	0.37	36.13	27.9	31.06	1.32	0.78
(RE) at 18 km	Range	6.3-7.8	0.02-3.2	11-196	25-30	1-165	0-8	0-5

The amount of suspended solids was measured in terms of turbidity. Turbidity values in the estuary (sampling points V and P) were very high on certain days mainly due to sand mining and soil erosion. Conversely, the turbidity values recorded in the other locations were moderate. This was expected as the basin watershed is in fairly good condition, with balanced land use. However, small pockets of sparsely used cropland and grasslands in the middle catchment have given rise to increased turbidities on rainy days especially near towns. The chloride concentrations too show exceptionally high values in the lower reaches. Main factor for this high chloride levels is the more saline water intrusions during low flow conditions because of severe sand mining. The saline water even reach the Ambatale drinking water intake during periods of extreme drought interrupting water supply to the city on several occasions. Hence now sand mining in this stretch of the river is prohibited. Also, the observed levels of dissolved Cr and Pb indicate extensive industrial pollution. Therefore the proposed increased abstraction in future at Ambatale will certainly reduce flushing the estuary further, resulting in higher Cr and Pb in the river bed, worsening conditions.

Table 2. Maximum, Minimum and Average water quality parameters II

Location		DO (mg/l)	COD (mg/l)	BOD (mg/l)	NO_3 (mg/l)	PO_4 (mg/l)
Seethawaka	Average	6.87	12.33	2.03	0.381	0.023
(S) at 55 km	Range	4.7-8.5	2-30	1-4	0-2.5	0-0.16
Glencorse	Average	7.09	9.47	1.90	0.442	0.031
(G) at 52 km	Range	5-8.4	1-33	1-4	0-1.51	0-0.28
Hanwella	Average	6.75	11.40	2.00	0.319	0.045
(H) at 34 km	Range	2.6-8.7	2-39	1-5	0-1.43	0-0.4
Ambatale	Average	6.63	9.60	2.20	0.324	0.056
(A) at 16 km	Range	4.9-8.3	2-30	1-5	0-2.43	0-0.38
Victoria bridge	Average	5.53	22.63	6.37	0.532	0.196
(V) at 2.5 km	Range	0.8-7.5	2-108	1-66	0-2.06	0-1.61
Paliyagoda	Average	6.02	29.03	3.80	0.553	0.091
(P) at 1 km	Range	3.2-7.4	2-207	1-18	0-3.15	0-0.6
Eswathu Oya	Average	6.74	11.20	1.93	0.424	0.046
(EO) at 51 km	Range	4.3-8	2-36	1-4	0-1.5	0-0.49
Pugoda Ela	Average	6.65	13.07	2.07	0.518	0.058
(PE) at 45 km	Range	4.7-8.1	2-39	1-5	0-2.1	0-0.56
Wak Oya	Average	6.77	14.63	2.27	0.317	0.041
(WO) at 35 km	Range	5-8	2-42	1-6	0-1.35	0-0..36
Pusseli Oya	Average	5.66	15.37	2.67	0.499	0.043
(PO) at 31 km	Range	2.8-7.8	2-39	1-10	0-2.72	0-0.28
Maha Ela	Average	4.10	18.29	4.92	0.369	0.048
(ME) at 20 km	Range	2.3-7.8	4-54	1-10	0-2.15	0-0.3
Raggahawatte	Average	4.51	28.13	6.83	0.944	0.123
(RE) at 18 km	Range	1.2-7.7	5-99	1-25	0-3.8	0-1.06

Table 2 data shows that the pollution in the river is considerable. There is minor decrease in some waste concentration between Hanwella and Ambatale but, significant increase in concentration between Ambatale and Victoria Bridge. These results were rather unexpected because of lesser human activities upstream of Hanwella. However it must be noted that, for this comparison, the waste concentrations are considered rather than loads. There are two possible reasons for this variation. Firstly in comparison there are little discharges of urbane or industrial wastes in this stretch. Secondly it is at Hanwella that the river begins to be influenced by tidal effects, thus reducing flow velocity especially during dry season resulting in significant degradation to take place along this fairy long reach.

The nutrient concentrations (table 2) observed in the river is smaller than what was anticipated considering the loads of applied fertilizer in the catchment. It is even more unusual to obtained low Nitrate nitrogen (NO^{-}_3 –N) levels as it is very mobile in soils and also is the most highly oxidized form of nitrogen usually abundant in streams. The low Phosphorus concentrations were as expected. Usually runoffs carry lower phosphorous amounts, as they are retained in soil and a further potion can readily bind with suspended particles in the river water and settle to the stream bottom. Another reason for these lower nutrient levels can be nutrients trapping by the large number of hydropower reservoirs constructed in the upper catchment. Nevertheless these low concentrations are far adequate for eutrophication to occur (Muller, 1999) and periodic algal blooms observed in the shallower areas especially in stagnant waters during periods of low flow are a good example. On the other hand, the PO_4 load under low flow conditions increases very steeply below Ambatale. This increase cannot account only to increased urban pollution. It may also be due to discharges of phosphate containing detergents, especially by the textile factories. The observed decrease below Victoria bridge may be due to sedimentation and uptake by organisms.

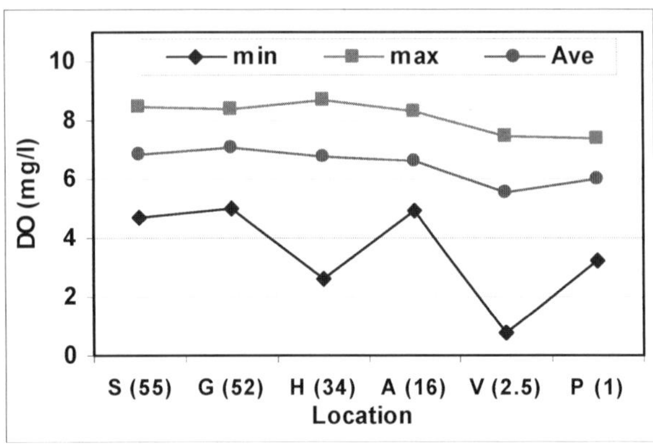

Figure 2. DO variation in main river

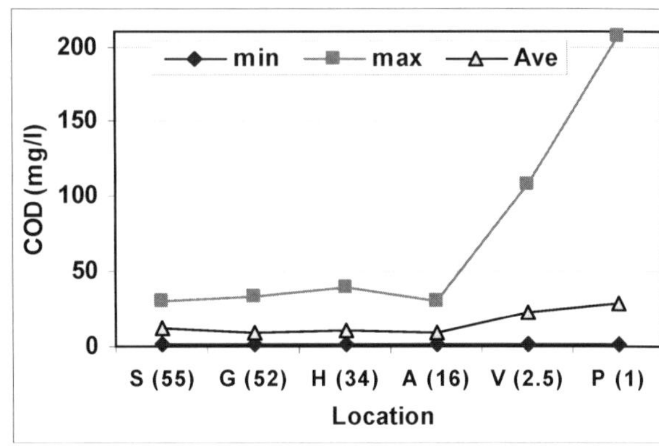

Figure 3. COD variation in main river

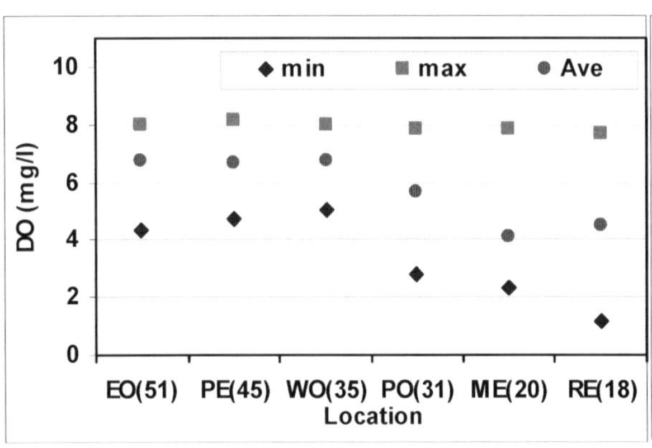

Figure 4. DO variation in tributaries

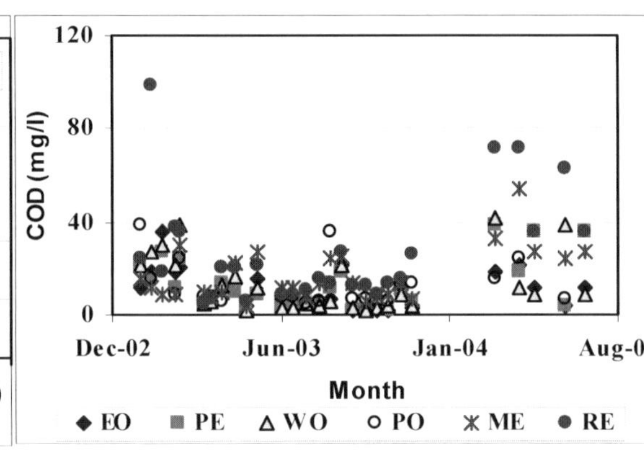

Figure 5. Observed COD in tributaries

The measured DO, BOD_5 and COD levels especially at Ambatale and further downstream are often in excess of the Sri Lankan inland surface water standard required for bathing or as source water for drinking with complete treatment. Current Sri Lanka standard for this is 5 mg/l for DO, 4 mg/l for BOD_5, and 20 mg/l for COD respectively. The observed maximum, minimum and average variations of DO and COD along the main river are presented in figure 2 and figure 3 respectively. Figure 4 show the maximum, minimum and average DO values and figure 5 show all observed COD values in the tributaries.

Estuary, hydropower reservoirs, the main body of the river and its tributaries are the main components of the ecosystem in the Kelani River. The low DO (occasionally below 1 mg/l) levels in the estuary and tributaries is a good indication for possible complete destruction of their ecosystems. The estuary may be stratified especially during low flows, leading to a complete depletion of its DO levels in the lower layers and the consequent formation of toxic gases.

Analysing the concentrations of organic material indicate that the obtained values at Hanwella too are significant compared to the levels at Ambatale. A reason could be the frequent discharges from textile and rubber industries above Hanwella. Even with approximately equal contributions from urban and industrial sources to the BOD load another aspect for these variations is that there appears to be significant degradation of organic material within the river reach especially during periods of low flow. This is due to the river flowing full even at low discharges influenced by the sea level and as the extent and the deepness of the channel- approximately 100m wide and 2m to 3m deep with flow velocity in the order of 0.1m/s. At this low velocity the travel time between Hanwella and Ambatale will be about 2.5 days during which a significant degradation of BOD can be expected.

Figure 3 and table 2 results show a rapid increase in the COD levels at the lower reach of the river. This could be due to larger COD loads form industries and the highly polluted nature of the river estuary, which can be atrophic under low flow conditions as discussed. However the reason for the BOD not increasing so rapidly may be due to the nature of the pollution or the presence of toxic materials that inhibit the natural degradation of the organic matter. Further the figure 5 shows that the COD levels in the tributaries appear to be increasing, even though there is a considerable scattering of the results. The reason may be the significant increase in COD load from the industrial parks.

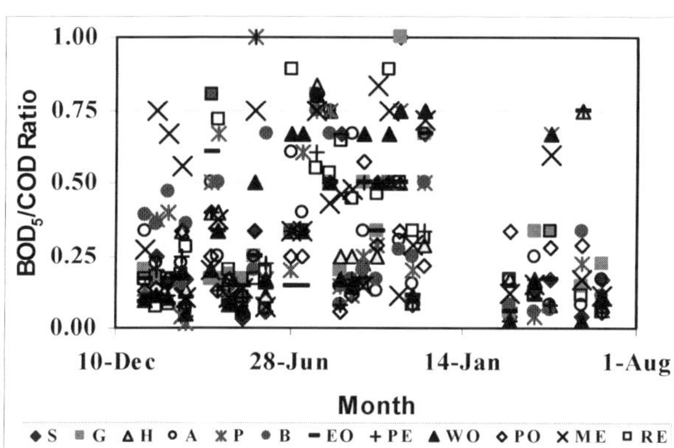

Figure 6. Obtained BOD_5/COD ratio

It is common that water bodies especially rivers are at greater risk in developing countries as they are commonly used for untreated sewage disposals (Dissanayake et.al. 1982, Pearce et. al. 1998). Conforming this, the observed coiliform levels during this study too was quite high. Coliform levels over 15,000 cfu/100ml were common in most stations exceeding the required level of drinking water for conventional treatment too.

The BOD/COD ratio can be used to assess the relative importance of industrial pollution as against "natural" organic pollution. The ratio is approximately 0.5 for "natural organic materials while it ranges from 0.5 to 0.1 for industrial effluents. The calculated ratios for this study are plotted in figure 6 above and it shows more pollution contribution from industrial sources compared to urban sources.

4. Conclusions

Most of the time, the water quality at Ambatale is found to be within the standard for drinking water with conventional treatment but not suitable as a source of drinking water with simple treatment. The water quality data obtained so far show that industrial and urban effluents pollute the river above the Greater Colombo water intake at Ambatale. The river below Ambatale is more polluted, by both urban and industrial sources, and the water quality is poor endangering the aquatic life and severely effecting ecology of the estuary. Analysis of the organic loads show that the contributions from industrial sources are much higher compared to urban load. Also it appeared that significant degradation of organic material occur between

Hanwella and Ambatale during all seasons. The nutrient loads measured in the river are somewhat less than the values expected compared to the applied fertilizer load to the catchment. Trapping of nutrient by hydropower reservoirs may be the main reason for this difference. Further the obtained fairly high dissolved Cr and Pb concentration necessitates an investigation of heavy metals and toxic organic matter on this use of water.

Given the current and future importance of the Ambatale intake, the construction of a salinity barrier arrears to be a priority. The intake is currently protected from further large industrial effluent discharges by a government policy not to site such industries above Ambatale. However, the river water quality should be monitored and modeled so as to understand the limits of future development

5. References

Blind M. and Aaiderink H. (1998). Designing a monitoring network for long term trend detection, *Wat. quality international*, September/October, 43-46.

Danish Hydraulic Institue (1999). Kelani Ganga basin detailed basin assessment, working document C.

Dissanayake C.B., Senaratne A., Weerasooriya S.V.R. and De Silva S.H.G. (1982). The environmental pollution of Kandy lake: A case study from Sri Lanka. *Env. Int.*, 7, 343-351.

Eaton A.D., Clesceri L.S. and Greenberg A.E., (1995). Standard methods for examination of water and wastewater: American public health association (Eds), Edition 19, United book press, Maryland, US.

Muller D. and Helsel D. (1999). Nutrients in the Nation's Water-Too Much of a Good Thing, US Geological Survey circular 1136.

Ongley E D (1998). Modernisation of water quality programmes in developing countries: issues of relevancy and cost efficiency, Wat. quality int., September/October 1998, pp 37-42.

Payne, A. I. (1986); The ecology of tropical lakes and rivers, John Wiley & Sons, UK, pp 301.

Pearce G.R., Chaudhry M.R and Ghulam S. (1998). A simple methodology for water quality monitoring: HR Wallingford in collaboration with International Waterlogging and Salinity Research Institute,Lahore.

Pollution of Some Toxic Metals in Canal Water Leached Out from Acid Sulphate Soils in the Mekong Delta, Vietnam

Nguyen My Hoa[1], Tran Kim Tinh [2], Mats Aström [3] and Huynh Tri Cuong [4]

[1] Soil Science & Land Management Department, Cantho University, Vietnam
3/2 street, Cantho city, Vietnam.
Email: nmhoa@ctu.edu.vn.
[2] Soil Science & Land Management Department, Cantho University, Vietnam
3/2 street, Cantho city, Vietnam.
Email: tktinh@ctu.edu.vn.
[3] Department of Biology and Environmental Science, Kalmar University, Sweden
Barlastgatan 11, 39182 Kalmar, Sweden.
Email: mats.astrom@hik.se
[4] Soil Science & Land Management Department, Cantho University, Vietnam
3/2 street, Cantho city, Vietnam.
Email: htcuong@ctu.edu.vn.

Abstract

It has been known that water leached out from actual Acid Sulphate Soil (ASS) acidified water in canals in the Mekong Delta. However, the leaching of toxic metals from these ASS is poorly known and unquantified. This study aims at accessing pH and amounts of 9 toxic metals (Al, As, Cd, Cu, Fe, Mn, Ni, Pb, Zn) in first, second and third-order canals, and in ground water in two canal systems in severe and moderate ASS in the Long Xuyen Quadrangle in the Mekong Delta during 5 periods in comparison to those in an non-ASS area.

Results showed that pH was lowest, and concentration of Al, Cd, Cu, Fe, Mn, Ni, Zn in canal water in the severe ASS was higher than that in moderate ASS. Concentration of these metals was often highest in ground water, high in the third-order (the smallest canal) and second-order canal (the intermediate canal), and lowest in the first-order canal (the largest canal). The highest concentration of these metals was mostly observed at the beginning of the rainy season during May and July. Concentration of As, and Pb was found not related to the occurrence of ASS. Results of this study clearly showed that ASS is a source of contamination of these toxic elements to water environment, and hence further drainage, reclamation of severe ASS for crop cultivation should now be seriously considered.

Keywords: acid sulphate soils, acid water, Mekong Delta, toxic metals, water environment

1. Introduction

In the Mekong Delta (MD), Acid Sulphate Soils (ASS) occupied about 1.7 millions ha, mainly located in Long Xuyen Quadrangle in Northern West, Plains of Reeds in the Northern East and the central of Camau Peninsular in Southern part of the Mekong Delta. Since 1980, many severe ASS in the MD have been reclaimed for cultivation of rice and other crops such as pineapple, yam, cassava, sugar cane, etc..., resulted in artificial drainage of the areas for crop production. Dikes have been made, which prevented the leaching of acid and toxic substances via yearly flood. The acid water was accumulated in the area and then was pumped out to canals, rivers, causing severe water pollution. The consequence of leaching of ASS on acidification and accumulation of Fe^{2+} and Al^{3+}in canal water has been known. However, little information on concentration of other toxic metal elements in canal water in ASS in the MD is available.

Sundstrom et al. (2002), reported that the median concentration of Al, Cd, Co, Mn, Ni and Zn in the draining ditches in Acid Sulfate soils in Finland was found to be 135-1044 times higher than the median concentration of Finnish headwater stream. The diffuse leakage of metals (Al, Cd, Co, Mn, Ni, and Zn) from acid sulfate soil was also found to be a massive supplier in comparison to the present-day industrial discharges. This alarms an environmental contamination, which is very much likely to exist also in other ASS areas than those in Finland.

This study is therefore, in the first stage, aimed at investigating the concentrations of the 9 metal elements (Al, As, Cd, Cu, Fe, Mn, Ni, Pb, Zn) in the first, second and third-order canals and ground water in the ASS Long Xuyen Quatrangle during 5 periods: at the beginning and at the end of the dry season, and at beginning, the middle and at the end of the rainy seasons 2003.

2. Materials and methods

2.1. Area descriptions

Investigation was carried out in ASS Long Xuyen Quatrangle at two first- order canal systems: Tam Ngan and Kinh Moi canal systems. The Tam Ngan canal system located on severe ASS (Typic Sulfaquepts). From Tam Ngan first-order canal, 2 second-order canals in Khoai Mi state farm and Cong Ranh area were selected. Khoai Mi state farm is a closed-dike area, where dike were made around the whole area to prevent flood for cassava and rice cultivation. Cong Ranh area was an open area, where there are several third-order canals connected to Cong Ranh canal for drainage of acid water to Western sea via Tam Ngan canal. In Cong Ranh area, cajuput forest and double rice cropping are the main land use type. The Kinh Moi canal system located on moderate ASS (Sulfic Humaquepts). From Kinh Moi canal, Ben Bo and Ben Dau second-order canals were selected. These two canals were ended by small mountain area; the only mountain area remained in the Mekong Delta. Therefore water flow in these two canals was slow. Rice is the main crop in the area. Canals in Binh Thuy non-ASS area were also selected for comparison.

2.2. Sampling method and chemical analysis

In Tam Ngan area, water samples were taken from Tam Ngan main canal, two second-order canals in Khoai Mi state farm and in Cong Ranh area, and one third-order canal at each second-order canal. Similarly, in Kinh Moi area, water samples were taken from Kinh Moi main canal, two second-order canals in Ben Bo and Ben Dau, and one third-order canal at each second-order canal.

Water samples were collected in duplicate at the headwater and at the outlet of each first- and second-order canal, and at the middle of each third-order canal. The order of canal were named following the name for irrigation system, meaning that first order canal is the biggest canal and third-order canal is the smallest that bring water to the field, and the second-order canal is the middle canal that connects the first to the third-order canal. Ground-water samples were collected at the four fields where the third-order canal located. Five sampling periods were: at beginning and at the end of the dry season (January and March, respectively) and at the beginning, the middle and the end of the rainy season (May, July and November, respectively).

Ground and canal water samples were analyzed for pH, EC in the same day of collection, then acidified to pH<2 by addition of 0.2 ml concentrated HNO_3 in 100 ml of water sample and filter through 0.45 μm polycarbonate screen filters for analyzing of 9 toxic metal elements (Al, As, Cd, Cu, Fe, Mn, Ni, Pb, Zn) by ICP-MS and by ICP-AES for overanged samples.

3. Results and discussion

3.1. Variation of pH and EC in different ASS areas during different time periods
Results in Figure 1.a showed that water pH was high in the first order canal, low in the second order canal and lowest in the third order canal, reflecting the lowest acidity in canal water which is directly drained from ASS, and the dilution of these acid water in the first-order canal by non-acid water from river. Water pH in the third order canal was often lower in Tam Ngan than in Kinh Moi systems. This reflects the variation of acidity in water leached out from different ASS areas. The more severe the ASS was, the more acid the water from canals in these areas was. The hydrology and the management of ASS in the areas also affect the acidity of canal water in these areas. In the Khoai Mi state farm where dikes were made around the whole area to prevent flood, acid water leached out from the dikes, bunds and from the soil was accumulated into the third-order canal. Therefore, pH in this third-order canal was often lower than in Cong Ranh area where acidity was leached out of the soil during flooding period. Water pH was low in third order canal in Ben Bo and Ben Dau areas, because of the stagnant condition of water in these canals. pH was almost below 3 in most of the periods in the second and third order canals in all areas. pH in the third-order canal in May and July are 2.61-2.7, 3.02-2.88, 2.9-2.87, for Khoai Mi, Cong Ranh, and Ben Bo and Ben Dau, respectively. During May and July, pH in the first order canal was even below 3.5, whereas it was often about 5-7 in other months. This shows that at the beginning of the rainy season (in May and July), rain water have dissolved and leached H^+ forming during the dry season into the canal, resulted in severe acidity during this period. pH in ground water was as low as that in the third-order canal. EC was highest in ground water and in third-order canal water, especially in May and July, and was in accordance with variation in pH. EC was high when pH was low and vise versa.

Very low pH of canal water leached out from ASS was mainly due to acidification of the soil, which caused by oxidation of sulphidic material during formation of actual ASS in potentially ASS area, and the reclamation of ASS for agricultural production such as artificial drainage canal excavation, constructing closed-dike to prevent flood, etc…in actual ASS areas. The low water pH in ASS closed-dike area in comparison to ASS open-area showed that management of ASS is important in controlling the acidification in canal water.

3.2. Variation of concentration of metals in different ASS areas during time periods
The concentration of Al, Cd, Cu, Fe, Mn, Ni, Zn was often highest in severe ASS in the close-dike Khoai Mi state farm, high in Cong Ranh area and lowest in moderately ASS Ben Bo and Ben Dau areas (Fig 1.a &1.b)). The results again showed the high contamination of toxic

metals in severe ASS in comparison to moderate ASS and the effect of ASS management on contamination of toxics metals in canal water leached out from ASS. The construction of dike around the area to prevent flood in Khoai Mi state farm made acid and toxic substances unleached by yearly flood. In general, highest concentration of metals was observed in May and July and this tendency was mostly observed in Cong Ranh open area, but was not clear in Khoai Mi, Ben Bo and Ben Dau areas because of slow flows in these areas.

Al concentration in third-order canal during May and July was highest in Khoai Mi (59.6-89.4 mg/l), lower in Cong Ranh (4.5-10.7 mg/l), and lowest in Ben Bo and Ben Dau (4.1-10.2 mg/l). Concentration of Al in ground water in May and July was highest in Khoai Mi (72.5-123 mg/l) lower in Cong Ranh (23.6-28.5 mg/l), Ben Bo (22.4-56.7 mg/l) and Ben Dau (33.3-38.1 mg/l).

As concentration was low (< 10 μg/l) in canals and ground water in the 4 locations, except for high value in ground water Cong Ranh in May (40.3μg/l) and November (44.1μg/l).

Cd concentration was highest in ground water (1.75μg/l in Khoai Mi, 1.16 μg/l in Cong Ranh and < 1 μg/l in Ben Bo and Ben Dau), and lower in all canals (<1 μg/l).

Maximum Cu concentration in third order canal was found in May and July ranging from 24.1 μg/l to 55.4 μg/l. Cu concentration reaches maximum value in ground water in Khoai Mi state farm (113μg/l), and lower in Cong Ranh (95.2μg/l), Ben Bo (37.2 μg/l) and Ben Dau (22.4 μg/l).

Fe and Mn concentrations were mostly highest in third order canal and ground water in May and July. Fe concentration in third order canal was highest in May and July in Khoai Mi (72.2-102 mg/l), and lower in Cong Ranh (3.6-14.3 mg/l), Ben Bo (8.63- 9.07 mg/l) and Ben Dau (4.38 mg/l). Highest Fe concentration was in ground water (64.1-182 mg/l in Khoai Mi; 66.7-125 mg/l in Cong Ranh; 57.7-83.1 mg/l in Ben Bo; and 22.1-34.9 mg/l in Ben Dau during May and July).
Maximum concentration of Mn in second and third order canal water in May and July was 5.05 mg/l in Khoai Mi, 2.23 mg/l in Cong Ranh, 3.93 mg/l in Ben Bo and 1.87 mg/l in Ben Dau, and in ground water in May-July was 4.49-5.19 mg/l in Khoai Mi, 1.79-1.92 mg/l in Cong Ranh, 3.68-5.49 mg/l in Ben Bo and 2.08-2.11 mg/l in Ben Dau.

Maximum Ni concentration was found highest in Khoai Mi state farm (444 μg/l) in second-order canal, lower in Cong Ranh (96.3 μg/l) and lowest in Ben Bo (70.1 μg/l) and Ben Dau (52.1 μg/l) in third-order canal.

Maximum concentration of Pb in the third-order canals in Khoai Mi, Cong Ranh, Ben Bo and Ben Dau was 12.3μg/l, 3.4μg/l, 3.6μg/l and 1.4 μg/l, respectively. However, Pb concentration in ground water in May and July was highest in Cong Ranh, while for other metals, they were often highest in Khoai Mi state farm.

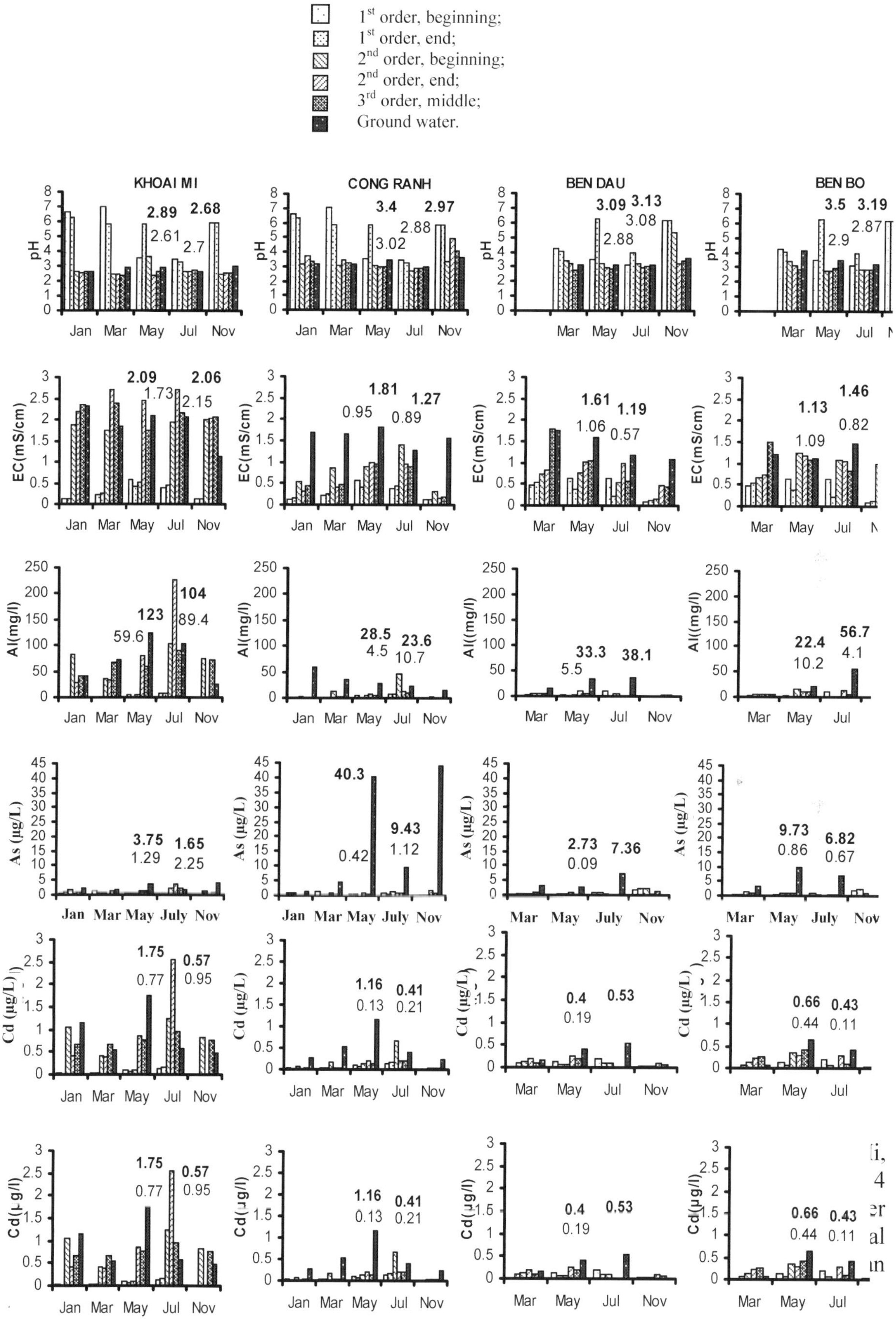
1st order, beginning;
1st order, end;
2nd order, beginning;
2nd order, end;
3rd order, middle;
Ground water.
KHOAI MI
CONG RANH
BEN DAU
BEN BO
pH
EC(mS/cm)
Al(mg/l)
As (µg/L)
Cd (µg/L)
Cd(µg/l)
Jan Mar May Jul Nov
2.89 2.68 2.61 2.7
3.4 2.97 2.88 3.02
3.09 3.13 3.08 2.88
3.5 3.19 2.87 2.9
2.09 2.06 1.73 2.15
1.81 1.27 0.95 0.89
1.61 1.19 1.06 0.57
1.13 1.46 1.09 0.82
123 104 59.6 89.4
28.5 23.6 4.5 10.7
33.3 38.1 5.5
22.4 56.7 10.2 4.1
3.75 1.65 1.29 2.25
40.3 9.43 0.42 1.12
2.73 7.36 0.09
9.73 6.82 0.86 0.67
1.75 0.57 0.77 0.95
1.16 0.41 0.13 0.21
0.4 0.53 0.19
0.66 0.43 0.44 0.11

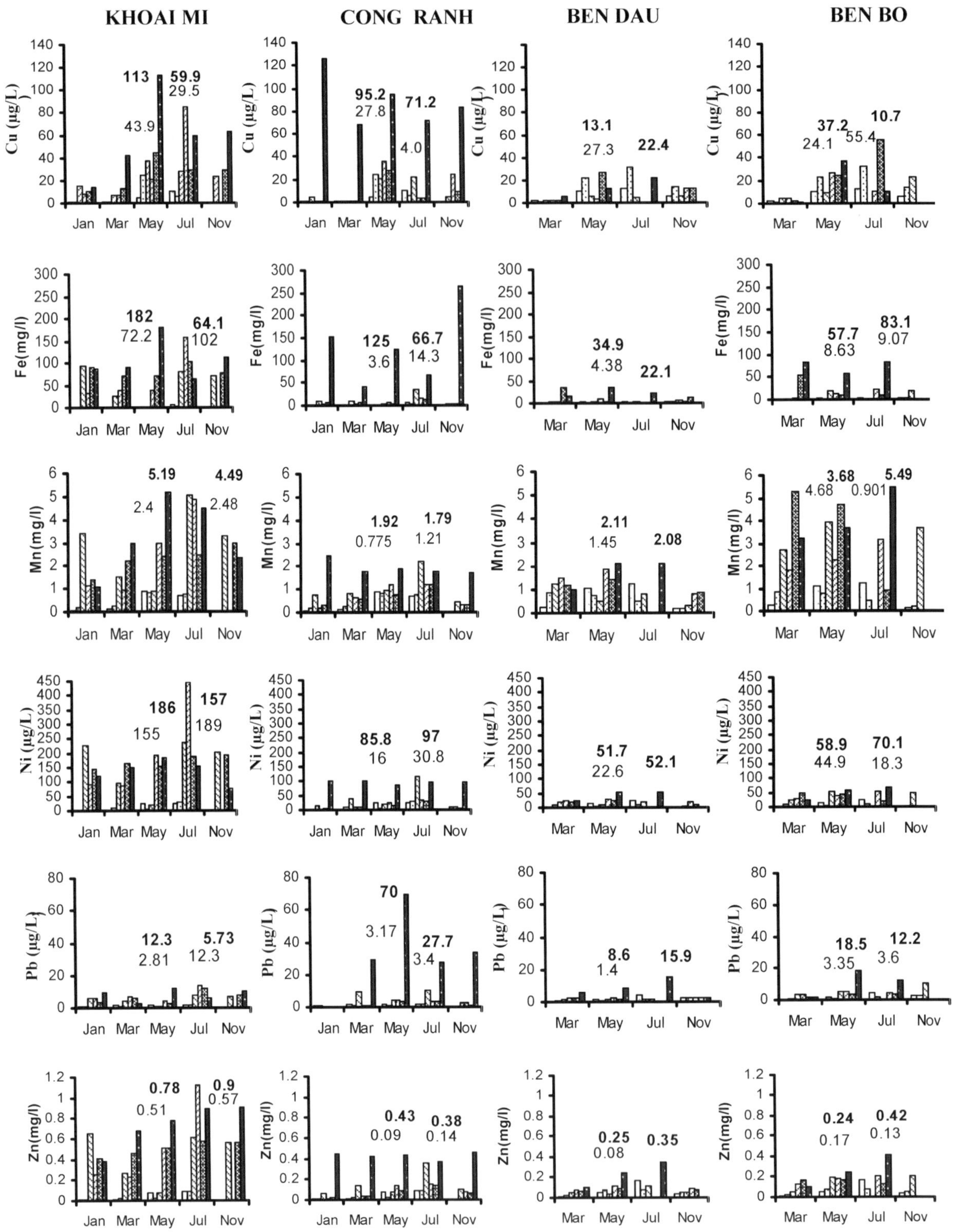

Figure 1b. Concentration of Cu, Fe, Mn, Ni, Pb and Zn at four second-order canals (Khoai Mi, Cong Ranh, Ben Bo and Ben Dau) and at first and third order canals connected to the 4 second-order canals during 5 periods (graphs were named by the name of the 4 second-order canals). Numerical data shown in the graph were for ground water (bold) and third-order canal (not bold) in May and July only. Some data in July and November are missing; hence it can not be seen in the graphs. Legends in the graphs are the same as in Fig. 1a.

Zn concentration was high in second-order, third-order and in ground water and was almost below 0.5 mg/l.

3.3. Comparison of concentration of pH, EC and toxic metals in ASS and non-ASS areas.
It was found that the concentration of Al, Cd, Cu, Fe, Mn, Ni, and Zn in the first, second and the third-order canal, and in ground water in ASS was 17 - 296, 4.5 -37.0, 3.0 – 15.3, 5.3-58.5, 1.9 – 35.8, 12 - 233, and 6.8 – 36.7 times higher than those in Binh Thuy non-Acid Sulphate Soils (non-ASS), respectively (Table 1). These results showed that the removal of these elements from soils to canals was related to acidification of the soil, which is caused by oxidation of sulphidic material during formation of actual ASS and artificial drainage during reclamation of ASS for agricultural production. Concentration of As, and Pb was found comparable in canal and ground water in ASS and non-ASS. Concentration of As and Pb in ground water in non-ASS was even higher than in ASS (Table 1). Therefore, it may be that concentration of these elements was unrelated to extensive ASS occurrence as found by Åström (2001). pH in ground and canal water in ASS was very much lower than in non-ASS and was over limited range for surface water quality used for living activity according to Vietnamese criteria. Concentration of Al, Fe, and Mn in ASS canals was also higher than limited range for surface water quality used for living activity according to Vietnamese criteria.

4. Conclusions

The results clearly showed that ASS is source of acid and of Al, Cu, Fe, Mn, Ni and Zn toxic elements which contaminated the aquatic environment. Water in canal leached out from severe ASS contains more toxic metals than moderately ASS. High concentration of Al, Cd, Cu, Fe, Mn, Ni, Zn in canal system in ASS showing that the removal of these elements from soils to canals was related to acidification of the soil, and the reclamation of ASS for agricultural production such as artificial drainage canal excavation, constructing closed-dike to prevent flood, etc…in actual ASS areas. High concentration of toxic metals and low water pH in close-dike area in comparison to ASS open-area showed that management of ASS is important in controlling contamination of these metals to the environment. Since concentration of these toxic elements was often in accordance with low pH, controlling the removal of these elements from top soil by liming is perhaps important in limiting the contamination of these elements to the environment, which should be studied further. Another important action which should be considered is that no further deterioration of water quality will occur, and according to suggestion by Sunström et al., no drainage of actual/potential ASS field should be allowed.

5. References

Sundström, R, Åström, M. and Österholm P. (2002). Comparison of the metal content in acid sulphate soil runoff and industrial effluents in Finland. *Environ. Sci. and Tech.,* 36, 4269-4272.

Åström, M. (2001). The effect of acid soil leaching on trace element abundance in a medium-sized stream, W. Finland. Appl. geochem. ,73 (2001),181-191.

Table 1. Average concentration of pH, EC and the nine toxic metals in the first, second and third order canals in ASS and in Binh Thuy non-ASS in comparison to Vietnamese criteria for surface and ground water quality, and ratio of these concentrations in ASS and non-ASS.

Sampling site	pH	EC (mmhos/cm)	Al (mg/l)	As (µg/l)	Cd (µg/l)	Cu (µg/l)	Fe (mg/l)	Mn (mg/l)	Ni µg/l)	Pb (µg/l)	Zn (mg/l)
ASS canal system											
First order	5.06	0.3	2.5	0.8	0.07	9.5	1.6	0.554	9.7	1.5	0.05
Second order	3.07	1.1	26.6	1	0.37	14.6	23.4	1.790	70	4.6	0.21
Third order	2.95	1.2	23.7	0.9	0.34	18.3	35.5	1.816	67.2	3.6	0.22
Ground water	3.25	1.6	43.9	9.2	0.58	51.6	92.5	2.703	90.8	18.4	0.45
non-ASS canal system											
First order	7.68	0.15	0.07	1.32	0.01	1.3	0.3	0.03	nd	1.60	0.010
Second order	7.17	0.15	0.09	1.51	nd [1]	1.7	0.4	0.05	nd	0.99	0.010
Third order	7.18	0.16	0.1	4.04	nd	1.2	0.9	0.26	nd	1.48	0.006
Ground water	6.27	0.43	2.6	36.4	0.13	17.1	14.5	1.42	7.6	22.50	0.066
ASS/ non-ASS	Ratio of average concentration of metals in ASS and in non-acid soil										
First order	na [2]	2.0	36	0.6	7.0	7.3	5.3	18.5	32 [3]	0.9	5.0
Second order	na	7.3	296	0.7	37.0 [3]	8.6	58.5	35.8	233 [3]	4.6	21.0
Third order	na	7.5	237	0.2	34.0 [3]	15.3	39.4	7.0	224 [3]	2.4	36.7
Ground water	na	3.7	17	0.3	4.5	3.0	6.4	1.9	12	0.8	6.8
Source of water	Vietnamese criteria for surface and ground water										
Surface water	6.0-8.5		-	50	10	100	1	0.1	-	50	1
Ground water	6.5-8.5		-	50	10	1000	1 - 5	0.1 - 0.5	-	50	5

[1] : not detected at lower limit, (detection limit was 0.01 µg/l for Cd, and 0.3 µg/l for Ni).
[2] : not applied
[3] : detection limit of Cd and Ni was used as values for non-ASS canal water in the calculation.

Water Pollution in Inner-city Canals in Ho Chi Minh City, Viet Nam

Nguyen Thi Thuong[1], Takizawa Satoshi[1], Nguyen Phuoc Dan[2], Bui Dan Truong[2], Oguma Kumiko[1] and Katayama Hiroyuki[1]

1) Department of Urban Engineering, The University of Tokyo
7-3-1 Hongo, Bunkyo-ku, Tokyo 113-8656 JAPAN
e-mail: thuong@env.t.u-tokyo.ac.jp
2) Department of Environment, Ho Chi Minh City University of Technology
268 Ly Thuong Kiet St., Dist. 10, Ho Chi Minh City, VIET NAM
e-mail: dan-np@hcm.vnn.vn

Abstract

Ho Chi Minh City (HCMC) is the biggest and most populous city in Viet Nam. The city has a dense network of canals, which receives and conveys both storm water and untreated wastewater from municipal sewerage. Furthermore, illegally dumped domestic solid waste to the canals and unsatisfactory dredging of the drainage canals aggravate remarkably the water pollution and bad odor. This deteriorated water quality condition creates a public health hazards. The problem has become worsening as the urbanization was rapidly expanded without development of sufficient infrastructure. This paper presents the current status and to elucidate mechanism of pollution in HCMC of Viet Nam in terms of dissolved oxygen (DO), oxidation-reduction potential (ORP), ammonia (NH_4^+), dissolved organic carbon (DOC) and major anions.

Keywords: canal, Ho Chi Minh city (HCMC), water Pollution, water quality, wastewater

1. Introduction

Ho Chi Minh City (HCMC) located in the southern part of Viet Nam is the fastest growing region and the most populous city. The city produced 18% of the national GDP in 1999, and is a primary economic center of the country. Population is over 5 million, of which 71% are classified as urban. Annual population grew at 3.2%, half of which was attributed to immigration.

Network of canals in the city is relatively dense. Total length is nearly 100 km and the water slope of most canals is very small. Similarly to the status in other cities in Viet Nam, these canals receive and convey both storm water and wastewater generated from urban activities and discharge to Saigon river and then to Nha Be river. Apart from the partial treatment of toilet wastes in septic tanks, wastewater treatment is practically absent, resulting in the discharge of untreated wastewater to canals. There are approximately 36,189 households with 181,924 residents inhabit along and on the canals. Considering only 70 km of urban canals, this surface water environment has to receive around 550,000 m^3 of untreated municipal and industrial wastewater and 450 tones of solid waste (P. N. Dang, 2000). Consequently, all of the canals have been severely polluted. The special feature of canal network in HCMC is that it is strongly influenced by tides and some canals are influenced from more than one direction. As a result, the pollutants remain in the canals and have been accumulating.

The situation has become worsening as the development of sufficient infrastructure failed to keep pace with the current rate of the urbanization. The main objectives of this paper are to present the current status and trend in contamination and to elucidate the mechanism of water pollution in the canals in HCMC of Viet Nam.

2. Materials and methods

Sampling was carried out three times in September of 2003, March and May of 2004. A total of 12 sites from Tan Hoa-Lo Gom canal, Sai Gon River and canals in Binh Thanh district was selected for water sampling (see Figure 1). Except for the samples taken in May it was raining, the others were quite sunny.

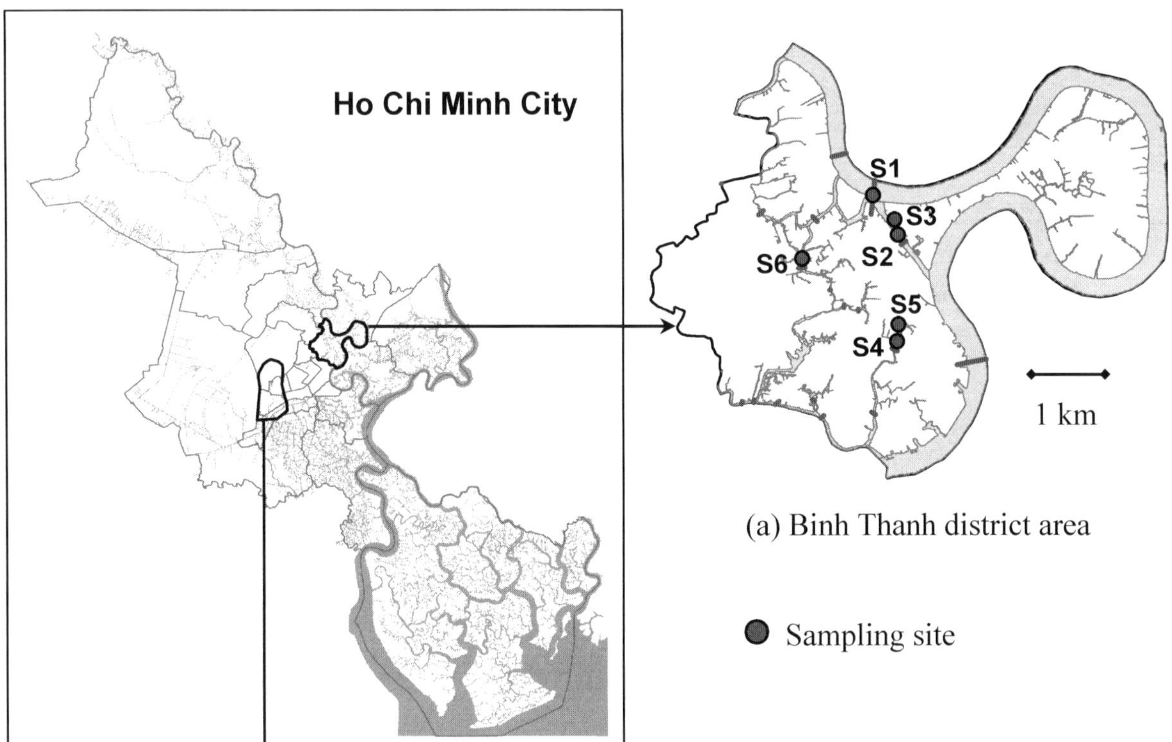

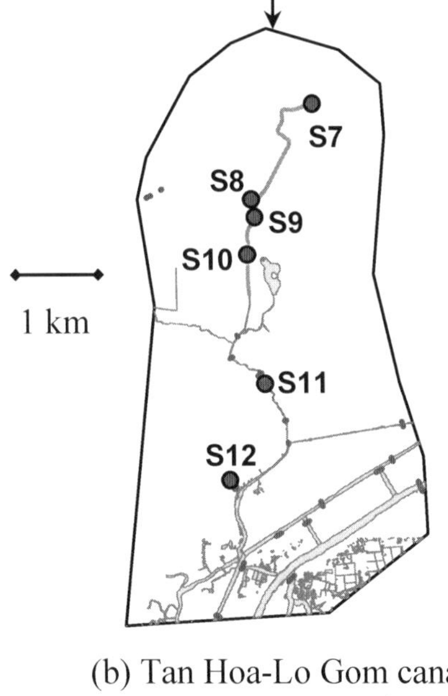

ID No.	Sampling site
S1	Thanh Da, Sai Gon river
S2	Cau Kinh Thanh Da bridge, on the bridge
S3	Cau Kinh Thanh Da, bridge side, wastewater outfall
S4	Hang Xanh, on the Van Thanh bridge
S5	Hanh Xanh, beside Van Thanh bridge, wastewater outfall
S6	Chu Van An bridge
S7	Huong Lo 2
S8	Bridge at Trinh Dinh Thao street
S9	Same location as S8, outfall
S10	Hoa Binh bridge
S11	Phu Lam bridge, wastewater outfall
S12	Pham Van Chi bridge

Figure 1. Study area and sampling sites.

Several water quality parameters measured at the field by Water Quality Monitoring System U-20XD (Horiba) included dissolved oxygen (DO), oxidation-reduction potential (ORP) and ammonia (NH_4^+). Dissolved organic carbon (DOC) was analyzed in laboratory along with major anions comprising of chloride (Cl^-), nitrate (NO_3^-), phosphate (PO_4^{3-}), sulfate (SO_4^{2-}). Since the surface water usually has high turbidity, whole of the samples were filtered with the 0.45 μm pore sized membrane. The samples used for determination of DOC were acidified by chlohydric acid (HCl) thereafter. All the samples were stored at 4°C. The analysis was performed in compliance with Standard Method of Analysis of Water and Wastewater (1994, AWWA). DOC was analyzed by Shimadzu TOC-5000, whereas major anions were measured by ion chromatographic method, using Yokogawa IC-7000RS.

As this paper is mainly aimed at presenting the situation and trend of pollution, some relevant statistical analyses such as boxplot, histogram and correlation methods have been employed in order to illustrate the factual phenomena.

3. Results and discussion

3.1 Water quality profile

Boxplot of the parameters are shown in Figure 2, which visualize the overall differences of water quality among 12 sampling sites. The bold line across the box represents the median of the entire data set. The bottom and top sides of the box show the value of the first and third quartiles.

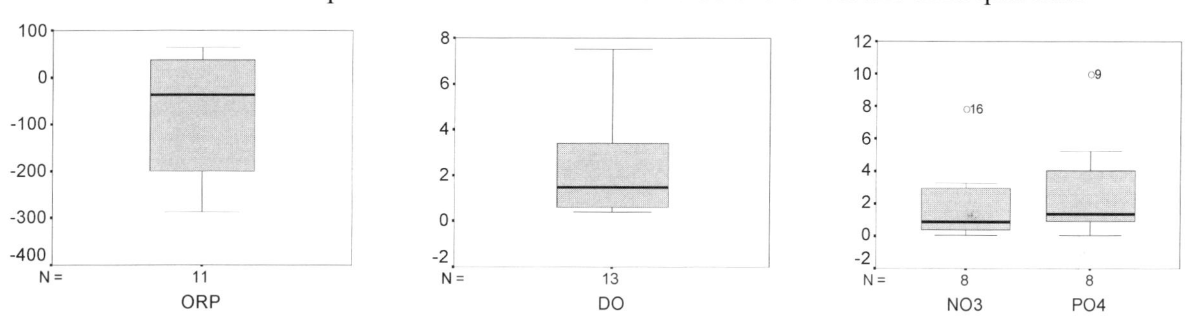

(a): ORP (mV) (N=11) (b): DO (mg/l) (N=13) (c): NO_3^- (mg/l) and PO_4^{3-} (mg/l) (N=8)

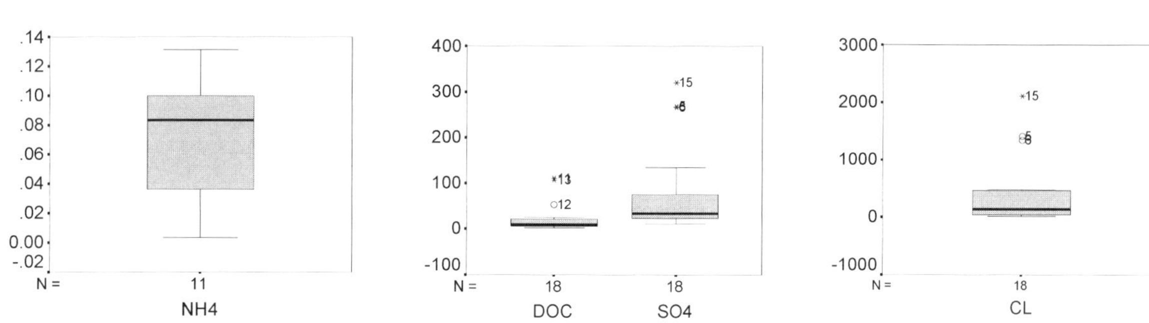

(d): NH_4^+ (mg/l) (N=11) (e): DOC (mgC/l) and SO_4^{2-} (mg/l) (N=18) (f): Cl^- (mg/l) (N=18)

Figure 2. Boxplot of parameters in surface water.

Figure 2(a) presents a large variation of ORP in surface water, revealing redox conditions quickly changed within a short distance in the area. Most sites had a negative value up to

(-288) mV, suggesting the water in the areas was under anaerobic condition.
The contents of DO, NO_3^- and PO_4^{3-} are low and vary in quite a short range (see Figure 2(b) and Fig. 2(c)). The concentration distribution of these three parameters is skewed to the right. DO, NO_3^- and PO_4^{3-} fell from 0.4 to 7.5 mg/l, 0.04 to 3.27 mg/l and from 0.05 to 5.25 mg/l, respectively. However, there is an outlier among the concentrations of NO_3^- and PO_4^{3-} as well. The Figure 2(d) also exhibits the narrow variation of NH_4^+, which ranges from 0.007 mg/l to 0.131 mg/l.

DOC concentration varies fairly less as shown in the Figure 2(e), of which the third quarter has the greatest variation. However, DOC was found to have extremely high values at some points. Also in Figure 2(e), the variation of SO_4^{2-} presents relatively large range, of which some concentration values were considered as outliers. The situation of Cl^- variation (see Figure 2(f)) is similar to that of SO_4^{2-} at which the three outliers in concentration originated from the same sampling sites.

Specifically, the histograms were constructed as an attempt to identify the frequency of the each parameter as concentration.

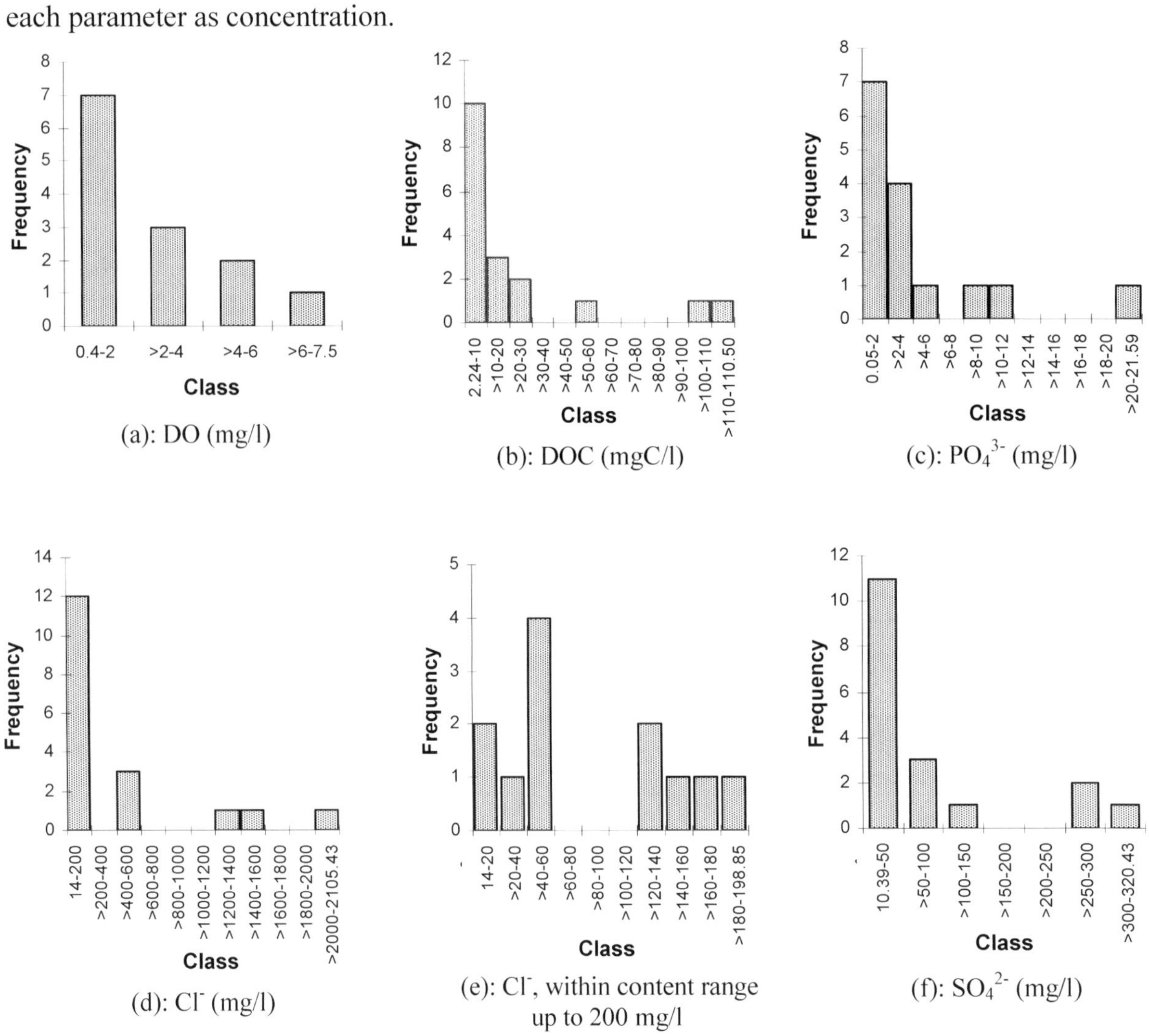

Figure 3. Histogram of parameters in surface water.

In general, all of parameters have the high frequencies within the portion of low concentration. The frequencies trend to fall down in the various ranges of higher concentration.

DO, as shown in Figure 3(a), has more than 50% of observations fell into the portion where the content is up to 2 mg/l. That means approximately the rest half has exceeded the maximum allowable concentration (Viet Nam Standard for Surface Water Quality TCVN 5942-1995, column B). Besides, there are several frequencies at which DOC reaches the concentration of higher than 10 mg/l (see Figure 3(b)), implying the sign of organic compound pollution.

Figure 3(c) presents PO_4^{3-} reached to 8 observations with the concentration over 2mg/l, of which the unusual value of more than 20 mg/l occurred once.

The case of another anion, namely Cl^-, seems a little while complicated as it has quite a large variation. Hence, Figure 3(e) was performed to give more details of how the frequency distribution within only the content range up to 200 mg/l is. Figure 3(d) exhibits the frequency of Cl^- for the whole of observations. From the concentration of more than 200mg/l, the distribution of Cl^- extremely scattered. However, at value of 200 mg/l and under, the frequency number at concentration of lower than 60 mg/l is a little while greater than of those, which is higher than 120 mg/l.

In most cases, the content of SO_4^{2-} is under 150 mg/l, except some outliers of more than 250 mg/l in concentration have been identified (see Fig. 3(f)).

3.2 Correlation between parameters

In order to examine the correlation between parameters in the surface water, we performed correlation analysis of the available data. The details of the result are presented in Table 1.

Table 1. The Correlation coefficient

	DO	ORP	NH_4^+	DOC	Cl^-	NO_3^-	PO_4^{3-}	SO_4^{2-}
DO	1							
ORP	-0.05	1						
NH_4^+	0.13	-0.50	1					
DOC	0.15	-0.66	0.40	1				
Cl^-	-0.30	0.22	-0.64	-0.19	1			
NO_3^-	0.31	0.86	-0.32	-0.41	0.42	1		
PO_4^{3-}	0.02	-0.69	0.35	0.37	-0.08	-0.42	1	
SO_4^{2-}	-0.35	0.27	-0.63	-0.19	0.98	0.43	-0.17	1

The results show the existence of positive correlation between ORP and NO_3^-, and Cl^- and SO_4^{2-} as the correlation coefficients are positive and high value. In addition, there is also the negative correlation between ORP and DOC, ORP and PO_4^{3-}, NH_4^+ and Cl^-, and NH_4^+ and SO_4^{2-}. In this case, the correlation coefficients are minus and far away from zero.

Indeed, the following graphs support the finding that the large values of one parameter set are associated with large values of the other. Figure 4(a) showed the correlation between ORP and NO_3^- contents. The data is slightly scattered, however they show a positive correlation between the ORP and NO_3^- contents. The lower boundary is given by the data obtained from S5, and the upper boundary is given by the data obtained from the site of S7. The ORP values fell into a positive range of 26, 63 and 48 mV at the sampling sites of S2, S1 and S7, respectively, which are either near the confluence with river, or on the river, or at canal's upstream.

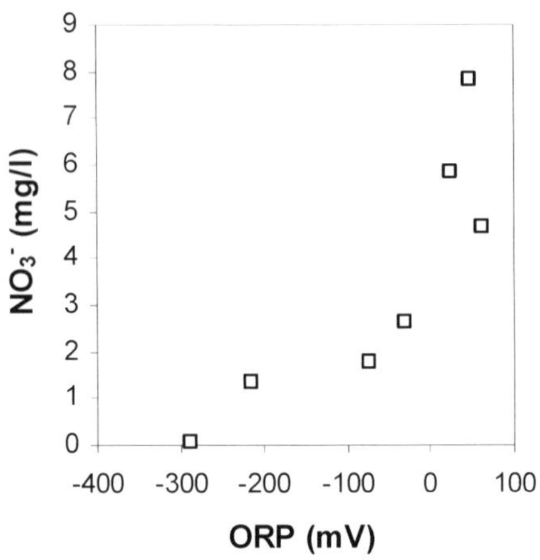

(a): ORP and NO_3^-

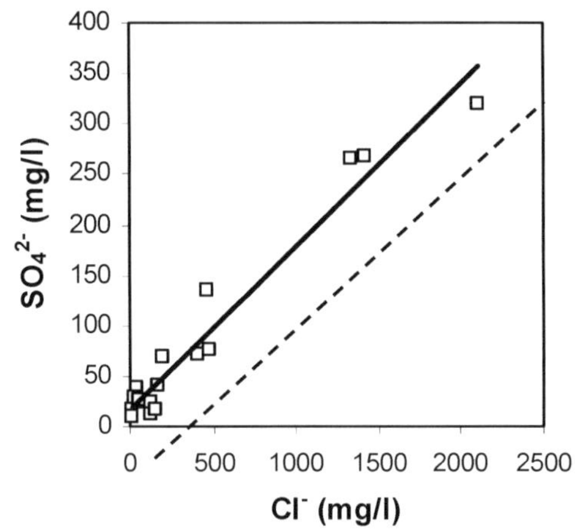

(b): SO_4^{2-} and Cl^-

------ Relation between concentration of SO_4^{2-} and Cl^- in seawater

Figure 4. Correlation between parameters in surface water.

Cl^- and SO_4^{2-} contents can be correlated as shown in Figure 4(b). The graph exhibits a clear increasing trend between Cl^- and SO_4^{2-}. Cl^- varied between 14.00 to 2,105.43 mg/l, whereas SO_4^{2-} ranges within 10.39 to 320.43 mg/l. The trend line of these two parameters is relatively similar to the relation line between concentration of SO_4^{2-} and Cl^- in the seawater. This findings implies SO_4^{2-} and Cl^- in the canals are under the influence from the sea altogether.

3.3 Variation pattern in water quality

In the March of 2004, the water sampling was carried out in 10 sites, including a sample of river water, the last were taken from the various locations in the canals inner HCMC. Figure 5 summarizes the contents of DOC, Cl^- and PO_4^{3-} from that sampling time.

In general, the content of Cl^- is quite high in most sampling points with the value of more than 100 mg/l. In fact, the concentration of Cl^- included inputs from human activity in Asia is only 10 mg/l (J. David Allan, 2000). Cl^- reached to the amazing value of thousands at 2 sites in Binh Thanh district, namely S1, S2, and S12 on the Tan Hoa-Lo Gom canal. Locating quite close to S1, and S2, however, the concentration of Cl^- at S3, which is the outfall, is not as great as in those points. The situation is similar to Van Thanh area comparing the sample taken on the mid of canal (S4) and outfall (S5). The lower concentration of Cl^- in the outlets than in the mid-canal sites plus the weak flow in the canal once again suggest the urban streams receive Cl^- from the sea water but not the municipal wastewater.

Meanwhile, the concentration of DOC trends to decrease at the sites where the concentration of Cl^- comes to high values. Higher values in the sampling sites along the Tan Hoa-Lo Gom canal (from S7 to S12) are due to effects of seasonal variations in the rainfall and drastic tidal

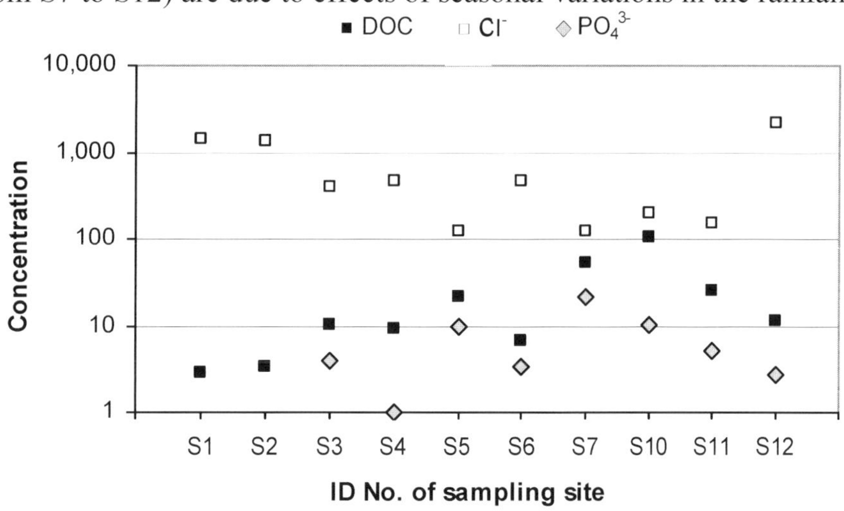

Figure 5. Concentration of DOC (mgC/l), Cl^- (mg/l), and PO_4^{3-} (mg/l) in the surface water (Data from March 2004 sampling).

regimes; these are discussed hereinafter.

The trend in DO, DOC, Cl^- and NO_3^- along the Tan Hoa-Lo Gom canal (the data presented is from March and May of 2004) is presented in Figure 6. The graphs show the remarkable variation in the parameters between two sampling times. As mentioned at the beginning, the sampling in May was carried out on a rainy day. The large dilution offered by the rainfall strongly decreased the concentration of most of parameters, including DOC, Cl^- in May. Also in most cases, the concentration of parameters trends to increase from upstream to the site of S10 and fall down as getting closer to the downstream. Due to the tidal influence, the wastewater in the downstream of canal is diluted as water from Can Gio River goes into canal, but upstream is stagnant for many days. But the primary affect of the tides on the water quality is the increase of salinity due to mixing of fresh and salt water. This may explain the reason why the Cl^- content has risen up at the point of S12 as shown in Figure 6(a). In another respect, the result also concretizes the assumption from previous section that Cl^- and SO_4^{2-} are influenced by a process, namely tidal regime.

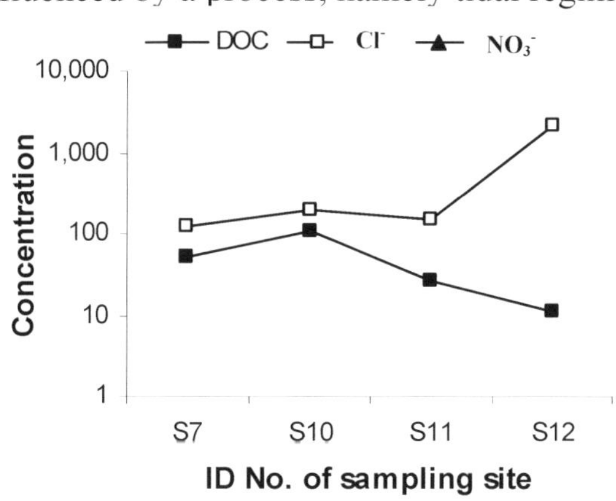

(a): March 2004

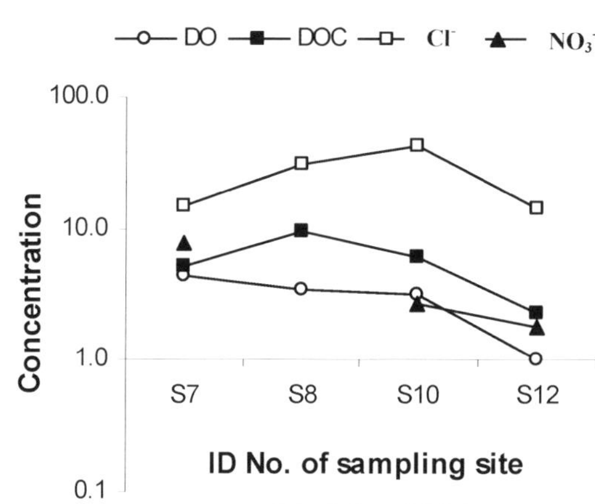

(b): May 2004

Figure 6. Concentration of DO (mg/l), DOC (mgC/l), Cl^- (mg/l) and NO_3^- (mg/l) along the Tan Hoa-Lo Gom canal.

4. Conclusions

The urban streams in HCMC have the problem of water pollution. High organic compound levels in urban surface water are caused by huge discharge of untreated municipal wastewater and urban drainage in to the water bodies. DO in most sampling sites is very low, of which around 50% of observations fell into portion of $\leq$ 2 mg/l. Approximately, there are 50% frequency which DOC and PO_4^{3-} have the value more than 10 mg/l and 2 mg/l, respectively. The concentration of parameters in water might decrease clearly from the range of hundreds to tens, resulting from the dilution of heavy rainfall. The influence from tide regime is important in changing water quality by increasing the salinity and causing stagnant water in the upstream of the Tan Hoa-Lo Gom canal. A combination of measures on municipal drainage improvement that takes the tide pattern into account as well as enhancement of public awareness should be implemented to deal with the present situation of water quality in the city.

5. Acknowledgement

The authors wish to acknowledge the financial assistance by RR2002 project. The graduate scholarship for this study was supported by Japan International Cooperation Agency.

6. References

American Public Health Association, American Water Works Association, Water Environment Federation (1995) Standard Methods for the Examination of Water and Wastewater 19th ed.

Edward A. Laws (1993) Aquatic Pollution: An Introductory Text 2nd ed., John Wiley & Sons, Inc.

General Statistics Office, Viet Nam (2003) Statistical YearBook 2002. Statistical Publishing House.

Japan International Cooperation Agency (1999) The Study on Urban Drainage and Sewerage System for Ho Chi Minh City in the Socialist Republic of Viet Nam. Final report.

J. David Allan (2000) Stream Ecology: Structure and Function of Running Waters. Kluwer Academic Publishers.

Ministry of Science Technology and Environment, Viet Nam (1995) Viet Nam Standards.

Neil A. Weiss (2002) Introductory Statistics 6th ed., Addison Wesley publisher.

Pham Ngoc Dang (2000) Management of Urban and Industrial Environment (in Vietnamese). Publisher of Construction.

Sunil Kumar Karn, Hideki Harada (2001) Surface Water Pollution in Three Urban Territories of Nepal, India, and Bangladesh. *Environment Management,* **28**(4), 483-496.

World Bank (2000) Ho Chi Minh City Environmental Sanitation Project - Nhieu Loc-Thi Nghe Basin: Environmental Impact Assessment. Final report.

World Bank (2002) Feasibility Study on Sanitation and Urban Upgrading of Tan Hoa-Lo Gom Canal Basin.

Optical Model for Remote Sensing in Turbid Waters Based on the Ship Observations Conducted in the Upper Gulf of Thailand

Hiroshi Kobayashi[1], Mitsuhiro Toratani[2], Hajime Fukushima[3]
Satsuki Matsumura[4] and Absornsuda Siripong[5]

1) Interdisciplinary Graduate School of Med. and Eng., University of Yamanashi
4-3-11 Takeda, Kofu, 400-8511, Japan
e-mail: kobachu@yamanashi.ac.jp
2) School of High-Tech. for Human Welfare, Tokai University
317 Nishino, Numazu, 410-0395, Japan
e-mail: tora@fksh.fc.u-tokai.ac.jp
3) School of High-Tech. for Human Welfare, Tokai University
317 Nishino, Numazu, 410-0395, Japan
e-mail: hajime@fksh.fc.u-tokai.ac.jp
4) Faculty of Sci., Chulalongkorn University
254 Phayathai Rd., Bangkok, 10330 ,Thailand
e-mail: smatsu@sc.chula.ac.th
5) Faculty of Sci., Chulalongkorn University
254 Phayathai Rd., Bangkok, 10330 ,Thailand
e-mail: absornsuda.S@Chula.ac.th

ABSTRACT

Spectral measurements of downward irradiance and upward radiance were conducted in the upper Gulf of Thailand to construct an optical model for ocean color remote sensing in turbid waters. Simultaneously, chlorophyll-a concentration; suspended solid concentration; and absorption coefficients of chlorophyll-a, particulate matter, and colored dissolved organic matter (CDOM) were measured. An inorganic suspended solid (ISS), which exhibits a high scattering property, was also measured as a residue after ignition. The measured subsurface reflectance, which was derived from spectral radiation measurements, was compared with that calculated using the optical model to evaluate the optical model. On the other hand, the estimated reflectance was calculated using the optical model on the basis of inherent optical properties, such as absorption and backward scattering coefficients of in-water substances. The ISS backward scattering coefficient was estimated by the measurement of its absorption coefficient and size distribution on the basis of the Mie theory. The estimated reflectance primarily agrees with the measured reflectance except in some cases. We have to perform additional observations including the improvement in the estimation method of ISS optical properties to obtain a suitable optical model.

Keywords: Ocean color remote sensing, optical model, turbid waters, inorganic suspended solid

1. Introduction

Coastal waters are affected by human activities and suffer from many environmental problems. On the other hand, they have a high productivity. Therefore, the monitoring of coastal regions is important. Ocean color remote sensing is an effective monitoring method. Many satellite remote sensing algorithms exist for the open ocean (e.g. for OCTS (Kishino *et al.* 1998), for SeaWiFS (O'Reilly *et al.* 1998)). However, it is difficult to apply these algorithms to coastal waters because the high concentration of suspended solids (SS) and colored dissolved organic matter (CDOM) causes optical complexities. Therefore, algorithms for turbid waters are being developed on the basis of a theoretical optical model (Schiller and Doerffer 1999; Tanaka *et al.* 2004). Till such time, optical properties such as absorption and backward scattering coefficients of suspended and dissolved matter in water need to be relied upon. In coastal waters, nonliving particulate matter, which mostly comprise mineral particles or inorganic suspended solids (ISS), are dominant due to scattering processes in water. However, a few studies regarding the SS or ISS optical properties were reported; on the other hand, many researchers conducted studies on phytoplankton and CDOM. Babin and Doerffer (1996) reported the wavelength dependence of the ISS backward scattering coefficient and Kronfeld (1988) reported the relationship between the SS scattering coefficient and the SS concentration; however, they did not include the SS absorption. In this study, the ISS backward scattering coefficient taking into account absorption was derived from the measurement of the ISS absorption coefficient and size distribution on the basis of Mie theory. We evaluated the optical model for turbid waters, comparing surface reflectance measured using submersible radiometers with the reflectance calculated using an optical model with inherent optical properties of phytoplankton, SS, and CDOM. The ship measurement was conducted in the upper Gulf of Thailand where the water exhibits high turbidity and optical diversity.

2. Materials and methods

The observation was carried out in the upper Gulf of Thailand on May 2004 (Figure 1). Spectral upward radiance and downward irradiance in water at each depth were measured with submersible radiometers (TriOS optical sensors; RAMSES ACC/ARC) equipped with depth meters. The measured radiance and irradiance were calibrated with the immersion factors (Zibordi and Darecki, 2006). The upward radiance just below the surface $L_u(0^-,\lambda)$ and downward irradiance just below the surface $E_d(0^-,\lambda)$ were extrapolated from the profile measurements of $L_u(z,\lambda)$ and $E_d(z,\lambda)$, respectively. The upward irradiance just below the surface is defined as

$$E_u(0^-,\lambda) = QL_u(0^-,\lambda) \quad (1)$$

where the upward radiance field was assumed to be isotropic in water and Q was equal to π. The irradiance reflectance below the surface referred to as subsurface reflectance is defined as

$$R(0^-,\lambda) = \frac{E_u(0^-,\lambda)}{E_d(0^-,\lambda)} \quad (2)$$

Water samples from each depth were collected using a Van Dorn sampler. The water sample was filtered onto a glass fiber filter (Whatman, GF/F). The concentration of chlorophyll-a was determined fluorometrically with a fluorometer (Turner Designs; 10-000R) after extracting the pigments in N,N-Dimethylformamide (Suzuki and Ishimaru 1990). The SS concentration was derived by subtracting the weight of the glass fiber filter (Whatman,

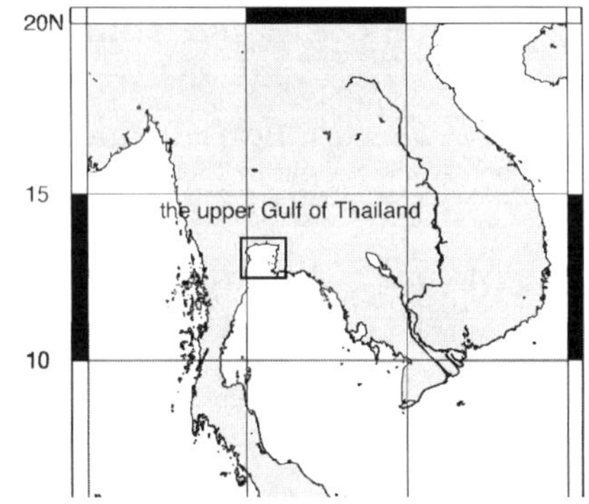

Figure 1 Location of the upper Gulf of Thailand.

GF/F) before filtration from its weight after the filtering of the sample water. The GF/F filters were preprocessed at 500 °C for 2 h to remove organic components. After measuring the total SS weight, the GF/F filters were heated again under the same conditions as earlier, following which the ignition loss was measured. The difference between the total SS and the ignition loss was considered as the ISS amount. The absorption coefficients of phytoplankton and particulate matter were calculated from the measured optical density of filter samples using the pathlength amplification factor β (Cleveland and Weidemann 1993; Kishino *et al.* 1985). The optical density of particles including phytoplankton was measured with a spectrophotometer (Nihon Bunko; V-550) equipped with an integrating sphere (Nihon Bunko; ISV-469) (Mitchell 1990). The optical density of nonliving particulate matter was measured after extracting the pigments on the filter by methanol (Kishino *et* [illegible] measured using the same procedure after the glass fiber filter was heated at 500 °C for two h. The absorption coefficient of CDOM was measured using the spectrophotometer with a 10-cm long cell from the water sample filtered using a polycarbonate filter with a pore size of diameter 0.2 μm (Whatman; Nuclepore). The ISS size distribution was measured using a Coulter Counter (Beckman Coulter; Multisizer 3) equipped with an aperture of diameter 70 μm. The water sample was filtered onto a low-ash membrane filter (Toyo Roshi; Cellulose acetate), following which it was heated at 500 °C for two h. The remaining particles were suspended in a Na_3PO_4 solution.

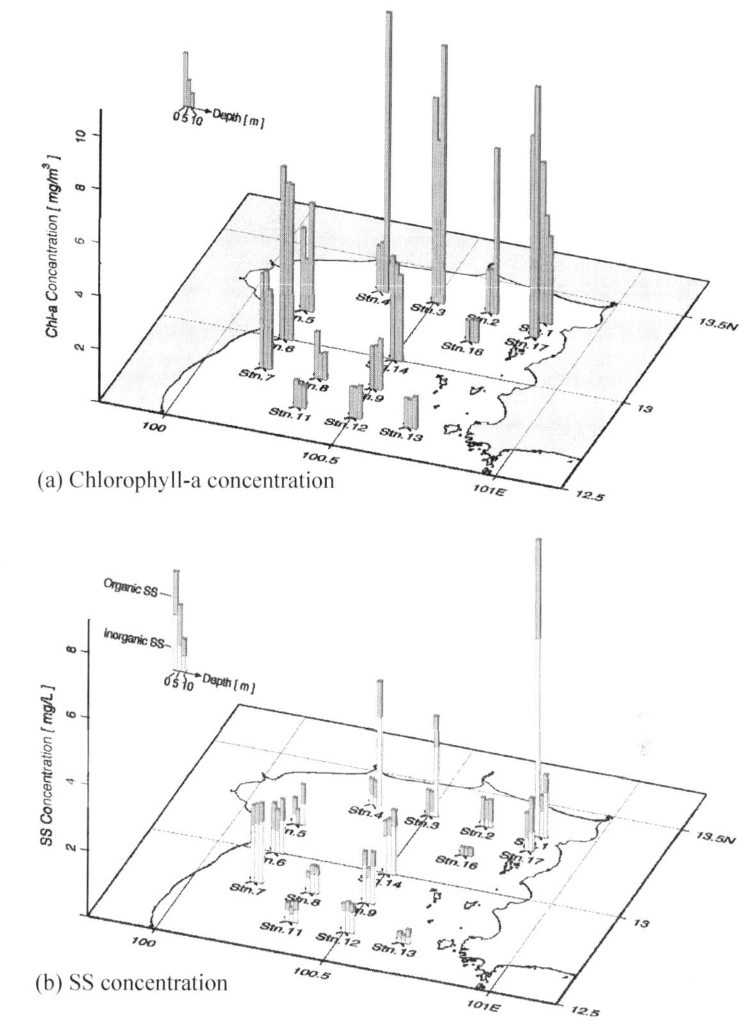

Figure 2 Chlorophyll-a and SS concentrations at each observation site in the upper Gulf of Thailand. The SS concentrations for organic and inorganic components are shown separately.

The subsurface reflectance is calculated from the backward scattering coefficient $b_b(\lambda)$ and the absorption coefficient $a(\lambda)$ on the basis of an optical model (Morel and Prieur 1977) and can be expressed as

$$R(0^-,\lambda) = f'(\lambda)\frac{b_b(\lambda)}{a(\lambda)+b_b(\lambda)} \quad (3)$$

where $f'(\lambda)$ varies with the solar altitude, chlorophyll concentration, wavelength, and the inherent optical properties (Morel and Gentili 1991). The value of f(λ) was simply assumed as

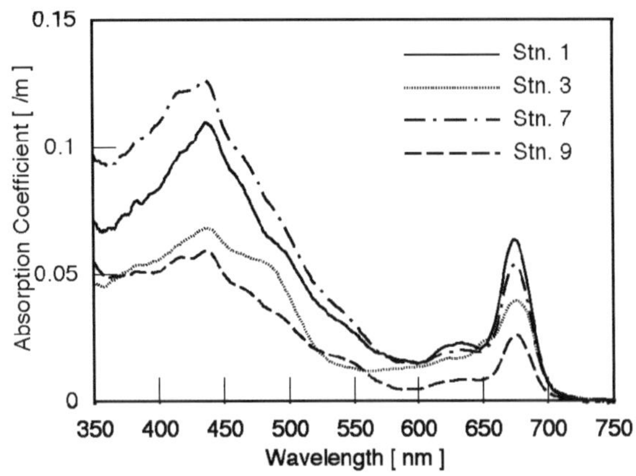

Figure 3 Absorption coefficients of phytoplankton at the surface water at each sampling station.

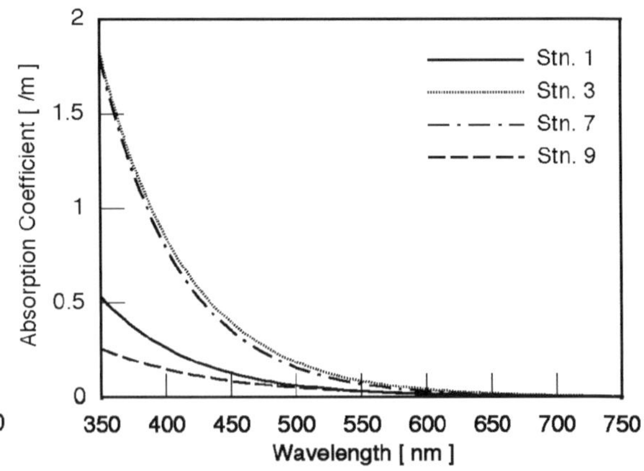

Figure 4 Absorption coefficients of CDOM in the surface water at each sampling station.

0.33 independent of wavelength. The backward scattering and absorption coefficients are expressed as the sum of their components:

$$b_b(\lambda) = b_{bw}(\lambda) + b_{bph}(\lambda) + b_{bISS}(\lambda) \tag{4}$$

and

$$a(\lambda) = a_w(\lambda) + a_p(\lambda) + a_{CDOM}(\lambda) \tag{5}$$

where the subscript w represents pure seawater; ph, phytoplankton; and p, total particulate matter.

The absorption and backward scattering coefficients of pure seawater are taken from previous studies (Morel 1974; Pope and Fry 1997). Although the absorption coefficients of particulate and dissolved matter were measured, the backward scattering coefficients of particulate matter could not be measured directly. Therefore, the backward scattering coefficient of phytoplankton is calculated as follows (Loisel and Morel 1998):

$$b_{bph} = 0.005\left(\frac{660}{\lambda}\right)^{0.407} C^{0.795} \tag{6}$$

The ISS backward scattering coefficient was estimated from the measurements of the ISS absorption coefficient and size distribution. First, the imaginary part of the refractive index of the ISS was derived from these measurements using the Mie theory with the real part assumed as 1.16. Next, the ISS backward scattering coefficient was calculated from the assumed real part of the refractive index, its estimated imaginary part, and the measured size distribution using the Mie theory.

3. Results and discussion

The results of measurements of chlorophyll-a and SS concentrations are shown in Figure 2. The chlorophyll-a concentration was higher in the northern and western regions of the upper Gulf of Thailand. The organic SS concentration, which is related to phytoplankton, shows the same pattern as chlorophyll-a. On the other hand, the ISS concentration showed a different pattern. The ISS concentration at the sea surface around the mouth of the upper Gulf was higher than that in the northern region. Although the ISS concentration at the sea surface is low in the northern region of the upper Gulf, the ISS concentrations are high at a depth of 10 m at stations 3 and 4. At these points, the depth of the seafloor is around 10 m; therefore, the high ISS concentration is caused by the resuspension of sediment on the seafloor.

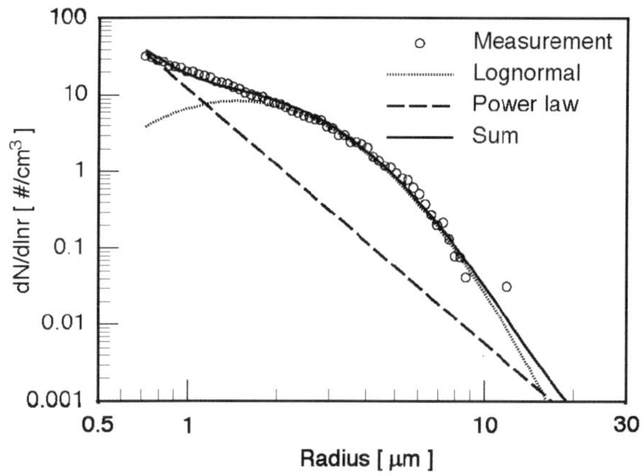

Figure 5 ISS size distribution at Station 7. The size distribution is fitted on the sum of a lognormal distribution and a power-law distribution

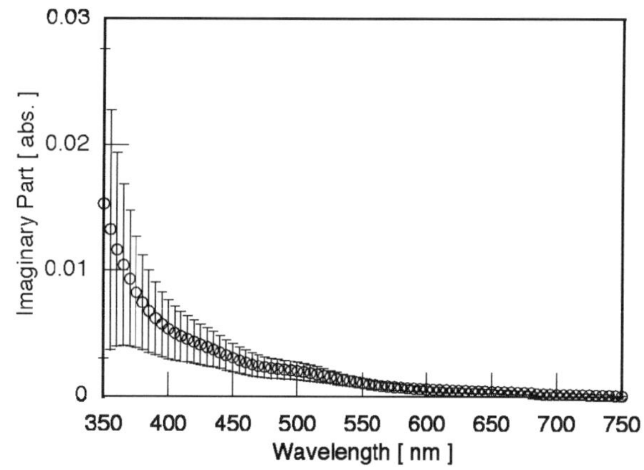

Figure 6 Average of the estimated imaginary parts of the refractive index of the ISS. The error bar shows the standard deviation of all the estimated results

Hereafter, we only show the results of stations 1, 3, 7, and 9 because the subsurface reflectance at these stations was derived from a good correlation between the depth and measurements of downward irradiance and upward radiance. The measured spectra of the absorption coefficients of phytoplankton and CDOM are shown in Figures 3 and 4, respectively. The spectral shapes of phytoplankton are the same except for those at station 3. The measured ISS size distributions fit fairly well with the sum of a power-law distribution and a lognormal distribution (Figure 5). The spectral imaginary part of the refractive index of the ISS at each sampling station was estimated from the curve-fitted size distribution and the measured absorption coefficient by using the Mie theory. The estimated imaginary part of the ISS is averaged (Figure 6). The value increases in the shorter wavelength region and it has hinge points around 380, 460, and 560 nm. The backward scattering coefficient at each sampling station was derived from the average of the results of the estimated imaginary part and size distribution (Figure 7). The backward scattering coefficient decreases at shorter wavelengths because the imaginary part increases around that region, i.e., the high absorptivity of the ISS weakens its scattering property.

Figure 8 shows the comparison between the spectral subsurface reflectance measured using radiometers with that calculated from the optical model with inherent optical properties. The results using the other ISS backward scattering coefficient (Babin and Doerffer 1996; Kronfeld 1988) are also shown to compare them with our results. Although the best result implies that the calculated reflectance agrees with the measured reflectance perfectly, the minimum requirement is the agreement between the relative shapes of their subsurface reflectance with discrepancies in their absolute values. This is because general algorithms of ocean color remote sensing are constructed using the ratio of reflectance in each band or wavelength. Therefore, the results normalized to 550 nm are also shown. The result at stations 1 and 3 are in good agreement. In particular, shapes of the reflectance at station 1 agree perfectly (Figure 8(a')). In the wavelength region shorter than 520 nm, the relative result obtained using the previous b_{bISS} equation is higher. It suggests that this

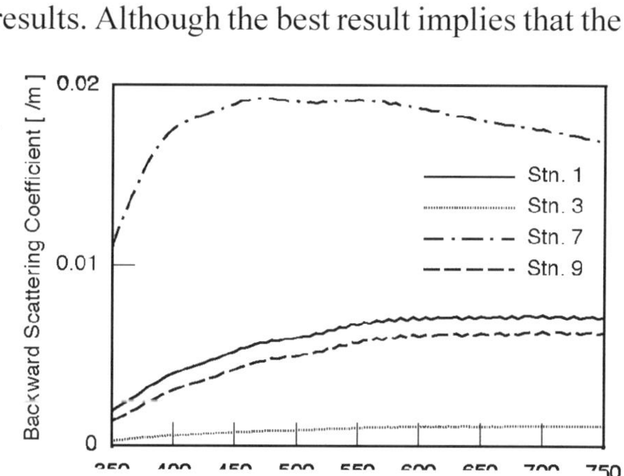

Figure 7 Estimated backward scattering coefficients of the ISS at each station.

coefficients of the ISS at each station.

overestimate is caused by neglecting ISS absorption in this wavelength region. At station 3, both calculated results almost agree (Figure 8(b)) because the ISS concentration is as low as 0.13 mg·L^{-1} and
the used optical properties except for that of ISS are completely identical. Therefore, this discrepancy is caused due to the error in the optical properties of phytoplankton or CDOM. At station 7, the reflectance calculated using our model was underestimated in the wavelength region shorter than 600 nm. On the other hand, the result underestimated the entire wavelength region in the case of the previous b_{bISS} equation. Further, these shapes disagreed with the measurement (Figure 8(c')). The wavelength region in which the result used our model underestimated is an absorption band of a phytoplankton. Although the chlorophyll-a concentration at station 7 was 3.53 mg·m^{-3}, which was lower than that at station 1 (5.99

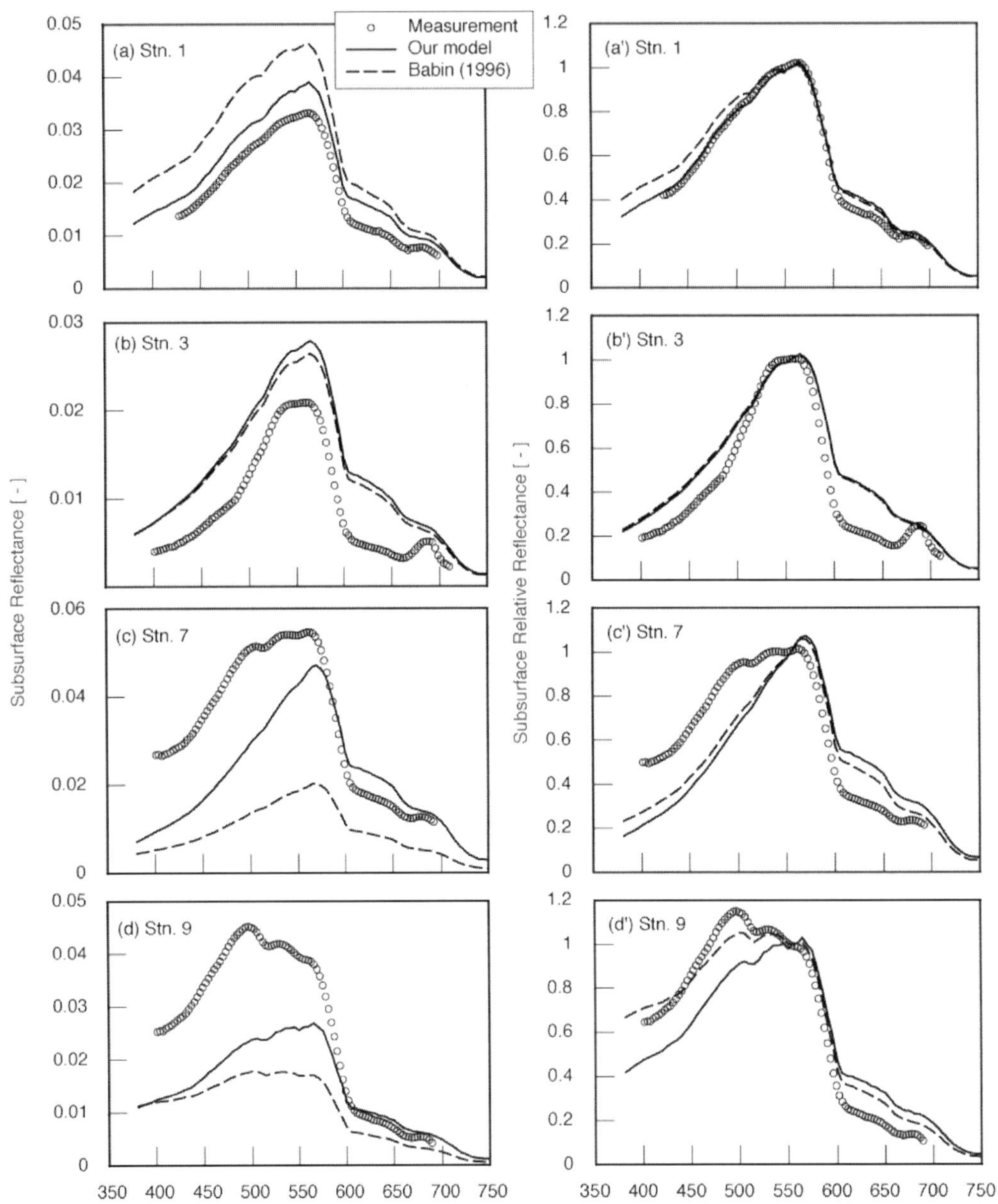

Figure 8 Comparison between the subsurface reflectance measured using submerged radiometers and the one estimated from the optical model on the basis of the measured and derived inherent optical properties at each station (left). For a relative comparison, the results were normalized to 550nm (right).

mg·m^{-3}), the measured absorption coefficient at station 7 was higher than that at station 1 (Figure 3). If this higher absorption coefficient was caused due to the error in its measurement, one of the reasons of the underestimation of the calculated subsurface reflectance is this overestimation of the absorption coefficient. At station 9, the absolute reflectances did not agree with each other, but there was a slight agreement in their shapes (Figure 8(d')). In addition, the relative reflectance used in the previous b_{bISS} equation was closer to the measurement as compared with our result. In this case, the result indicates that the ISS absorptivity was overestimated. It is considerable that the average ISS imaginary part of the refractive index used in our model was larger than its actual value at station 9. On the other hand, the ISS absorption coefficient necessitates sensitive measurement due to the scattering of the ISS particles, whose real part of the refractive index is high, resulting in our overestimation of the absorption coefficient while measuring the optical density. We have to perform additional observations to obtain a suitable optical model for turbid waters.

4. Conclusions

The subsurface reflectance estimated using the optical model on the basis of the inherent optical properties was compared with the reflectance measured with radiometers to evaluate an optical model for turbid waters. The required ISS backward scattering coefficient was estimated from measurements of the size distribution and the absorption coefficient. The results of the comparison were in good agreement. Based on the comparison of our model with the previous model, the result obtained using our model agreed to the measurement made at station 1; on the contrary, the result obtained using the previous ISS equation was more close to that at station 9. These results indicate that the estimation method of the ISS optical properties including the measurement of the ISS absorption coefficient needs to be improved. Therefore, we have to perform additional observations to obtain a suitable optical model. Furthermore, we will construct an algorithm for ocean color remote sensing for turbid waters.

5. Acknowledgement

This research was partially supported by the Ministry of Education, Science, Sports and Culture, Grant-in-Aid for Young Scientists (B), 15760406 and COE Research, 1699997. The authors thank Mr. K. Fujita and Mr. H. Yoshimura for assistance in the analysis. We also thank Prof. J. Ishizaka, Ms. P. Singhruck, and Mr. C. Bawornkittiwong for the measurement and providing the CDOM data, and Dr. T. Hirawake and Mr. T. Hirano for their help in the measurement of the chlorophyll-a concentration. We are grateful to the Captain and crew of the R.V. Kasetsart 1 and research group of Chulalongkorn Univ. and of Kasetsart Univ. for their willing help during the cruise.

6. References

Babin, M., and Doerffer, R. (1996). Specifications for case II coastal water reference model. *Algorithm Theoretical Basis Document for MERIS (ATBD 2.12)*, ESA-ESTEC, 61–73.

Cleveland, J. S., and Weidemann, A. D. (1993). Quantifying absorption by aquatic particles: A multiple scattering correction for glass-fiber filters. *Limnol. Oceanogr.*, **38**, 1321–1327.

Kishino, M., Ishimaru, T., Furuya, K., Oishi, T., and Kawasaki, K. (1998). In-water algorithms for ADEOS/OCTS. *J. Oceanogr.*, **54**, 431–436.

Kishino, M., Takahashi, M., Okami, N., and Ichimura, S. (1985). Estimation of the spectral absorption coefficients of phytoplankton in the sea. *Bull. Mar. Sci.*, **37**, 634–642.

Kronfeld, U. (1988). *Die optischen Eigenschaften der ozeanischen Schwebstoffe und ihre Bedeutung fuer die Fernerkundung von Phytoplankton*, Ph.D. Dissertation, GKSS Forschungszentrum, Fachbereich Geowissenschaften, Universitaet Hamburg.

Loisel, H., and Morel, A. (1998). Light scattering and chlorophyll concentration in case 1 waters: A reexamination. *Limnol. Oceanogr.*, **43**, 847–858.

Mitchell, B. G. (1990). Algorithms for determining the absorption coefficient of aquatic particulates using the quantitative filter technique (QFT). *Ocean Optics*, 137–148.

Morel, A. (1974). Optical properties of pure sea water. Optical Aspects of Oceanography, N. G. Jerlov and S. Nielsen, eds., Academic Press, New York, 1–24.

Morel, A., and Gentili, B. (1991). Diffuse reflectance of oceanic waters: its dependence on Sun angle as influenced by the molecular scattering contribution. *Appl. Opt.*, **30**, 4427–4438.

Morel, A., and Prieur, L. (1977). Analysis of variations in ocean color. *Limnol. Oceanogr.*, **22**(4), 709–722.

O'Reilly, J. E., Maritorena, S., Mitchell, B. G., Siegel, D. A., Carder, K. L., Garver, S. A., Kahru, M., and McClain, C. (1998). Ocean color chlorophyll algorithms for SeaWiFS. *J. Geophys. Res.*, **103**(C11), 24937–24953.

Pope, R. M., and Fry, E. S. (1997). Absorption spectrum (380-700 nm) of pure water. II. Integrating cavity measurements. *Appl. Opt.*, **36**(33), 8710–8723.

Schiller, H., and Doerffer, R. (1999). Neural network for emulation of an inverse model operational derivation of Case II water properties from MERIS data. *Int. J. Remote Sensing*, **20**(9), 1735–1746.

Suzuki, R., and Ishimaru, T. (1990). An improved method for the determination of phytoplankton chlorophyll using N,N-Dimethylformamide. *J. Oceanogr. Soc. Japan*, **46**, 190–194.

Tanaka, A., Kishino, M., Doerfeer, R., Schiller, H., Oishi, T., and Kubota, T. (2004). Development of a neural network algorithm for retrieving concentrations of chlorophyll, suspended matter and yellow substance from radiance data of the Ocean Color and Temperature Scanner. *J. Oceanogr.*, **60**, 519–530.

Zibordi, G., and Darecki, M. (2006). Immersion factors for the RAMSES series of hyper-spectral underwater radiomters. *J. Opt. A: Pure Appl. Opt.*, **8**, 252–258.

Macroinvertebrate Pilot Study for Ecological Health Monitoring in the Lower Mekong Basin

Bounnam Pathoumthong[1)] and Chanda Vongsombath[2)]

1) Centre of Environment and Development, National University of Laos
Postal Address: P.O. Box: 73 22, Vientiane, Lao PDR
e-mail: bounnamp@hotmail.com
2) Biology Department, National University of Laos
Postal Address: P.O. Box: 73 22, Vientiane, Lao PDR
e-mail: vongsombath@yahoo.com

Abstract

The priority aim of the study is to select the most appropriate method used for the ecological health monitoring program in Lower Mekong Basin. Two apparent sampling approaches were applied (kick method and sweep method). Due to the physical difference of substrates at the individual sampling sites, however, the kick method cannot be applied uniformly. Collected macroinvertebrate data were summarized in term of diversity and abundance, and in a biotic index (SIGNAL value). The sites also demonstrated differences by clustering. A total of 8483 individual macroinvertebrates were collected from 13 sites in the Mekong River and its tributaries (Nam Ou, Nam Moun). Representatives of 21 orders, 113 families, 202 genera, and 218 species were sampled. The diversity of macroinvertebrates increases along the up river sites. Similarly, when the data were assessed by using SIGNAL value, the up river sites have shown high value which means less stress than lower sites. Overall, based on this assessment, water in study sites is relatively clean, and only some down stream sites show some stress from pollution.

Keywords: macroinvertebrates, biomonitoring, biotic index, pollution

1. Introduction

The Mekong river has the third highest fish richness in the world after the Amazon and the Zaire. There are many endemic species of crabs, snails and fish which have high ecological values as well as economic values for the people living in the region. In order to keep these resources for long term benefits a "precautionary principle" should be applied to new development activities. This principle may best be implemented through the concept of "river health". There are many ways to assess river health, but this study applies a biological assessment approach by focusing on using macroinvertebrates as stress indicators of the water.

The advantages of biomonitoring versus physical or chemical monitoring are: (1) biomonitoring reflects overall ecological integrity (i.e., physical, chemical and biological); (2) it provides a holistic measure of the environmental condition by integrating stresses over time (Rosenberg and Resh 1993) ; and (3) the public better understands living organisms as other measures of a "healthy" environment (Plafkin et al. 1989). Generally, macroinvertebrates are often varied as the assemblage of choice. They have several characteristics that make them particularly useful for biomonitoring: (1) macroinvertebrates occur in almost all types of freshwater habitats; (2) There are many taxa of macroinvertebrates, and among these taxa there is a wide range of sensitivity to pollution and environmental stress; (3) They have mostly sedentary habits so they are likely to be exposed to pollution or environmental stress; (4) Their life cycles are sufficiently long that they are likely to be exposed to pollution and environmental stress, and the community will not recover too quickly so that the impact will go undetected; (5) Sampling the benthic macroinvertebrate assemblage is relatively simple and does not require complicated devices or great effort; (6) Taxonomic identification is almost always easy to the family level and usually relatively easy to the genus level. (Voshell *et al.* 1997).

Under the Agreement On the Cooperation for the Sustainable Development of the Mekong River Basin the four signatory countries agreed to *"protect the environment, natural resources, aquatic life and conditions and ecological balance of the Mekong River Basin*" (Article 3), They also agreed to "*make every effort to avoid, minimize and mitigate harmful effects that might occur to the environment, especially the water quantity and quality, the aquatic (eco-system) conditions, and ecological balance of the river system*" (Article 7).
In order to ensure that these objectives are met the MRC-coordinated monitoring programme for the Mekong will need to be broadened to include biological or ecological health information. For example, from the existing chemical monitoring data it is possible to say whether salinity at a site is increasing or decreasing, but what it is impossible to say is whether the change that has been detected as having an effect on the fish and other aquatic life in the river. That is a significant omission because what the people of the basin are concerned about changes in the fish stock, and the consequences of changes in water chemistry rather than the changes in water chemistry per s. This also means that we cannot tell whether we are meeting the ecological objectives mentioned above.

A difficulty in developing this monitoring program is that there have not yet been successful biological monitoring programmes implemented for any river the size of the Mekong. Many programmes are operating in smaller rivers, but there are reasons for believing that parameters that work in smaller rivers may not be successful as indicators in the Mekong. Therefore, in order to set an appropriate criterion, pilot studies and base line data also must be provided.

As the macroinvertebrate study is a part of Ecological Health Assessment of Lower Mekong Basin, the study aims to the following objectives:

- to evaluate the applied methods and select the most appropriate for the Mekong river;
- to develop baseline information about macroinvertebrates in the Mekong river;
- to evaluate site stress by using invertebrates as an indicator; and
- to develop a possible strategy by using macroinvetebrates for ecological heath assessment in the Lower Mekong basin.

2. Materials and methods

2.1 Study sites

The sampling activities started on 19/5/03 and finished 13/6/03. The sampling was conducted within 4 countries (Laos, Thailand, Cambodia and Vietnam) with 13 sites including: 5 sites in Laos, 3 sites in Thailand, 3 sites in Cambodia, and 2 sites in Vietnam (Figure 1). Due to the large size of the study area; the sampling sites are physically and environmentally different.

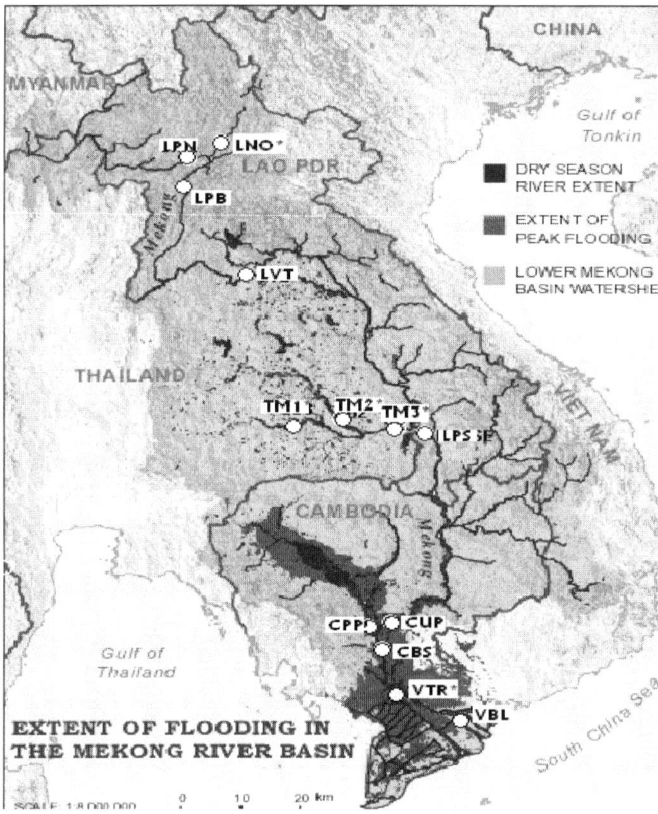

Site Code	Name of location
LPN	Pak Nam Ou, far from Nam Ou Mouth stream 1.5 Km (Mekong main river)
LNO*	Nam Ou, far from mouth stream 4 Km, (Tributary of Mekong river)
LPB	It is Luoang Pa Bang site (Mekong main river)
LVT	Vientiane Site (Mekong main river)
LPS	Pakse site (Mekong main river)
TM1*	Mun river (Tributary of Mekong river)
TM2*	Mun river (Tributary of Mekong river) Kaeng sa peu
TM3*	Mun river (Tributary of Mekong river) Pak Nam Moun
CUP	Mekong River Upper Phnompenh 4 km.
CPP	Phompenh port (the river is connecting with Tonglesarp River)
CBS	Basac River (one branch of Mekong river before close to Vietnam)
VTR*	Ten River (it also one branch of MK, but the site is Bentre Province VN)
VBL	Hong river (it also one branch of MK, but the site is Benlai Province VN)

Note: * refers to the tributary of the Mekong river

Figure 1 Sampling sites

Some have muddy, some sandy, and some rocky substrates. Furthermore, the sites in Vietnam are estuarine sites

2.2 Sampling methods

Sampling was under taken in sampling zones of 100 m at each site. Within this zone we randomly placed sampling 5 sampling plots on each side of the river bank, each plot 10 m long. Due to the variance of substrate in the Mekong river (especially, upriver vs. downriver), two types of net sampling, kick sampling and sweep sampling, were applied for this pilot study. The evaluation of the two methods would then provide the basis for setting the standard in further studies.

2.3 Macroinvertebrate collecting

The macroinvertebrates were picked by using forceps and pipettes. Picking was done in a zig-zag manner from one end of the tray to the other and from side to side. The samples were stored in jars of 70% ethanol, labeled with site name, location code, date, and replicate number.

2.4 Laboratory analysis

The identification was made by using Stereo-microscope and several references (Merritt and Cummins 1996; Snomouang 1997; and Pinder *et al.* 2000). All the collected specimens were preserved in jars and kept for references at the Biological Laboratory, National University of Laos.

2.5 Data analysis

Macroinvertebrate data were interpreted in 3 forms: (1) diversity and abundance, (2) Biodiversity index: SIGNAL value, and (3) Clustering.
The diversity of invertebrates in selected sites were summarized in different taxa as orders, families, genera and species, and also the number of individuals counted for each site.
The use of biotic indices based on macroinvetebrates to assess water pollution and other human inputs on rivers has long history (Washington 1984). This study applied the biotic index: SIGNAL value. *SIGNAL* (Stream Invertebrate Grade Number Average Level) is a simple biotic index for river macroinvertebrate, which was firstly developed for application in the Hawkesbury-Nepean river system, eastern Australia (Chessman 1995). It was adopted from the Average Score Per Taxon (ASPT) version of the Biological Monitoring Working Party (BMWP) system using in Great Britain (Hawkes 1997). The SIGNAL is calculated by averaging the pollution sensitivity grade number of the families present, which may range from 10 (most sensitive) to 1 (most tolerent) (Chessman 2003).

Due to the fact that macroinvertebrates of the Mekong River were not yet studied inrelation to their pollution sensitivity, there is no grade given specifically to the maccroinvertebrate families in the Mekong river system. However, grades for many important families were awar from several authors from different regions, so the pollution sensitivity grades for the macroinvertebrate families in this calculation will be used in modification of BMWP and other authors in Chessman (1995). Chessman also recommended that in some case the grades are necessarily a compromise, either because of variation in the sensitivities of species within a family or because some families are sensitive to certain types of pollutant but relatively tolerant of others.

Signal value: collected data from each site were calculated by summing up the grades of all families present in the site than the total is divide by the number of families to provide an average grade per family.

$$\text{SIGNAL} = \frac{\Sigma \text{ grade of each family presents in the site}}{\Sigma \text{ number of families in the site}}$$

The calculation of average grade is designed in order to reduce possible bias associated with variation in stream size and sampling technique (Chessman 1995).

The state of sampling sites is classified according to the SIGNAL value, i.e. greater than 6 = water status is classified as clean water; 5-6 = doubtful quality, possible mild pollution; 4-5 = probable moderate pollution; and less than 4 = probable severe pollution.

Clustering: the clustering is employed for demonstrating the similarity of the site, according to the occurring of invertebrates of each site. The cluster will group the similar site together and at the same time the degree of similarity was also graphically shown.

3. Results and discussion

3.1 Diversity and Abundance

A total of 8483 individual macroinvertebrates were collected from 13 sites in the Mekong

River and its tributaries (Nam Ou, Nam Moun). Representatives of 21 orders, 113 families, 202 genera, and 218 species were sampled (Table 1). Generally, the number of macroinvetebrate species in the upriver sites is higher then of the downriver sites, except LPN site which is diffcult to access.

Table 1 Overall richness and abundance of macroinvertebrates in each site

No site	Code	No. of Order	No. of Family	No. of Genus	No. of Species	No. Individual
1	LPN	10	19	22	22	81
2	LNO*	14	52	92	99	1824
3	LPB	13	50	65	69	843
4	LVT	9	26	37	38	413
5	LPS	13	29	43	44	1733
6	TM1*	10	19	21	21	304
7	TM2*	10	30	48	48	695
8	TM3*	10	21	25	25	163
9	CUP	8	14	16	17	159
10	CPP	11	28	38	39	670
11	CBS	10	26	33	33	873
12	VTR*	13	24	26	26	622
13	VBL	7	11	12	12	103
Total		21	113	202	218	8483

3.2. Site comparison

Sweeping was applied to all sites but in some sites sweeping was done at both sides, in others only on one side. Sweeping on the left side of each site was used in this analysis because only this method was applied to all sites.

Diversity and abundance: LPS was the most abundant with 736 individuals in total. Site VTR* represented the next most abundant with 382 individuals. These two sites represented more than one third of the total number of individuals sampled. This high number of individuals resulted from some dominant species in the sites (Table 2).

Table 2 Comparison of diversity and abundance between sites

No site	Code	No. of Order	No. of Family	No. of Genus	No. of Species	No. Individual
1	LPN	8	15	16	16	52
2	LNO*	11	21	32	34	194
3	LPB	10	18	20	22	79
4	LVT	9	19	28	28	261
5	LPS	10	18	25	26	736
6	TM1*	7	12	12	12	92
7	TM2*	9	23	30	31	347
8	TM3*	8	13	15	15	67
9	CUP	8	12	14	14	63
10	CPP	10	20	27	27	374
11	CBS	8	19	23	23	358
12	VTR	9	15	17	17	382
13	VBL	7	12	12	12	103
Total						3,108

In terms of diversity, the Nam Ou site (LNO*) was the most diverse with 34 species. The second most diverse is TM2* with 31 species. The lowest diversity sites are VBL and TM1* with 12 species for each site (Table 2)

SIGNAL value: according to the signal value, the overall water quality of the Mekong river is classified as medium quality, and some pollution is possible. If the sampling sites are considered differently, only the three uppermost sites are classified as clean water namely Nam Ou, Pak Nam Ou and Luangphrabang sites. Five sites possibly show mild pollution and four sites are classified as showing moderate pollution (see Table 3).

Table 3 Comparison between sites by SIGNAL value

Site	Sum of grade	Number of family	SIGNAL value
LNO*	132	20	6.6
LPN	88	14	6.3
LPB	102	17	6.0
LPS	88	15	5.9
LVT	104	18	5.8
TM2*	125	22	5.7
CBS	99	18	5.5
CPP	98	19	5.2
TM3*	60	12	5.0
TM1*	60	12	5.0
VTR	28	6	4.67
CUP	52	12	4.3
VBL	52	12	4.3
Average			5.4

Clustering: the data from sweeping were applied for graphing clusters between sites. The sites were divided into two big groups: first group is composed of two sites, Luangphrabang and Nam Ou site; and the other 11 sites are in the second group. The second group is divided into two subgroups of which one of them again is divided into small groups (Figure 2).

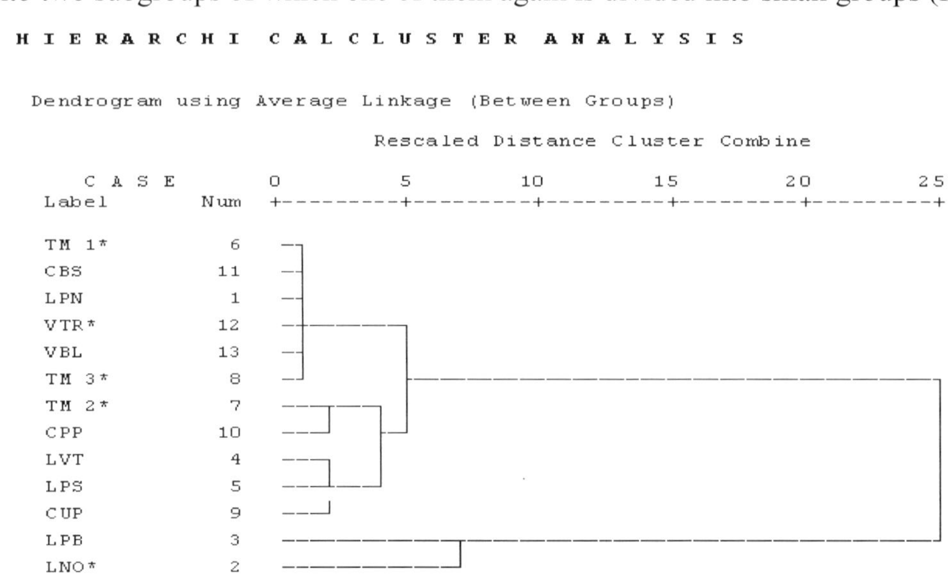

Figure 2 Clustering of similar sites based on macro invertebrates

The use of macroinvertebrates in assessing water quality or ecological health will be more effective when appropriate methods are selected. So, one of the important objectives of this study is to test which the used method is appropriate for the study location. Two methods of sampling (sweeping and kicking) were applied in this study. However, these two methods could not be applied at each site equally due to the characteristic habitat differences. Therefore, this analysis described overall taxa of all sites and for the comparison purposes, the analysis was made separately according to the method used for sampling.

Two methods of sampling were applied in this study (kick and sweep), but only sweeping can be used for analyzing comparisons between sites since it can be applied to all sites. However, the data from kick sampling is still useful for more specific sites where it can be applied. The data from kick sampling also shows more diverse and interesting data to study in more detail. If every site had a location where the kick method could also be used, the results would be more comprehensive and reliable.

Data analysis can also be done statistically but for the purpose of easy understanding, a biotic index was applied. In this case the SIGNAL value is considered more appropriate since we were not able to conduct a complet quantitative sampling (only semi quantitative). The use SIGNAL value used in this study is more appropriate for the sites from LPN site to TM3* since other sites in Cambodia and Vietnam show large difference in habitat characteristics and conditions.
In addition, using the SIGNAL value applies the scores from different sources which are not exactly developed for this region. Therefore, a systematic study can help in developing this kind of criteria for assessment.

4. Conclusion

According to the species richness and SIGNAL values of macroinvertebrates sampled at 13 sites, these indicators show high value in upper site and low values in down river sites. This means the water in the lower sites has some stress from pollution. However, according to the SIGNAL value, overall water quality of Mekong river is classified as medium quality, it is possible that some pollution occurs in some sites.
As the main objective of this study the focus was on selecting appropriate sampling methods for developing a monitoring protocol used in the monitoring program of MRC and local monitoring programs in Lower Mekong basin countries, some considerations have been recommended for further improvement:

- The study covers quite large areas which have different characteristics of habitats, so it makes no sense to compare across the sites. The sites should be monitored and compared in time series.
- Sites with similar substrates must be selected with the most similar patterns of habitat (i.e. have similar weed, algae or bed rocks) because from this study we found that the characteristics of habitat have great influence on macroinvertebrate assemblage.
- Rapid assessment methods, involving simple but precise protocols, coupled with easily calculated quantitative indices of environmental quality are widely used. However, most of biological indices have been developed in the temperate regions. Thus, carrying out this pilot study is an important step to provide baseline data and to develop biological indices and a standardized protocol for this region.

- For further improvement of this method more comprehensive programmes of sampling need to be developed in the Mekong river Basin.

5. Acknowledgement

The macroinvertebrate pilot study for ecological health monitoring in the Lower Mekong River Basin was funded by MRC as part of the Environmental Monitoring Program. I am grateful to Dr. Ian Campbell, Senior Environmental specialist, Environment Division, MRC Secretariat for organizing this study; Professor Vince Resh and Dr. Bruce Chessman for their constructive advices. Also thanks to all in the team of ecological health monitoring pilot study for their enthusiastic helps during the field-sampling period. Special thanks to Dr. Clemens Grunbuhel for proof reading of this paper.

6. References

Chessman, B. C. (1995). Rapid assessment of rivers using macroinvertebrates: a procedure based on habitat-specific sampling, family level identification and biotic index. *Australian J. of Ecol.*, **20**, 122-129.

Chessman, B. C. (2003). New sensitivity grades for Australian river macroinvertebrates. *Marine and Freshwater Research,* **54**, 95-103.

Hawkes, H. A. (1997). Origin and development of the Biological Monitoring Working Party Score Systems. *Water Research,* **32**, 964-968.

Merritt, R. W. and Cummins, K. W. (1996). *An introduction to aquatic insects of North America*, 3rd ed. Kendall/Hunt, Dubuque, IA.

Pinder C. and Tilling, S., Nguyen Xuan Quynh and Mai Dinh Yen. (2000). *Biological surveillance of freshwaters, using macroinvertebrates*. A practical manual and Identification Key for use in Vietnam. Field studies council.

Plafkin, J. L., Barbour, M. T., Porter, K. D., Gross, S. K. and Hughes, R. M. (1989). *Rapid bioassessment protocols for use in streams and rivers: benthic macroinvertebrates and fish.* U. S. Environmental Protection Agency, Office of Water. Washington, DC.

Rosenberg, D. M. and Resh, V. H. (1993). *Freshwater Biomonitoring and Benthic Macroinvertebrates*. Chapman & Hall, Inc., New York.

Snomouang, L. (1997). *Invertebrate* (Thai). Zoology Deparment, KhonKaen University.

Voshell, J. R., Smith, E. P., Evans, S. K. and Hudy M.. (1997). Effective and scientifically sound bioassessment: opinions and corroboration from academe. *Human Ecol. Risk Assess.*, **3**, 941-954.

Washington, H.G. (1984). Diversity, biotic, and similarity indices: a review with special relevance to aquatic ecosystems. *Water Research,* **18**, 653-694.

Effect of Heavy Metals on Oxygen Utilization Rate of Nitrifying Mixed Cultures and Its Application to the Biosensing of Toxicity

Jaruwan Tantasut[1)], Hiroyasu Satoh [2)], Preeda Parkpian[1)] and Skorn Mongkolsuk[3)]

1) Environmental Engineering and Management, Asian Institute of Technology,
P.O. Box 4, Klong Luang, Pathumthani, 12120, Thailand
e-mail: Jaruwan.Tantasut@ait.ac.th or preeda@ait.ac.th
2) Institute of Environmental Studies, The University of Tokyo,
5-1-5 Kashiwanoha, Kashiwa, Chiba 277-8563, Japan
e-mail: hiroyasu@k.u-tokyo.ac.jp
3) Laboratory of Biotechnology, Chulabhorn Research Institute,
Bangkok, 10210, Thailand
e-mail: skorn@tubtim.cri.or.th

Abstract

This study investigated the effect of heavy metals on the response of nitrifiers while having the future goal to develop a biosensor using nitrifiers for toxicity monitoring. Nitrifying mixed cultures were used instead of purified nitrifying culture because of the easiness of the culture development. The toxic effect of heavy metals on the cultures was detected as the reduction of oxygen utilization rate or OUR. When ammonia oxidizing mixed culture (AOMC) was employed, with 15 min exposure time, it was found that cadmium and nickel have stronger inhibitory effect on AOMC than copper and zinc. In the tests with low heavy metal concentrations either OUR inhibition or enhancement was observed. However, percentage of OUR reduction increased with higher heavy metal concentrations. The similar study was performed with nitrite oxidizing mixed culture (NOMC) and it was found that NOMC is not effective to detect heavy metals toxicity in comparison to AOMC. The results from this study illustrated that AOMC can be used to monitor heavy metals. The response of the cultures affected by heavy metal concentration and exposure time length. Generally, the extended exposure time improved the toxic response of AOMC to cadmium. Therefore, in the toxicity test with low heavy metal concentrations, the extended exposure time would be recommended.

Keywords: ammonia oxidizing mixed culture, biosensor, heavy metals, nitrifiers, nitrite oxidizing mixed culture, oxygen utilization rate

1. Introduction

Significant amount of heavy metals are used in industrial activities as well as in our daily lives. They are finally discharged to receiving water bodies and cause adverse effect on environment and ecosystem eventually. It is needed to set up proper countermeasure to control the emission of these heavy metals. One of the most effective measures is to set up regulations and another effective measure is the introduction of effective wastewater treatment processes. In both measures, the monitoring of toxic chemicals in water is essential. Without monitoring methods, regulation cannot be implemented, and the effectiveness of wastewater treatment processes cannot be achieved.

Several monitoring methods are available for the detection of toxic chemicals in water and wastewater. Biosensor is one of monitoring methods based on biological mechanisms being used for toxicity monitoring (Grunditz *et al.*, 1998). Nitrifiers are known to be sensitive to toxic chemicals (Blum and Speece, 1991) and the effects of toxic chemicals on nitrifiers have been reported by several researchers (Inui *et al.*, 2002; Tanaka *et al.*, 2002; Juliastuti *et al.*, 2003). But only a few studies have been performed seriously on the response of nitrifiers to toxic chemicals. Inui *et al.* (2002) developed a biosensor using *Nitrosomonas europaea*, and reported its response to different kinds of heavy metals. Blum and Speece (1991) also reported the effect of toxic chemicals on the inhibition of ammonia consumption of *Nitrosomonas*. Both of such studies used pure culture as biosensing element. While the results from single species may be repeatable and unambiguous, it is difficult to extrapolate to other species. Further, it could be possible that the use of several species rather enhance the accuracy of the bioassay test. In addition, it might beneficial in lowering the cost for biosensor preparation. Therefore, it is of great interest to use nitrifying mixed cultures as biosensing element. A new effort in biosensor development is worth to be extensively conducted to realize biosensor using nitrifiers.

In this study, nitrifying mixed cultures were developed in the laboratory and then applied for the toxicity tests in which the cultures were exposed to heavy metals and the reduction of oxygen utilization rate was monitored as the indicator of the toxic response. Further, the toxic response and sensitivity of the cultures to different heavy metals were discussed and defined.

2. Materials and methods

Experimental approach: In this study, ammonia oxidizing and nitrite oxidizing mixed cultures were developed using ammonia and nitrite as the nitrogen source respectively. The developed cultures were then subjected to the toxicity test, or a batch experiment to observe the response to different heavy metals with different concentrations. In the toxicity test, the effect of heavy metals on the cultures was detected as the reduction of oxygen utilization rate in the presence of heavy metal.

Nitrifying mixed cultures enrichment: 5L batch reactors continuously supplied with air were employed for the acclimatization of nitrifying mixed cultures. Different compositions of the enrichment solution was fed to each reactor to enrich the cultures. Ammonia oxidizing mixed culture was enriched with ammonium sulfate as the sole energy source in a mineral salts solution. The compositions of enrichment solution for the cultivation of ammonia oxidizing mixed culture from sewage was as follows: $(NH_4)_2SO_4$ 6.6 g, K_2HPO_4 0.44 g, $MgSO_4.7H_2O$ 1.9 g, $CaCl_2.2H_2O$ 0.1 g, $Na_2MoO_4.2H_2O$ 0.5 g (Madigan *et al.*, 1999). On the other hand, sodium nitrite was used to enrich nitrite oxidizing mixed culture and the compositions of the enrichment solution for the cultivation of nitrite oxidizing mixed culture from sewage was as follows: $NaNO_2$ 1.1 g, K_2HPO_4 0.44 g, $MgSO_4.7H_2O$ 1.9 g, $CaCl_2.2H_2O$ 0.1 g, $Na_2MoO_4.2H_2O$ 0.5 g (Madigan *et al.*, 1999). Everyday, 30mL of the enrichment solution was supplied to each reactor. The pH of the enrichment solution was adjusted to 7.5 with 0.5M K_2CO_3, which also had a role as the carbon source. The reactor was operated under the room temperature of 27-30°C. Once a week, aeration was stopped, the biomass was allowed to settle, and supernatant of 4L was removed by decantation, and the same volume of deionized water was filled into the reactor. In the reactor, pH was maintained at 8.5 by the addition of 10%$NaHCO_3$ solution. The pH, temperature and dissolved oxygen in the reactors were routinely monitored. After one month of operation, biomass from each reactor was collected and used for the toxicity test.

The microbial population used as the seed in this study was obtained from the mixed liquor of an activated sludge process located in Thammasat University Rangsit Campus in Thailand treated of domestic wastewater from the university campus.

Tested toxic chemicals: Heavy metals considered in this study were cadmium sulfate, copper sulfate, nickel sulfate and zinc chloride. They were obtained from BDH Laboratory Supplies (AR grade).

Toxicity test: The toxicity test to observe the toxic effect of heavy metals on nitrifying mixed cultures was carried out in a sealed 100mL bottle installed with an oxygen electrode connected to a volt meter (Figure 1). In the toxicity test with ammonia oxidizing mixed culture, the toxicity test was conducted as follows. To the bottle, 25-30mL of the ammonia oxidizing mixed culture containing final biomass of 100 mg dried weight, ammonia solution of final concentration of 70 mg/L NH_4-N, and different concentrations of heavy metal were added. Deionized water was added up to eliminate head space in the bottle with final volume of 105mL, then the bottle was closed with a cap equipped with an oxygen electrode. The pH of the tested solution was adjusted to 8.5 using 0.1N NaOH and 0.1N H_2SO_4. A magnetic stirrer at 120 rpm was employed for the mixing and the testing temperature was in between 24-26°C. The inert oxygen utilization rate was determined in the similar test without the addition of heavy metal and ammonia. The maximum oxygen utilization rate was determined in the similar test with ammonia but without heavy metal. Toxicity test with nitrite oxidizing mixed culture was conducted in the same way, but the culture used was nitrite oxidizing mixed culture instead of ammonia oxidizing mixed culture, and nitrite solution of final concentration of 8.14 mg/L NO_2-N was used instead of ammonia.

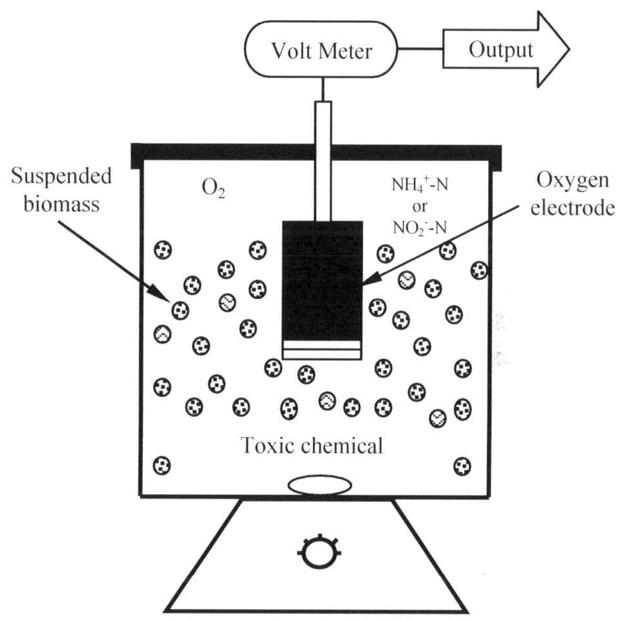

Figure 1 Toxicity test

In the toxicity test to observe the effect of exposure times on the response of ammonia oxidizing mixed culture to cadmium, cadmium and ammonia were added at time 0, then, the OUR was monitored during 0 to 15 min. The mixture was reaerated for 10 min, and the OUR during 25 to 40 min was monitored, reaerated again for 10 min and the OUR during 50 to 65 min was monitored. At the beginning of the test, the pH was adjusted to 8.5 using 0.1N NaOH and 0.1N H_2SO_4. The pH during the test was not controlled, but was observed to be between 8.0 and 8.5.

The inhibition of OUR was calculated as follows.

$$\%\,\text{inhibition} = \frac{OUR_{max} - OUR_{tested\ solution}}{OUR_{max}} \times 100$$

Where; OUR_{max} : Oxygen utilization rate in the test with ammonia or nitrite but without the addition of heavy metal.

$OUR_{tested\ solution}$: Oxygen utilization rate of tested solution with the addition of ammonia or nitrite and heavy metal.

3. Results and discussion

The effect of heavy metals on ammonia oxidizing mixed culture (AOMC): The effect of cadmium, nickel, copper, and zinc, on ammonia oxidizing mixed culture was studied. Six different concentrations, 0, 1.0, 2.5, 5.0, 10 and 20 mg/L of each heavy metal were tested. The oxygen utilization rate was measured after the addition of each heavy metal within 15 min exposure time, and the results of the tests are given as Figure 2.

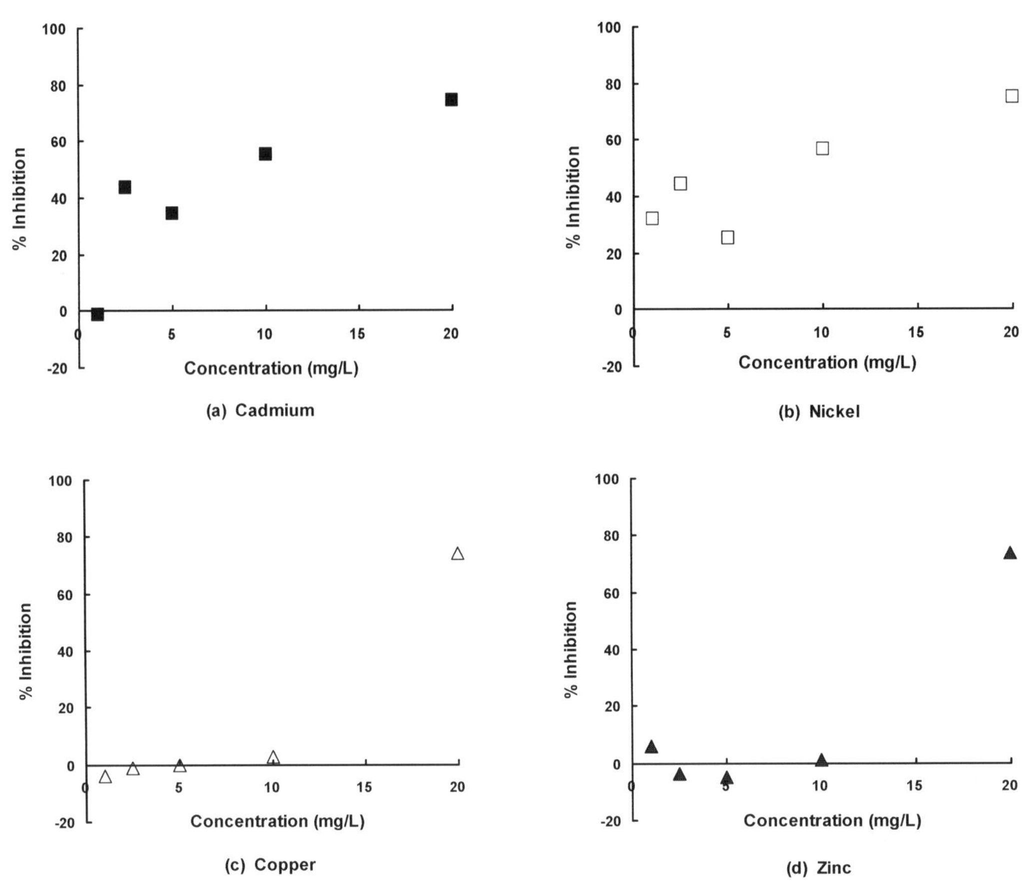

Figure 2 Effect of heavy metals on oxygen utilization rate of AOMC

For cadmium and nickel, it was found that in some cases the increase of heavy metal concentration did not result in the increase of %inhibition, or rather, %inhibition was reduced. Such cases were observed especially at the concentration range lower than 5.0 mg/L.

For example, in toxicity test with cadmium, while %inhibition was around 44% at 2.5 mg/L, it was reduced to around 38% when cadmium concentration was increased to 5.0 mg/L. However, linearity of the response and %nhibition can be clearly observed at higher concentrations.

In the experiment with copper and zinc, it was found that the inhibition of OUR of the tests was low in comparison to the tests with cadmium and nickel. This indicated that cadmium and nickel have stronger inhibitory effect on AOMC than copper and zinc. One of the possible explanation for the difference of toxic effects of these heavy metals would be that copper is a micronutrient and play structural role in various enzymes, while cadmium is not. Indeed, Ensign *et al*. (1993) reported that the addition of $CuCl_2$ to *N. europaea* cell extracts resulted in 5-to15-fold stimulation of ammonia-dependent O_2 consumption. But this explanation cannot be applied to nickel, as it is also micronutrient but the reduction of OUR was observed in it existence.

The effect of heavy metals on nitrite oxidizing mixed culture (NOMC): The effect of cadmium, nickel, copper, and zinc, on nitrite oxidizing mixed culture was studied with different heavy metal concentrations. The OUR was measured after the addition of each heavy metal within 15 min exposure time, and the results of the tests are given in Figure 3.

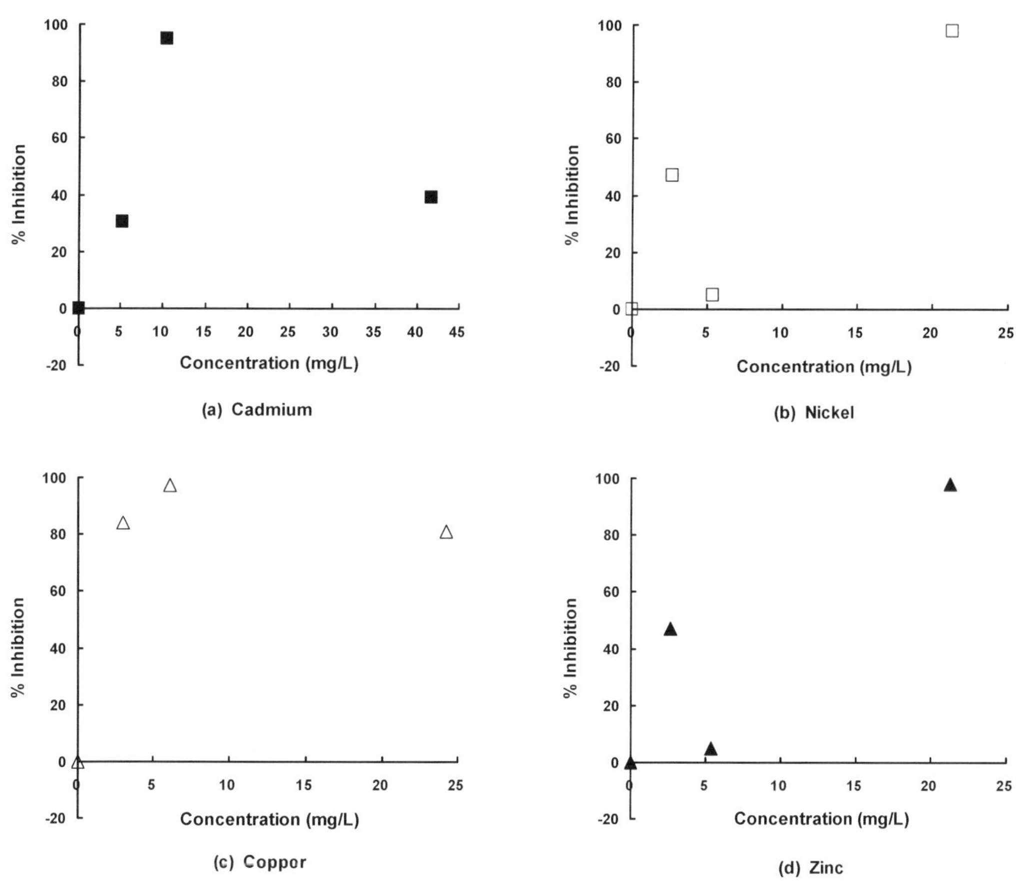

Figure 3 Effect of heavy metals on oxygen utilization rate of NOMC

The results in Figure 3 showed that the correlation between heavy metal concentrations and %inhibition of the toxicity tests with NOMC was not high. Further, it was found that the maximum oxygen utilization rate of nitrite oxidizing mixed culture was about 10 times less than that of ammonia oxidizing mixed culture as shown in Figure 4. Therefore, it was expected that it would not be easy to apply NOMC for the detection of heavy metals or attain high sensitivity in comparison to AOMC.

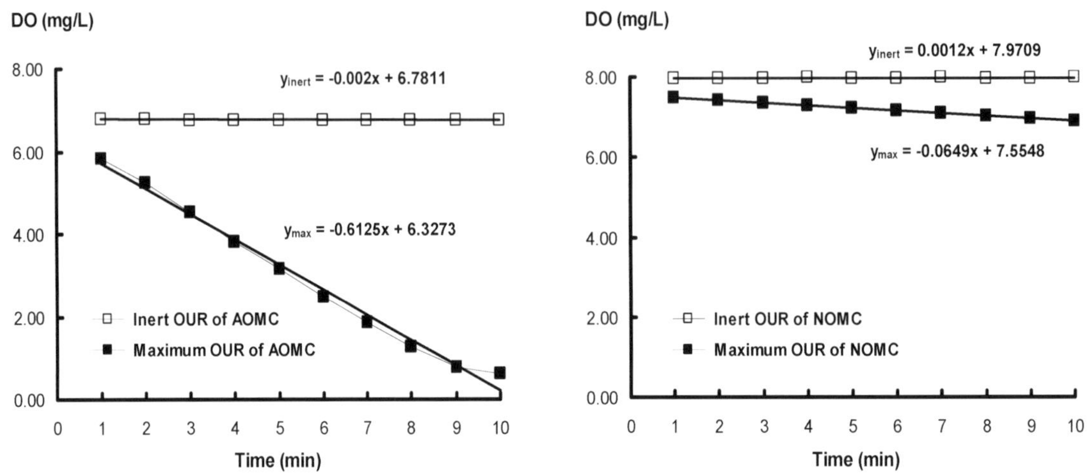

Figure 4 The inert and maximum oxygen utilization rate of AOMC and NOMC

The effect of exposure times on the response to cadmium: The effect of exposure times on the response pattern of ammonia oxidizing mixed culture to cadmium was investigated in an attempt to improve the response of AOMC to heavy metals. Six different concentrations of 0, 1.0, 2.5, 5.0, 10 and 20 mg/L of cadmium were tested. The oxygen utilization rate was measured after the addition of cadmium during three exposure time periods: 0-15, 25-40 and 50-65 min. The toxicity tests were performed with two replications and the results of the toxicity tests are given as Figure 5.

It was found that at the shorter exposure time (Figure 5a), the response of AOMC to low cadmium concentrations (1.0, 2.5 and 5.0 mg/L) was unpredictable (either inhibit or enhance the oxygen utilization rate). However, the reduction of oxygen utilization rate or %inhibition to all tested concentrations can be observed more clearly when the longer exposure times of 25-40 and 50-65 min were applied (Figure 5b,5c).

It is not easy to explain why at shorter exposure time and at the lower cadmium concentrations the fluctuation of OUR was observed and why such fluctuation was eliminated by applying longer exposure time. One of the possible explanations is that at a lower concentration range, some microorganisms initiated resisting or detoxifying metabolisms in response to the dosed cadmium, which resulted in the increase of OUR rather than its decrease. Another possible explanation could be that it takes time for the cadmium molecules to reach the target site within a cell to complete the toxic reaction.

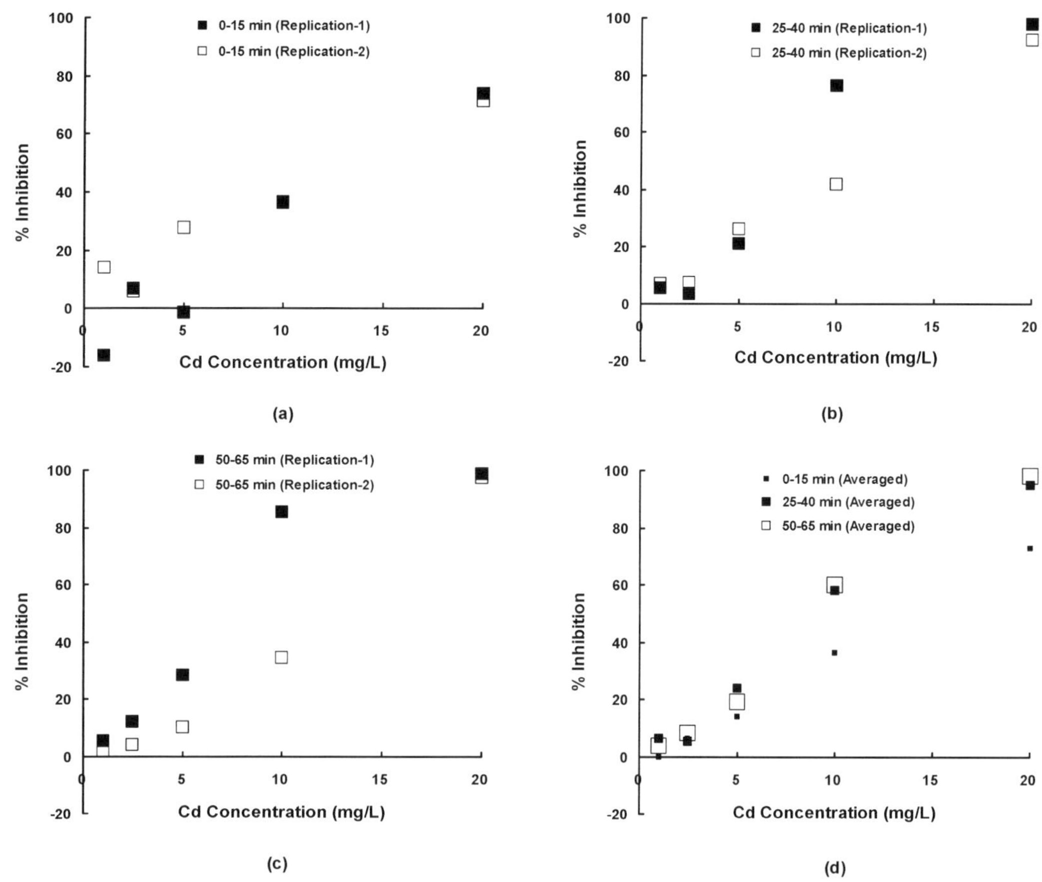

Figure 5 Effect of exposure times on the response of AOMC to cadmium

When the averaged values of the two replications were applied (Figure 5d), it was illustrated that generally the sensitivity of AOMC to cadmium was improved by the extending exposure times to 25-40 and 50-65 min. In other words, the reduction of OUR was clearly observed when the longer exposure time was applied. As can be seen that %inhibition at 0-15 min was lower than at 25-40 and 50-65 min. Further, at higher cadmium concentrations (10, 20 mg/L), the longest exposure time give the highest %inhibition. This observation is consistent with the study by Inui *et al*. (2002). They reported that for the detection of toxicity of heavy metals by their sensor using *N. europaea*, longer exposure time was required to achieve the maximum reduction of respiratory activity.

4. Conclusions

In this study, the response of ammonia oxidizing and nitrite oxidizing mixed culture to heavy metals were studied in terms of the inhibition or reduction of oxygen utilization rate. When ammonia oxidizing mixed culture was applied, it was found that the effect of low heavy metal concentrations on the culture was unpredictable and the relationship between heavy metal concentrations and oxygen utilization rate was less linear, either OUR inhibition or enhancement was observed. However, linearity relationship can be clearly observed at the higher concentrations. Further, it was found that the toxic response or %inhibition at low cadmium concentrations (1.0, 2.5, 5.0 mg/L) was improved by applying extended exposure times to 25-40 and 50-65 min. In the toxicity tests with nitrite oxidizing mixed culture, it was expected that it would not be easy to attain high sensitivity in comparison to AOMC.

The results from this study illustrated that ammonia oxidizing mixed culture can be used to monitor heavy metals with rapid response for water that might contain tens of mg/L of heavy metals. For water that have heavy metals in the order of a couple of mg/L, longer exposure time would be recommended to improve the sensitivity of the test.

5. Acknowledgement

This research work was supported by the grant of the Post-Graduate Education, Training and Research Program in ESTM through Higher Education Development Project of the Ministry of University Affairs, Thailand. The authors also thank Southeast Asian Center for Water Environment Technology (SACWET) for financial support. The authors would also like to thank to Fuji Electric Co. Ltd. for the donation of DO electrode.

6. References

Blum D.J.W. and Speece R. E. (1991). A database of chemical toxicity to environmental bacteria and its use in interspecies comparisons and correlations. *Research Journal of the Water Pollution Control Federation,* **63**(3), 198-207.

Ensign S.A., Hyman M.R. and ARP D.J. (1993). In vitro activation of ammonia monooxygenase from *Nitrosomonas europaea* by copper. *Journal of Bacteriology*, **175**(7), 1971-1980.

Grunditz C., Gumaelius L. and Dalhammar G. (1998). Comparison of inhibition assays using nitrogen removing bacteria: Application to industrial wastewater. *Water Research*, **32(**10), 2995-3000.

Inui T., Tanaka Y., Okayasu Y., and Tanaka H. (2002). Application of toxicity monitor using nitrifying bacteria biosensor to sewerage systems. *Water Science and Technology*, **45**(4-5), 271-278.

Juliastuti S.R., Baeyens J., Creemers C., Bixio D. and Lodewyckx E. (2003). The inhibitory effects of heavy metals on the net maximum specific growth rate of the autotrophic biomass in activated sludge. *Journal of Hazardous Materials*, **B100**, 271-283.

Madigan M.T., Martinko J.M. and Parker J. (1999). Brock Biology of Microorganisms, Eighth edition, Prentice Hall, New Jersey, USA.

Tanaka Y., Taguchi K. and Utsumi H. (2002). Toxicity assessment of 255 chemicals to pure cultured nitrifying bacteria using biosensor. *Water Science and Technology*, **46**(11-12), 331-335.

Metallothionein, a Biomarker for Copper Toxicity in Mussel

Phanit Ratasuk [1)], Panida Navasumrit [2)] and Preeda Parkpian [3)]

[1)] Water Quality Management Bureau, Pollution Control Department
92 Soi Phaholyothin 7, Samsennai, Bangkok, 10400 Thailand
e-mail: ratasuk@hotmail.com
[2)] Environmental Toxicology Laboratory, Chulabhorn Research Institute
54 Moo 4 Laksee, Bangkok, 10210 Thailand
e-mail: panida@tubtim.cri.or.th
[3)] School of Environmental Resources and Development, Asian Institute of Technology
P.O. Box 4, Klong Luang, Pathumthani, 12120 Thailand
e-mail: preeda@ait.ac.th

Abstract

Effect of copper on metallothionein (MT) induction was studied. Experimental studied were carried out on green mussel *Perna viridis* exposed in aquarium for 14 days to different concentrations of $CuCl_2$. Metallothionein quantitation was used as a marker of metal exposure. Results demonstrated that copper accumulation in mussel tissue was concentration and time dependent particularly at high level as 50 μg/l. MT expression in mussel exposed to 25 and 50 μg/l were generally higher than those exposed to lower copper concentrations. This result demonstrated positive dose-response relationship between MT expression and copper concentration in the experimental tanks. However, significant MT expression in exposed mussels was not observed when compared to that of control. Copper accumulated in the mussel tissue and MT expression in the digestive gland as a function of time showed the inverse relationship demonstrating the important role of MT in binding with metals in the detoxification process. Results from field study showed that mussel collected from the Samut Prakarn coastal area had higher level of MT than level found in the experiment. The possible reason may be due to longer exposure period in the field as well as multi-exposure from other metals. The overall results demonstrated the possibility of using metallothionein induction as a biomarker for copper exposure however the basal level of MT need to be identified in mussel cultured in the pristine area

Keywords: biomarker, copper, green mussel, metallothionein, Samut Prakarn coastal area

1. Introduction

In Thailand, environmental monitoring of coastal areas consisted of the measurement of physical and chemical variables and only occasionally were biological variables incorporated. The use of biomarker as an application to obtain information on the state of the environment and the effects of pollution on living biological resources have been adopted in international monitoring programs, especially in aquatic ecosystems. Among specific biomarkers, the contents of metallothionein are widely used in heavy metal pollution monitoring. The induction of metallothionein is implicated in cellular protection and play a crucial role in metal detoxification (Olsson, 1993). The potential value of metallothionein as an indicator of metal exposure, in mussels commonly found in European waters, *M. edulis* and *M. galloprovincialis* was reported. This study aims to study the effect of copper exposure on the induction of metallothionein in the local mussel, *Perna viridis* and to evaluate the potential of using the induction of this special protein called metallothionein as a biomarker in the environmental monitoring program.

2. Materials and methods

2.1 Field study

Green mussels were randomly collected from Samut Prakarn coastal area in September 2002. Thirty mussels were used for the determination of copper accumulation and metallothionein level in the mussel. The digestive glands were rapidly removed from the soft part of the mussel and stored in -80°C freezer for the analysis of metallothionein expression. Other soft parts were taken for the analysis of copper content.

2.2 Laboratory study

The effect of exposure was studied at 10, 25 and 50 µg/l of copper in which copper accumulation and metallothionein expression were determined. This range of copper concentration was within the range of copper levels found in the green mussel farms of Thailand (this study). In the time-course study, the effect of copper exposure on copper accumulation and metallothionein expression in mussel digestive gland at 10, 25 and 50 µg/l of copper were studied at the various time intervals. These were the first day, the third day, the seventh day and the fourteenth day. Green mussel *Perna viridis* Linnaeus 6-8 cm long, collected at Sriracha Fishery Research Station were used. The mussels were acclimatized in aerated natural seawater for one week before testing. The conditions during the acclimatization period were temperature 28°C, salinity 30 ppt, dissolved oxygen 4.3-4.8 mg/l and pH 7.5. During this time, all mussels were fed with algae cultures.

Three aquarium tanks with the size of 20 X 30 X 20 inch, each containing 50 mussels, were filled with 40 liters of filtered seawater. Each of the tanks was spiked with copper in the form of $CuCl_2$ (analytical grade). The copper concentration in each tank was adjusted to reach 10, 25 and 50 µg/l, respectively. A fourth tank with 50 mussels and 40 liters of filtered seawater without copper added was used as a control unit. Both contaminated and control mussels were fed with the *Tetraselmis* sp. daily. Seawater was changed daily and copper was added fresh everyday. Ten mussels from each experiment tank along with the control were harvested at day 1, 3, 7 and 14. The digestive glands were rapidly removed from the soft part of the mussel and stored in -80°C freezer for the analysis of metallothionein expression. Other soft parts were taken for the analysis of copper content.

2.3 Copper accumulation in mussel

The soft parts of mussel were homogenized in a glass blender . The sample weight 1 g was then digested with 5 ml of 65% HNO_3 and 2 ml of 30% H_2O_2 in the microwave digestion unit. Determinations were carried out by flame atomic adsorption spectrophotometry.

2.4 Metallothionein expression

RNA was isolated using TRIZOL Reagent and isolation procedure as described by the manufacturer (Invitrogen, US). To remove DNA, the total RNA was incubated at room temperature for 15 minutes with DNase I (Invitrogen, US) in 1x DNase I reaction buffer, in a final volume of 10 µl. 25 mM EDTA was added and followed by incubation at 65 °C for 10 minutes, for inactivation of DNase I.

Complementary DNA synthesized reaction contained DNase I-treated RNA, Random hexamer (Promega, US) and were then incubated at 65°C for 10 minutes and placed on ice immediately. After incubation, 5X EXPAND™ reverse transcriptase buffer, 5mM dNTPs, 100 mM DTT solution and EXPAND™ reverse transcriptase (Roche, Germany) were added into the mixture, in the final volume of 20 μl. The reaction was incubated at 30°C for 10 minutes, followed by at 42°C for 48 minutes and 95°C for 2 minutes to inactivate the primer-template binding.

The primers for metallothionein were designed from mRNA of *Perna viridis* : 5' TGC ACG AAC AAC TGG AGT CAC 3' and 5' ATC TGT GGT ACT GGG TGC AGC 3'. The PCR product of metallothionein was 185 bp. The amplification reaction of metallothionein gene consisted of 3 μl template cDNA, 5 mM dNTP, 4 μM MT primers, 10X PCR buffer, 50 mM $MgCl_2$ and 1 unit *Taq* polymerase in a total volume of 50 μl. The reaction was run at 94°C for 2 min. followed by 35 cycles at 94°C (1 min), 62°C (1 min), 72°C (1 min) and 72°C for 5 min. PCR product was then electrophoresed on an agarose gel, and stained with ethidium bromide.

3. Results and discussion

3.1 Dose response relationship study

The dose response study has been designed to evaluate the effect of copper concentration on the accumulation in soft tissue and MT expression in the digestive gland.

3.1.1 Copper accumulation in the tissue

In this study, the copper accumulation in green mussel can be observed shortly after 1 day exposure to copper at 25 and 50 μg/l but not statistical significant (Figure1). Generally, results indicated that there was a positive dose-response relationship between copper accumulation in the tissue of mussel and copper concentrations in the experimental tanks. Mussel exposed to 25 μg/l of copper show significant accumulation of copper in the tissue ($p < 0.05$) at the seventh day. Whereas exposed to higher concentration of copper at 50 μg/l significantly accelerated the copper accumulation at the third day of exposure. At the longer exposure period up to 14 days, the accumulation increased significantly compared to that of control at all three concentrations. When compared to control, the elevated concentration of copper in tissue was detected in all mussels exposed to 25 and 50 μg/l as well as most of the mussel exposed to 10 μg/l. Thus, results demonstrated that *P. viridis* can not regulate Cu^{2+} at these trace exposure concentrations.

3.1.2 Metallothionein expression in the digestive gland

MT expression in Cu-exposed mussel, expressed as a percentage of control values was presented in Figure 2. MT expression in mussel exposed to 50 μg/l were generally higher than those exposed to lower copper concentrations, particularly at the first day, seventh and fourteenth day of exposure. Mussel exposed to 25 μg/l also showed slightly higher level of MT than those exposed to 10 μg/l. This result demonstrated generally positive dose-response relationship between MT expression and copper concentration in the experiment tanks. However, significant MT expression in exposed mussels was not observed at all copper concentrations (10, 25 and 50 μg/l) when compared to that of the control.

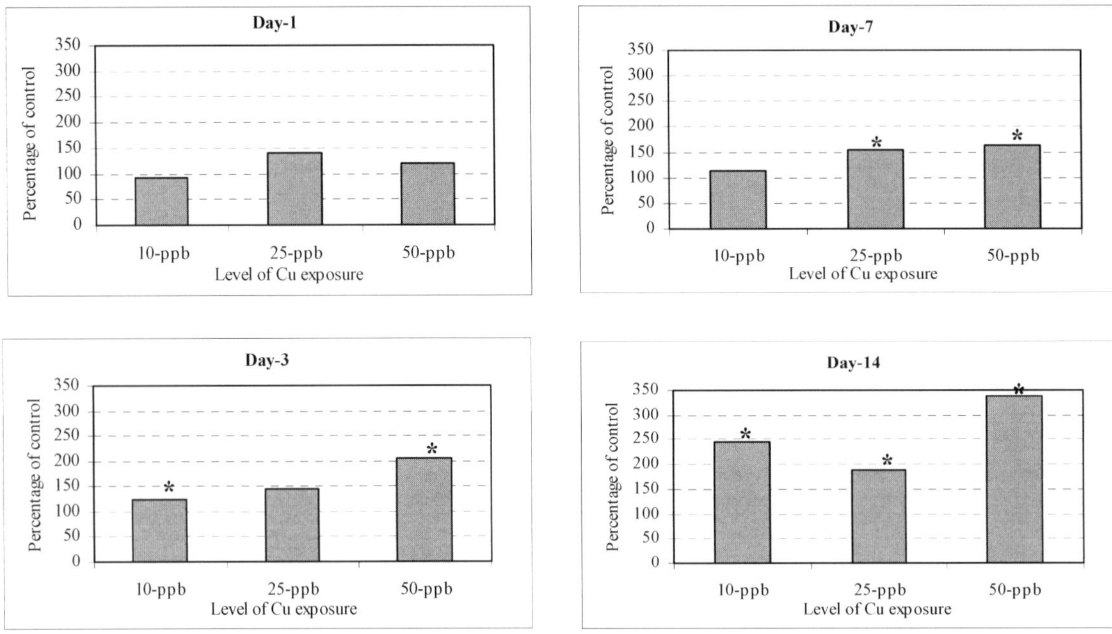

Figure 1. Copper accumulation in the tissue of *Perna viridis* exposed to 10, 25 and 50 μg /l(ppb) after 1, 3, 7 and 14 days of exposure. All the results are expressed as percentage of the control. Asterisk indicates significant difference ($p < 0.05$) from the control value.

It should be noted that the basal level of MT in control group has raised up after the first day of exposure. We assumed that the basal level of MT was influenced by both internal and external factors. In aquatic animals, MT levels have been shown to vary with time of year (season), reproductive state and water temperature. MT levels are elevated during sexual maturation and spawning. Capture-stress when taking animal from the field could also causes a small elevation in MT level (George and Langston, 1994). MT induction in the marine mussel was also shown to be size dependent and increased significantly with temperature. Large mussels exhibited higher MT levels than smaller one. MT expression was significantly higher at 18 and 25 °C than 5 °C (Serafim et al, 2002). Mussels used in this study were in the adult stage, captured during the spawning period (November 2002) and cultured in water temperature around 25 – 28 °C. These conditions may be the possible explanation for the high basal level of MT in the control mussel and this existing MT level was sufficient enough in binding with excess Cu^{2+} without a significant induction of MT mRNA expression. Therefore, the exposed-mussel to trace copper concentrations could not induce MT in a significant level.

3.2 Time course study

3.2.1 Copper accumulation in the tissue

An increase in Cu^{2+} concentration in tissue (excluding digestive gland) as a function of exposure time at 10, 25 and 50 μg/l as shown in Figure 3. Copper accumulation in the control mussel was gradually decreased with the duration of study. This decreasing trend could be due to the depuration. Placing mussel in the clean seawater for 3 days could result in the lost of biological incorporated contaminant where the half-life is low e.g. copper (Lobel et al.,1991). Our assumption was supported by the slightly increasing of copper concentration in the control tank after 24-h exposure period.

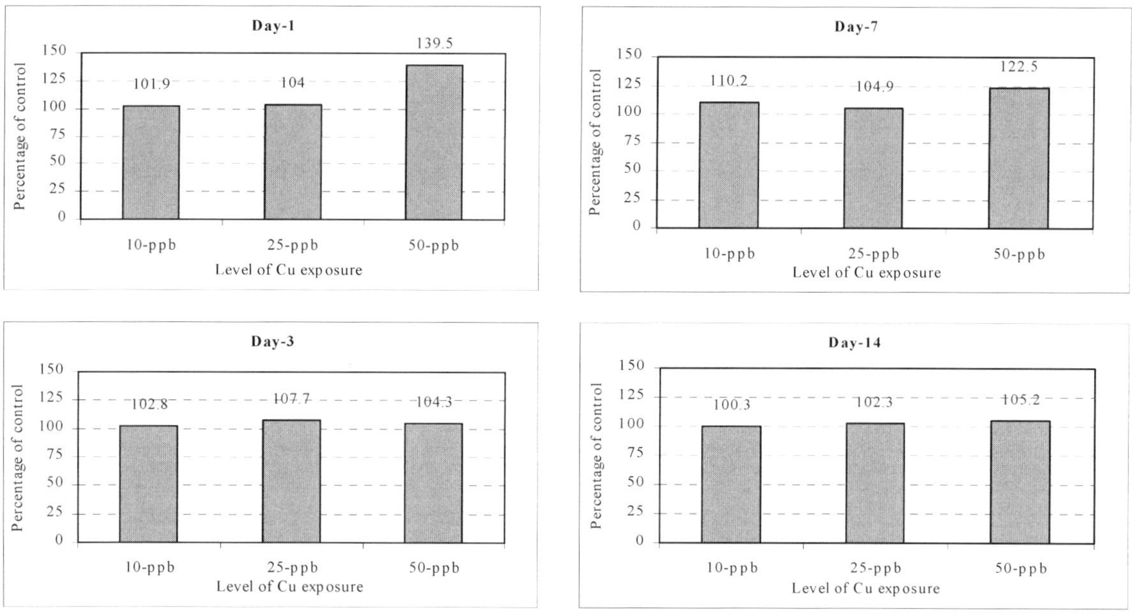

Figure 2. Relative MT expression in the digestive gland of *Perna viridis* exposed to copper at 10, 25 and 50 μg/l (ppb) after 1, 3, 7 and 14 days of exposure. All the results are expressed as percentage of the control.

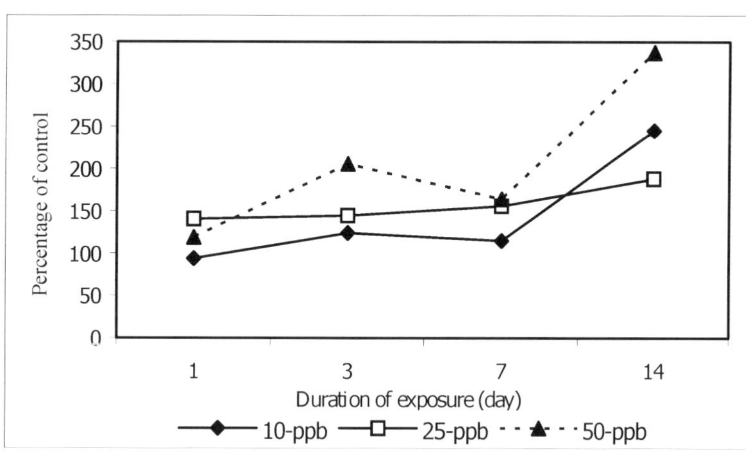

Figure 3. Copper uptake by *P. viridis* exposed to various water concentrations of Cu^{2+} (as $CuCl_2$). All the results are expressed as percentage of the control.

At 10 μg/l, the level of copper accumulation gradually increased at the thirdth day and slightly decreased at the seventh day, but then sharply rose at the fourteenth day. Copper level in the mussel tissue exposed to 25 μg/l concentration were slightly increased up to the fourteen day. We found the similar pattern of variation in copper accumulation between mussel exposed to 10 and 50 μg/l. However a more pronounced time-dependent copper accumulation was recorded in mussel exposed to 50 μg/l. The depletion of copper accumulation in the mussel tissue at the seventh day is due to a very short biological half-life of copper. Copper has a biological half-life of about 9-10 days, a very short period when compared with other metal such as cadmium (Viarengo et al., 1985).

3.2.2 Metallothionein expression in the digestive gland

Figure 4 shows that at 50 μg/l concentration, MT expression started to be induced shortly after one day exposure and then sharply decreased at the third day of experiment. Induction of MT appeared again at the end of the seventh day and drop down at the end of experiment. This fluctuation in MT expression as observed at 50 μg/l concentration was also found in mussel exposed to 10 μg/l but in the lesser extent. Whereas MT expression in the 25 μg/l concentration during 2 weeks experimental period showed slightly change in MT level.

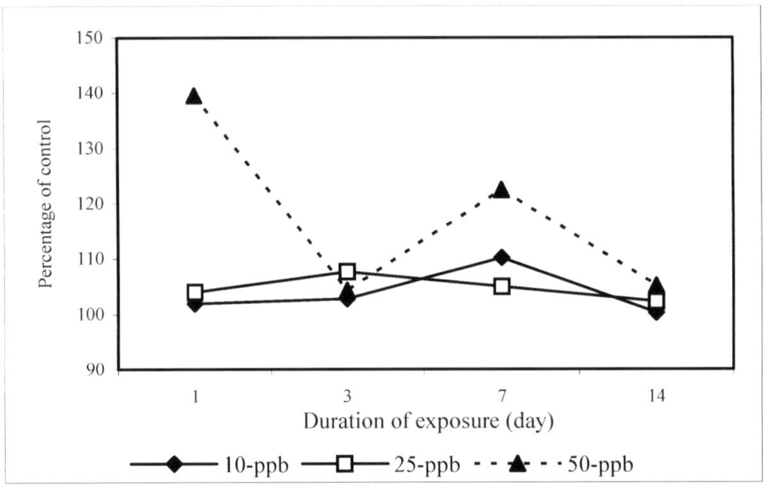

Figure 4. MT expression in the digestive gland of *P. viridis* exposed to various water concentrations of Cu^{2+} (as $CuCl_2$). All the results are expressed as percentage of the control

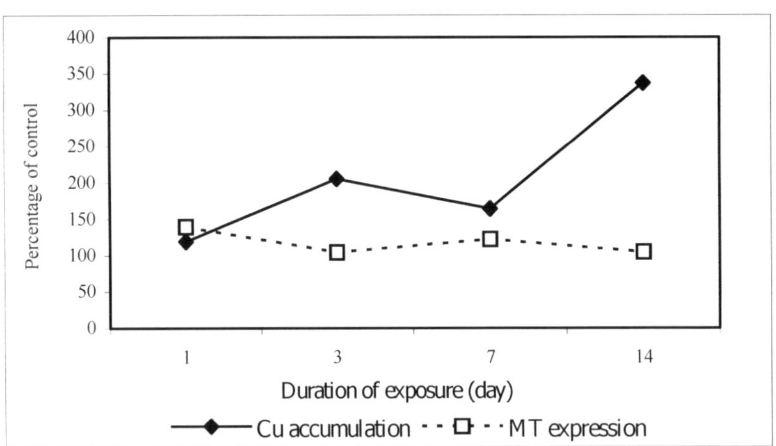

Figure 5. The inverse relationship between copper accumulation in the tissue and MT expression in the digestive gland of mussel exposed to copper at 50 μg/l

Copper accumulated in the tissue (Figure 3) and MT expression in the digestive gland (Figure 4) as a function of time showed the inverse relationship as clearly seen at 50 μg/l concentration (Figure 5). After three days exposure period, copper accumulation in the tissue increased while level of MT expression decreased. At seventh day, a lower level of copper accumulation in the tissue was in contradict to a higher level of MT expression. This inverse relationship between copper accumulation and MT expression was also observed at the fourteenth day of experiment. At lower copper concentrations as 10 and 25 μg/l, the inverse relationship between copper accumulation in the tissue and the level of MT expression was exhibited in the similar manner with a lesser extent. This kind of relationship demonstrated that MT in the digestive gland play an important role in binding with the excess copper resulting in less copper accumulated in the tissue. At a longer exposure period, copper concentration may exceed the capacity of MT in metal sequestration. As a result, copper is accumulated more in the tissue. However due to a very short biological half-life of copper, MT become available again in binding with copper at the seventh day.

3.3 Field study

The mean concentration of copper in soft tissue of mussel was 11.24 ± 1.65 mg/kg dry wt. The mean relative intensity of MT was 3.38 ± 0.75 (Table 1). MT expression in mussel from study area was significantly higher than that found in the laboratory study using mussel collected from Sriracha. We assumed that the difference may be due to two reasons. First is the duration of exposure. In the laboratory study, mussels were exposed to copper at trace concentrations of copper for maximum 2 weeks. Whereas mussel in the field study were definitely longer exposure in the natural condition with copper concentration about 10 μg/l. Second reason should be contributed to multiexposure to other metals in the study area due to the fact that Samut Prakarn is an industrial zone. As well as it has been demonstrated that MT synthesis may also be induced by other metals such as cadmium and zinc in mollusce (Roesijadi, 1992). Therefore, a high induction of MT expression in the field study may be due to simultaneous exposure to copper and other metals.

Table 1 Copper accumulation in the tissue and MT expression in the digestive gland of *P. viridis* collected from Samut Prakarn.

No.	Copper concentration (mg/kg dry wt)	Relative intensity of MT
1	12.59	2.87
2	11.72	3.02
3	9.40	4.24
	11.24 + 1.65[a]	3.38 ± 0.75

[a] Data are reported as mean ± SD from 3 pooled sample (10 mussels/pool)

4. Conclusions

Copper accumulation in the mussel tissue was concentration and time dependent particularly at high level as 50 μg/l. MT mRNA expression in the mussel digestive gland demonstrated generally concentration dependent but not statistical significant. The inverse relationship between copper accumulation in the tissue and MT expression in the digestive gland of mussel was observed as a function of time demonstrated an important role of MT in detoxification process. MT expression in the field study showed higher level than laboratory study probably due to longer exposure period as well as multi-exposure to other metals in the field.

Iinduction of metallothionein in mussel need further validation in the field study. Mussel should be collected from various sites including polluted and non-polluted sites. The basal level of metallothionein need to be identified in mussel collected from the pristine area. It is interesting to conduct the metallothionein study using other sensitive species which are not metal accumulator as green mussel in order to find out the more sensitive bioindicator of metal pollution in coastal environment.

5. References

George, S.G. and Langston, W.J. (1994). Metallothionein as an indicator of water quality-assessment of the bioavailability of cadmium, copper, mercury and zinc in aquatic animals at the cellular level. In: *Water Quality and Stress Indicators in Marine and Freshwater System*, Sutchliffe (Ed). Linking Levels of Organization. Freshwater Biological Association.

Lobel, P.B., Longerich, H.P., Jackson, S.E. and Belkhode, S.P. (1991). A major factor contribution to the high degree of unexplained variability of some element concentrations in biological tissue: 27 elements in 5 organs of the mussel *Mytilus* as a model. *Arch. Environ. Contam. Toxicol.*, **21**: 118-125.

Olsson, P.E. (1993). Metallothionein in fish: induction and use in environmental monitoring. In: *Toxicology of Aquatic Pollution. Physiological, Cellular and Molecular Approaches*, E.W. Taylor (Ed). Cambridge university press, London.

Roesijadi, G. (1992). Metallothioneins in metal regulation and toxicity in aquatic animals. *Aquat. Toxicol.*, **22**: 81-114.

Serafim, M.A., Company, R.M., Bebianno, M.J. and Langston, W.J. (2002). Effect of temperature and size on metallothionein synthesis in the gill of *Mytilus galloprovincialis* exposed to cadmium. *Mar. Environ. Res.*, **54**: 361-365.

Viarengo, A., Palmero, S., Zanicchi, G., Capelli, R., Vaissiere, R., and Orunesu, M., (1985). Role of metallothioneins in Cu and Cd accumulation and elimination in the gill and digestive gland cells of *Mytilus galloprovincialis Lmk. Mar. Environ. Res.*, **16**: 23-36.

Relief of Inhibitors for the PCR-based Monitoring of Occurrence and Fate of *Cryptosporidium* in Wastewater and Surface Water

Alfredo Anceno [1)], Oleg Shipin [1)], Buyan Chuluun [1)] and Hiroyuki Katayama [2)]

1) Environmental Engineering and Management, Asian Institute of Technology, Klong Luang, Pathumthani 12120, Thailand
e-mail: ajancenoii@ait.ac.th, oshipin@ait.ac.th, buyanc@ait.ac.th

2) Department of Urban Engineering, The University of Tokyo, 7-3-1 Hongo, Bunkyo-ku, Tokyo 113-8656 Japan
e-mail: katayama@env.t.u-tokyo.ac.jp

Abstract

The effectiveness of several DNA extraction methods for the PCR based detection of *Cryptosporidium parvum* oocyst spiked in negative samples was compared. The effect of post-lysis treatments and use of different PCR amplification enhancing additives for the relief of PCR inhibitory substances was investigated. Towards the end of the study, a low-cost direct DNA extraction method was proposed for use in the quantification of the pathogen in surface water and wastewater. This would serve to promote staging of affordable routine pathogen occurrence and fate monitoring in water environments in the developing world where infection is more prevalent.

Keywords: *Cryptosporidium parvum*, realtime PCR quantification, PCR inhibitors, wastewater, pathogen fate and occurrence

1. Introduction

Several techniques have been proposed for the detection of *Cryptosporidium* in wastewater, surface water and drinking water. These techniques are particularly employed to evaluate the effectiveness of wastewater treatment processes in reducing oocyst loads and in various health or infection risk assessments. Conventional methods are based on three key steps including concentration (filtration, floculation and centrifugation), purification (percoll-sucrose flotation, fluorescence activated cell sorting and immonomagnetic capture) and detection (immunofluoresence microscopy and PCR). The latest USEPA-validated method (Method 1623) is based on filtration, immunomagnetic separation (IMS) and immunofluoresence assay (IFA) with vital dye staining using 4',6-diamidino-phenylindole (DAPI) and differential interference contrast (DIC) microscopy (USEPA, 1999). Disadvantages of this method include being low-throughput, the use of expensive microscopy and non-specificity for human pathogenic *C. parvum* (Fontaine and Guillot, 2003). On the othe hand, the occurrence of *Cryptosporidium* in surface waters is in the range of 0.5-5,000 oocysts/100 L while outbreak preventive action level is only 10-30 oocysts/100 L that demands a very sensitive detection or quantifiation technique such as realtime PCR (Guy *et al.*, 2003).
Several PCR and realtime PCR based techniques for the detection and quantification of the parasite in the past employed immunomagnetic capture for oocyst purification. Due to the

prohibitive cost of commercial IMS kits, the full promise of employing high-throughput, PCR based assay for both routine assays and comprehensive surveys on the occurrence and quantification of the parasite in the environment could not be realized in the third world. Recently, the use of less expensive commercial DNA extraction kits in place of IMS coupled with a variety of treatments to alleviate the effects of PCR amplification inhibitors in DNA samples directly extracted from water concentrates were attempted (Guy *et al.*, 2003; Jiang *et al.*, 2005). Considering the cost of these commercial DNA extraction kits and the subsequent realtime PCR, the affordability of routine assay for quantification remains an impediment in the developing world. In the present study, we attempted to develop a cost effective method of extracting DNA from *Cryptosporidium* oocysts without the use of commercial kits. Several treatments were compared to relieve the effects of PCR-inhibitory substances during DNA extraction stage. The use of additives that reportedly enhance PCR amplification was also attempted.

2. Materials and methods

2.1 Cryptosporidium oocysts. Viable *C. parvum* oocysts (Iowa isolate) produced by passage in calves was purchased from Waterborne Inc. (New Orleans, LA, USA). The oocysts were supplied as a purified suspension in phosphate-buferred saline (50 mM sodium phosphate buffer, pH 7.2, 0.7% NaCl) supplemented with antibiotics and stored at 4°C. Dilutions containing 10^3-10^5 oocysts used in spiking experiments were prepared by counting using a hemocytometer, sedimentation, washing and suspension in PBS.

2.2 DNA extraction from purified oocysts. DNA yields across DNA extraction methods or modifications used in this study were compared using a fixed oocyst count. Aliquots of 10^4 oocysts were pipeted into 1.5-mL eppendorf tubes, sedimented, washed and resuspended in 50 μL sterile PBS. DNA was extracted from oocysts using DNeasy Tissue Kit (Qiagen, Hilden, Germany) and Wizard Genomic DNA Prurification Kit (Promega, Madison, WI, USA) according to manufacturer instructions. Alternatively, 0.5 mL 5% guanidine thiocyanate or 1% Triton X-100 was added on other treatment lots. Samples were subjected to different combinations of five rounds of freeze-thawing (5 min at -40°C and 5 min at 95°C), proteinase K treatment (~12 mAU, 55°C, 1 hr) and six 10-s bursts of sonication (20% amplitude) using a 130-W Cole-Parmer probe sonicator. DNA was purified either according to manufacturer instructions, otherwise alcohol precipitation was employed. Surface moisture from DNA pellets was removed by brief freeze drying. DNA pellets were resuspended in 10 μL Tris buffer (10 mM Tris-HCl, pH 8.0). One μL DNA sample portion was subjected to realtime PCR. Treatments giving realtively lower threshold (Ct) values were assumed to have yielded more DNA and therefore used in subsequent experiments.

2.3 DNA extraction from samples. Aliquots of 10^3 oocysts were seeded into 0.5-mL portions of oocyst-free 5-mL concentrate of 2-L wastewater sample lots tested at least five times by PCR. DNA was extracted using half of all methods or treatment combinations above that previously gave a relatively higher DNA yield based on Ct-values. A combination of Dynabead anti-*Cryptosporidium* Kit (Dynal, Lake Success, NY, USA) and DNeasy Tissue Kit as the reference DNA extraction and purification method was also tested. After sonication and debris removal by centrifugation, lysates in all treatments were treated with Chelex 100 to 20% final concentration, polyvinylpyrrolidone (PVP 360) to 2% final concentration or a combination of both. Lysates were incubated at 50°C for 30 min, centrifuged at 10,000 x g for 10 min, transferred to fresh tubes and DNA was purified according to each method. DNA pellets were resuspended 50 μL Tris buffer and 5 μL portions were used in realtime PCR.

2.4 Realtime PCR amplification. PCR amplification of oocyst wall protein (COWP) was carried out according to Guy *et al.* (2003) with modification. The 20-μL reaction mixture contained 1X iQ SYBR Green Supermix (BioRad, Hercules, CA, USA), 0.5 μM of COWP P702 forward and reverse primers, and DNA template. Thermocycling using an iCycler (BioRad, Hercules, CA, USA) consisted of 10-min initial denaturation at 95°C and 45 cycles of 30-sec denaturation at 95°C, 1-min annealing at 60°C and 30-sec extension at 72°C.

2.5 Relief of inhibitors during PCR and sensitivity studies. The best method above, based on relative Ct-values, was used to extract DNA from 0.5-mL portions of oocyst-free wastewater concentrate spiked with 10^3 purified oocysts. The ability of previously established PCR enhancers to reduce the effect of PCR inhibitors present in DNA extracts was investigated. PCR additives including 1% PVP 360, 5% dimethylsulfoxide, 500 ng/μL bovine serum albumin (BSA) and combinations thereof were incorporated in PCR reaction mixtures. The most effective treatment was used subsequently. On the other hand, sample dilutions containing 10^5, 10^4, 10^3, 500, 100, 50, 10, 5 oocysts either in pure or spiked preparations were used to evaluate the sentivity of the chosen method.

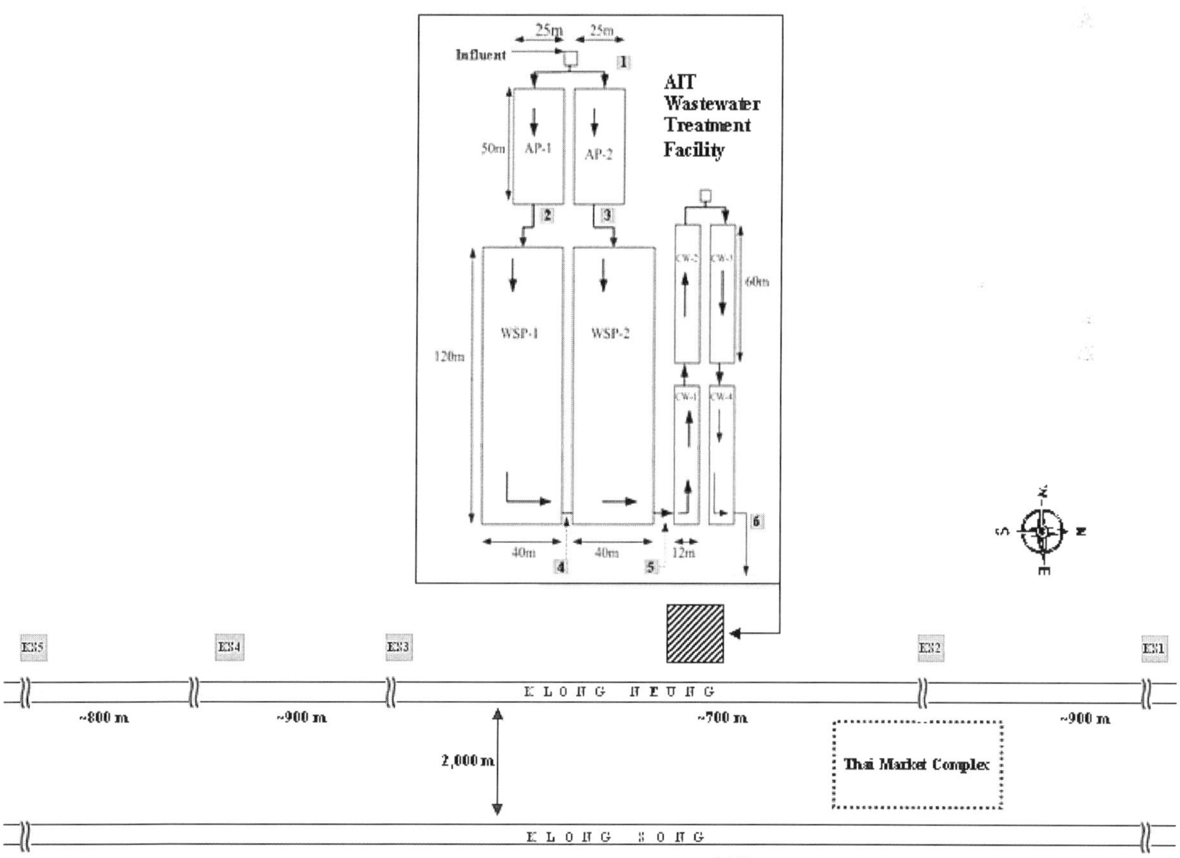

Figure 1. Sampling points within AIT wastwater treatment facility and nearby waterways: (1) Influent outlet, (2) Outlet of anaerobic pond 1, (3) Oulet of anaerobic pond 2, (4) Outlet of facultative pond 1, (5) Outlet of facultative pond 2, (6) Outlet of constructed wetland, (KX1-KX5) Sampling points 1-5 either for Klong Neung or Klong Song.

2.6 Quantification of *Cryptosporidium* in field samples. The occurrence and fate of *C. parvum* at the Asian Institute of Technology wastewater treatment facility located in Pathumthani, Thailand was investigated. As the facility discharges its effluents to a nearby waterway, Klong Neung, samples upstream and downstream of this klong were examined to assess the impact of the released treated wastewater to open water environment. Water samples were taken from five points 700-900 m apart (Fig. 1). Upstream of Klong Neung is a Thai market complex that releases wastewater on its banks and downstream is a heavy residential area. Samples were also taken in a parallel waterway, Klong Song, at the same distance interval for comparison. The vicinity of this waterway is rather less populated. Wastewater samples (2 L) were directly collected from the outlet of each treatment ponds and also 20 cm below the surface in the midpoint of a waterway. Samples were transported to the laboratory within 2 hours for analysis. Two liter samples were equally portioned among 200-mL centrifuge tubes and centrifuged at 4,000 x g for 10 min at 4°C. Supernatants were discarded and sediments were pooled to a total of 5 mL. Triplicates of 0.5-mL portions of each concentrated sample were subjected to DNA extraction and realtime PCR assay.

3. Results and discussion

The use of polymerase chain reaction (PCR) for the rapid, specific and sensitive detection of *Cryptosporidium* is well documented (Wiedenmann *et al.*, 1998; Quintero-Betancourt *et al.*, 2000). PCR based techniques allows the determination of the genus, species, genotype and presence and quantity of the parasite in various food, clinical and environmental samples including surface water and wastewater. While endpoint PCR assays may be useful in determining the presence or identity of the parasite, real-time PCR allows accurate quantification and could arguably reduce cost when more than one target organism or assemblage is to be detected or quantified in a multiplex assay.

Table 1. Comparison of DNA yields using different methods of direct DNA extraction from purified *C. parvum* oocysts as inferred from Ct-values

	DNA extraction methods [a]				Ct [b]
M1	Qiagen lysis buffer + F/T	Proteinase K	Sonication	Qiagen purification system	26.8 ± 0.8
M2	Qiagen lysis buffer + F/T	Proteinase K		Qiagen purification system	31.5 ± 1.1
M3	Promega lysis buffer + F/T	Proteinase K	Sonication	Promega purification system	25.7 ± 1.2
M4	Promega lysis buffer + F/T	Proteinase K		Promega purification system	28.7 ± 1.9
M5	5% Guanidine thiocyanate + F/T	Proteinase K	Sonication	Isopropanol precipitation	24.9 ± 1.7
M6	5% Guanidine thiocyanate + F/T	Proteinase K		Isopropanol precipitation	30.2 ± 2.1
M7	5% Guanidine thiocyanate + F/T		Sonication	Isopropanol precipitation	27.9 ± 0.9
M8	5% Guanidine thiocyanate + F/T			Isopropanol precipitation	31.8 ± 1.4
M9	1% Triton X-100 + F/T	Proteinase K	Sonication	Isopropanol precipitation	25.6 ± 1.2
M10	1% Triton X-100 + F/T	Proteinase K		Isopropanol precipitation	32.4 ± 0.8
M11	1% Triton X-100 + F/T		Sonication	Isopropanol precipitation	29.3 ± 2.1
M12	1% Triton X-100 + F/T			Isopropanol precipitation	34.5 ± 1.8

(a) F/T = freezing and thawing
(b) Mean ± standard deviation to nearest tenths of triplicate DNA extractions from 10^4 oocysts per method

Several published works on PCR detection and quantification of the parasite including the USEPA Method 1623 employ expensive IMS in oocyst purification. In later studies, this is obviated by replacement with less expensive commercial kits for direct DNA extraction although the affordability of this approach remains a question in the third world where infection by the parasite is more prevalent. In the present undertaking, we started with comparing the DNA yields of different direct extraction methods on purified oocysts as reflected by Ct-values (Table 1). We observed that, across DNA extraction modifications, both incubation with 12 mAU Proteinase K at 55°C for 1 hr and use of sonication increased DNA yields as reflected by lower Ct-values.

In another experiment, six of the methods (M1, M3, M5, M7, M9 and M11) that gave relatively higher Ct-values were chosen to directly extract DNA from spiked samples. These methods were compared with the use of IMS coupled to direct DNA extraction using Qiagen DNeasy Tissue Kit as reference method after Guy *et al.* (2003). Lysates were treated to remove PCR inhibitors after lysis or before DNA purification. Table 2 shows that the use of both IMS and DNeasy Tissue Kit is better than the use of the latter alone. On the other hand, among samples without post-lysis treatment, those extracted using method 5 gave the most promising result. PVP 360 appears to have delayed PCR amplification in all samples across treatments where it was applied. Lower Ct-values in samples treated with Chelex 100 indicated the relief of PCR inhibitors.

Table 2. Effect of post-lysis treatment on the reduction of PCR inhibitors and the subsequent realtime PCR amplifiability

Post-lysis treatment	**Ct-values for different DNA extraction methods**[a]						
	IMS + M1	M1	M3	M5	M7	M9	M11
Untreated	29.4 ± 0.7	32.5 ± 0.7	29.9 ± 0.9	27.7 ± 1.3	29.5 ± 0.7	33.2 ± 1.2	na
PVP 360[b]	30.5 ± 1.1	35.3 ± 1.3	33.1 ± 1.8	31.7 ± 0.8	31.9 ± 1.4	na	na
Chelex 100[c]	28.4 ± 1.2	31.3 ± 1.1	26.2 ± 1.2	25.9 ± 0.9	29.1 ± 1.6	31.9 ± 1.7	na
PVP 360 + Chelex 100[d]	30.8 ± 0.9	33.7 ± 1.5	27.6 ± 0.8	27.8 ± 1.1	30.6 ± 0.8	na	na

(a) Mean ± standard deviation to nearest tenths of duplicate DNA extractions from 10^3 oocysts per method
(b) 2% final concentration, (c) 20% final concentration, (d) both at the same concentrations

DNA extracted by Method 5 above — that included the use of 5% guanidine thiocyanate during freeze-thawing, ~12 mAU proteinase K, sonication and Chelex 100 — was subjected to realtime PCR assay with the addition of PCR amplification enhancing compounds such as PVP 360, DMO, BSA and their combinations. With the use of BioRad iQ SYBR Green Supermix, it was observed that these additives did not further enhance the sensitivity of the PCR assay (Table 3). Hence, in subsequent experiments these additives were not employed.

Table 3. Effect of different PCR additives on PCR amplification

	Ct-values for different PCR additives[a, b]						
No additive	PVP 360	DMSO	BSA	PVP 360 + DMSO	PVP 360 + BSA	DMSO + BSA	PVP 360 + DMSO + BSA
27.4 ± 0.8	28.6 ± 1.1	28.4 ± 0.9	28.2 ± 1.3	28.9 ± 1.0	29.3 ± 0.8	29.6 ± 0.8	30.2 ± 0.8

(a) Mean ± standard deviation to nearest tenths of triplicate PCRs per treatment
(b) PVP 360, DMSO and BSA at 1%, 5% and 500 ng/μL final concentration, respectively.

Dilutions containing 10^5, 10^4, 10^3, 500, 100, 50, 10, 5 *C. parvum* oocysts were subjected to direct DNA extraction using Method 5 with Chelex 100 treatment and realtime PCR. Both purified and spiked preparations were used to check the sensitivity of the chosen method. Figure 2 indicates that the method could detect up to 10 oocysts in purified preparations and 50 oocysts in spiked samples, although the latter method have shown to detect up to less than 10 oocysts in subsequent tests on field samples. It was noted that, perhaps due to the presence of PCR inhibitory substances, the sensitivy of the method was reduced by approximately two threshold cycle units when detecting spiked oocysts compared to pure preparations. Coincidentally, portions of field samples with possibly dilute amounts of inhibitors but happen to contain high oocyst loads could have possibly given positive but weak signals.

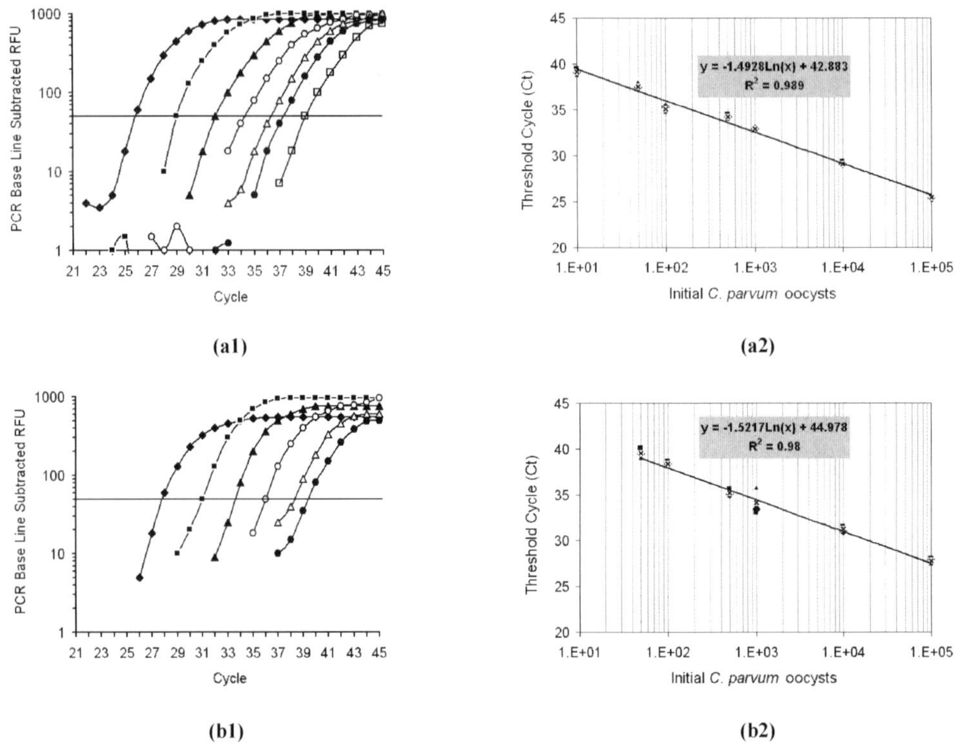

Figure 2. Sensitivity of *C. parvum* realtime PCR quantification using DNA extraction Method 5 above. (a) Purified oocyst preparation;, (b) Wastewater concentrate spiked preparation; (1) Realtime amplification profiles for 10^5, 10^4, 10^3, 500, 100, 50, 10, 5 *C. parvum* oocysts in the same order; (2) Corresponding standard curves.

The number of oocyst from fields samples was calculated using the line equation in Fig. 2-b2. This was obtained from realtime PCR assays of different oocyst dilutions spiked into 0.5-mL portions of oocyst-free wastewater concentrates (5 mL from 2 L). The following working equation was used in ^{MS}Excel calculations:

$$x = \left[\exp\left(\frac{y - 44.978}{-1.5217} \right) \right] x\ 5 \qquad [1]$$

Where, **x** is the number of *C. parvum* oocysts per L of sample and **y** is the threshold cycle (Ct) value and **5** is a conversion factor calculated by dividing 5 mL (concentrated sample) by 0.5 mL (portion used) and again dividing this by 2 L (original sample size).

The presence of pathogens in wastewater and surface water does not directly imply prevalence of disease but indicates potential health risk. The study of the occurrence of a pathogenic microorganism is indispensable for the evaluation of the actual risks of potential waterborne disease outbreaks. On top of that is the need of a reliable, sensitive and cost-effective detection technique that could assess the extent of pathogen occurrence. In that regard, the present quantification method was employed to evaluate the efficiency of the integrated AIT wastewater treatment system in reducing *C. parvum* oocyst loads. Water samples from Klong Neung, a canal adjacent to the facility and into which the facility discharges effluents, were examined to assess impact of effluent released to open water environment. As there is an extensive waste-generating activity taking place upstream and many other local klongs, it is assumed that the klongs play a substantial role in pathogen removal.

C. parvum was detected in raw influent in numbers ranging from 80 to about 100 oocysts/L while the number reduced to about 5- to 10-fold as treated wastewater exits the anaerobic ponds. Oocysts were not detected in facultative pond and constructed wetland effluents. It appears that a series of ponds could offer a satisfactory treatment capacity for the removal of the protozoan parasite while more data in succeeding months need be gathered for more statistically reliable conclusions. Also, from Table 4, it appears that the quality of canal water in Klong Neung in terms of *C. parvum* oocyst load is not compromised by effluents released from the AIT wastewater treatment facility. In fact, higher oocyst loads were detected in water samples taken upstream and downstream the facility; we refer to sampling points KN2 and KN3 where the wasterwater treatment facility lie in between. In contrast, a less predictable occurrence of the parasite was observed in Klong Song.

Table 4. Quantity of *C. parvum* in AIT wastewater treatment facility and nearby waterways

Sampling Points	May 15, 2005		June 26, 2005	
	Ct[a]	Oocysts[b] / L	Ct[a]	Oocysts[b] / L
Klong Neung 1	41.1	63.9	42.9	19.6
Klong Neung 2	40.2	115.5	41.6	46.0
Klong Neung 3	40.8	77.9	43.8	10.8
Klong Neung 4	42.8	20.9	na	na
Klong Neung 5	na	na	na	na
Klong Song 1	na	na	42.4	27.2
Klong Song 2	na	na	na	na
Klong Song 3	41.3	56.1	na	na
Klong Song 4	na	na	na	na
Klong Song 5	42.3	29.1	na	na
Influent Outlet[c]	40.4	101.3	40.6	88.8
Anaerobic Pond Outlet 1[c]	42.9	19.6	na	na
Anaerobic Pond Outlet 2[c]	44.1	8.9	43.1	17.2
Facultative Pond Outlet 1[c]	na	na	na	na
Facultative Pond Outlet 2[c]	na	na	na	na
Constructed Wetland Effluent[c]	na	na	na	na

(a) Average to nearest tenths of triplicate DNA extractions on one sample
(b) Calculated using the line equation in Figure 2-b2
(c) Constituents of AIT wastewater treatment facility; see Figure 1

4. Conclusions

A modified method for direct extraction of DNA from *C. parvum* oocysts was developed in this study. It consists of three cycles of freezing-thawing of 0.5-mL sample portions of surface water and wastewater concentrates suspended in 5% guanidine thiocyanate, treatment with proteinase K, sonication and purification by isopropanol precipitation. Since the released DNA need not bind to selective matrices for purification, unlike those employed in commercial kits, the yield will be higher. The use of 20% Chelex 100 was quite satisfactory in relieving the effects of PCR inhibitors while the use of manufacturer-optimized iQ SYBR Green Supermix eliminates adding PCR amplification enhancers. Nevertheless, the compatibility of the present DNA extraction method and other PCR mastermixes should be tested. For a 0.2-L actual sample size, the present protocol could detect up to 50 *C. parvum* oocysts/L in less than 5 hours. This length of time is necessary to substantially remove inhibitors. The method has been successfully employed in quantifying the amounts of the target pathogen in environmental water samples and wastewater samples. A more extensive application of this protocol on field samples shall be attempted.

5. Acknowledgement

Financial support provided by SACWET is gratefully acknowledged.

6. References

Fontaine M. and Guillot E. (2003). An immunomagnetic separation-real-time PCR method for quantification of *Cryptosporidium parvum* in water samples. *J. Microbiol. Methods* **54**(1), 29-36.

Guy R., Payment P., Krull U. and Horgen P. (2003). Real-time PCR for quantification of *Giardia* and *Cryptosporidium* in environmental water samples and sewage. *Appl. Environ. Microbiol.* **69**(9), 5178-5185.

Jiang J., Alderisio K., Singh A., Xiao L. (2005). Development of procedures for direct extraction of *Cryptosporidium* DNA from water concentrates and for relief of PCR inhibitors. *Appl. Environ. Microbiol.* **71**(3), 1135–1141.

Quintero-Betancourt W., Peele E. and Rose J. (2002) *Cryptosporidium parvum* and *Cyclospora cayetanensis*: a review of laboratory methods for detection of these waterborne parasites. *J. Microbiol. Methods* **49**(3), 209-224.

US Environmental Protection Agency (USEPA), (1999). Method 1623: *Cryptosporidium* and *Giardia* in water by Filtration / IMS / IFA. Office of Water. EPA-821-R-99-006.

Wiedenmann A., Kruger P. and Botzenhart K. (1998). PCR detection of *Cryptosporidium parvum* in environmental samples—a review of published protocols and current developments. *J. Ind. Microbiol. Biotechnol.* **21**, 150–166.

The Application of Fecal Sterols and their Ratios as Fecal Contamination Indicators in Water Environment around Bangkok

Hoa Thai Nguyen[1)], Kaisai Zang[2)], Thammarat Koottatep[1)] and Hiroyasu Satoh[2)]

1) School of Environment, Resources and Development,Asian Institute of Technology
P.O.Box 4, Klong Luang, Pathumthani 12120, Thailand
phuongthaohoa@yahoo.com
2) Dept. of Urban Engineering, The University of Tokyo
7-3-1 Hongo, Bunkyo-ku, Tokyo 113-8656, Japan
kaisai@env.t.u-tokyo.ac.jp
3) School of Environment, Resources and Development, Asian Institute of Technology
P.O.Box 4, Klong Luang, Pathumthani 12120, Thailand
thamarat@ait.ac.th
2) Institute of Environmental Studies, The University of Tokyo
#674, Env. Building, Kashiwanoha 5-1-5, Kashiwa, Chiba 277-8563, Japan.
hiroyasu@k.u-tokyo.ac.jp

Abstract

In this study, fecal sterols such as coprostanol (5β-cholestan-3β-ol), cholesterol (cholest-5-en-3β-ol) and cholestanol (5α-cholestan-3β-ol) in the sediments of nine water bodies in and around Bangkok and in one urban wastewater treatment plant in Bangkok were analyzed. Fecal sterols were determined by gas chromatography-mass spectrometry (GC/MS) analysis after solvent extraction and concentration while omitting the saponification, clean-up, and derivatization steps. The careful observation of the mass spectra obtained in sample analyses indicated that clean-up can be omitted for the samples analyzed in this study. The coprostanol concentration ranged from less than 0.01μg/g-dry to 278μg/g-dry. Many of the samples gave the ratios of coprostanol to the sum of coprostanol and cholestanol (5β/(5β + 5α)) higher than 0.5 but less than 0.7 and these sites are most probably affected by fecal contamination. Fecal sterols were also found in sewage, whereas in Bangkok feces are supposed to be treated or trapped by septic tanks. Fecal sterols proved itself to be a useful indicator for the monitoring of fecal contamination.

Keywords: Bangkok, cholesterol, coprostanol, fecal contamination, sewage, sterols

1. Introduction

Water pollution has been a matter of concern as the population growth and urbanization proceeds very rapidly without establishing proper wastewater management systems. Water pollution caused by feces is thought to be one of the most serious water pollution, because it means higher risk of water borne diseases. Determining the extent of fecal pollution is of primary concern to control water borne diseases, and thus there is a strong need for proper indicator for fecal pollution. Many researchers as can be seen from the fact that it is included in the "Standard Methods" have long used fecal coliform group. On the other hand, indigenous soil microorganisms are better suited to grow in soil, and fecal bacteria are able to establish a non-fecal environmental source of these bacteria (Byappanahalli et al., 1998). In addition, samples for the analysis of bacterial indicator cannot be stored for a long time; this is a difficulty especially in developing countries where analytical laboratories are scarce (Isobe et al., 2002).

As an alternative or an additional indicator of fecal contamination, fecal sterols have been attracting the interest of researchers. Sterols and related compounds have been widely used to detect sewage pollution, with most work having focused on coprostanol. Murtaugh and Bunch (1967) first recognized the potential usefulness of coprostanol as chemical indicator for fecal pollution. Fecal sterols are found in considerable amount in human and higher animal feces. And in human feces approximately 60% of fecal sterols is coprostanol (Leeming et al., 1996) In aquatic environments coprostanol is microbially degraded under aerobic conditions, while it is known to be stable under anaerobic condition. Coprostanol is hydrophobic and can easily be associated with particulate matter in sewage and water columns, and it has been reported that more than 80% of sterols existed in the particulate phase (Isobe et al., 2002). That is, coprostanol can be preserved for a long time under anaerobic condition (Isobe et al., 2002). The presence of coprostanol in environment is only significant from human and higher animal feces. Thus coprostanol has been proposed as a powerful molecular marker for fecal pollution monitoring and has been applied to several environmental matrices. Furthermore, concurrent measurement of coprostanol together with structurally related sterols can give useful information for source identification of fecal pollution (Leeming et al., 1996). This is another advantage of chemical versus microbial indicators and particularly important in monitoring in the tropical Asian region where human and domestic animals often share residences.

In this study, the concentrations of fecal sterols (coprostanol and cholestanol) and coprostanol ratio in sediment of nine selected water bodies and an urban wastewater treatment plant (WWTP) in Bangkok and its surrounding areas were studied.

2. Materials and Methods

2.1 Chemicals

The following chemicals were used in this study. As standard, dichloromethane (DCM) HPLC grade (J. T. Backer); methanol (MeOH) , HPLC grade (J. T. Backer); 5α – cholestane as the internal standard (Sigma); 5β -cholestan - 3β - ol or coprostanol (Sigma); cholest - 5 - en -3β – ol, or cholesterol (Sigma); 5α -Cholestan - 3β – ol, or cholestanol (Sigma). The standard solutions with concentrations around 100 mg/l were prepared by dissolving fecal sterol compounds into dichloromethane.

2.2 Site selection and sample collection

Sediment samples were collected from nine water bodies as shown in Figure 1. Selected sampling sites covered both urban and suburban areas, where fecal contaminations were suspected varying at different levels are as follows:

- Central Bangkok: (1) Klong Phra khanong, (2) Klong Ong Ang, (3) Klong Bang Lamphu, (4) Klong Saen saep
- Suburban area: (1) Klong Nueng,

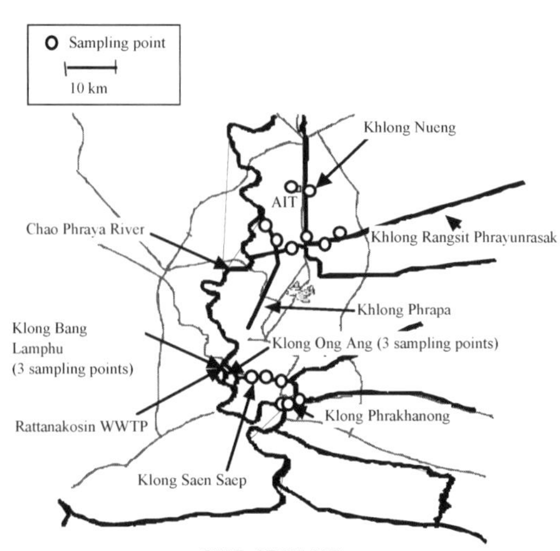

Figure 1. Sampling Site

(2) Klong Rangsit Prayunrasak
-Others: (1) Upper stream of Chao Phraya River, (2) Klong Prapa (drinking water source),
(3) Pond in AIT campus

Here, "Klong" means canals or artificial waterways in Thai. In this study, 47 surface sediment samples in the canals and the river above were collected in June 2004. The samples were transported to the laboratory within 5 hours, where they were stored at - 30^{0}C until analysis, normally within 3 days. All glassware were immersed in DCM 24 hours and baked in the oven prior to use.

The suspended solids in the influent and effluent of the primary sedimentation tank of Rattanakosin WWTP were also collected in June 2004, and fecal sterols in the suspended solids were analyzed.

2.3 Fecal sterols analysis

As it has been reported that significant part of fecal sterols stay in association with sediments or suspended particles rather than in dissolved form in water, in the present study, fecal sterols in dissolved form were not analyzed (Isobe et al., 2002).

The analytical procedure of fecal sterols is essentially composed of sample drying, extraction, saponification, clean-up, derivatization, and determination by GC or GC/MS (Sundin et al., 1999; Seguel et al., 2001; Isobe et al., 2002). In this study, as the sample drying method, freeze drying method has been adopted widely, but as a freeze dryer was not available, samples were dried at room temperature. Room temperature drying under aerobic condition could cause biological degradation of fecal sterols, but has been employed by Seguel et al. (2001). The steps for saponification was omitted, as esterified fecal sterols are reported to be rather minor fraction contributing less than 20% of total (McCalley et al., 1981). As the extraction method, the method reported by Isobe et al. (2002) was adopted, which employs successive extraction with MeOH, MeOH/DCM (1:1 v/v), and DCM. Clean-up procedure was omitted but the purity of the peaks was checked by carefully checking the mass spectra. For derivatization, trimethylsilylation with Bis (trimethylsilyl) trifluoroacetamide (BSTFA) was first tried and found did not work effectively most probably because of inteference from humidity of samples. Based on our observation that even without derivatization enough sensitivity was obtained and the peaks seemed to be separated well for quantification, the authors decided to omit derivatization. The method finally adopted is as described below.

A weighed sediment sample (0.2-4 g-dry) was dried under room temperature while paying care so that the drying can be completed rapidly, and placed in a 50-mL glass centrifuge tube. Here, due care was paid to reduce the time for drying, as fecal sterols may be biologically degraded while drying as the condition is aerobic. Into the tube, 30 mL of MeOH was added and fecal sterols were extracted for 10 min under ultrasonication. Then, the sediment and the extract were separated by centrifugation at 3700rpm for 7min, and the extract phase was collected in a 200-ml pear-shaped flask. The same extraction procedure was repeated twice from the residual sediment sample in the tube but first time with 30ml MeOH/DCM (dichloromethane) (1:1, v/v) and the second time 30ml DCM. All the extracts were collected and combined into the same 200-ml pear-shaped flask. Then the combined extract was concentrated to about 2ml using a rotary evaporator at 35^{0}C with applying adequate vacuum. To the concentrated extract, a 50μl of 50mg/l 5α-cholestane was added as the internal standard and centrifuged at 12000 rpm for 5 min prior to GC/MS analysis. The GC/MS analytical condition was as follows. The GC/MS machine employed was GC-17A, GC/MS-QP5050 (Shimadzu, Japan). A fused silica capillary column CP SIL 8 CB, 50m, 0.25 mm i.d.

and 0.12μm film thickness (Varian-Chrompack) was employed. Helium was used as carrier gas at 123 kPa and the flow rate 1 ml/min. The injection was by the splitless mode at a temperature of 280°C with a 1μl injection volume. The GC oven program was; 100°C 1 min, then to 280°C at a rate of 20°C/min and then to 310°C at a rate of 2°C/min. The mass spectrometer was operated under the selective ion-monitoring (SIM) mode after initial solvent delay of 8 min. The selected ions for SIM mode analyses are listed in Table 1. Each sample was analyzed under SIM mode for quantification and under SCAN mode for the check of the purity of the peak.

Table1: Target compounds and selective major ion fragments

Sterol	Common name	Selective major ion fragment [a] (m/z)
5α – Cholestane*	5α - Cholestane	<u>217.20</u>, 357.40, 372.45
5β-Cholestan-3β-ol	Coprostanol	<u>233.25</u>, 373.40, 355.30, 388.50
Cholest-5-en-3β-ol	Cholesterol	<u>386.40</u>, 275.35, 301.35, 353.40
5α-Cholestan-3β-ol	Cholestanol	<u>215.20</u>, 233.20, 373.40, 388.45

[a] The underlined ions were used for quantification (SIM mode)
* 5α – Cholestane was used as the internal standard.

Recovery was tested by analyzing environmental samples with and without the addition of known amount of targets before drying. The obtained average recoveries of the 3 sterols were from 81%, 76% and 87% for coprostanol, cholesterol and cholestanol respectively. The detection limit signal / noise = 3 (S/N = 3) for individual sterols by the current SIM mode were found to be 4.0 ng/L, 1.6 ng/L, and 2.8 ng/L in the DCM solution to be injected to GC/MS for coprostanol, cholesterol and colestanol respectively.

3. Results and discussion

3.1 Purity of the peaks

As the clean-up step was omitted in this study, the purity of the peaks must carefully be checked. Examples of the total ion chromatograms obtained under scan mode analyses are shown in Figure 2. As can be seen, distinct peaks were observed for the internal standard, coprostanol and cholesterol except for the sample from the pond in AIT campus. The mass spectra of coprostanol and cholesterol are shown in Figures 3, 4 and 5. The mass spectra of the coprostanol and cholestanol from the samples were almost identical to that from standard solution. The mass spectra of cholesterol from the samples were also found to be similar to that from standard solution, except that some of the mass spectra had ions of m/z=430 and 316, indicating that the peaks were overlapped with other unknown chemicals. While the ion intensity ratios of the monitored ions in SIM mode analyses were found to be more or less constant, it was indicated that the peaks of cholesterol is not pure and thus there is a fear of over estimation of it concentration.

3.2 Determined sterol concentrations

The determined concentrations of the fecal sterols in the sediments and the suspended solids samples are shown in Figure 5. All of the examined water bodies, except for AIT pond and the samples from the Rattanakosin WWTP gave fluctuation of coprostanol and cholesterol concentrations ranging around one order or more. One of the cause of the fluctuation must be that the samples were taken at multiple locations in the same water body.

Suspended solids in the influent to and effluent from the primary sedimentation tank of the Rattanakosin WWTP had 69μg/g-dry and 250μg/g-dry of coprostanol, 37μg/g-dry and 147μg/g-dry of cholesterol, and 95 μg/g-dry and 64 μg/g-dry of cholestanol respectively. Unfortunately, the influent and the effluent samples were taken on different days, and thus the comparison of these values is not easy.

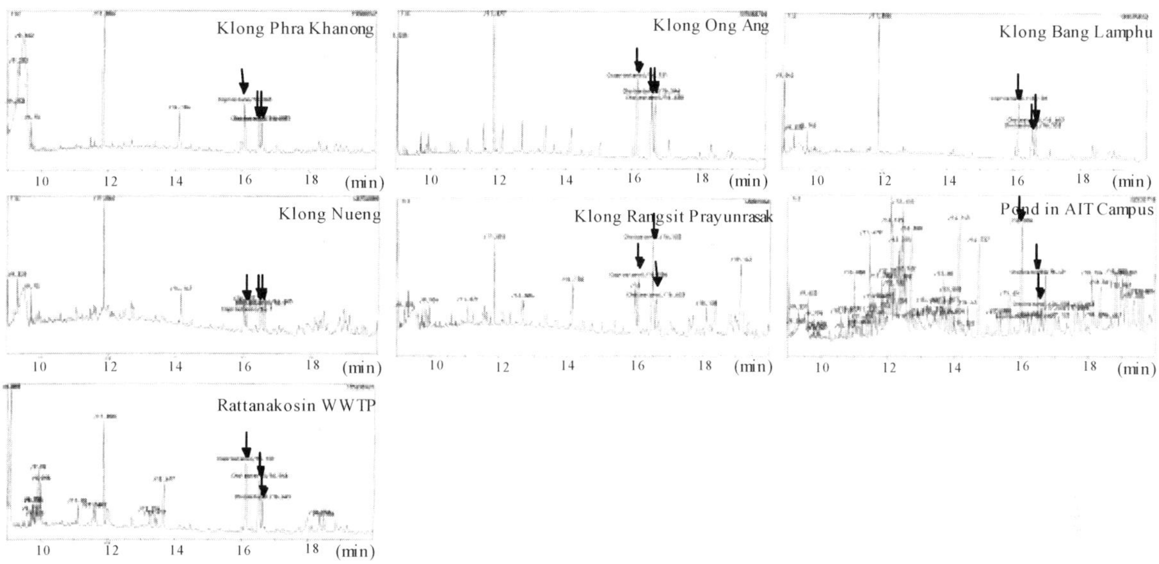

Figure 2. TIC Chromatograms of the samples obtained under SCAN mode analyses. (Peaks highlighted with arrows are coprostanol, ~16.1, cholesterol, ~16.5 min, and cholestanol, ~16.5min)

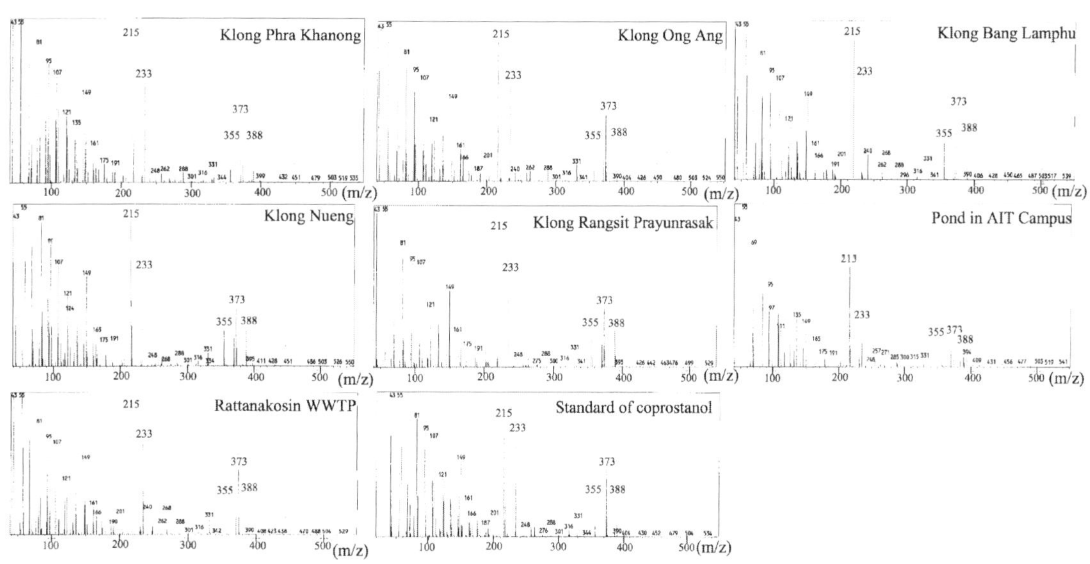

Figure 3. Mass spectra of coprostanol.

Isobc ct al. (2002) reported that influent into an urban wastewater treatment plant in Tokyo contains around 300μg/L of coprostanol. Suppose that influent suspended solids is in the order of 200mg/L, then, the coprostanol concentration in the influent suspended solids is estimated to be 1,500μg/g-dry. The coprostanol content in the sewage in Rattanakosin WWTP was found to have only around one tenth of that in Tokyo. One but the most influential cause for the difference is that feces are collected as sewage in Tokyo whereas feces are intercepted by septic tanks in Bangkok. But the detection of fecal sterols in the sewage in Rattanakosin WWTP indicates that some parts of feces are collected as sewage.

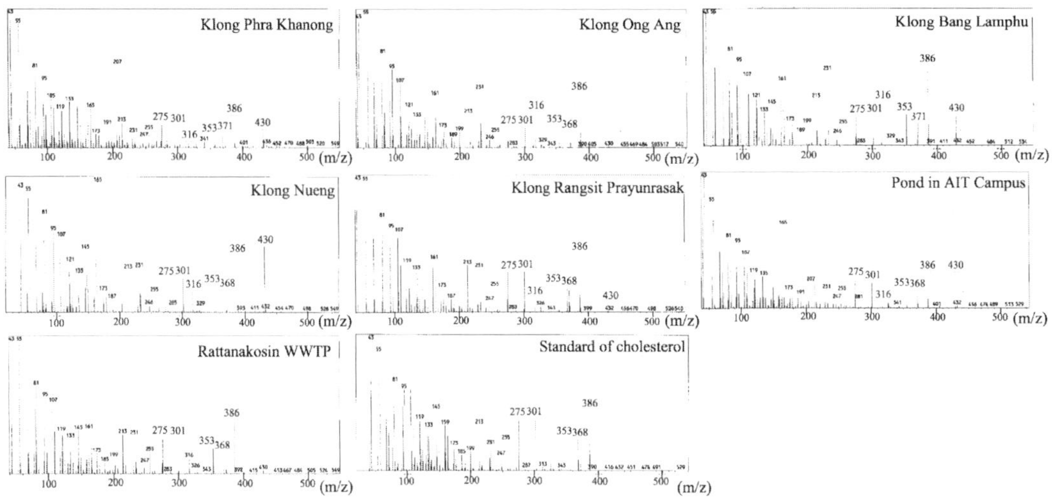

Figure 4. Mass spectra of cholesterol.

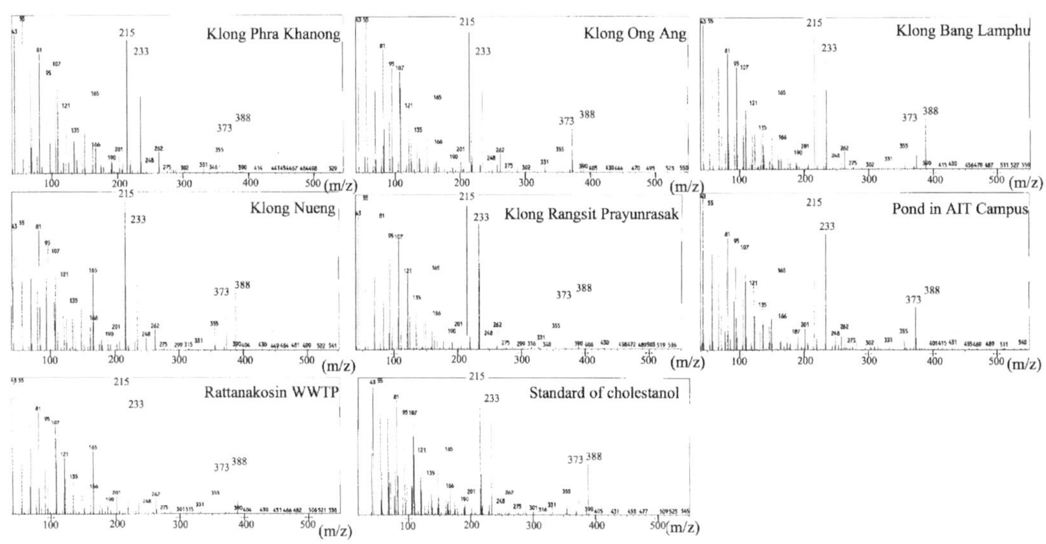

Figure 5. Mass spectra of cholestanol.

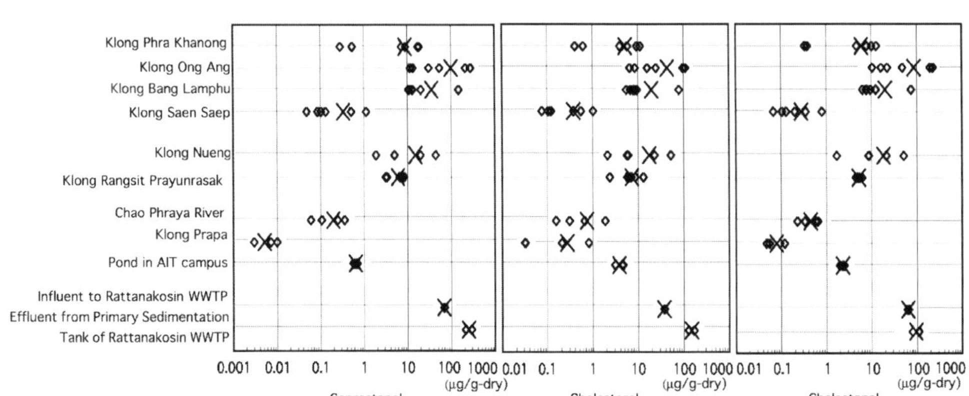

Figure 6. Concentrations of coprostanol,cholesterol and cholestanol in the samples. (Open boxes are for the samples, and the crosses are the average for the water bodies)

Of the water bodies examined, except for the AIT pond and the Rattanakosin WWTP, Klong Ong Ang gave the highest concentrations for both coprostanol and cholesterol, then followed by Klong Bang Lamphu, Klong Nueng, Klong Phra Khanong, Klong Rangsit Prayunrasak. Klong Rangsit Prayunrasak was followed by Klong Sean Saep and Chao Phraya River followed, and then by Klong Prapa. The concentration levels of Chao Phraya River and Klong Prapa could be thought to be background level. Klong Prapa is for the water conveyance from Chao Phraya River for drinking water supply, and it was good to find the lowest concentration of coprostanols in it. Some of the samples from Klong Ong Ang and Klong Bang Lamphu gave as high concentration as that in Rattanakosin WWTP. These canals may be severely affected by sewage, or could be affected by feces directly discharged by some means into these water bodies. Suburban canals, Klong Nueng and Klong Rangsit Prayunrasak were found to have significantly higher concentrations than those in Chao Phraya River and Klong Prapa. These canals may also be affected by fecal contamination. In the field observation, on the bank of these canals, many households were found with septic tanks installed in and along the canal bed.

3.3 Fecal sterol ratio

Fecal sterol concentrations in sediments are known to be affected by the nature of the sediment, such as particle size distribution and organic contents. To overcome such effect, Grimalt et al. (1990) in troduced the stanol ratio of $5\beta/(5\beta + 5\alpha)$, or the ration of coprostanol to the sum of coprostanol and cholestanol. They proposed that the value being higher than 0.7 is a positive indicator for fecal contamination, and less than 0.3 as background level.

The ratios $5\beta/(5\beta + 5\alpha)$ were calculated for the samples examined in this study, and are presented in Figure 7. The effluent from the primary sedimentation tank of Rattanakosin WWTP had the highest value of 0.72, whereas that of influent into the primary sedimentation tank was 0.52. Most of the samples from Chao Phraya River, Klong Prapa, and AIT pond had the values of less than 0.3, indicating these water bodies are less contaminated by feces. All other samples gave the ratio between 0.3 and 0.7. Isobe et al. (2002) studied fecal sterol concentrations in the water environment of Malaysia and Vietnam, and even in supposedly contaminated site the maximum ratio fell at 0.54 to 0.59. They suggested that the criteria value of 0.7 proposed by Grimalt et al. (1990) may be too high for tropical climate countries. Similar observation has been reported by Chan et al. (1998). Considering the result that influent into the primary sedimentation tank of Rattanakosin WWTP had the ratio of 0.52, as the criteria, it seems more rational to adopt the value of 0.5 for severe contamination in the case of Bangkok. But to set up such criteria, more careful study would be needed. At least, it seems that the ratio of $5\beta/(5\beta + 5\alpha)$ and coprostanol concentrations are reflecting the degree of fecal contamination of the studied water bodies.

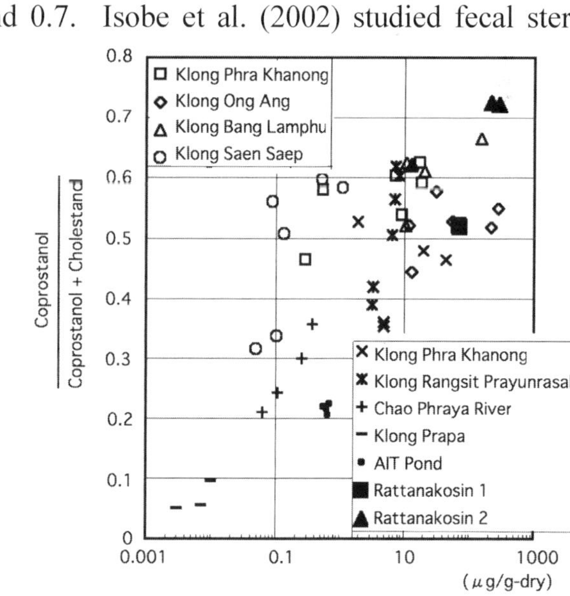

Figure 7. Fecal sterol ratios. (Rattanakosin 1 and 2 are for the influent to and effluent from the primary sedimentation tank of Rattanakosin WWTP)

4. Concluding remark

In Thailand, feces are thought to be trapped and treated by septic tanks, which are installed in all buildings. But grasping the actual situation of their efficiencies is not easy. Indeed, in the present study, while the degree of contamination cannot be mentioned, fecal contamination was found in many of water bodies. Fecal sterols could be a useful indicator as an aid for the management of water bodies, as it can detect leak of feces to water environment. Also, in the case of Bangkok, it could be used in the transition from the current system in the coexistence of septic tanks with sewage system to the elimination of septic tanks in assessing the amount of feces discharged to sewage.

5. Acknowledgements

The authors would like to thank to SACWET for the financial support. Sincere gratitude is addressed to Environmental Research and Training Centre (ERTC) of Thailand for their kind support of GC/MS. The authors would also like to express their appreciation to Ms. S.Phunsiri and other laboratory staffs in EEM, AIT for their technical and logistical support.

6. References

Byappanahalli, M. N., Fujioka, R.S. (1998). Evidence that tropical soil environment can support the growth of escherichia coli. *Wat. Sci. Tech.*, 38 (12), 171-174.

Chan, K. H., lam, M. H. W., Poon, K. F., Yeung, H. Y., Chiu, T. K. T. (1998). Application of sedimentary fecal stanols and sterols in tracing sewage pollution in coastal waters. *Wat. Res.*, 32 (01), 225-235.

Grimalt, J. O., Fernandez, P., Bayona, J. and Albaiges, J. (1990). Assessment of fecal sterols and ketones as indicators of urban sewage inputs to coastal waters. *Environ. Sci. Technol.*, 24, 209-211.

Isobe, K. O., Tarao, M., Zakaria, M. P., Chiem, N. H., Minh, L. Y., Takada, H. (2002). Quantitative application of fecal sterols using Gas Chromatography-Mass spectrometry to investigate fecal pollution in tropical waters: Western Malaysia and Mekong Delta, Vietnam. *Environ. Sci. Technol.*, 36 (21), 4497-4507.

Leeming, R., Ball, A., Ashbolt, N., Nichols, P. (1996). Using faecal sterols from humans and animals to distinguish faecal pollution in receiving waters. *Wat. Res.*, 30 (12), 2893-2900.

McCalley , V.M., Cooke, M., Nickless, G. (1981) Effect of sewage treatment on fecal sterols. *Wat. Res.*, 15, 1019-25.

Seguel, C. G., Mudge, S. M., Salgado, C., Toledo, M. (2001). Tracing sewage in the marine environment: Altered signatures in Concepcion bay, Chile. *Wat. Res.*, 35 (17), 4166-4174.

Sundin, K. A., Leeming, R. L., Stenstrom, T. A. B. (1999). Degradation of fecal sterols in urine for assessment of fecal cross-contamination in source-separated human urine and urine storage tank sediment. *Wat. Res.*, 33 (09), 1975-1980.

Section 2

Groundwater: Pollution and Management

Groundwater Protection Zoning by Groundwater Vulnerability Mapping in the Lower Central Plain, Thailand

Somkid Buapeng[1)] and Panchote Kulvanit[2)]

1) Bureau of Groundwater Conservation and Restoration
Department of Groundwater Resources, Bangkok, 10400, Thailand
E-mail address: somkid@dgr.go.th
2) Metrix Associates Company, Limited
177 Rama VI Rd., Rong Muang, Pathumwan, Bangkok, 10330, Thailand
E-mail address: pku@metrixassociates.com

Abstract

Due to rapid economic development, careless land use, and unconcerned groundwater users in the lower central plain, Thailand, where most of exported rice is cultivated, major industrial goods are produced, and capital city is located, groundwater in this region is highly possible to be contaminated from community, agriculture, and industry activities. In order to conserve and sustain this water resource in this area, groundwater quality protection policy and planning must be set. Groundwater protection zoning which indicates priority of areas needed to be protected is a basis for such planning. To define which area should be first concerned and protected, factors effecting groundwater contamination include groundwater vulnerability and possible contamination load must be studied. Therefore, DRASTIC, vulnerability assessment method, developed by U.S.EPA, was applied for this study. The scale of some parameters i.e., A (aquifer media) and I (impact of vadose zone media) were adjusted according to the specific condition of the study area and available data. The groundwater vulnerability was assessed in form of 50x50 m^2 grid cell and shown in Geographic Information System (GIS). Then DRASTIC index of each grid cell was processed to each subdistrict. The result of vulnerability assessment shows that there is high groundwater vulnerability in the western part of the study area (part of Nakornpathom, part of Karnjanaburi, and part of Supanburi province) compared to other parts. Most significant factors effecting groundwater vulnerability in this region are depth to groundwater (D) and impact of vadose zone media (I). Vulnerability assessment result of each sub district was presented on map and GIS showing 3 levels of groundwater vulnerability and it then be used as groundwater protection map indicating protection area levels.

Keywords: Groundwater protection zone, Groundwater vulnerability, Groundwater quality protection, DRASTIC, The Chao Phraya river basin

1. Introduction

Groundwater is considered as one of the most important natural resources since it provides high quality drinking water at low price to meet the needs of community, agriculture, and industries. Effective groundwater protection is the most appropriate method to obtain high quality of this raw water source and ensure an availability of supply in order to response to the needs arisen from agricultural and industrial sectors. Previously, people did not pay much attention to the groundwater. Also, no effort has been contributed to the improvement of groundwater quality. Therefore, it is necessary to find an effective practical measure to maintain and sustain high-quality groundwater supplies.

Accordingly, The Study of Groundwater Quality Protection in the Central Plain is considered as a crucial step in developing guidelines and measures to guarantee the quality of groundwater based on the standard as well as setting up a Groundwater Protection Zone. The main objectives of such guidelines and measures are to control land use and monitor users' activities that will be taken place in order to ensure that the quality of groundwater will not be affected and contaminated by these unaware behaviors. Such measures try to tackle problem and its root causes, not the end-of-pipe solution likes groundwater treatment. Undoubtedly, at present, there is a requirement for a clear and applicable standard measures and well-organized groundwater protection zone.

2. Area description

Under the Study of Groundwater Quality Protection in the Central Plain, the study area covers 18,332 km^2 of the lower Chao Phraya River basin, from Chainat Province to the Gulf of Thailand, as shown in Figure 1. The study area covers Bangkok, Samut Prakan, Samut Sakhon, Nonthaburi, Pathum Thani, Nakhon Pathom, Phra Nakhon Si Ayutthaya, Suphan Buri, Ang Thong, and Sing Buri province, including some area of Chai Nat, Nakhon Nayok, Chachoengsao, Samut Songkhram, Ratchaburi, and Kanchanaburi province.

Topography: The topography of the Central Plain is flood plain, comprising thick sedimentation layers, with undulating terrains and monadnocks in some parts. In the past, the majority of the area was below the sea level. Later on, the sea level descended while the ground level ascended. This area was also affected by the force of many rivers, which were the Chao Phraya River and some other main rivers, namely the Noi River, Suphan Buri River, Pa Sak River, Lop Buri River, and Bangpakong River. These rivers flowed from surrounding highlands to the Gulf of Thailand, forming a flood plain, which is two metres above MSL at Bangkok. The flood plain has an incline slope toward the north with an elevation of four meters above MSL in Phra Nakhon Si Ayutthaya and 20 metres above MSL in Chainat. The boundaries of the lower plain area in central region are connected to: Mae Klong Basin to the west; the Korat Plateau to the east (separated by Phetchabune Mountain Range); the upper plain area in central region to the north (separated by a small mountain range in Nakhon Sawan and Chainat); and the Gulf of Thailand to the south.

Most of the area is used for agriculture, accounting for 16,244 km^2 or 88.61% of the study area. The study area is utilized for rice paddy field, which is cultivated more than twice a year using irrigated water. In the areas without irrigation, however, cultivation takes place only in rainy season. Sugar cane and corn crops are also found in the area. Apart from agriculture, forest accounts for four percent of the study area. High-density population areas are found in Bangkok, lower part of Pathum Thani, and middle part of Samut Prakan and make up five percent of the study area.

Geology and hydro-geology: There are two types of aquifers in the lower central plain: unconsolidated rocks, which originate from river and marine deposits with alternate sand/gravel deposits and clay; and consolidated rocks, where water is stored in the voids or cracks of limestone from Permian period, which contains alternate sandstones and shale. In this study, the aquifer is divided into 4 parts within the depth of 200 m. from the ground level. The criteria used are based on the characteristic of the rock bed and the sand/gravel deposits layer expanding vertically and horizontally at different levels including transmissivity and storage coefficient of the aquifer. Characteristics of each aquifer are as follows:

The first aquifer is found approximately 30 – 50 m. from the ground level with a thickness of approximately 50 m. It consists of sand sediments and gravel that has poor to good grading. This aquifer is connected to Bangkok Aquifer in the southern part of the study area, from Phra Nakhon Si Ayutthaya to the Gulf of Thailand.

The second aquifer lies at the depth of some 60 – 80 m. from the ground level with a thickness of 20 – 50 m. It consists of sand and gravel with intervention of thin clay layer. This aquifer is connected to Phra Pradaeng Aquifer in the southern part of the study area, located at the depth of 60 – 120 m. from the ground level. This aquifer has high storage capacity and the groundwater quality from this aquifer includes good, brackish, and salty.The third aquifer is found at the depth of 100 – 140 m. from the ground level with a thickness of approximately

50– 70 m. This aquifer consists of sand and gravel layers that have mediumsegregation together with thin layers of clay. This aquifer is connected to Nakhon Luang Aquifer in the southern part of the study area, located at the depth of 120 – 180 m. from the ground level. This aquifer provides good amount of water with good quality except for the groundwater in the southern part of Thonburi and Bangkok which is brackish or salty.

The fourth aquifer lies at the depth of 150 – 200 m. from the ground level with a thickness of 30 – 70 m. This aquifer consists of sand and gravel layers together with thin layers of clay providing good amount of water with good quality. This aquifer is connected to Nonthaburi Aquifer in the southern part of the study area, located at the depth of 180 – 280 m. from the ground level.

Groundwater Quality : The groundwater quality in the study area is analyzed by means of groundwater maps of each province; related reports; groundwater well records; groundwater quality data from the Division of Agricultural Hazardous Substances; and additional field surveys. In general, it is found that groundwater quality in the central region is in good standard except in the lower part of the study areas where the groundwater is brackish or briny. These areas include Sing Buri and Suphan Buri down to the river mouth areas covering Bangkok, Samut Prakan, and Chachoengsao. However, the saline groundwater sources are limited to only certain areas and found only in shallow aquifers but moreover in middle and deep aquifers such as those in Bangkok, yield fresh water too. In addition, iron and manganese concentrations in groundwater are high in the upper part of the study area (e.g.Suphan Buri, Ang Thong and Chai Nat) and some areas in the lower part of the study area.

3. Methods

The vulnerability mapping in this study use DRASTIC concept and GIS (Geographic Information System) techniques. DRASTIC index was developed by Aller, L. *et al.* (1987) for the U.S. Environmental Protection Agency to be one of the standard methods for evaluating groundwater vulnerability to pollution. As shown in table 1, 7 hydro-geological factors will be considered in this method. Each of the parameter is classified into ranges and assigned a subjective rating (r). Weight multiplier (w) is then used for each factor to balance and enhance its importance.

GIS techniques are applied to manage all data in raster format with 50 x 50 sq.m.-grid cells. The prepared 7 weighted hydro-geological was assigned to each grid. Then the DRASTIC index can be computed from equation 1. The resulting DRASTIC index represents a relative measure of groundwater vulnerability. The higher DRASTIC index is greater relative to pollution potential. Figure 2 shows the parameters of DRASTIC.

$$D_rD_w + R_rR_w + A_rA_w + S_rS_w + T_rT_w + I_rI_w + C_rC_w = \text{DRASTIC Index} \quad (1)$$

The hydro-geological parameters in equation 1 are described below:

Depth to Aquifer (D): Depth to Aquifer determines the thickness of the material through which contaminated water must travel before it reaches groundwater. In general, deeper aquifer usually means longer travel time of pollutants. This parameter was considered from type of aquifer, water level, accompanying with geological characteristic and human activity effecting top soil such as river and sand mining, respectively.

Net Recharge (R): Net Recharge is the amount of water per unit area that infiltrates the ground surface and reaches the aquifer. The net recharge was calculated by multiplying the amount of rainfall in each area and infiltration rate, based on the soil map of the Land Development Department.

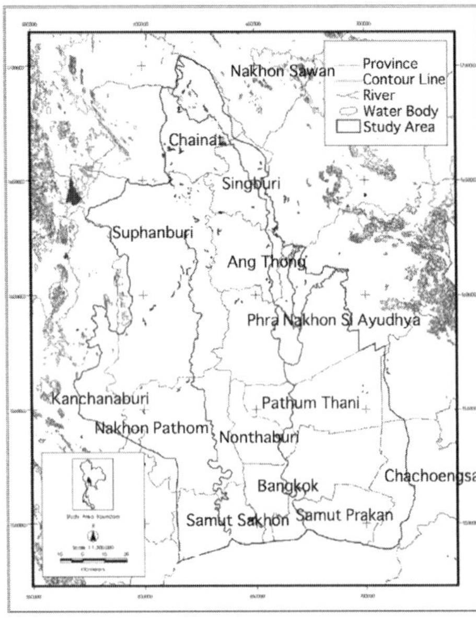

Figure 1. Study Area – Lower Chao Phraya basin

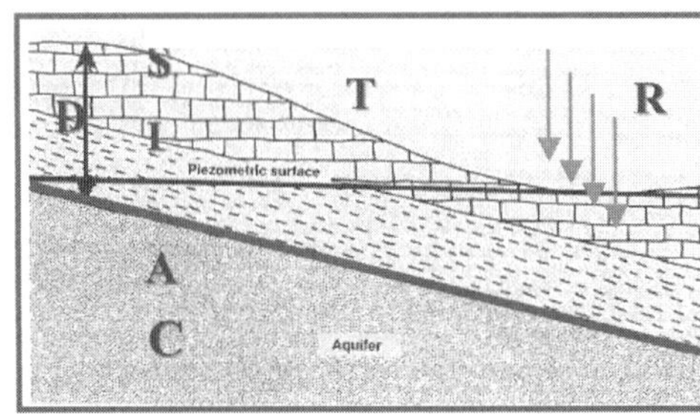

Figure 2: Parameters of DRASTIC

Aquifer Media (A): Type of aquifer in the study area, consolidated rocks or unconsolidated rocks, was classified. The larger grain size and the more fractures, the permeability, can tell the vulnerability of the aquifer. Information on aquifer media was obtained from the study of hydrogeology, data of groundwater exploration of 733 wells and hydrogeology map (Department of Groundwater Resource).

Soil Media (S): Soil is commonly considered the upper weathered zone of the earth. The presence of fine-textured and/or non-swelled materials can decrease relative soil permeability and restrict contaminant migration. Soil data was obtained from soil map, conducted by the Land Development Department.

Topography (T): The topography of the land affects groundwater vulnerability because the slope of the land is an important factor in determining whether the contaminant released will become run-off or infiltrate the aquifer. With a low slope, the contaminant is less likely to become run-off and therefore more likely to infiltrate the aquifer.

Impact of Vadose Zone Media (I): The vadose zone is defined as the zone above the water table which is unsaturated. The texture of the vadose zone determines how long the contaminant will travel through it which can be assessed by averaging. Source of data includes log data (Department of Groundwater Resource) and hydrogeology map.

Hydraulic Conductivity of the Aquifer (C): Hydraulic conductivity refers to the ability of the aquifer material to transmit water. The Hydraulic conductivity is positively correlated with the vulnerability rating. Data is available from hydrogeology map and data of groundwater exploration (Department of Groundwater Resource)

The weighting factor of DRASTIC parameters that been used in this study is shown in Table 1. The factor scale is adjusted to satisfy the hydro-geologic condition of the study area that some parameters data ranges is small (ie, aquifer media (A) and impact of vadose zone media (I).

4. Results and discussion

Vulnerability Mapping: As shown in Figure 3 and 4, the result of the study indicates that both Depth to Aquifer (D) and Impact of Vadose Zone Media (I) are the most important factors, controlling the degree of aquifer vulnerability in the study area. This finding implies that in the southern part of the study area, in which the top layer is mostly contained with marine clay, has aquifer vulnerability lower than the north and the west. Moreover, the relatively deep aquifer found in Samut Sakhorn, Samut Prakarn, Bangkok, Nonthaburi and some parts of Pathumthani result these areas the least vulnerable. On the other hand, the study found that the study area in Nakhon Pathom, especially in Tambon Huay Kwang, Thungbua, Thunglooknok, Gumphangsean, Lumhuy, Huayphra, Donthum, Nonggyuleum in Muang District, are the most vulnerable due to the spreading sand mining in these areas.

In summary, based on the study and evaluation, aquifer vulnerability is directly related to geological and hydrogeological factors. There is a slight vulnerability in the groundwater in the lower part of the study area, in which the top layer in covered by marine clay and is regarded as the confined aquifer. In contrast, the study found that the area in the upper part, such as in Ang Thong, Sing Buri, and Chai Nat province, have higher degree in aquifer vulnerability due to the semi-confined aquifer and the thickness of the sand layer. Furthermore, the high level of aquifer vulnerability could be resulted from the removal of soil for sand mining. This case is happened in some parts of Nakhon Pathom, Kanchanaburi and southwest of Suphan Buri province.

Groundwater Protection Zoning: Groundwater quality protection can be done in many levels depending on area size and protection objectives.

The aquifer protection prevents groundwater from pollution sources on the whole. The emphasis is made on the areas that are prone to groundwater contamination. At the same time, groundwater quality monitoring programmed is undertaken to evaluate the situation such that mitigation can be done promptly should problems arise. The results of this study will lead to regional and national guidelines and policies of groundwater resource conservation as well as groundwater source conservation.

Table 1. Weighting factor of hydro-geologic parameter used in this study

Parameter	Data Range	Weight
D-Depth to Aquifer (Dw = 5)	0-1.5	10
	1.5-4.5	9
	4.5-9.0	7
	9.0-15.0	5
	15.0-22.5	3
	22.5-30.0	2
	>30.0	1
R-Net Recharge (Rw = 4)	0-22.5	0.45
	22.5-45.0	0.9
	45.0-67.5	1.8
	67.5-90	2.7
	90-112.5	3.6
	112.5-135	4.5
	135-157.5	5.4
A-Aquifer Media (Aw = 3)	Sand / Clay	5
	Sand	6
	Sand / Gravel	7
	Gravel / Sand	8
	Gravel	9
S-Soil Media (Sw = 2)	Thin or Absent	10
	Gravel	10
	Sand	9
	Peat	8
	Aggregated Clay	7
	Sandy Loam	6
	Loam	5
	Silty Loam	4
	Clay Loam	3
	Muck	2
	Clay	1

Parameter	Data Range	Weight
T-Topography (Tw = 1)	0-2	10
	2-6	9
	6-12	5
	12-18	3
	>18	1
I-Impact of Vadose Zone Media (Iw = 5)	0-0.06	0.2
	0.06-0.11	0.4
	0.11-0.17	0.6
	0.17-0.23	0.8
	0.23-0.28	1
	0.28-0.34	2
C-Hydraulic Conductivity of the Aquifer (Cw = 3)	0-10	1
	10-30	2
	30-70	4
	70-100	6
	100-200	8
	200+	10

The process of setting aquifer prevention and monitoring guideline started with assessing aquifer vulnerability to pollution in order to establish a groundwater protection area according to its vulnerability level. The result was then considered together with current activities and their associated risks to the area, leading to establishment of acceptable/unacceptable activities for particular groundwater protection area and groundwater monitoring system.

The establishment of the groundwater protection area was based on its vulnerability and employs the boundary of Tambol administration. The vulnerability was used to divide areas into three levels, which were red area, yellow area, and green area. The red area has the highest vulnerability followed by the yellow area and the green area, respectively. It should be noted that these vulnerability levels are comparative within the study area only. The groundwater protection areas of the study area based on Tambol administration are shown in Figure 5.

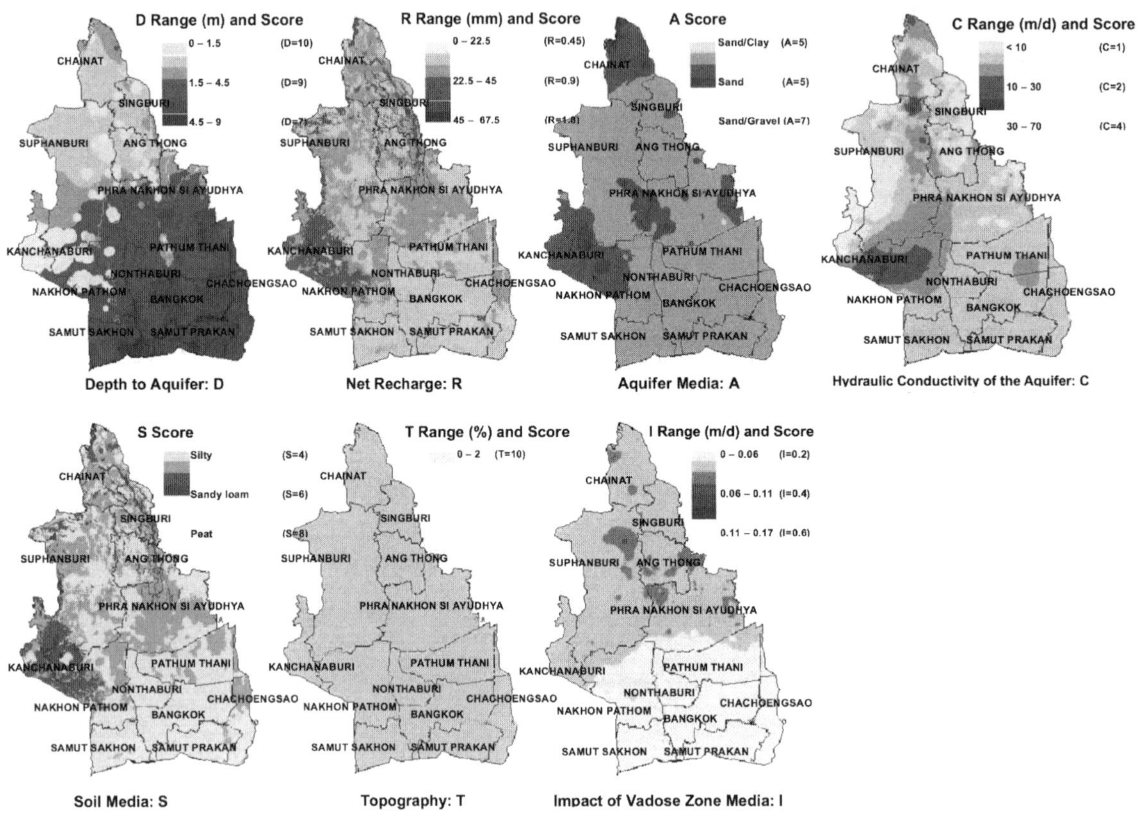

Figure 3. Weighting result of each hydro-geological parameter in DRASTIC

5. Conclusions and recommendations

In this study, the groundwater protection zone was proposed by the means of the ground water vulnerability mapping. The groundwater vulnerability mapping was created by the DRASTIC concept and GIS techniques. The result of vulnerability mapping showed that the southern part of the study area (Nonthaburi, Bangkok, Samut Sakon and Samut Prakan) has aquifer vulnerability lower than the north and the west area (part of Nakornpathom, part of Karnjanaburi, and part of Supanburi province). Most significant factors effecting groundwatervulnerability in this region are depth to groundwater (D) and impact of vadose zone media (I). There are some recommendations for the improvement of ground water protection zoning study

1. Recommendation on groundwater quality protection and assessment methods

In order to evaluate and protect the groundwater quality, it is vital to select and develop an evaluation method as well as a protection measure that are reliable and follow international standards. At the same time, they should be consistent with the area characteristic and the database of the study area. Most importantly, they must be practicable.

2. Recommendation on the database related to groundwater quality protection and evaluation
In order to obtain the best study result, it is necessary to develop relevant database and analysis method. Emphasis should be made on data compilation and processing so that the data from different sources can be linked together systematically and readily used for the future studies. Moreover, additional data collection is recommended for the study area and other areas in the country.

3. Recommendation on practical use of the results from the groundwater quality protection study
A good protection guideline must be practical and efficient in protecting groundwater quality. It must be impartial and suitable to socio-economic condition. Hence, the concept of Polluter-Pay-Principle (PPP) should be employed. Besides, the guideline must be able to be inspected, so that preventive and mitigation measures can be promptly used should the pollution occur. Moreover, the aquifer protection and wellhead protection scheme should be implemented in a holistic way, such that groundwater quality protection can be fully realised.

6. Acknowledgement

The authors greatly appreciate The Department of Groundwater Resources, The Royal Irrigation Department, The Meteorological Department, The Land Development Department and The Department of Environmental Quality Promotion for access to all hydrological and hydro-geological information.

7. References

Aller, L., Bennett, T., Jay, H.L., Rebecca, J.P. and Hackett, G., (1987). *DRASTIC: A Standradized System for Evaluating Ground Water Pollution UsingHydrogeologic Settings*, NWWA/EPA Series, p.455.

Castany, G., Groba, E., and Romijn, E., (1994). Guidebook on Mapping Groundwater Vulnerability, *Internation Association of Hydrologists*, Volume 16, p. 131.

Chalmers. (2003). *Regional Follow-Up Seminar 2003-Groundwater Supply for Urban Area*, Chiangmai, p. 53.

Foster, S., Hirata, R., Gomes, D., D'Elia, M. and Palis, M. (2002). *Groundwater Quality Protection, The World Bank*, GW-MATE Briefing Note 8. p. 103.

Foster, S.S.D. and Skinner, A.C. (1995). *Groundwater Protection: The Science and Practice of Land Surface Zoning, Groundwater Quality: Remediation and Protection.* (Proceeding of the Prague Conference, May 1995), Num.225, pp.471-482

Gupta, A.D. (1997). *Groundwater Protection*, Third Regional Training Workshop on Groundwater Contamination, Cambodia.

Jarupongsakul, S., (1990). *Geomorphology of The Chaophraya Delta, Thailand*, Kyoto University, Japan, p.86.

Khantaprab, C., and Boonnop, N. *Urban Geology of Bangkok Metropolis: A Preliminary Assessment*, Department of geology, Chulalongkorn University, Bangkok, pp. 107-135.

U.S. Environmental Protection Agency. (1991). *Protecting Local Ground-Water Supplies Through Wellhead Protection*, Office of Water.

Zaporozec, A., 2002. *Groundwater Contamination: A Methodological Guide*, Series on Groundwater, Num.2.

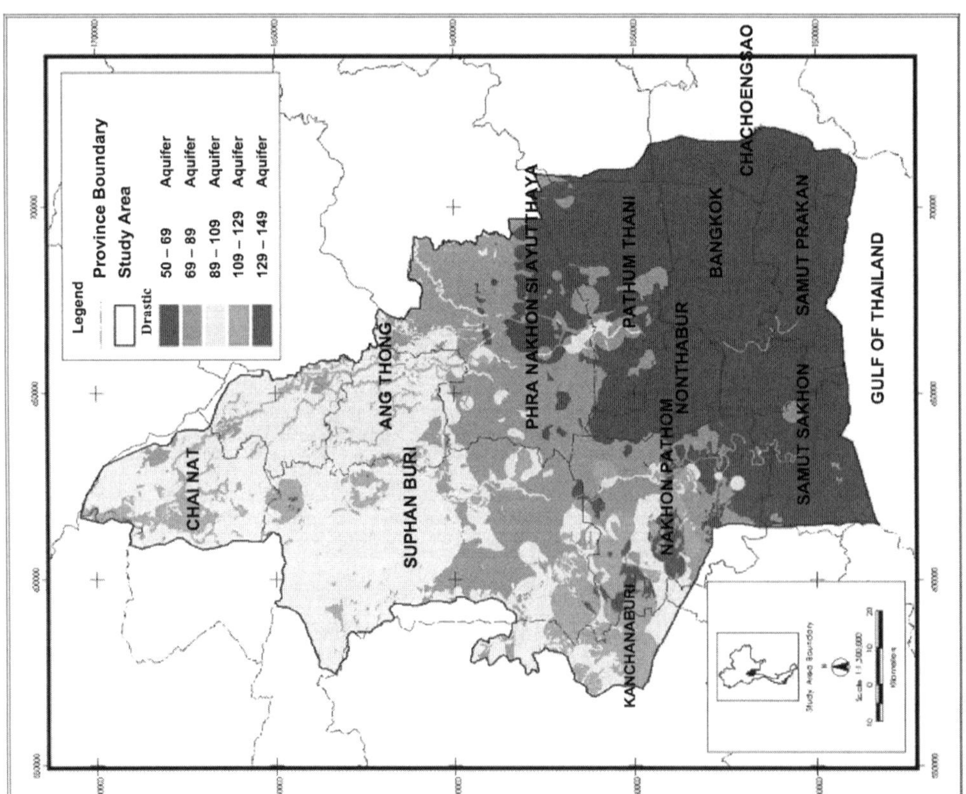

Figure 4. Groundwater vulnerability mapping

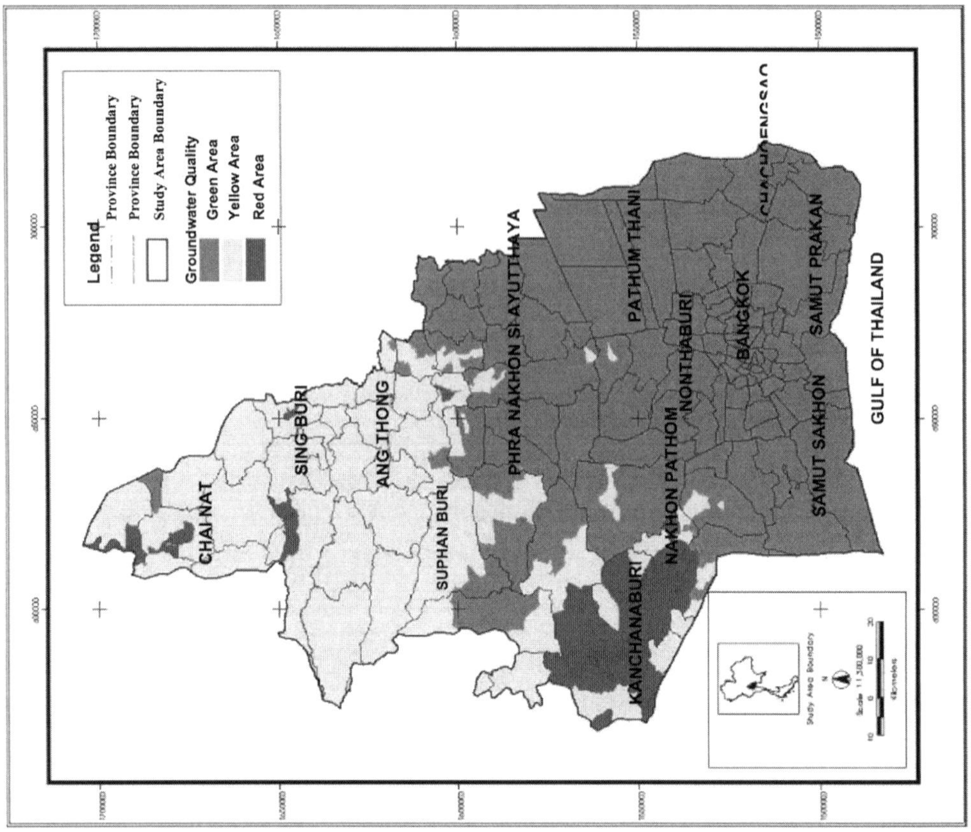

Figure 5. Aquifer protection zone map

Tracing Nitrate Transport in Groundwater Aquifers by Isotope Techniques—a Case Study in the Western Kofu Basin, Japan

Takashi Nakamura [1)], Sangam Shrestha [1)], Hiroshi Satake [2)] and Futaba Kazama [1)]

[1)] Dept of Eco-social System Engineering, University of Yamanashi
4-3-11 Takeda, Kofu, Yamanashi, JAPAN
E-mail: g05de001@ccn.yamanashi.ac.jp; kfutaba@ccn.yamanashi.ac.jp; sangam@ccn.yamanashi.ac.jp
[2)] Faculty of Science, Toyama University, Toyama
3190 Gofuku, Toyama, JAPAN
E-mail: satake@sci.toyama-u.ac.jp

Abstract

The objective of this paper is to identify the source and transport of nitrate in the groundwater aquifers of alluvial fans by stable isotopic techniques. Water samples collected from the precipitation, groundwater, and river water at the western Kofu basin were analysed for the major cations, anions, and isotopes ($\delta^{18}O$ and $\delta^{15}N$). Through the cluster analysis of the cations and anions, two groundwater aquifers formed on the Midai and Kamanashi alluvial fans were identified in the basin. Higher nitrate concentrations were observed in the groundwater aquifers of the Midai alluvial fan, and the values of $\delta^{15}N$ suggest that the major sources of nitrate are the inorganic and organic fertilizers applied in the orchards and paddy fields. The values of $\delta^{18}O$ show that precipitation and river water are sources of recharge in both the aquifers. A strong positive correlation between the nitrate concentration and the $\delta^{18}O$ values of both the groundwater aquifers shows that precipitation water acts as a carrier and river water facilitates dilution in the nitrate loading from terrestrial catchments. The mixing ratio and nitrate concentration were used to develop models to estimate the loading of nitrate into the groundwater aquifers of both the alluvial fans; for the models, the contribution of the precipitation and river water to the recharge was used.

Key words: Nitrate, Stable isotopes, Alluvial fans, Mixing ratio, Groundwater aquifers

1. Introduction

The widespread contamination of groundwater with nitrate is a matter of global concern to both environmentalists and public health officials. A recent study reported that the groundwater aquifer of the Kofu basin was contaminated by nitrate (Kazama and Yoneyama, 2002) because of the excessive fertilizer application in agricultural areas. In the past, very few studies have been carried out to investigate the groundwater quality and groundwater systems and the origin and effect of groundwater on pollution loading in the Kofu basin. Generally, the shallow groundwater in alluvial fans is recharged by two sources, namely, rivers and rainwater (e.g. Mizutani et al., 2001). In particular, rainfall is one of the principle sources of groundwater recharge in alluvial fans, and the infiltrated rainfall is regarded as the carrier of nitrate nitrogen into groundwater systems (e.g. Ogawa et al., 2001). However, in the case of the western Kofu basin, the riverbed elevation is higher than the surrounding area because of the long-term deposition of flow materials. The nitrate concentration and distribution are affected by the behaviour of both the groundwater sources. Therefore, it is very important to know the contribution of precipitation and river water to nitrate loading for the proper management of the groundwater quality of the shallow groundwater aquifers of alluvial fans.

The understanding of the hydrological cycle was markedly improved by analysing and interpreting the hydrogen (δD) and oxygen ($\delta^{18}O$) isotope ratios of ocean water, precipitation, ice cores, surface water, and groundwater. Isotope ratio determinations have also been used to identify the sources of dissolved constituents in precipitation, surface water, and groundwater, such as sulphate ($\delta^{34}S$), nitrate ($\delta^{15}N$), and dissolved inorganic carbon ($\delta^{13}C$), among others. (Ging et al., 1996; Kendall et al., 1997). However, the usefulness of $\delta^{15}N$ alone in identifying these sources is limited by the overlapping ranges of $\delta^{15}N$ values (Kendall and McDonnell, 1998). This is because the isotope ratio also includes the physical (e.g. mixing), chemical (e.g. dissolution of salts), and/or microbial processes (e.g. denitrification, bacterial sulphate reduction) in the hydrosphere (e.g. Rock and Mayer, 2002; Rock and Mayer, 2004). Therefore, in this study, we tried to use both $\delta^{15}N$ (nitrate) and $\delta^{18}O$ (water) with the available indexes to identify the sources and transport of nitrate in the groundwater aquifers of the western Kofu basin, Japan.

2. Materials and methods

2.1 Description of study area

The Kofu basin, located in the central part of Japan (Figure1), covers an area of approximately 100 km^2. The study area lies from 35°34' to 35°41'N latitudes and from 138°32' to 138°24'E longitudes, and the altitude varies from 200 to 500 m. The average annual precipitation is 1425 mm in the upper part of the basin. This area is a compound alluvial fan formed by two rivers, namely, the Kamanashi and Midai Rivers. The riverbed elevations are higher than these alluvial fans.

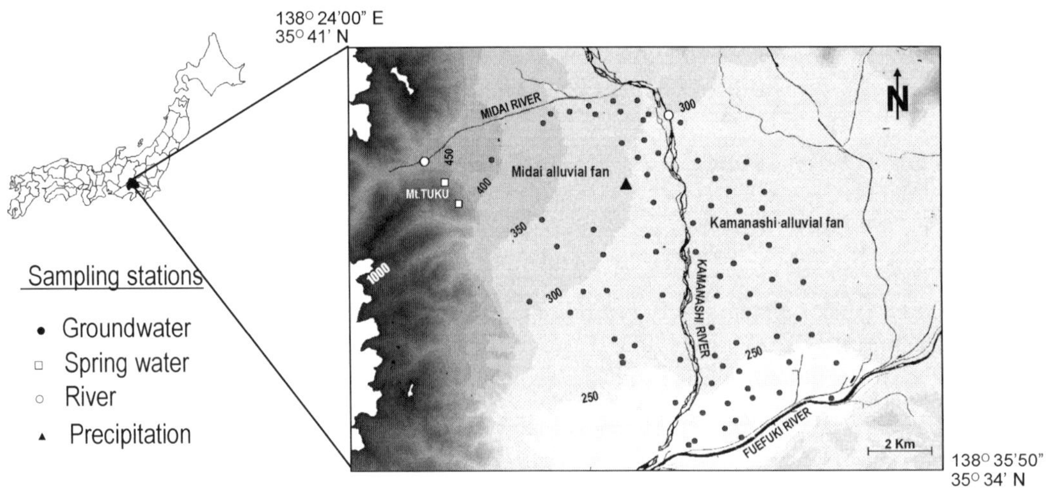

Figure 1. Location map and sampling stations in the study area

Groundwater is a major source of drinking water in the Kofu basin. The centre and upper part of the Kamanashi alluvial fan is made of gravel having a depth of 75~95 m, which was transported by the Kamanashi River. Furthermore, the lower part is composed of clay soil, which is covered by 15~25 m deep gravel (MLIT, Japan, 2001). The Midai alluvial fan is made of gravel that was transported by the Midai River, and this part has no clay soil. The dominant land use types of the Midai alluvial fan and the Kamanashi alluvial fan are orchards (>64%) and paddy fields (>51%), respectively. These land use types are uniformly distributed in the fans. The average rate of nitrogen fertilizer application in an orchard and paddy field is 12 kgN/ha/yr and 2.5 kgN/ha/yr, respectively.

2.2 Sampling and method of analysis

Groundwater samples (n = 80) were collected from domestic and farm wells distributed over an area of 100 km^2 from 22–29 May 2003. River water samples (n = 14) were collected bimonthly from the Midai River and the Kamanashi River from May 2003–May 2004. Precipitation samples (n = 12) were collected monthly from August 2003–July 2004 from the upland area of the basin (Fig.1). Spring water samples (n = 14) were collected bimonthly from June 2003–June 2004 at two locations on Mt. Tuku (Fig.1).

All the water samples (n = 123) were analysed for their water $\delta^{18}O$ composition. First, the samples were equilibrated with CO_2 gas and extracted using standard cryogenic separation techniques. The measurements were performed using an isotope ratio mass spectrometer (IRMS) (Micromass, Prism Mass, UK), and these values were normalized with Vienna standard mean ocean water (VSMOW) (Taylor, 1973). The concentrations of the major cations—sodium (Na^+), ammonium (NH_4^+), potassium (K^+), calcium (Ca^{2+}), and magnesium (Mg^{2+})—were determined using an ion chromatograph (TSK 1/2HR, Shodex, Japan). Similarly, the concentrations of the major anions—fluoride (F^-), chloride (Cl^-), nitrate (NO_3^-), and sulphate (SO_4^{2-})—were determined using another ion chromatograph (YK-G, Shodex, Japan). The concentration of bicarbonate (HCO_3^-) was determined immediately after sampling by acid-base titration using an MR-BCG indicator. The compositions of $\delta^{15}N\text{-}NO_3^-$ were determined using an elemental analyser coupled to the IRMS (PDZ-Europe, ANCA-SL, UK) according to the method described by Sakata (2000).

2.3 Separation of groundwater aquifer

The spatial pattern of an aquifer plays an important role in the fate and transport of nutrients into a groundwater system. Therefore, in order to identify the spatial variation of an aquifer in the study basin, we used a hierarchical agglomerative clustering technique by means of Ward's method, using the Euclidean distances as a measure of similarity. The major variables used for the clustering are fluoride (F^-), chloride (Cl^-), nitrate (NO_3^-), sulphate (SO_4^{2-}), sodium (Na^+), ammonium (NH_4^+), potassium (K^+), calcium (Ca^{2+}), and magnesium (Mg^{2+}). The linkage distance is reported as Dlink/Dmax, which represents the quotient of the linkage distance for a particular case divided by the maximum distance; the quotient is multiplied by 100 for standardizing the linkage distance represented on the y-axis (Wunderlin et al., 2001).

3. Results and discussion

3.1 Groundwater aquifer

The cluster analysis rendered a dendrogram, where all the sampling sites of the groundwater aquifer were grouped into two statistically significant clusters at (Dlink/Dmax) × 100 < 40 (Figure 2a). The clustering procedure generated two distinct and clear sites. Clusters 1 and 2 correspond to the Midai alluvial fan and the Kamanashi alluvial fan, respectively (Figure 2b).

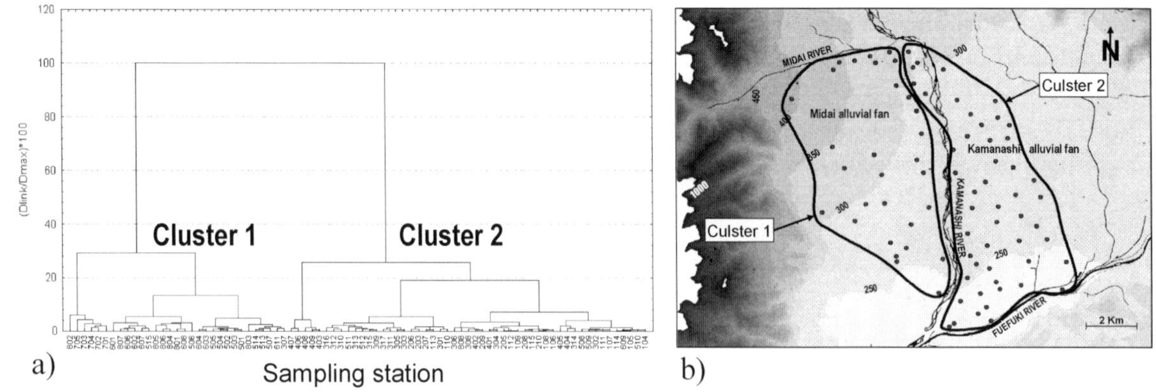

Figure 2. a) Dendrogram showing the clustering of the sampling stations and b) its spatial extent on the basis of the major anion and cation values.

3.2 Nitrate concentration and its source identification

The nitrate concentrations of the precipitation, river water, and groundwater are summarized in Table 1. The mean values of the nitrate concentrations of the precipitation and river water are much lower than those of the groundwater aquifers. The highest nitrate concentrations are observed in the groundwater aquifer of the Midai alluvial fan. The dominant land use type of this area is orchard plantations. Furthermore, farmers in this area usually apply a higher dose of fertilizer as compared to the Kamanashi alluvial fan where paddy fields are the dominant land use type. A wide range of nitrate concentrations was observed in each aquifer although the land use types are uniformly distributed. This suggests that apart from the land use types, dilution and physical mixing processes also affect the loading of nitrate into the groundwater aquifers.

Table 1. Nitrate concentrations of different sources

Source	*n*	Mean ± SD (mg/l)	Range (mg/l)
Precipitation	12	1.0 ± 0.6	0.0–2.0
Rivers			
Midai	7	1.6 ± 0.3	1.1–1.9
Kamanashi	7	3.6 ± 0.4	3.0–4.3
Aquifers			
Midai	30	17.0 ± 12.5	0.0–46.8
Kamanashi	50	6.3 ± 4.1	0.0–18.3

The $\delta^{15}N$ values can be used to distinguish between the different sources of nitrate. The Ministry of Environment, Japan, 2002, has published a range of $\delta^{15}N$ values for different kinds of sources within Japan (shown in Figure 3 as a range). The histogram of the $\delta^{15}N$ values analysed for 34 groundwater samples is within the range of the values for inorganic and organic fertilizers. This suggests that the major sources of nitrate are the inorganic and organic fertilizers applied in the orchards and paddy fields of the Midai and Kamanashi alluvial fans, respectively.

3.3 Origin of groundwater

In order to identify the origin of groundwater in both the Midai and Kamanashi aquifers, the weighted mean values of $\delta^{18}O$ of the precipitation, river water, and spring water at Mt. Tuku were compared (Table 2). The bimonthly $\delta^{18}O$ values of the river water of both the alluvial

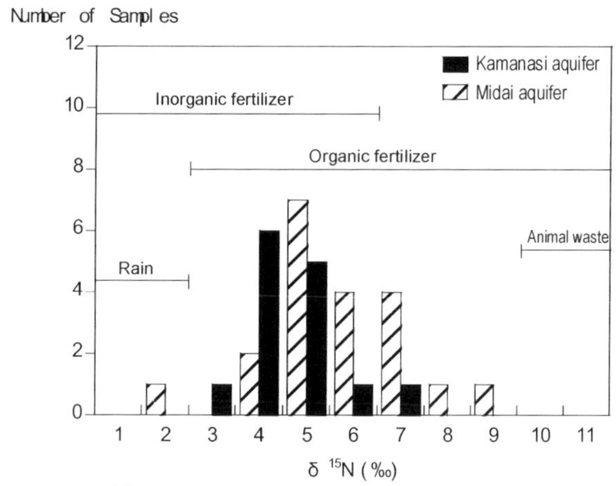

Figure 3. Histogram of the $\delta^{15}N$ values of the groundwater in the Midai and Kamanashi alluvial fans as compared with the available indexes from the MOE, Japan, 2002

fans did not show any wide variation, suggesting a constant source of river discharge. The mean values of $\delta^{18}O$ of the spring water at Mt. Tuku were similar to those of the precipitation samples. This indicates that precipitation is a source of spring water at Mt. Tuku. However, the contrast between the wide range values of the precipitation samples and the narrow range values of the spring water samples at Mt. Tuku suggests that mixing occurred during the infiltration of the groundwater. A similar finding was also reported by Mizutani et al. (2001).

Table 2. $\delta^{18}O$ values of water from different sources

Location	*n*	Mean ± SD (‰)	Range (‰)
Midai River	7	−11.1 ± 0.4	−11.5~−10.3
Kamanashi River	7	−11.0 ± 0.3	−11.3~−10.5
Spring water at Mt. Tuku	14	−9.7 ± 0.1	−9.9~−9.5
Precipitation*	12	−9.5 ± 2.2	−14.1~−6.8

*Weighted mean by volume

3.4 Contribution of precipitation to groundwater and its effect on nitrate loading

The relationship between the nitrate concentration and $\delta^{18}O$ values of the groundwater was investigated in order to determine whether precipitation and river water have an effect on the loading of nitrate into groundwater. In both the Midai and Kamanashi aquifers, a strong positive correlation was observed as shown in Figure 4. The nitrate concentration increased with the $\delta^{18}O$ values. In this figure, the values in the lower left corner correspond to the $\delta^{18}O$ values of river water, and the values in the upper right corner correspond to those of the spring water samples at Mt. Tuku. In the former case, the nitrate concentration was observed to be diluted by the river water during loading of nitrate into the groundwater. On the other hand, in the latter case, nitrate seems to have been transported to the groundwater through the infiltration caused by precipitation. This suggests that the distribution of nitrate concentration is clearly affected by the groundwater sources and the mixing ratio at every sampling station.

It is also observed that the slope in the case of the Midai aquifer is almost double that of the Kamanashi aquifer. This can be attributed to the land use type of the respective aquifers. The dominant land use type of the Midai aquifer is orchard plantations, and the farmers generally apply a higher dose of fertilizers (i.e. 12 kgN/ha/yr). On the other hand, the dominant land use

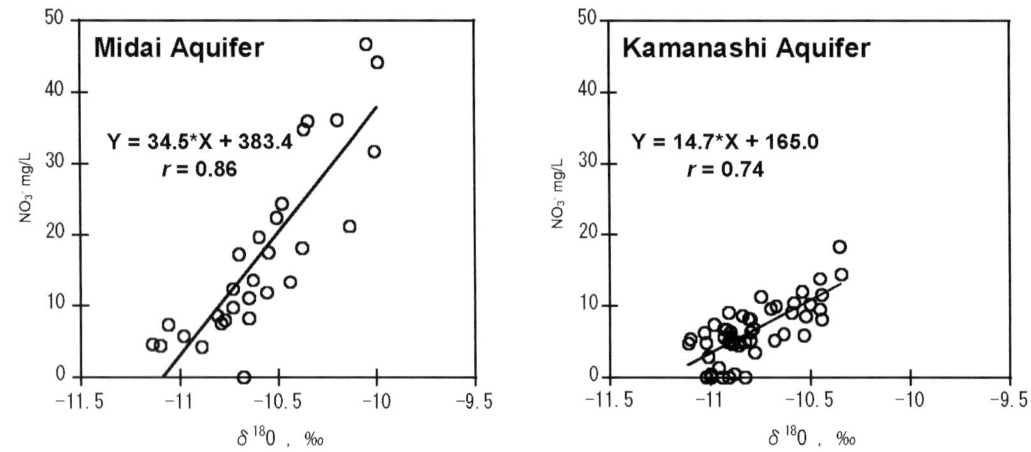

Figure 4. Relationship between the $\delta^{18}O$ values and nitrate concentration of the groundwater in the Midai and Kamanashi aquifers

type of the Kamanashi aquifer is paddy fields, and the farmers apply a lower dose of fertilizers (i.e. 2.5 kgN/ha/yr) than the orchard plantations. Therefore, it can be concluded that the nitrate loading was not only affected by the land use type but also the mixing ratio of the groundwater aquifers.

In order to generalize the scope of this study, the contribution of precipitation to the groundwater (mixing ratio) and the nitrate concentrations is used to develop a model to estimate the loading of nitrate into the groundwater aquifers of this area.

The mixing ratio was deduced using the analysed values of $\delta^{18}O$ and the following equation:

$$X_p = \frac{C_{gw} - C_r}{C_p - C_r} \qquad (1)$$

where X_p is the mixing ratio of groundwater and C_{gw}, C_p, and C_r are the $\delta^{18}O$ values of groundwater (*gw*), precipitation (*p*), and river water (*r*), respectively.

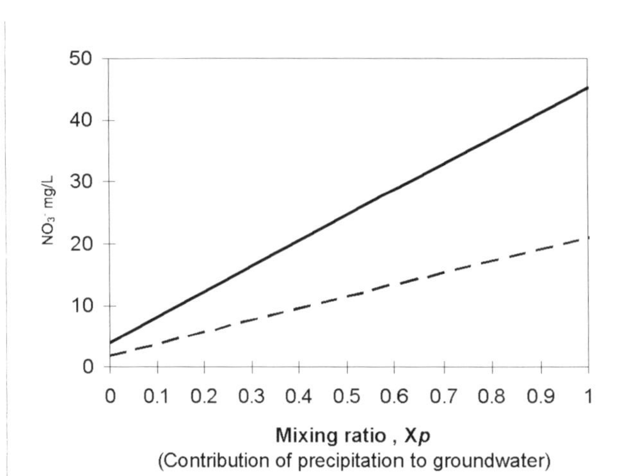

Figure 5. Nitrate loading models developed from the contribution of precipitation to the groundwater (mixing ratio) and the nitrate concentrations of the groundwater in the Midai and Kamanashi aquifers

Two equations were obtained for the Midai aquifer and the Kamanashi aquifer using the precipitation mixing ratio and the nitrate concentrations (Figure 5). These equations can be considered as general models to estimate the loading of nitrate into the groundwater in the study area.

4. Conclusion

In this study, the source and transport of nitrate into the groundwater aquifers of the Midai and Kamanashi alluvial fans were investigated by isotopic techniques. The nitrate concentration was found to be higher in the groundwater aquifer of the Midai alluvial fan. The values of $\delta^{15}N$ indicate that the major sources of nitrate are the inorganic and organic fertilizers applied in the orchards and paddy fields. Similarly, the isotopic values of $\delta^{18}O$ indicate that precipitation and river water are sources of recharge in both the aquifers. The relationship between the $\delta^{18}O$ values and the nitrate concentration shows that the precipitation acts as a carrier and the river water facilitates dilution in the loading of nitrate into the groundwater aquifers. This suggests that the loading of nitrate into the groundwater aquifers is not only affected by the land use type but also the recharge pattern of the precipitation and river water. Therefore, this study shows that for the management of groundwater quality, the groundwater recharge system together with the land use type should be appropriately considered.

5. Acknowledgement

We would like to express our sincere thanks to the staff of S-J laboratory (Faculty of Science, Toyama University) for helping with the measurements of $\delta^{18}O$ and the major dissolved ions. We are also very thankful to Mrs. Yuki Hiraga for her help in sampling and analysing and Dr. Tomoya Iwata and Mr. Bipin Pathak for their constructive criticisms during the compilation of this paper. We would also like to extend our special thanks to the regional inhabitants for their significant assistance in the water sampling.

6. References

Fukada, T., Hiscock, K.M., Dennis, P.F. and Grischek, T. (2003). A dual isotope approach to identify denitrification in groundwater at a river-bank infiltration site. *Water Res.*, **37**, 3070–3078.

Ging, P.B., Lee, R.W. and Silva, S.R. (1996). Water chemistry of Shoal Creek and Waller Creek, Austin, Texas, and potential sources of nitrate, *U.S. Geol. Surv. Water Resour. Invest. Rep. 96–4167.*

MOE, Japan. (2002). Measure of water pollution for nitrate and nitrite. In: A guide book the measure of groundwater pollution by nitrate-nitrogen, pp. 3–162, Chap. 1. Ministry of Environment, Japan (in Japanese).

MLIT (2001). A report of boring investigation in the Fuji River-bed, Ministry of Land, Infrastructure and Transport, Japan. (in Japanese, unpublished)

Kazama, F. and Yoneyama, M. (2002). Nitrogen generation in the Yamanashi prefecture and its effects on the groundwater pollution. *Int. Environmental science*, **15**(4), 293–298. (in Japanese)

Kendall, C. (1998). Tracing nitrogen sources and cycling in catchments. In: Kendall, C. and McDonnell. J.J (Eds.), Isotope tracers in catchment hydrology. Elsevier Science B.V., Amsterdam, pp. 517–576, Chap. 16.

Mizutani, Y., Satake, H., Yamabe, A., Miyaci, H., Mase, N., and Yamamura, K. (2001). Hydrogen and oxygen isotope ratios of groundwaters in shallow aquifers beneath the alluvial fan. *Japan Association of Groundwater Hydrology*, **43**(1), 3–11. (in Japanese)

Ogawa, Y., Nisikawa, M., Nakasugi, O., Ii, O. and Hirata, T. (2001). Determination of the abundance of $\delta^{15}N$ in nitrate ion in contaminated groundwater samples using an elemental analyzer coupled to a mass spectrometer. *Analyst*, **126**, 1051–1054.

Rock, L. and Mayer, B. (2002). Isotopic assessment of sources and processes affecting sulphate and nitrate in surface water and groundwater of Luxembourg. *Isotopes in Environmental and Health Studies*, **38**(4), 191–206.

Rock, L. and Mayer, B. (2004). Isotopic assessment of sources of surface water nitrate within the Oldman River Basin, Southern Alberta, Canada. *Water, Air and Soil Pollution: Focus*, **4**, 545–562.

Sakata, M. (2000). A simple and rapid for $\delta^{15}N$ determination of ammonium and nitrate in water samples. *Geochemical Journal*, **35**, 271–275.

Taylor, H.P.J. (1973). $^{18}O/^{16}O$ evidence for meteoric-hydrothermal alteration and ore deposition in the Tonopah, Comstock Lode, and Goldfield mining districts, Nevada: In Economic Geology. **68**, 747–764.

Wunderlin, D.A., Diaz, M.P., Ame, M.V., Pesce, S.F., Hued, A.C. and Bistoni, M.A. (2001). Pattern recognition techniques for the evaluation of spatial and temporal variations in water quality. A case study: Suquia river basin (Cordoba-Argentina). *Water Res.* **35**, 2881–2894.

Groundwater Arsenic Pollution: Impacts and Sustainable Solutions for Bangladesh

Md. Bayzidur Rahman[1)], Md. Salequzzaman[2)] and Abul Hasnat Milton [3)]

1) Doctoral Research student, School of Public Health, The University of Sydney, Australia; Phone: +61 02 93514382, Mobile: +61 0403277520; E-mail: bayzidr@health.usyd.edu.au. Location and Postal address: Room 128 (C), Edward Ford Building (A27), Camperdown Campus, The university of Sydney, NSW-2006, Australia

2) Professor and Head, Environmental Science Discipline, Khulna University, Khulna 9208, Bangladesh, Phone: +880 41 813239, Mobile: +88 0176 111 075, Email: salek_uz@yahoo.com / msalequzzaman@hotmail.com / M.Salequzzaman@murdoch.edu.au

3) Senior Lecturer, Centre for Clinical Epidemiology and Biostatistics, The University of Newcastle, Newcastle, NSW 2308, Australia. Phon: +61 2 49236200, Mobile: 61 412141971, E-mail:milton.hasnat@newcastle.edu.au

Abstract

This paper describes the occurrence of Arsenic in ground water of Bangladesh above the safe level, its impacts on reproductive and other health aspects of human life along with recommendation of some sustainable solutions. A cross-sectional study was conducted to determine the effect of arsenic poisoning on reproductive outcomes, through interviewing 533 eligible women. The study detects excess risks for spontaneous abortion, stillbirth and neonatal death among the participants exposed to increasing concentrations of arsenic in drinking water. The odds ratio (OR) for spontaneous abortion, stillbirth and neonatal death were 2.5 (95% confidence interval = 1.5 – 4.3), 2.5 (95% CI = 1.3 – 4.9) and 1.8 (95% CI = 0.9 – 3.6), respectively. Melanosis and keratosis were observed among 51 (10%) and 43 (8%) participants, respectively. Main source of this contamination is ingestion of arsenic contaminated water. A sustainable solution to this problem requires availability of pure drinking water either through purification of arsenic contaminated water or alternative affordable drinking water supply. The research finally recommends some feasible alternative sources of affordable and drinkable pollutants/contaminants free water as well as a policy framework on how the proposed solution could reach to the community people.

Key Words: Arsenic pollution, arsenicosis, spontaneous abortion, stillbirth, neonatal death, Bangladesh

1. Introduction

The majority of the villagers in Bangladesh used to drink water from unsafe water sources such as hand-dug wells or natural ponds until 1970's. This condition exposed the people to endemic and deadly diseases like diarrhea, dysentery and other intestinal ailments. These water-borne diseases at that time used to take a huge toll of lives every year. To overcome this situation, Bangladesh government with technical and financial assistances from different donor agencies extensively installed thousands of hand tube wells that abstracts microbiologically safe water from underground . It has been estimated that approximately 97% of the rural population of Bangladesh have had access to safe drinking water through these tube wells (Smith *et al.*, 2000). Ironically, the use of ground water for drinking purposes revealed as a PBTC (Planned for Blessings but Turned to Curse) syndrome for Bangladesh due to arsenic (As) contamination in such water.

The first reported case of arsenic-contaminated ground water (greater than 50 μg As L^{-1}) from the Bengal Basin was recorded in 1978 in West Bengal and the first case of arsenicosis was diagnosed there in 1983 (Acharrya *et al.*, 2000). Subsequently a number of studies conducted by the relevant agencies including the School of Environmental Studies (SOES) at Jadabpur University of West Bengal, India, confirmed that the Bengal Basin consists of high arsenic content in its groundwater. Later on, a national survey conducted jointly by the DPHE (Department of Public Health and Engineering) of Bangladesh and British Geological Survey revealed that 53 districts out of 64 districts are affected with arsenic contamination and approximately 28 to 35 million people of Bangladesh out of 120 million are at potential risk of arsenic poisoning (British Geological Survey, 2001).

Inorganic arsenic is mainly responsible for arsenic related toxicity on human health. It is also a known carcinogen and has the potential to cause cancer to the human skin and other internal organs (IARC, 1980). Studies have shown that arsenic also causes several non-malignant adverse health effects including weakness, edema, conjunctival congestion, diabetes mellitus, hypertension and adverse respiratory effects (Mazumdar *et al.*, 2002, Milton and Rahman, 1999, Milton *et al.*, 2001, Rahman *et al.*, 1998, Rahman *et al.*, 1999). Different studies report that development of skin lesions and other symptoms broadly depends on the concentration and duration of arsenic exposure including the nutritional factors.

Until now, minimum attention has been shown to the research on potential impact of arsenic poisoning on human reproduction. Few studies from different parts of the world reported that high arsenic exposure may result in spontaneous abortion, low birth weight baby, perinatal death, pre-eclampsia and congenital malformation (Nordstrom *et al.*, 1978a; Nordstrom *et al.*, 1978b ; Nordstrom *et al.*, 1979a ; Nordstrom *et al.*, 1979b). Ferm and Carpenter (1968) through experiments with lower animals such as rats have mentioned that arsenic can cross placental barrier and bears potential risk of producing adverse pregnancy outcomes.

This research attempted to determine the effect of arsenic on human reproduction process through a cross-sectional study. To understand the full impact of arsenic poisoning/contamination, the research also concentrates on other social, environmental and economic impacts in the society. Finally the research identifies few community-based, low-cost, locally available solutions to provide safe drinking water. Furthermore, the paper elaborates on how pond-sand filter, protected pond, rainwater harvester and kalshi filtration act to cover sustainability in affording community based arsenic free drinking water. Finally, to achieve sustainability in the rural and urban communities, an implementable policy framework has been proposed.

2. Methods and Study area

This research comprises a cross-sectional study to determine the association between chronic arsenic exposure and adverse pregnancy outcomes coupled with previous research and professional experiences. The professional experience includes the study on heavy metals conducted in Tokyo University of Japan during the period of 1994-1996, and surveys of different projects addressing the problems of drinking water problem in Bangladesh. An extensive literature review was also carried out to access the available information related to arsenic problems in Bangladesh.

The epidemiological study to determine the effect of arsenic on pregnancy outcomes were conducted in three thanas (sub districts) named as Homna under Commilla district, Saharastri under Chandpur district and Damurhuda under Chuadanga district. These three thanas were randomly selected from the unpublished database of NGO Forum for Drinking Water Supply & Sanitation (a national NGO in Bangladesh). The unpublished database contained information on tube well details. The obtained data on tube well details included information on age, depth and concentration of arsenic in ppb (parts per billion) level measured by either ultraviolet / visible spectrophotometry or flow injection hydride generation atomic absorption spectrometry (FIHG-AAS) method. As the study population, 533 eligible participants were selected who were ever-married women aged between 15-49 years with at least one prior pregnancy, no history of smoking, who had been living in the study area since their marriage and drank water from the respective tube well prior to and during the pregnancy. A structured pre-tested interviewer-administered questionnaire was used to collect information on socio-demographics, drinking water use, occupation and adverse pregnancy outcomes i.e., spontaneous abortion, stillbirth and neonatal death. In this study, any pregnancy termination within 28 weeks of gestation that occurred naturally without deliberate action and intention to terminate the pregnancy was regarded as spontaneous abortion (ICDDR,B, 1996). A stillbirth was considered to be any birth after 28th completed week of gestation when the baby did not breathe or show any sign of life (Dutta, 1994). Neonatal death was regarded the death of the new born within 28 days after birth (Dutta, 1994).

Arsenic exposure was assessed individually. For all subjects, only a single well-water measurement obtained in one point in time was used to characterize chronic arsenic exposure. Duration of individual arsenic exposure was calculated from the period of using the particular arsenic contaminated tube well by the participant. We analysed the epidemiological data of this study by Logistic regression method using the software STATA-7. Other related impacts of arsenic were derived from the interviews and case studies. Low cost technologies along with some sustainable options as alternative water sources were explored through observation of different projects dealing with arsenic problems and others including the Community based arsenic mitigation project in Jhikargachha thana of Jessore district arsenic mitigation project and CARE-RVCC (Reducing Vulnerability to Climate Change) Project. Analyzing the opinion of arsenic affected people and considering their alarming future, a policy framework is recommended.

3. Results and Discussion

Impact of Arsenic on Pregnancy outcomes: A total of 533 eligible women were interviewed to determine the pregnancy outcomes. Their mean age was 31.22 years (± 6.44) and the average duration of arsenic exposure of the participants was 8.11 years (± 7.32). The majority of them (51%) got married before the age of 16-years. Mean number of pregnancies was 3.72 (± 1.79) and mean number of live births was 3.48 (± 1.65). Occurrence of melanosis, leucomelanosis and keratosis observed among the study population was 51(10%), 55 (11%) and 43 (8%) respectively. Mean and median arsenic concentration of the tube well drinking water of the study participants' was 279 (± 355) and 116 μg/L respectively, ranging from non-detectable level to 1710 μg/L. Overall incidence of spontaneous abortion, stillbirth and neonatal death were found as 59.61 /1000 live birth, 35.55 /1000 live birth and 49.31 /1000 live birth respectively. Results regarding the risk (OR) of arsenic concentration and adverse pregnancy outcomes are given in Table 1. An increased risk is ovserved for spontaneous abortion and still birth for arsenic exposure over 50 $\mu g/L$. Milton and his associates found that for most levels of arsenic concentration over 50 $\mu g/L$, the risks are higer for longer maternal exposure (Milton, et al., 2005). Findings of the study suggest that chronic arsenic exposure may increase the risk of fetal and infant death. Chronic arsenic exposure has been reported to be associated with human adverse pregnancy outcomes in a few studies (Hopenhayn *et al.*, 2000), and the evidence is suggestive not conclusive. Animal model experimentation reported the toxicity of several forms of arsenic on pregnancy outcomes including malformations (Nemec *et al.*, 1998).

Table 1: Effect of Arsenic exposure on reproductive outcomes: Field survey 2002.

Pregnancy outcome	Frequency	Strength of Association (95% CI)		
		Odds Ratio	Lower limit	Upper Limit
Spontaneous abortion	70 /1000 live birth	2.5	1.5	4.3
Neonatal Death	67 / 1000 live birth	2.5	1.3	4.9
Still birth	48 /1000 live birth	1.8	0.9	3.5

Social Impacts of Arsenic: Skin lesions due to arsenic poisoning also results in some fallacies which trigger several social delinquencies. In the case study area, three women were found to be divorced as they developed arsenicosis after marriage. In addition, a large number of people specially the mothers, who are the victims of arsenic contamination, are now worried about their breast-fed babies. It is a matter of apprehension, whether the breast milk of an arsenic affected mother contains arsenic to affect the baby. Most of the rural people compare arsenicosis with leprosy, so they avoid any type of relation with the arsenic victims. Even in the national job market, the young generation bearing skin lesion is being disqualified for getting any job.

Economic Impacts of Arsenic: The economic impact of arsenic contamination of groundwater has not encountered in terms of monetary value, which at first, requires the actual figure of arsenicosis patients in the whole country, but such type of information is yet non existent. However, this study assumes both direct and indirect costs from the menace. The direct costs are associated with acute sickness of skin lesion, lung and bladder cancer, diabetes, bronchitis and some carcinogenic effects through expenditure for treatment and emergency hospital bed services. The indirect costs are incurred by abstinence from work, decline in average life expectancy, reduction in work efficiency due to various complications of the arsenic toxicity. Neurologic, dermatologic and carcinogenic effects bear the prominent responsibility to this type of cost. Any of the above mentioned adverse pregnancy outcomes may pose a psychological upset to the victim and her associates and ultimately contributes to the indirect cost.

Sustainable practices to combat arsenic contamination: Sustainable solution to arsenic problem is meant by the community access to safe drinking water for present and future generation through simultaneous social, environmental and economic viability for the options without any risk for facing adversity in the times to come. This concept highlights two options: removal of arsenic from the contaminated water and availing arsenic free water resources. Several non government and government agencies introduced such type of practices but could not satisfy the sustainability approach as a whole. Some of the practices are analyzed below:

Kalshi Filter: A simple three-pitcher (Kalshi) filtration system (Figure.1) made up of entirely from available local materials have proven to remove arsenic, other trace metals and anions from the groundwater without using any additive chemicals. The filtered water contained about 10 mg/L As (total), no detectable As (III) exceeding the guideline of US EPA (United States Environmental Agency), WHO (World Health Organization) and Bangladesh standard, and significantly reduced major, minor and trace metals (Munir, et al., 2001). The raw materials required to develop the system includes cast iron turnings, sand, wood charcoal and Kalshi. With a maximum costs of US $ 5.0- 6.0, the system is capable to supply five people for about five months at a consumption rate of 50 L/day (Munir, et al., 2001). This option may satisfy the sustainability criteria up to some extent; however the lack of awareness and motivation may not provide the expected benefit.

Figure 1. Kalshi-filter system in Bangladesh.

Pond Sand Filter (PSF): PSF is a semi engineered system to purify pond water for community service, where water is collected in a big chamber from a nearby protected pond with the help of a tube well and subsequently it is passed through a constructed filter bed. This system is mainly used for removal of bacteria and other disease bearing agents from surface water, which is an alternative water source to avoid the arsenic contamination. According to different studies, the construction costs for this system is around US$ 650. Several projects (such as community based arsenic mitigation project in Jhikargachha thana of Jessore district) are running in rural Bangladesh to adopt PSF through participatory management process, where the construction costs is recovered from community people by taking a trivial amount (01.00 taka/month/user). However, this option is not sustainable for the areas where surface water is contaminated by salinity as PSF does not remove saline from water.

Rain Water Harvesting: Rain water harvesting (RWH) is a viable alternative for supply of safe drinking water of superior quality being naturally soft and devoid of dissolved minerals or salts. The RWH system vary from as simple as a small container placed under a rain gutter downspout for drinking purpose of a small household or as complex as an engineered, multi tank, pumped and pressurized construction to supply residential and irrigation needs. The components of this system are: (1) a catchment area usually a roof of a house, made up of tiles, slates or corrugated iron sheeting, (2) gutters and downspouts to channel water from the catchment to storage, (3) storage system for receiving and storing water. According to several studies, the cost of a medium sized RWH system to provide 3-5 families is approximately US$ 5. It has been estimated from data on average roof area and annual rainfall in Bangladesh that about 122 liters of rainwater is harvestable per day. This could provide an amount that is on average 50% in excess of the requirements in dry season (Ahmed, 2002). Main hindrance to sustainability of this option is the storage consideration for dry season elapsing from 6 to 7 months. However, RWH is very much effective for the salinity prone area (such as the coastal region) of Bangladesh. Several NGOs like NGO Forum for DWSS is working to familiarize this system.

Protected Pond: This concept speaks of returning to the classical source of drinking water in rural Bangladesh that is the protected rural pond. RVCC project of CARE (Co-operative American Relief Everywhere) - Bangladesh is introducing this system by adopting the policy to re-excavating an existing pond to hold sufficient water for the surrounding community. By elevating the natural levee and fencing the edge, the pond is protected from intervention of pollutants and polluting agents. These protected ponds are restricted from other uses such as bathing, washing, cultivating fishes using chemical fertilizers, etc. A pond management committee is being formed to look after the system. Thus, it could be proven as a sustainable solution to arsenic problem.

Recommended Policy Framework along with the existing Practices:

Improved awareness programs: The field survey reveals that a large population in rural Bangladesh is reluctant to believe on arsenic contamination and its adverse health effects. Thus, they continue to drink arsenic contaminated water from tube wells remaining under substantial risk of developing related health hazards. More over, low literacy rate and limitation of proper knowledge to adopt alternative water option aggravates the situation. Therefore, a mandatory awareness campaign coupled with arsenic literacy program is strongly recommended. Relevant governmental and non-governmental agencies may share the responsibilities of raising mass awareness on arsenic mitigation maintaining a particular sustainable process.

Search for low-cost, locally available and appropriate alternative sources of drinking water: Shallow tube well has been the main source of drinking water in most of the arsenic affected regions as in the study areas such as Comilla and Chandpur district. The tube well screening program red marked (marked as arsenic contamination) a large proportion of functioning tube wells. As a consequence, users are in a confused state regarding their drinking water sources. So, an immediate action should be taken to provide them with best feasible alternative drinkable water resources accessible through locally available and low cost technologies. This requires firstly determination of suitable alternative water resources, choice for low-cost, locally available, and affordable technological and financial selection and then adaptation of the best option.
Conduct epidemiological research: Epidemiological studies (e.g. cohort, case-control studies) need to be carried out to determine the association between different health effects with arsenic exposure. For instance, the cross–sectional study discussed in this research to

determine the effect of chronic arsenic exposure to pregnancy outcomes is not conclusive yet. To confirm the association, a case-control study is recommended.

Develop reliable techniques for Arsenic screening: Millions of dollar is invested in arsenic mitigation project through tube well screening for arsenic contamination, which is carried out by using field kits. Those with arsenic levels >50 micrograms per liter (μg/L) were painted red to indicate that the tube well water is unsafe for drinking, and those with levels <50 μg/L were painted green to indicate that the tube well water is safe. Most experts agree that the latest developed kit works better for samples containing arsenic levels <10 μg/L or >100 μg/L, but not in between. Dipankar Chakraborti and his associates in Jadavpur university, Calcutta, conducted a study to verify accuracy of the field kits in comparison to FI-HG-AAS and discovered that for samples with arsenic levels <50 µg/L, 45% were mislabeled (Chakraborti, 2000). For samples with higher arsenic levels, the percentage of wells that were mislabeled previously was very low. Thus a reliable detection technique is essential for sustainable calculation and its subsequent management. On the other hand, arsenic level in the aquifers may not remain constant throughout the year. Furthermore, a green tube well once found safe may get arsenic contaminated in future. So, a continuous tube well surveillance mechanism needs to be developed locally, specially to test the green colored tube wells as well as newly installed ones.

4. Concluding remarks

The problem of arsenic contamination in Bangladesh is one of the most unfortunate events for the 21st Century. It has posed a substantial threat on the public health situation of Bangladesh. In addition to many other health effects, adverse effects on pregnancy outcomes due to chronic arsenic exposure as observed in this study add to the severity of the calamity. Most of the Bangladeshi population remains unaware of this fact and they are falling victim to this veiled slayer. To find a sustainable solution of this problem, low-cost locally available remedial measures have to be taken and to ensure that further contamination of domestic water supplies do not occur. To this end, various explanations being put forward for the source of arsenic and its release mechanism should be investigated to determine the single most important source. To effectively combat the situation, it is likely that a collaborative effort is needed with participatory approach.

5. References

Acharrya S.K., Lahiri S., Raymahashay B.C. and Bhowmik A. (2000). Arsenic toxicity in groundwater in parts of Bengal basin in India and Bangladesh: the role of Quaternary Stratigraphy and Holocene sea-level fluctuation. In: *Environmental Geology*, Springer-Verlog, pp1127-1137.

Ahmed M.F. (2002). Alternative water supply options for arsenic affected areas of Bangladesh (a theme paper for discussion). International workshop on arsenic mitigation in Bangladesh. Dhaka, Bangladesh: ITN-Bangladesh, Centre for Water Supply and Waste Management.

British Geological Survey (2000). Arsenic contamination of groundwater in Bangladesh. 2001. http://www.bgs.ac.uk/arsenic/bangladesh.html (access on 23 September 2003).

Chakraborti D. (2000). Groundwater Arsenic Contamination in Bangladesh. Book published jointly by School of Environmental Studies, Jadavpur University of Calcutta, India and Dhaka Community Hospital of Bangladesh.

Dutta. D.C.(1994). Textbook of Obstetrics Including Perinatology and Contraception 3rd edn. Kolkata, India: New Central Book Agency P Ltd.

Ferm V.H. and Carpenter S.J. (1968). Malformations induced by sodium arsenate. *Journal of Reproductive Fertilization*, vol. 17, pp. 199-201.

Hopenhayn R.C., Browning S.R. and Hertz-Picciotto. (2000). Chronic arsenic exposure and risk of infant mortality in two areas of Chile. *Environmental Health Perspective*, vol. 108, pp. 667-673.

IARC (International Agency for Research on Cancer) (1980). Monographs on the evaluation of the carcinogenic risk of chemicals to humans: some metals and metallic compounds. *Lyon: International Agency for Research on Cancer*, vol. 23, pp. 39-141.

ICDDR,B (International Centre for Diarrhoeal Diseases Research, Bangladesh) (1996). Abortion in Rural Bangladesh: Evidence From the MCH-FP Extension Project. Bangladesh. Working paper no. 121:4

Mazumder D.N., Haque R. and Ghosh N. (2000). Arsenic in drinking water and the prevalence of respiratory effects in West Bengal, India. *International Journal of Epidemiology*, vol. 29, pp.1047-1052.

Milton A.H. and Rahman, M. (1999). Environmental pollution and skin involvement pattern of chronic arsenicosis in Bangladesh. *Journal of Occupational Health*, vol. 41, pp. 207-208.

Milton A.H.; Hasan, Z. and Rahman, M. (2001). Chronic arsenic poisoning and respiratory effects in Bangladesh. *Journal of Occupational Health*, vol. 43, pp. 136-140.

Milton A.H., Smith W., Rahman B, et al. (2005). Chronic arsenic exposure and adverse pregnancy outcomes in Bangladesh. Epidemiology, vol. 16, pp.82-6.

Munir A.K.M., Rasul S.B., Habibuddowla M., Alauddin M., Hussam A. and Khan A.H (2001). Evaluation of performance of Sono 3-Kolshi filter for arsenic removal from groundwater using zero valent iron through laboratory and field studies. International workshop on technologies for arsenic removal from drinking water. Dhaka, Bangladesh

Nemec M.D., Holson J.F. and Farr C.H. (1998). Developmental toxicity assessment of arsenic acid in mice and rabbits. *Reproductive Toxicology*, vol. 12, pp. 647-658.

Nordstrom S. Beckman L. and Nordenson I. (1978a.). Occupational and environmental risk in and around a smelter in Northern Sweden-I: Variations in birth weight. *Hereditas,* vol. 88, pp. 43-46.

Nordstrom. S., Beckman L. and Nordenson I. (1978b). Occupational and environmental risk in and around a smelter in Northern Sweden-VI: Congenital malformations. *Hereditas,* vol. 90, pp. 297-302.

Nordstrom S., Beckman L., and Nordenson I. (1979b). Occupational and environmental risk in and around a smelter in Northern Sweden-III: Frequencies of spontaneous abortion. *Hereditas,* vol. 88, pp. 51-54.

Nordstrom S., Beckman L. and Nordenson I. (1979a.) Occupational and environmental risk in and around a smelter in Northern Sweden-V: Spontaneous abortion among female employees and decreased birth weight in their offspring. *Hereditas, vol. 90, pp. 291-296.*

Rahman M., Tondel M. and Ahmad S.A. (1998). Diabetes mellitus associated with arsenic exposure in Bangladesh. *American Journal of Epidemiology*, vol. 148, pp. 198-203.

Rahman M., Tondel M. and Ahmad S.A. (1999). Hypertension and arsenic exposure in Bangladesh. *Hypertension*, vol. 33, pp. 74-78.

Smith A.H., Lingas E.O. and Rahman M. (2000). Contamination of drinking-water by arsenic in Bangladesh: a public health emergency. *Bulletin of World Health Organization*, vol. 78, pp. 1093-1103.

Natural Occurrence of Arsenic in the Organic-rich Aquifer in Hanoi City: Sources and Mobilization Processes

Tran Thi Viet Nga[1)] and Satoshi Takizawa[2)]

1) Environment and Sustainable Development, United Nations University
53-70, Jingumae 5-chome Shibuya-ku, Tokyo 150-8925, Japan
E-mail: nga@hq.unu.edu
2) Department of Urban Engineering, Graduate School of Engineering, University of Tokyo
7-3-1 Hongo, Bunkyo-ku, Tokyo 113-8656, Japan
E-mail: takizawa@env.t.u-tokyo.ac.jp

Abstract

High concentration of arsenic (up to 112 μg/l) was found in groundwater from the South region of Hanoi City. Arsenic speciation and other chemical parameters, e.g. pH, redox potential, HCO_3^-, DOC, NH_4^+, Fe have been studied and showed that arsenic is natually leaching in reducing condition. The spatial distribution of elevated aqueous arsenic concentrations (>50 μg/l) well correlated with the distribution of peat. In oder to further investigate the source and mechanism of arsenic leaching, core-samples were taken from four boreholes dug in the well-field area. The ambient geochemial concentration and groundwater quality mornitoring both suggested that the degradation of organic matter plays a role in arsenic moilization. Arsenic bound to organic matter and ferric oxyhydroxides are considered to be released into groundwater when soil organic is degraded to reduce Fe(III) to Fe(II).

Keywords: groundwater, contamination, arsenic, organic matter, Hanoi

1. Introduction

The arsenic is well recognized for its high toxicity, complex hydrochemistry and difficulty in removal. Elevated concentration of arsenic in natural waters associed As-rich materirals and arsenic then being release through variuos mechanisms depending on local environment. Among numerous mineral containing arsenic have been reported, iron oxyhydroxides are considered important solid phases for arsenic. Various pathways aslo have been proposed for arsenic mobilization. Some have suggested that arsenic may have been released through sedimentary sulfide oxidation induced by the extensive pumping in dry-season. Some have argued that arenic is mobilized by the reduction of iron oxyhydroxide or sorbed arsenate by biodegradation of peat deposits. Other mechanisms also involved to explain the arsenic contamination in groundwater are displacement of arsenic by carbonate or by phosphate.

Groundwater in Hanoi city generally met the requirements for Vietnam Drinking Water Standard except manganese and iron. However, groundwater pollution due to arsenic, iron, ammonia and organic matter has recently been reported (HWBS, 2000, Berg et al. 2001, Nga et al., 2001). These reports raised concern among public on the safety of their drinking water and as a result groundwater contamination has become an immediate issue in Hanoi city. Although numerous research studies on groundwater quality have been carried out in the area, the mechanisms controlling release of these contaminants remain unclear.

The research was carried out with the major objective is to identify sources and elucidate mechanisms of groundwater contamination due arsenic by characerization of the aqueous and solid phase-arsenic. Between August 2000 and September 2001, we collected goundwaters and sediment cores from well-fields of the Hanoi Water Business Company and analysed for aqueous arsenic and other water quality parameters as well as solid-phase arsenic and soil geochemical parameters. The correlations among different water and soil parameters were examined to estimate the sources and mechanisms of arsenic leaching. Our results indicate the peat is sources for sedimentary arsenic and show the evidences of two possible mechanisms of arsenic leaching, namely the anaerobic degradation of organic matter in the peat layers and the reduction of iron oxyhydroxides. These results would be beneficial for the development and management for water sources and water supply in Hanoi City.

2. Materials and methods

2.1 Water sampling and analysis

Groundwater samples were collected from 45 deep wells in 8 major water treatment plants (WTPs) over three sampling periods (August 2000, December 2000, and September 2001). A water sample was collected into three 100 ml-plastic bottles and brought back to the Environmental Engineering Laboratory, University of Tokyo for analysis. All samples were stored at 4oC. Acidified samples by HNO_3 (1ml of $HNO_3$1:1) were used to determine major cations, trace metals and arsenic species by ICP-EAS (Optima 3000, Perkin-Elmer) and ICP-MS (HP4500, Yokogawa). Non-acidified sample were used to analyze major anions by IC. Acidified samples by HCl (1ml of concentrate HCl) were used to determine NH_4^+, dissolved organic carbon DOC, and Total Nitrogen (TN). Arsenic speciation was performed with an liquid chromatography (Shimidzu LC-10A) coupled with ICP-MS to determine four As species: arsenous acid As (III) (H_3AsO_3, $H_2AsO_3^-$), arsenic acid As(V) ($H_2AsO_4^-$, $HAsO_4^{2-}$), methylarsonic acid MMAA ($(CH_3)AsO_2H_2$, $(CH_3)AsO_2H^-$) and dimethylarsenic acid (DMAA) ($((CH_3)_2AsO_2H$, $(CH_3)_2AsO_2^-$).

2.2 Cores analyses

Soil sampling was carried out in September 2001. Four bore cores were drilled within the Phap Van, Ha Dinh, Mai Dich and Luong Yen well-fields. The locations were chosen because groundwater quality in these well-fields was very much different. The bore cores were drilled next to the existing well which was frequently sampled in our investigations. Sediment cores were taken at different depths between 2-40m. Immediately after the sediment cores were collected, they were placed in plastic bags and then keep in a plastic box under anaerobic condition using AnaeroPark (Mitsubishi Gas Chemical), and were kept in dark 4^oC. A total of 68 sediment cores from four boreholes were collected. After brought back to Japan, a part of cores were crushed and dried for 4-5 days at the room temperature (22^oC). The air-dried cores were crushes to 2 mm in size, and homogenized by thorough mixing. The remained fresh cores were still kept in anaerobic box at 4^oC until further analysis.

The solid-phases of arsenic were determined by applying the sequential extraction procedure. The procedure modified from several reported references (Ahmann et al., 1997 and Zobrist, et al., 2000) was used to target different solid-phase arsenic as exchangeable, metal-oxide-bound, organic and sulfide-bound, well-crystallized (hydro)oxides of metal (Fe, Al, Mn)-bound and residuals. Sediment (1gram) was place in 50ml plastic bottle and 20ml of the extraction solutions were added sequentially. After every extraction step, the bottle containing the soil and the extractant were centrifuged for 5 minute at 3500rpm. The solution was filtered through 0.45 μm cellulose acetate filter paper and diluted with water to a volume of 30ml, then kept in a plastic bottle in 4°C prior to analysis by ICP-MS. The residual sediment was used for the sub-sequent extraction steps

3. Results and discussion

3.1 Arsenic in groundwaters

Arsenic levels in groundwaters ranged from less than the detection limit to 112 μg/l over the three sampling periods (Figure 1) . In the northern part, groundwaters collected from Mai Dich, Ngoc Ha, and Ngo Si Lien well-fields had low level of arsenic (less than 2μg/L). Groundwaters collected along Red River (Luong Yen and Yen Phu well-fields) had moderately high arsenic level (< 30μg/l). The rest, including groundwaters collected from Tuomg Mai, Phap Van and Ha Dinh well-fielda in the South had high arsenic concentrations of 43-112 μg/l which were above the WHO guideline for drinking water (10μg/l). The concentration ranges of arsenic detected in this study are about one-third of other report (Berg et al, 2001), though spatial distribution and seasonal variations of arsenic content have been observed in the lower aquifer was the same. One-way analysis of variance (ANOVA) was applied to the data set of three sampling periods shows that the concentration of arsenic was higher in rainy season (August 2000 and September 2001) than in dry season (December 2000) ($p>0.5$).

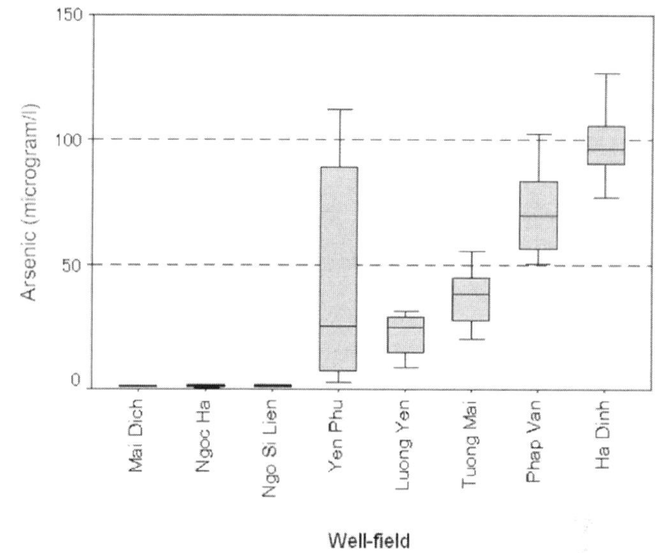

Figure 1. Box plots of arsenic concentration in eight well fields in Hanoi City. The line across the box represents the median. The bottom and top of the box show the location of the first and third quartile values. Data from the August 2000, December 2000 and September 2001 sampling.

Samples having arsenic concentration above 30 μg/l were selected for analysis of four arsenic species, namely arsenite As(III), arsenate As(V), MMA, and DMA. The total dissolved arsenic and the concentration of its species are given in Table 1. The main arsenic species detected in the groundwaters were As(III) and As(V), with the ratio of As(III) and total arsenic varied greatly from 0.49 to 1.00.

3.2 Arsenic in sediment cores

Arsenic concentration varied from less than 1 to 20 mg/kg with depth in stratigraphically different sediment layers in all four boreholes. The highest arsenic concentrations of 4-20 mg/kg were associated with the organic-rich peat samples, followed by greenish gray clay (3-12 mg/kg), and clayey silt and silt (1-9 mg/kg) and red brown clay and sand (up to 7 mg/kg). The core samples underwent a sequential extraction, and each extract was analyzed for arsenic to allow detection of a single solid-phase arsenic associated with sediments. The $(NH_4)_2SO_4$-extractable arsenic presumably associated with exchangeable arsenic, which is ionically bound to sediment particles, was found at low level as 0.08 mg/kg (max: 0.26; min:

0.02 mg/kg). Extraction with $NH_4H_2PO_4$ targeted arsenic bound to amorphous Fe oxyhydroxides, yielded about 0.03-5.34 mg/kg (mean: 1.43 mg/kg). Arsenic released by digestion with heated H_2O_2, intended to oxidize organic matter and sulfides and thus release associated species, were slightly less between 0.02-5.61 mg/kg (mean: 0.99 mg/kg). However, we found that the peats have higher content of arsenic-bound to OM when compared with the portions bound to iron oxyhydroxides. Oxalate-extractable arsenic, accounted for largest portions of extractable arsenic in the soils, which ranged from 0.01-9.33 mg/kg.

Table 1. Arsenic species in groundwater[1]

Well label (well-field name)	Fe^{2+} mg/l	NH_4^+ mg/l	ORP	As(III) μg/L	MMA μg/L	DMA μg/L	As(V) μg/L	T-As μg/L	As(III)/ ΣAs
Y20 (Yen Phu)	7.60	3.5	-98	67.4	ND	ND	57.1	99.6	0.54
Y22 (Yen Phu)	3.16	2.2	-65	61.1	ND	ND	62.7	85.7	0.49
P2 (Phap Van)	7.55	13.0	-60	39.7	ND	ND	15.1	56.3	0.73
P5 (Phap Van)	7.30	12.0	-100	36.0	ND	ND	6.7	51.1	0.84
P6 (Phap Van)	8.15	21.0	-65	37.7	ND	ND	7.7	50.6	0.83
P8 (Phap Van)	8.15	29.0	-92	81.3	ND	ND	ND	79.5	1.00
P9 (Phap Van)	8.10	19.0	-94	48.1	ND	ND	18.2	69.5	0.73
T11 (Tuong Mai)	6.85	4.5	-77	36.0	ND	ND	ND	55.5	1.00
T15 (Tuong Mai)	6.25	5.1	-77	23.8	ND	ND	ND	33.1	1.00
T16 (Tuong Mai)	7.80	11.0	-71	32.2	ND	ND	22.3	43.9	0.59
H6 (Ha Dinh)	20.00	9.4	-83	61.5	ND	ND	33.0	92.0	0.65
H10 (Ha Dinh)	13.65	7.5	-80	50.3	ND	ND	27.1	77.1	0.65
H17 (Ha Dinh)	8.20	5.0	-90	49.4	ND	ND	31.5	77.3	0.61
H18 (Ha Dinh)	10.25	7.2	-80	68.9	ND	ND	24.6	83.1	0.74
H14 (Ha Dinh)	8.45	6.6	-80	71.2	ND	ND	17.1	90.2	0.81

Note: ΣAs: As(III) + MMAA + DMAA + As(V), analysis by LC-ICP-MS
T-As : mean analysis by ICP-MS
ND: not detectable
1. Data from December 2000 sampling

3.3 Simultaneous contamination processes due to arsenic, iron, ammonia and organic matter in groundwater in South region of Hanoi City.

Nga et al (2003) had characterized the hydrogeochemistry of groundwater in Hanoi using multivariate data analysis and successfully identified three groundwater zones in the lower aquifers beneath the city, i.e. the northwest zone including Mai Dich, Ngoc Ha and Ngo Si Lien well-fields, the riverain zone including Yen Phu and Luong Yen well-fields, and the south zone cover the area of Tuong Mai, Phap Van and Ha Dinh well-fields. The northwest groundwater have very good quality and the riverain-groundwater quality shows similar with Red river water, the South-zone, however, have high concentration of iron, arsenic, ammonia and organic matter (Nga et al. 2003). In this report, we also analysis groundwater contamination by grouping of water quality data into three groups correspond to three distinguishable groundwater zones.

Reducing environment which indicated by negative ORP values, was observed in the aquifer along the Red River (Luong Yen and Yen Phu well-fields) and in the South (Phap Van, Ha Dinh and Tuong Mai well-fields). Hydrogen sulfide (H_2S), a pronounced indicator of reductive condition, was only detected in groundwaters collected along the Red River. No detection of both SO_4^{2-} and H_2S in the groundwaters from the South indicates that sulfate reduction is taking place under strongly anaerobic conditions and sulfide produced may combine with the a part of ferrous iron Fe(II) or dissolved As to form insoluble FeS or As sulfosalt. These processes might be the cause of no sulfide in groundwaters rich in arsenic and ammonia.

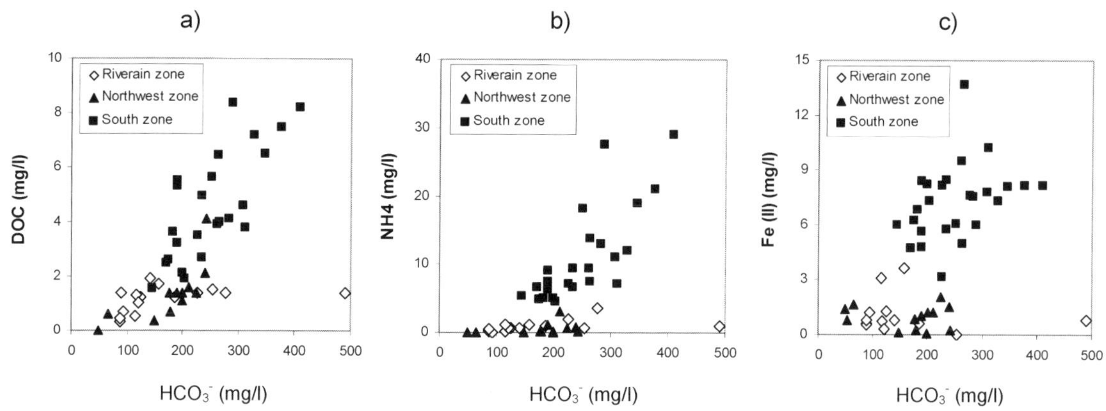

Figure 2. Correlation between groundwater quality parameters: a) Dissolved Organic Carbon (DOC) and bicarbonate (HCO^{3-}), b) ammonia nitrogen and bicarbonate (HCO^{3-}), and c) ferrous iron (Fe(II)) and bicarbonate (HCO^{3-}). Data sampled in December 2000 and September 2001 sampling periods

Elevated concentrations of NH_4-N (4.4-29 mg/l) and DOC (2.8-13.1 mg/l) were found in groundwaters of Tuong Mai, Phap Van and Ha Dinh well-fields in the South of the City. Figure 2a and 2b shows that HCO_3^- increase with the increase of DOC and NH_4-N. These good correlations may be explained by the anaerobic degradation of soil organic carbon that released DOC and NH_4-N. The DOC can be subsequently degraded to methane and bicarbonate in fermentation processes. As the anaerobic degradation processes of OM are slow, and some of DOC is not biodegradable, consequently it is usual to detect both DOC and HCO_3^-. Further, in most of the samples about 80% of the T-N was NH_4-N; and the range of NO_3-N in the groundwater was very low, between the undetected level to 0.22 mg/l. It reveals that under the strong reducing conditions, NH_4-N resulting from the degradation of OM would remain as a stable form.

The South groundwaters contain very high concentration of Fe (7-38mg/l) whereas very low Fe content was detected in other places. Of them, Fe(II) accounted for about 40-50 % (3.8-13.6 mg/l), and it increased with the total Fe concentration. These findings suggested that reduction of Fe(III) has occurred under a reducing condition. High concentration of Fe and degradation products of OM (DOC, NH_4-N and HCO_3^-), accompanied with good correlations among these parameters (Fig. 2a, 2b and 2c) strongly support that the Fe reduction process followed the model proposed by Nickson et al. (2000). In this model, reduction and dissolution of Fe are driven by the degradation of OM under reducing condition.

$$4FeOOH\ (limonite) + CH_2O + 7H_2CO_3 \rightarrow 4Fe^{2+} + 8HCO_3^- + 6H_2O \qquad (1)$$

3.4 Arsenic reduction and mobilization

Figure 3a shows that high arsenic concentration associated with high Fe(II) content in the groundwaters from the South of city, indicating that the reduction of Fe may have a link with the arsenic mobilization. These results are well consistent with many previous studies (Nickson et al., 2000). In addition, results from arsenic speciation showed that As(III) was predominant compared to As(V) in these groundwaters. We can conclude that the reduction of As(V) to As(III) has occured. As As(III) has been known to be more mobile spieces (Zobrist et al., 2000; Langner et al., 2000; Ahmann et al., 1997), the direct reduction of As(V) might play an important role in arsenic solubilization.

We have also found a good correlation between arsenic and some organic degradation-products, such as HCO3$^-$, P, and NH_4^+ (Fig. 3b). The strong correlation between dissolved arsenic and HCO_3^- concentration in the groundwater was also identified by many researchers

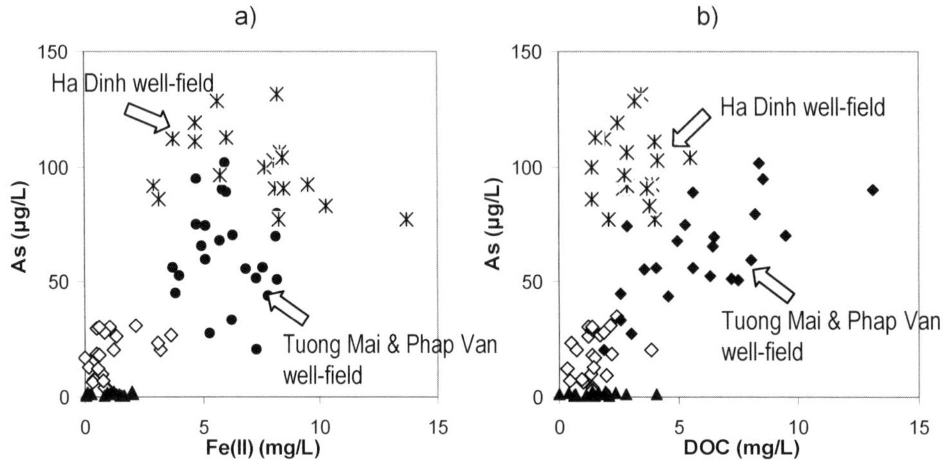

Figure 3. Correlation between groundwater quality parameters: a) Arsenic and Ferrous iron (Fe(II)), b) Arsenic and Dissolved Organic Carbon (DOC). Data sampled in August 2000, December 2000 and September 2001 sampling periods.

(Nickson et al., 2000; Kim et al., 2000). From Equation 1 the ratio of released Fe(II) and HCO_3^- is 2, but the ration found in our study is about 24, which is 10 time higher. It implies that ony small fraction of DOC in groundwater is used to reducce Fe(III) to Fe(II). Nickson et al. (2000) in their study of groundwater pollution in Bangladesh, also found that released HCO_3^- was by far greater than the concentration predicted by Equation 1, and they estimated that most of HCO_3^- came from natural mineral dissolution. Our results, instead, suggested that most of HCO_3^- came from the degradation of soil organic matter. These findings also suggest that degradation of oraganic matter plays an important roll in arsenic mobilization.

3.5 Peat as arsenic sources

Solid-phase arsenic and peat layer appear to be associated. Figure 4a shows that organic matter bound-arsenic increases with the carbon content and NH_4-N content in the cores taken from the Ha Dinh borehole. These results were consistently observed in all cores in four boreholes. Particularly, a high portion of As-bound OM was associated only with peat samples. These results suggest that organic materials in these soils are an important solid-phase for arsenic.

In a similar study on arsenic contamination in the Red River Delta, Berg et al. (2001) reported a good correlation between extracted arsenic and extracted Fe from sediments, suggesting that arsenic could be associated with Fe phases. Our results, however, did not show any significant correlation between these two parameters in all cores (Fig. 4a and 4b). The Fe contents ranged from 7.2-102.8 mg/g and were quite similar throughout the deep, with no apparent correspondence to the elevated arsenic level found in some sediment layers such as peat or green clay. These findings suggest that Fe minerals do not have significant influence on the distribution of soil arsenic.

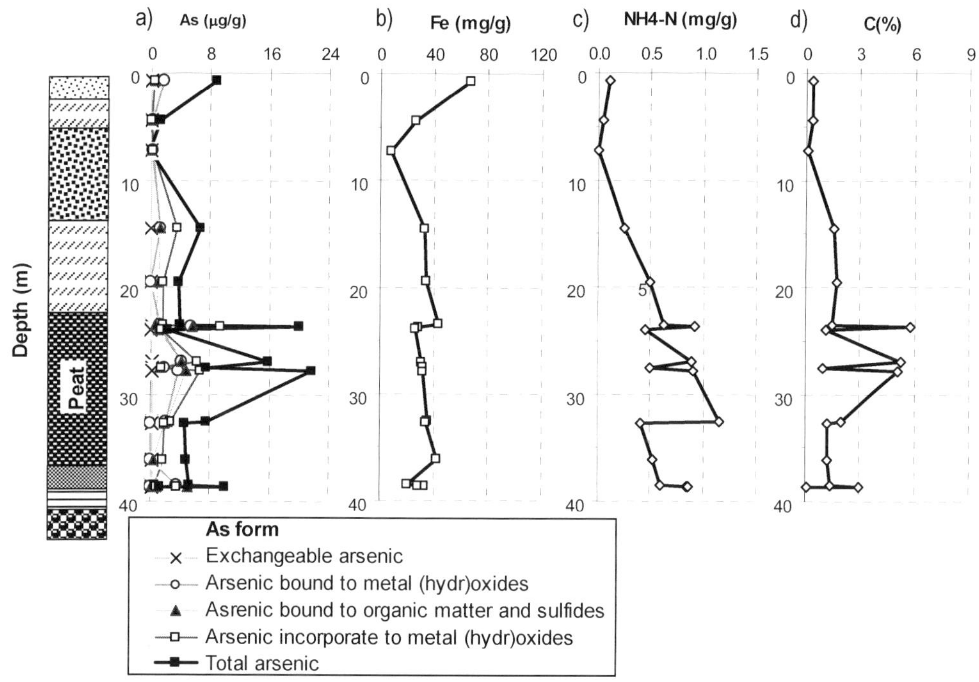

Figure 4. Vertical profiles of geochemical characteristics in Ha Dinh bore-hole: a)Arsenic dissolved by the sequence of extractions that target different solid phases, b) soil Iron, c) soil Ammonia Nitrogen, and charcoals, black mud, or leave and lignic) occurred at the depth from 24-36 m).

Further, results from sequential extraction show in the peat cores, 70-100% of the total sedimentary arsenic is comprised of Fe-bound and OM-bound arsenic. It indicates that peat is sources of sedimentary arsenic, i.e. peat sequester arsenic through two major solid phases: arsenic bound to OM and associated with Fe oxyhydroxides. Results obtained from water monitoring were consistent with our model, indicated by the coincidence of geographical distribution of arsenic rich-wells with the distribution of peat layer.

As mentioned earlier, both the dergadation of OM and reduction of Fe oxyhydroxides are responsible for the mobilization of arsenic. However, the correlation between different water paremeters do not show a similar trend when we look at each well-field independently (Fig. 3a and 3b). Figure 3b shows a good correlation between total arsenic and DOC for the groundwaters from Tuong Mai and Phap Van well-fields. Eventhough these groundwaters have relatively high level of Fe, the release of arsenic had a close link with the degradation of organic matter, and not with the reduction of iron . The groundwaters from Ha Dinh well-field, nevertheless, do not show such correlation. The spots were clustered, showing DOC and arsenic values vary within a narrow range. These results imply that the former or later mechanism will occur significantly depending on different hydrogeological conditions. Whille degradation of organic matter is the prequisite condition for arsenic mobilization, it also control the leaching of arsenic in the Phap Van and Tuong Mai well-field, whereas the reduction of Fe plays predominant roll in Ha Dinh well-field.

4. Conclusions

The geochemical concentration and water quality monitoring both suggested that the lower aquifer in the South of Hanoi City are undergoing simultaneous processes including the degradation of organic mater, the reduction and dissolution of iron, and the reduction and mobilization of arsenic in a strong reducing condition. Peat occurs extensively in the South and is consistent with the spatial distribution of arsenic-rich groundwater. Peat layer sequester arsenic through two major solid phases: arsenic bound to organic matter and associated with iron oxyhydroxides. Mobilization of arsenic can directly occur through the microbial fermentation of buried peat layers which are known to sequester arsenic, or be facilitated by the reductive dissolution of Fe oxyhydroxides and then the release of its sorbed arsenic.

5. References

Ahmann, D., Krumholz, L.R., Hemond, H. F., Lovley, D.R., and Morel, F.M.M. (1997). Microbial mobilization of arsenic from sediments of the aberjona watershed. In: *Environmental Science and Technology*, vol. 31, pp.2923-2930.

Berg, M., Tran, H. C., Nguyen, T. C., Pham, H. V., Schertenleib, R., and Giger, W. (2001). Arsenic contamination of Groundwater and drinking water in Vietnam: A Human Health Threat. In: *Environmental Science and Technology*, vol. 35, pp.2621-2626.

HWBC. (2000) *HWBC Monitoring data for 1989-1999*. Hanoi Water Business Company

Kim, M.J., Nriagu, J. and Haack, S. (2000). Carbonate ions and arsenic dissolution by groundwater. In: *Environmental Science and Technology*, vol. 34, pp.3094-3100.

Langner, H.W., and Inskeep, W.P. (2000). Microbial reduction of arsenate in the presence of ferrihydrite. In: *Environmental Science and Technology*, vol. 34, pp.3131-3136.

Nickson, R.T, Mcarthur, J.M., Pavenscroft, P., Burgess, W. G., and Ahmed, K.M. (2000). Mechanism of arsenic release to groundwater, Bangladesh and West Bengal. In: *Applied Geochemistry*, vol. 15, pp.403-413.

Nga, T.T.V., Inoue, M., Khatiwada, N.R. and Takizawa, S. (2001). Groundwater contamination due to arsenic, iron, ammonia, and organic matter in Hanoi City, Vietnam. In: *Proceedings of the IWA 2nd World Water Congress, October 2001, Berlin.*

Nga, T. T. V., Inoue, M., Khatiwada, N. R., and Takizawa, S., (2003). Heavy metal tracers for the analysis of groundwater contamination: case study in Hanoi City, Vietnam. In: *Water Science & Technology: Water Suppl,* vol. 3(1-2) pp 343-350, IWA Publishing 2003.

Zobrist, J., Dowdle, P.R., Davis, J.A., and Oremland, R.S. (2000). Mobilization of arsenite by dissimilatory reduction of adsorbed arsenate. In: *Environmental Science and Technology*, vol. 34, pp. 4747-4753.

Severity of Arsenic Concentration in Soil and Arsenic-rich Sludge of Bangladesh and Potential of their Biological Removal: a Novel Approach for Tropical Region

S.M. Atiqul Islam[1)], Kensuke Fukushi[2)] and Kazuo Yamamoto[3)]

1) Department of Urban Engineering, The University of Tokyo, 7-3-1 Hongo, Bunkyo-Ku, Tokyo 113-8656, Japan, e-mail: atiqul@env.t.u-tokyo.ac.jp
2) Environmental Science Center, The University of Tokyo, 7-3-1 Hongo, Bunkyo-Ku, Tokyo 113-0033, Japan, e-mail: fukushi@esc.u-tokyo.ac.jp
3) Environmental Science Center, The University of Tokyo, 7-3-1 Hongo, Bunkyo-Ku, Tokyo 113-0033, Japan, e-mail: yamamoto@esc.u-tokyo.ac.jp

Abstract

Arsenic toxicity in groundwater is a common problem in many countries including Bangladesh where the problem is very severe. Though lots of efforts have been paid to find safe drinking water in Bangladesh but no suitable outcome obtained till today. Moreover, disposal of arsenic-rich sludge from different arsenic removal units creates more problems. Due to use of contaminated irrigation water, arsenic concentration in soil increases. This study was conducted to assess the severity of arsenic concentration in soil and arsenic-rich sludge in Bangladesh and to confirm some processes of biotransformation of arsenic that can be involved in its remediation from soil/sludge. High-level arsenic concentration was observed in soil from agricultural land receiving arsenic-rich irrigation water and soil from disposal land receiving arsenic-rich sludge-water. This study confirmed anaerobic bioprocess of arsenate reduction and methylation conducted in mesophelic temperature by soil microorganisms collected from Bangladesh. Similar process in similar condition conducted by microorganisms present in cow-dung was also observed in a higher rate. Biomethylation/volatilization process has a very good potential in bioremediation in tropical region where normal atmospheric temperature is nearly mesophelic. Establishment of such bioremediation process is possible after necessary investigation. Thus it can be a fruitful solution for tropical climatic condition and may be perfectly suitable for affected countries in Southeast Asia.

Key Words: Arsenic methylating bacteria, arsenic reducing bacteria, arsenic-rich sludge, Bangladesh, cow-dung, soil

1. Introduction

Arsenic is a naturally occurring element and is assumed to be the 20th most abundant element in the biosphere (Ahmed, 2000). It has also several anthropogenic inputs such as agricultural and poultry applications in contrast to its biological toxicity. However, arsenic contamination has become a common problem in many parts of the world. Arsenic leaching problem has occurred from mine tailings in Australia, Canada, Japan, Mexico, Thailand, United Kingdom and the United States. Arsenic contamination from natural aquifers of water supply has occurred in Argentina, Bangladesh, Cambodia, Chile, China, Ghana, Hungary, India, Mexico, Nepal, New Zealand, Philippines, Taiwan, United States and Vietnam (Wilson, 2004).

Arsenic pollution has become in tremendous dimensions throughout Asia. Arsenic pollution crises are the most severe in Bangladesh and India (West Bengal) and significant in China, Nepal, Pakistan, Thailand and Viet Nam. It is estimated that more than 35 million people are consuming arsenic-polluted water in Bangladesh alone where underground water is used for drinking and cooking. In Guizhou, China, high arsenic concentrations in coal have caused air pollution with serious health casualties. Victims are also found among villagers near a geothermal power plant in Mindanao Island in the Philippines and among ex-workers and residents near Ronpibool mining sites in Thailand and Matsuo in Japan. (AAN, 2002).

The latest statistics indicates that it exceeds 50 µg/l (Bangladesh standard for Drinking Water) in many tube wells of 60 districts out of total 64 and about 75 million people are under the risk of arsenic poisoning in Bangladesh (Rahman *et al.*, 2002). The effects of drinking arsenic-contaminated water include skin lesions and cancers, gangrene and internal cancers of the bladder, kidneys and lungs. Hence lots of efforts have been made to find safe drinking water. Even today, no suitable solution has been established. Every method so far established to provide safe drinking water produces considerable amount of arsenic-rich sludge. Existing disposal options of this sludge are to through it on adjacent low land, ditches, which increases the possibility of further contamination of soil and groundwater. On the other hand, continued cropping with arsenic contaminated irrigation water increases the extent of contamination in agricultural land. Recent studies observed considerable arsenic concentration in rice and vegetables grown on arsenic polluted soil (Ali *et al.*, 2003a; Hironaka *et al.*, 2003). So, remediation of arsenic from soil is another challenge for the country.

The origin of arsenic and its fate in soil environment have not yet been known. A number of microorganisms, plants and invertebrates play important roles in the distribution and cycling of arsenic in soil environment (Tamaki *et al.*, 1992). Arsenic can be accumulated and transported by different processes like reduction, oxidation and methylation. These processes are the part of the arsenic transport cycle. Arsenic can be transferred to plants, invertebrates or microbiologically decomposed to volatile methylarsine gases (Mukhopadhyay *et al.*, 2002). The production of methylarsine gases from inorganic arsenic occurs as a result of yeast, fungal and bacterial methylation processes. Release of arsenic by natural low temperature biomethylation into the atmosphere was evidenced. It was estimated that as much as 210×10^8 g of arsenic is lost to the atmosphere in the vapor state annually from the land surface and major part of this is contributed by this biogenic activity (Tamaki *et al.*, 1992).

Microbial reduction of arsenate is the activity of dissimilatory arsenic-reducing bacteria (DAsRB). DAsRB couple the reduction of As (V) to the oxidation of an organic compound or H_2 and thereby conserve energy for growth. Bacterial methylation of inorganic arsenic has been studied extensively in methanogenic bacteria (McBride, 1971). Arsenate, arsenite and methanearsonic acid can serve as substrates in methylarsine formation. Inorganic arsenic methylation is coupled to CH_4 biosynthetic pathway and may be a widely occurring mechanism for arsenic detoxification (Tamaki *et al.*, 1992). Arsenic as inorganic arsenate is reduced and methylated to volatile methylarsine by DAsRB and Arsenic Methylating Bacteria (AsMB) in soil. This conversion is a multistep process, with arsenate initially being reduced to arsenite followed by a sequence of methylation and reoxidation. Arsenic volatilization may be used as a mechanism of arsenic remediation if it can be fully exploited (Valls *et al.* 2002).
Objectives of this study is to assess the severity of arsenic concentration in soil and arsenic-rich sludge produced from existing arsenic removal units in Bangladesh and to confirm the

microbial transformation of arsenic in contaminated soil/sludge for their potential application in bioremediation suitable for tropical regions. This study covers arsenic concentration analyses in different soil and arsenic-rich sludge samples of Bangladesh along with some reviews of similar data from other researchers. It also covers the process confirmation of microbial volatilization of arsenic via reduction and methylation in soil and cow-dung at mesophelic temperature. Considering one effective treatment of arsenic-rich sludge by anaerobic digestion with cow-dung, this study attempts to confirm the microbial processes of arsenic volatilization in cow-dung.

2. Materials and Methods

2.1 Sample collection

Soil samples from different arsenic-rich soil and arsenic-rich sludge water (flush water) from different AIRP (Arsenic Iron Removal Plant) were collected from 6 districts in Bangladesh. These locations were primarily selected according to high arsenic concentration in groundwater. Water samples from the source tube wells were collected to check the arsenic concentration. Arsenic-rich sludge-water samples were collected from different AIRP plants of two districts in Bangladesh. Soil samples were collected by 50 mm diameter polyvinyl chloride pipe of about 750 mm height. A 1.5 kg hammer was used to insert sampler to required depth. After withdrawing the sampler along with the soil core, both end of the pipe was sealed with tape (Ali *et al.* 2003a) and kept in plastic bottle for transportation to the University of Tokyo. Cow-dung samples were collected from Saitama Prefecture Agriculture and Forestry Research center from fresh cow dropping by cow of age 2-3 years old. Both soil and cow-dung slurry was prepared anaerobically for culture.

2.2 Quantification of arsenic and other metallic ions from soil

Total arsenic concentration (acid soluble) along with other metallic ions from soil, sludge and water samples were measured by ICP-MS (HP4500). Soil and Sludge was extracted with HNO_3 and HCl (1:3) following the procedure mentioned by Ali *et al.* (2003a).

2.3 Microbial transformation of arsenic

Strictly anaerobic batch culture precedure was followed to observe microbial processes of arsenate reduction and methylation in both soil and cow-dung. Duplicate batch cultures were conducted in scrum vials (30 ml in 50 ml vials) sealed with butyl rubber stopper using free headspace filled with NO_2:CO_2 (8:2). Cultures were grown in presence of arsenic as arsenate with different concentrations. Microbial transformation was confirmed by introducing control samples where bacterial inoculums were killed by autoclaving. Methanobacterium (ATCC Medium 1045) was slightly modified by lacking Na_2S and adding of 1 g sodium acetate, 2 g sodium formate per liter as substrate and media was reduced by addition of L-cysteine (0.3 to 0.5 g/L). Resazurin indicator was added to check the presence of oxygen (pink for oxygen). The culture was grown in dark in shaking waterbath (110 rpm) at 37^0C incubation.

Enrichments were analyzed for accumulation of different arsenic species in liquid and gas. In liquid samples, As(III), monomethylarsonic acid (MMAA), dimethylarsenic acid (DMAA), trimethylarsenic oxide (TMAO) were detected by ICP-MS coupled with HPLC (Shimadzu CTO-10A; column GL-IC-A15) with time of growth of bacteria. Volatile methylarsines (Monomethylarsine, MMA; Dimethylarsine, DMA; and Trimethylarsine, TMA) and arsine gas formation in headspace were determined by GC-MS (HP5890; Column DB-5MS). Gas production was quantified before release of gas from the bottle by inserting frictionless glass syringe (10-50 ml size equipped with 20-gauge needle) in the culture tubes.

3. Results and Discussion

3.1 Arsenic Concentration in Soil Sample

High level arsenic concentration (total arsenic) was observed in soil samples collected from close distance to the source tube wells in Comilla, Manikgonj and Munshigonj districts. Depth of soil samples varied from 0-30 cm and concentration varied from 46.9 to 127 mg/kg. It is obvious from the result that high arsenic concentration in source tube well resulted accumulation of high arsenic content in soil. However, no significant correlation between water-As and soil-As was observed from the results shown in Figure 1. Huq *et al.* also did not find any correlation between obtained water-As and soil-As from investigated samples collected from irrigation fields of different districts and some results are plotted in Figure 2. It indicates that the arsenic accumulation in soil not only depends on the source well but also depends on soil properties, especially the clay content (permeability), organic content (biological transformation), etc.

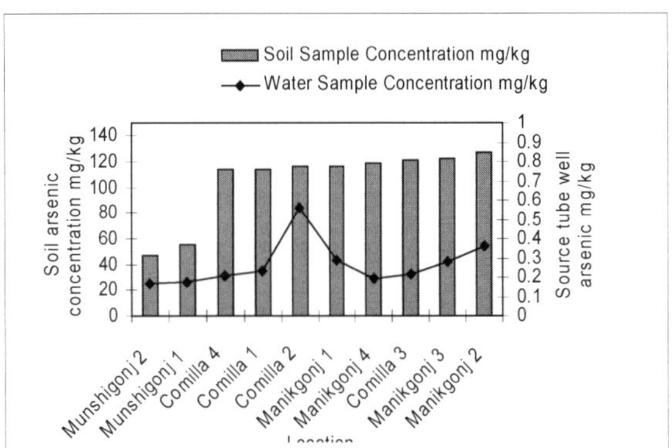

Figure 1. Arsenic concentration in soil and source tube well (Results plotted based on this study)

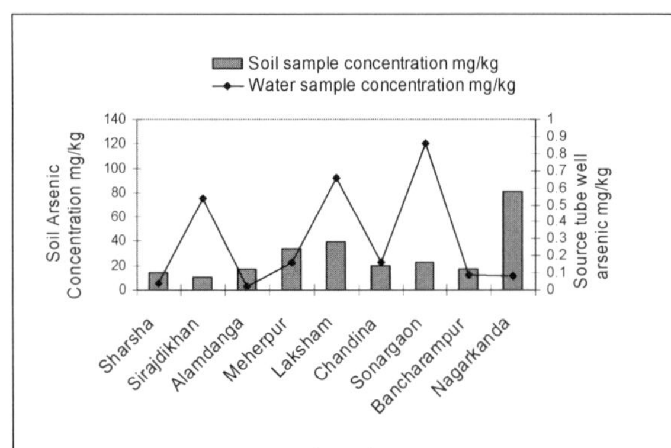

Figure 2. Arsenic concentration in soil and source tube wells (Results plotted based on *Huq et al., 2003*

Table 1: Arsenic concentration in agricultural soil along with other metallic concentration

Sample No.	Place, type and time of collection	As (mg/kg)	Fe (mg/kg)	Na (mg/kg)	Ca (mg/kg)	Mn (mg/kg)	Cr (mg/kg)	Ni (mg/kg)	Cu (mg/kg)	Zn (mg/kg)	Cd (mg/kg)	Pb (mg/kg)
BR-1	Nabinagar, B. Baria; Harvest time agricultural soil	9.8	16270.0	550.0	11170.0	2700.0	28.2	30.9	17.0	50.7	-	17.3
BR-2	Nabinagar, B. Baria; Harvest time agricultural soil	3.9	13325.0	0.0	13590.0	2100.0	24.3	29.0	16.0	45.0	-	30.4
Co-1	Barura, Comilla; Harvest time agricultural soil	40.4	12075.0	1200.0	9735.0	1950.0	14.3	16.8	6.2	28.1	-	6.7
F-2	Vanga, Faridpur; Harvest time agricultural soil	7.3	24480.0	1605.0	43800.0	200.0	25.9	33.3	36.8	49.5	0.2	33.4
MH-1	Shreenagar, Munshigonj; Harvest time agricultural soil	7.0	19100.0	1000.0	13750.0	4350.0	32.8	37.0	40.0	58.0	-	30.4
MH-2	Shreenagar, Munshigonj; Harvest time agricultural soil	3.9	18900.0	1845.0	14640.0	3550.0	26.4	29.6	31.7	42.6	-	29.0
MH-3	Shreenagar, Munshigonj; Harvest time agricultural soil	80.9	18565.0	1945.0	14410.0	3480.0	29.7	33.9	36.2	50.7	-	28.3
M-1-T	Manikgonj, Pre-harvest time agricultural soil	12.3	13750.0	0.0	34890.0	3350.0	17.0	20.5	32.8	12.7	0.3	24.2
M-1-TS	Manikgonj, post harvestime agricultural soil	60.1	15500.0	2150.0	26250.0	3500.0	26.0	31.6	34.4	85.8	0.8	28.8

agricultural fields irrigated by arsenic contaminated water. Results of soil analyses are shown in Table 1. Arsenic concentration in surface soil (0-30 cm) varied from 3.9 mg/kg to 80.9 mg/kg. In line with this study, Ali *et al.* (2003a) obtained high arsenic concentration in top soil (0-15 cm) with a decreasing profile along depth. In this study, similar to Ali *et al.* (2003a), obtained higher arsenic concentration at the end of irrigation period than the beginning. This study also obtained very high iron content in agricultural soil. Elevated concentration of iron and arsenic present in irrigation water certainly increases the concentration in soil. Though no significant correlation obtained between iron and arsenic content but fairly positive correlation between Fe-As and Na-As (Figure 3) obviously indicates that sources of these elements are anthropogenic. On the other hand almost 1:1 relationship between Cr-Ni and Cr-Pb certainly indicates their origin is natural.

Higher arsenic concentration in irrigation water and soil results in higher concentration of arsenic in roots, stem and leaves of rice plants (Abedin *et al., 2002;* Ali *et al.*, 2003a; Hironaka *et al., 2003*). Vegetables and other plants growing on arsenic contaminated soil exhibits higher arsenic concentration (Farid *et al.*, 2003). Arsenic concentration decreases considerably after harvesting due to flood and some bio-geochemical processes (e.g. bio-methylation, dissolution, desorption) occurring in the rice field (Ali *et al.*, 2003a).

Arsenic concentration in arsenic-rich sludge of arsenic removal units

Characterization was performed to know the quality of arsenic-rich sludge water obtained from Arsenic Iron Removal Plant (AIRP) and results are shown in Table 2. The sample was collected during back washing after one month operation of the plant and the source tube well arsenic concentration varied from 0.08 to 0.1 mg/l. The obtained arsenic concentration in flush water was not too high. Actually flush water is not concentrated sludge; it is just back washing water from the sedimentation tank and the filtration tank of AIRP. The volume of this water is around 30-50 liters in each wash. This sludge water is disposed in nearby low lands. However, soil collected from these lands showed very high arsenic concentration varied from 109 to 135 mg/kg.

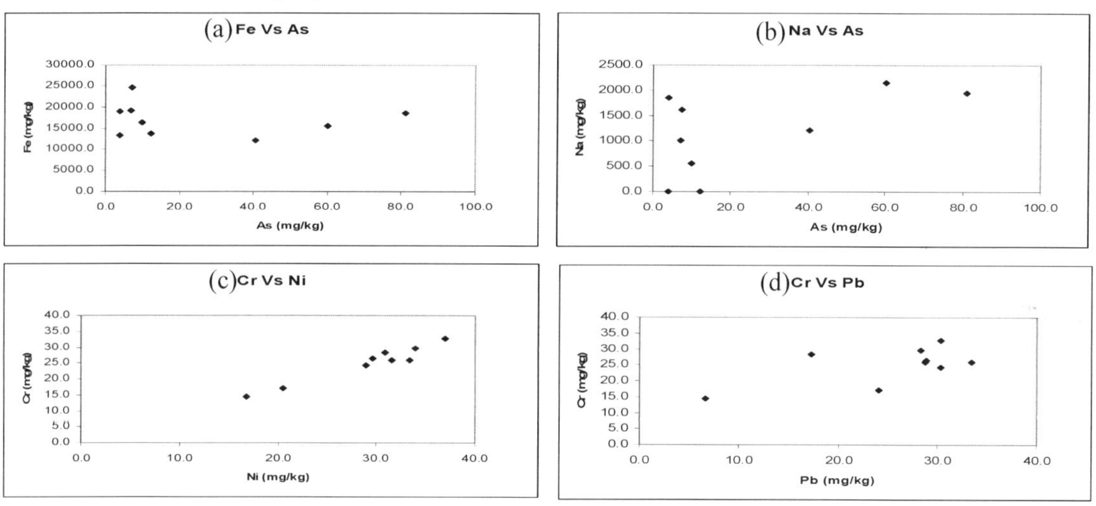

Figure 3: Correlation between different metallic ion concentrations in agricultural field soil (a) Fe Vs As; (b) Na Vs As; (c) Cr Vs Ni; (d) Cr Vs Pd

Table 2: Characterization of arsenic-rich AIRP flush water

Name of Sample	pH	TS (%)	As (mg/kg)	Fe (mg/kg)	K (mg/kg)	Na (mg/kg)	Ca (mg/kg)	Al (mg/kg)	Mn (mg/kg)	Mg (mg/kg)	Zn (mg/kg)
Putiazani 1: Collected from AIRP flush water	7.80	0.39	2.22	345.9	133.4	12.2	2690	0.02	0.35	525	10.5
Putiazani 2: Collected from AIRP flush water	7.88	0.37	2.54	376.1	145.1	13.5	2360	0.04	0.5	560	9.5
Pukuria 1: Collected from AIRP flush water	7.95	0.31	1.5	335.6	78.9	14.1	1570	0.01	0.6	226	3.8
Pukuria 2: Collected from AIRP flush water	7.98	0.23	1.1	277.5	82.6	11.5	1260	0.01	0.8	236	1.01

Besides, arsenic concentration in arsenic-rich sludge collected from other existing arsenic water treatment units of Bangladesh was found considerably high by Ali *et al.* (2003b) and some results are cited in Table 3. Large-scale use of these arsenic removal units may generate significant quantities of arsenic-rich sludge and indiscriminate disposal of this sludge may lead to serious environmental pollution (Ali *et al.* 2003b). One of the option of treatment of this sludge promoted by different forum is anaerobic digestion in nearby pond along with mud and/or cow-dung or domestic sewage (Badruzzaman, 2003). Ali *et al.* (2003b) conducted an anaerobic digestion study of sludge from STAR and BUET-UNU arsenic removal units in presence of cow-dung and found significant elimination of total arsenic from the sludge within few days of digestion. This method may lead to arsenic elimination of this sludge after conducting further research (Ali *et al.* 2003b). Presumably biomethylation of arsenic conducted by microorganisms of cow-dung is responsible for arsenic elimination, which is confirmed by this study and the results are presented in the following section.

Table 3: Arsenic concentration in arsenic-rich sludge of different arsenic removal units(summarized results from Ali *et al*., 2003b

Sample ID	Name of arsenic removal units	Type of sample	% Solids	Arsenic concentration (mg/kg)
TCLP-24	STAR BTU	Slurry waste	8.8	12.9
TCLP-26	STAR BTU	Slurry waste	9.2	38.5
TCLP-30	STAR BTU	Slurry waste	20.4	17.1
TCLP-32	STAR BTU	Slurry waste	16.6	30.3
TCLP-34	BUET-UNU	Slurry waste	2.6	0.9
TCLP-35	BUET-UNU	Slurry waste	2.4	0.93
TCLP-2	Shapla	Spent media	Solid waste	38.8
TCLP-14	Shapla	Spent media	Solid waste	30.2
TCLP-21	Shapla	Spent media	Solid waste	24.8
TCLP-9	SIDKO ARU	Spent media	Solid waste	798
TCLP-10	SIDKO ARU	Spent media	Solid waste	708
TCLP-22	BUET-UNU	Sand	Solid waste	37
TCLP-23	BUET-UNU	Sand	Solid waste	19.2

Microbial processes involved arsenic species transformation in soil

Very high rate of arsenate reduction to arsenite was observed by activity of soil microorganisms culturing in batch in presence of around 1 mM Arsenate, As(V), within 10 days of incubation at 37^{0}C under anaerobic condition. Results shown in Figure 4 indicates that around 59-92% of As(V) was converted to As(III) within 10 days of incubation for different culture samples while almost no transformation occurred in control samples (microorganisms were killed by autoclaving).

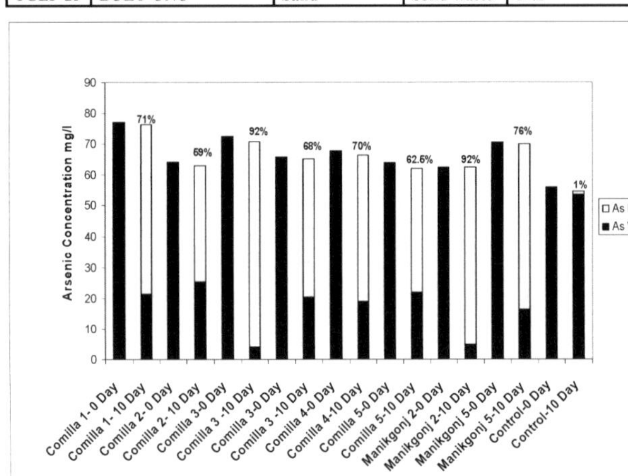

Figure 4: Arsenate reduction to arsenite by soil microorganisms

Microbial methylation/volatilization of arsenic was detected from different soil samples cultured in batch under anaerobic conditions at 37^{0}C using substrate as acetate and formate. Formation of monomethylarsine (MMA) and trimethylarsine (TMA) was observed up to 0.2 ppm within 10 days of incubation with initial arsenic concentration (AsV) of around 0.05 mM as presented in Table 4.

Table 4: Arsenic methylation/volatilization by soil microorganisms from different samples

Sample	MMA (ppm)	DMA (ppm)	TMA (ppm)	Arsine (ppm)
Manikgonj-1: From soil where AIRP flush water drains	0.02	ND	ND	ND
Manikgonj-2: From sediments collected nearby ditch	0.04	ND	ND	<0.005
Manikgonj 3: From sediments collected nearby ditch	0.05	ND	0.2	ND
Manikgonj 4: From soil near tube well	ND	ND	0.1	<0.005
Manikgonj 5: From soil near tube well	0.1	ND	ND	ND
Comilla 3: From soil near tube well	0.01	ND	ND	<0.005

*ND=Not detected (detection limit 0.001 ppm)

This transformation resulted by microbial activity was confirmed by control sample where no transformation occurred. Considerable methane formation was associated with this transformation (up to 30%), which indicates methanogens are dominating species involved in the microbial activity. Thus microbial methylation/volatilization of arsenic by soil microorganisms appears to be an important process eliminating considerable amount of arsenic from liquid/solid.

3.2 Methylation/volatilization of arsenic by microorganisms present in cow-dung

Significant amount of MMA (0.25 mg/l) and little amount of TMA (0.015 mg/l) were observed in batch culture of cow-dung (as seed) under anaerobic conditions incubation at $37^{0}C$ with initial arsenic concentration of 0.1 mM as AsV as shown in Figure 5. All samples tested in this experiment showed similar conversions. The rate of formation of volatile arsines is much higher than that of soil microorganisms. Rate of methane formation was found higher along with this transformation (up to 70%). The rate of conversion of volatile methylarsines depends on the initial arsenic concentration and initial biomass concentration.

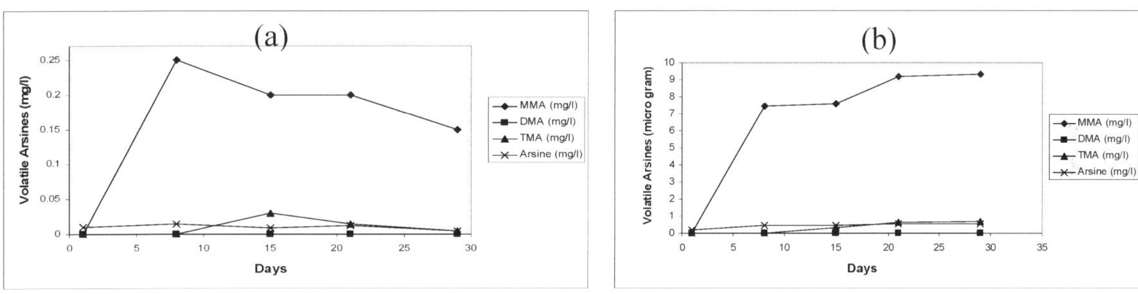

Figure 5: Volatile methylarsines production by cow-dung microorganisms in batch culture with a initial arsenic concentration of 0.1 mM (a) Concentration of arsines in gas phase (b) total arsines (weight) in gas phase

4. Conclusions

Arsenic concentration in agricultural soil in Bangladesh is increasing due to the continuing use of arsenic contaminated irrigation water. The amount of arsenic found in this study is as high as 80 mg/kg in some agricultural field soil (top soil) though the overall arsenic concentration in soil all over Bangladesh is not so high (Huq *et al.*, 2003). For agricultural soil, the regulatory limits of arsenic established for environmental health investigation level in Australia is 20 mg/kg and this suggests that exceeding this level requires soil clean up (O'Neil, 1990) while there is no such guidelines for Bangladesh. Considering increasing concentration in agricultural soil and its subsequent effect on plants, establishment of such guideline is necessary and thereby some soil may require clean up. Arsenic concentration in soil where arsenic-rich sludge is disposed was also observed very high. This study confirms that in anaerobic condition, microorganisms in soil and cow-dung is capable of methylation/volatilization of inorganic arsenic to organic arsenic (methylarsine) forms. Interestingly, organic arsenic has lower acute toxicity than that of inorganic ones. So, it can be concluded that this process can be involved in bioremediation of soil after necessary investigation. Using anaerobic microorganisms (certainly that of cow-dung), it is also possible to establish bioremediation process for elimination of arsenic from sludge before disposal.

Arsenic is a common problem around the world but in some Asian countries such as Bangladesh, India, Nepal, Vietnam, Thailand and China, this problem is quite significant. The possible microbial process (volatilization) for arsenic remediation confirmed in this study in mesophelic temperature, which is certainly favorable in tropical region. That is why, this process can commonly be used in all countries having tropical climate. Thus, this process can suitably be applied for many contaminated region in Southeast Asian countries.

5. References

Abedin M.J., Cresser M.S., Meharg A.A., Feldmann J. and Cotter-Howells J. (2002). Arsenic Accumulation and Metabolism in Rice. *Env. Sci. Technol.,* 36(5). pp 962-968.

Ahmed M.F and Rahaman M. M. (2000). Water Supply and Sanitation – Low Income Urban Communities, International Training Network (ITN) Centre, BUET.

Ali M.A., Badruzzaman A.B.M., Jalil M.A., Hossain M.D., Ahmed M.F., Masud A.A., Kamruzzaman M. and Rahman M.A. (2003a). Arsenic in Plant-Soil Environment in Bangladesh. In: *Fate of Arsenic in the Environment,* M. F. Ahmed, M. A. Ali & Z. Adeel (Ed.), BUET-UNU, pp 85-112.

Ali M.A., Badruzzaman A.B.M., Jalil M.A., Ahmed M.F., Kamruzzaman M., Rahman M.A., and Masud A.A. (2003b). Fate of Arsenic in Wastes Generated From Arsenic Removal Units . In: *Fate of Arsenic in the Environment,* Ahmed, Ali & Adeel (Ed), BUET-UNU, pp 147-159.

Badruzzaman A.B.M. (2003). Leaching of Arsenic from Wastes of Arsenic Removal Systems. In: *Fate of Arsenic in the Environment,* Ahmed, Ali & Adeel (Ed), BUET-UNU, pp 161-179.

Farid A.T.M., Roy K.C., Hossain K.M., and Sen R. (2003). A Study of Arsenic Contaminated Irrigation Water and its Carried Over Effect on Vegetables. In: *Fate of Arsenic in the Environment,* M. F. Ahmed, M. A. Ali & Z. Adeel (Ed), BUET-UNU, pp 123-130.

Hironaka H. and Ahmed S.A. (2003).Arsenic Concentration of Rice in Bangladesh. In: *Fate of Arsenic in the Environment,*M. F. Ahmed,M. A. Ali& Z. Adeel (Ed),BUET-UNU, pp69-84.

Huq S.M.I., Rahman A., Sultana N., and Naidu R. (2003). Extent and Severity of Arsenic Contamination in Soil of Bangladesh . In: *Fate of Arsenic in the Environment,* M. F. Ahmed, M. A. Ali & Z. Adeel (Ed), BUET-UNU, pp 69-84.

McBride B.C. and Wolf R.S. (1971). Biosynthesis of Dimethylarsine by Methanobacterium. *Biochemistry,*10(23). pp 4312-4317.

Mukhopadhyay R., Rosen B.P., Phung L.T., and Silver S. (2002). Microbial Arsenic: from geocycles to genes and enzymes. *FEMS Microbilogy Review,*26. pp 311-325.

O'Neil P (1990). Arsenic. In: *Heavy Metals in Soils,* B.J. Alloway (Ed), John Willey & Sons, New York.

Preface. Asia Arsenic Network (AAN) (2002). Miyazaki, Japan

Rahman M H, Rahman M M, Watanabe C, and Yamamoto K. (2002). In: *Arsenic Contamination in Groundwater -Technical and Policy Dimensions,* Proceedings of the UNU-NIES International Workshop, United Nations University (UNU) and Japan National Institute for Environmental Studies (NIES). pp 9-22

Richard Wilson (2004). Arsenic Project: Summary of the acute and chronic effects of arsenic and the extent of the world arsenic catastrophe. Harvard University, USA.

Tamaki S. and Frankenberger Jr W.T. (1992). Environmental Biochemistry of Arsenic. *Reviews of Environmental Contamination and Toxicology,* 124. pp 79-110.

Valls M. & Lorenzo V.D.(2002). Exploiting the genetic and biochemical capacities of bacteria for the remediation of heavy metal pollution.*FEMS Microbilogy Review,*26.pp 327-338.

Ground Water Shortage in Aquifer of Dhaka City and Land Subsidence: A Major Concern in Potable Water Supply in Dhaka Metropolis in 2030

Muhammad Abu Yusuf[1)], Satoshi Takizawa[2)] and Hiroyuki Katayama[2)]

1) Education Wing, Prime Minister's Office, Old Parliament Building, Tejgaon, Dhaka-1215, Bangladesh. E-mail: yusufjsps@yahoo.co.uk
2) Department of Urban Engineering, Graduate School of Engineering, The University of Tokyo, 7-3-1 Hongo, Bunkyo-ku, Tokyo 113-8656, Japan.
E-mail: takizawa@env.t.u-tokyo.ac.jp; katayama@env.t.u-tokyo.ac.jp

Abstract

Water supply in Dhaka metropolis is basically rely on ground water abstraction and a very small fraction treated water from surface water treatment plants. Dupe Tube Well (DTW) abstracted water from the aquifer under 50 meters and 120 meters of Dhaka City and supply water without any treatment, as the quality of the water is good. Due to excess abstraction ground water table of the aquifer under Dhaka City is declining 1.5 m per year. In the long run, it will not be possible to abstract water from the aquifer under Dhaka City. Moreover, Dhaka City experiences land subsidence which also force to find alternative sources of water for Dhaka City for different water uses. On the other hand, rapid urbanization due to migration, Dhaka City will be the 2nd largest city in the world housing 25.2 million population by 2030 which will put enormous pressure on DWASA to find alternative sources of water supply for Dhaka Metropolis. DWASA needs to shift to surface water treatment and reduce system loss to ensure sustainable water supply for city citizens of today and for population in 2030 which is discussed in this paper.

Keywords: ground water shortage, potable water supply, surface water treatment, water supply demand and deficit, water scarcity

1 Introduction

Dhaka is the capital and largest city in Bangladesh. According to UN Population Division (2003), Dhaka metropolitan is currently the 12th largest city in the world housing 11.6 million populations within an area of about 1529 sq. km (590 sq miles). It was estimated that Dhaka would be the 7th largest metropolis in the world with a population of 17.9 million by 2015 (UN Population Division, 2004). It was also estimated that Dhaka Metropolitan will be the 2nd largest city in the world by 2030 and it would be housing over 25.2 million populations. Dhaka Metropolitan area comprises with Dhaka City (Dhaka City Corporation area) and five other municipalities such as Savar, Narayangonj, Kadamrasul, Tongi and Gazipur lies and under three districts viz. Dhaka, Narayangonj and Gazipur. Dhaka City's water supply started first in 1874 by Nawab Abdul Gani through establishment of Dhaka Water Works (DWW) (Ahmed, 2001). Dhaka Water Supply and Sewerage Authority (DWASA) is responsible to ensure potable and other water uses in Dhaka City since its establishment in 1963. At present surface water treatment is practiced in a limited scale in the Dhaka City and lion share of water supply rely on ground water abstraction. Due to excess abstraction ground water table of the aquifer under Dhaka City is declining 1.5 m per year. In the long run, it will not be possible to abstract water from the aquifer under Dhaka City. Moreover, Dhaka City experiences land subsidence which also force to find alternative sources of water for Dhaka city for different water uses. On the other hand, rapid urbanization due to migration, Dhaka City will be housing 25.2 million population by 2030 and DWASA has to find sources of water supply for the world 2nd largest Dhaka Megalopolis urban agglomeration.

2 Materials and methods

Government published and unpublished documents, project documents and other publications were reviewed and analysed to write this paper. Existing status of drinking water supply, shortage of water supply, water scarcity, problem and challenges of water supply in Dhaka City were identified. An alternative source of potable water supply for Dhaka City is justified to ensure sustainable water supply for city citizens in 2030 is elaborated.

3. Results and discussion

3.1 Water Supply in Dhaka Metropolitan Area

It was mentioned earlier that DCC area and Narayangonj Municipality's water supply is basically based on major part ground water abstraction and small percentage surface water treatment. Dhaka City's water supply is overwhelmingly based on ground water as water quality of the aquifer is very good under Dhaka City. Dupe Tube Well (DTW) abstracted water from the aquifer under 50 meters and 120 meters of Dhaka City and supply water without any treatment, as the quality of the water is good. It was reported that in 2001, DWASA abstracted water through 336 DTWs. Moreover, in addition to DWASA's DTWs, 450 small diameters private DTWs have been installed by various organizations or persons in the Dhaka City and those produce 100 MLD of water to meet their own needs (Ahmed, 2001a; DWASA, 2001). Excess abstraction of water from the aquifer, ground water table of the aquifer under Dhaka City is declining 1.5 m per year (Ahmed, 2001). In the long run, it will not be possible to abstract water from the aquifer under Dhaka City. On the other hand, Dhaka City's land subsidence observed by the expert. DWASA, need to shift to surface water treatment.

3.2 Surface Water Treatment Plant to supply water in Dhaka Metropolitan

There are four surface water treatment plants exist in the Dhaka Metropolitan area. Among four surface water treatment plants, three located in the DCC area, while one is located in the Narayangonj Municipality. Until year 2000, the production capacity of two water treatment plants of DCC area and one of the Narayangonj Municipality were 85 million litres per day (MLD), but practical production was 40 million litres per day (MLD) of water which was less than 4% of the total water supply demand of the city (DWASA, 2000). In 2001, a new 225 MLD surface water treatment plant located in the Sayedabad area of the Dhaka City started operation. The volume of water treated by the surface treatment plants is small compare to total demand of water supply. Still DWASA rely for a significant volume of water supply on ground water abstraction. On the other hand, water quality of the Buriganga and Sitalakhya is deteriorating fast which will not be suitable for the treatment for potable water supply. DWASA is considering Megna and Padma Rivers will be the potential sources of water source for the water treatment plants to supply water in the Dhaka Metropolitan area. The water treatment plants generally disinfected water by chlorination. Chlorine and alum sulphate is using in the water treatment plants to treat the surface water by these three water treatment plants.

Dhaka Metropolitan areas lion share of water supply rely on ground water abstraction is already reported earlier. Prior to operation of Syedabad Water Treatment Plant, the sources of

Dhaka city's water production ratio were ground water and surface water treatment 97:3 (until 2001). Since Syedabad Water Treatment plant operation the ratio of sources of water turns to 82:18 (DWASA, 2001). Dhaka Water Supply and Sewerage Authority (DWASA) provided services to the citizens of DCC area and Narayangonj Municipality in terms of water supply is shown in Table 1. While, demand and deficit of water supply in the DCC area and Narayangonj is given in Table 2.

Table 1: Water Supply services provided by the DWASA in the Dhaka City Corporation area and Narayangonj Municipality

Services	1995	1996	1997	1998	2001
Deep Tube-well (No.)	197	216	225	237	336
Water lifting (MLD)	770	810	870	930	1220
Water Demand (MLD)	1260	1300	1350	1400	1600
Water Pipe line (km)	1404	1462	1475	1610	2127
Water connection (No.)	147,758	154,832	163,675	171,855	191,087

Sources: (Ahmed, 2001: DWASA, 2001); MLD: Million litres per day; Km: Kilometer

Table 2: Demand and Deficit of water supply in Dhaka City Corporation area and Narayangonj Municipality

Year	Population (million)	Running Deep Tube Well (number)	Water Demand (MLD)	Water Supply (MLD)	Deficit (MLD)
1963	0.85	30	150	130	20
1970	1.46	47	260	180	80
1980	3.03	87	550	300	250
1990	5.56	216	1000	510	490
1996	7.55	216	1300	810	490
1997	8.00	225	1350	870	480
1998	8.50	237	1400	930	470
2000	9.50	308	1500	1130	370
2001	10.00	336	1600	1220	380

(Sources: DWASA, 2001)

3.3 Shortage of Water Supply/Water Supply Deficit

The total volume of production of potable water by the DWASA was 1220 MLD as of June 2001. Total population of Dhaka Metropolitan area is about 10 million. The average daily use of water per person is 160 litres, the total demand of potable water in Dhaka Metropolitan area is account 1600MLD. As of June 2001, DWASA's water production was 1220 MLD which is 76% of the demand, the shortage of water supply in Dhaka City was stand 380 MLD, which is 24% of the demand of water supply (DWASA, 2001).

3.4 System Loss/Non Revenue Water of DWASA

DWASA supply water through a 2127 km long pipeline and connected to household level,commercial places and industries (DWASA, 2001). Until 2001, DWASA provided 191,087 water supply connections in Dhaka City. As it was mentioned earlier that a total of 336 deep tube wells and four water treatment plants were producing 1220 MLD of water out of 1600MLD water supply demand in Dhaka Metropolitan area, where shortage of water supply in Dhaka City is accounts 380 MLD, which is 26 % of water supply demand. On the other hand, the system loss in 2001 accounts 42.86 % of total volume of water production per day (DWASA, 2001). The trends of system loss of DWASA in different years are given in Table 3.

The major causes of system loss of DWASA's water are illegal connection, leakage of

Table 3: The percentage of system loss or non-revenue waters in Dhaka city

Year	Percentage of system loss of DWASA
1979	75%
1995	41%
1996	51%
2001	42.86%

(Source: DWASA, 2001)

pipeline, and corruption of DWASW's employees by changing meter reading with the connivance of water users. It was observed that there are many roadside tap connected to make sure water availability in the old city where population congestion is much more. In many incidences water supply is used to restrict for a limited period in residential area. People form the slum and even household level used to collect water from these roadside taps. Those connections were given with recommendation of DCC and these are directly connected with main pipe. But, who is paying for this public taps it is not clear or many incidences none is paying for these taps. Practically, DWASA used to treat these waters as system lose.

There are many other ways also water wastages in the city such as leakage or damage of supply pipes, main pipes and illegal connections with connivance or assistance of lower level of staffs of DWASA. These lower level staffs are involved in this process to earn illegal money. There is no efficient mechanism of monitoring leakage at this stage, as well as there is no community-based information sources exist in the city. That's why it was observed that water leakage continue over 2-3 days or even more. On the other hand, if the DWASA authority was informed in any way, even though mechanism is not that efficient to repair the pipe or lines immediately. It could be either due to inefficient monitoring and unaccountability or inefficiency of the staffs of DWASA. According to DWASA, due to shortage of manpower and absence of physical facilities, it is also difficult for the employees of DWASA to perform efficiently. In many instances, adequate fund is not available in DWASA to perform expensive repair. Moreover, DWASA still have to rely on Ministry's approval on many issues which is a lengthy process and due to inefficient bureaucracy it takes long to have the approval. Delaying of approval at the Ministry also create obstacle for the DWASA to provide services.

However, some corrupt employee of the DWASA engaged in corruption those ultimately ended up with system loss of water or non-revenue water. They used to do negotiation with household level water users and industrial water users. Generally, users were supposed to pay whatever the amount of water they used. Users don't have to pay the exact amount of bill (which is may be very high), if they pay a certain percentage of money to the corrupt employees of the DWASA and their bill used to reduce from exact one. DWASA's staffs are involved in changing meter readings. Through this process system loss is used to be a significant in the Dhaka City and DWASA is facing problem. To overcome this problem, modern technology need to be installed where meter readers cannot change the meter reading. Moreover, efficiency in management may help to resolve this problem. DWASA should approach in that direction to ensure adequate safe water supply to the city.

3.5 Water Scarcity in Dhaka Metropolitan area

A lion share of water supply in the Dhaka Metropolitan area depends on ground water abstraction through Deep Tube Well (DTW), besides water from surface water treatment plants. It has been observed that water table in the aquifer is lowering 1 to 1.5 meter every year under Dhaka City. On the other hand, during dry season water table usually become lower than other seasons. Moreover, during dry spell water consumption rate of the city dwellers is usually also increase, as more water is requires for cooling or bathing purposes. Due to hot weather, city dwellers use more water in the dry season compare to other seasons. According to DWASA, water use usually increased 10% in Dhaka City during summer (DWASA, 2001). But, DWASA continued abstraction at the same rate- like other seasons. In many instances water production disrupt due to electricity failure or load shedding of electricity.

Water supply has the direct link with electricity, as the pump in the dupe tube well and water treatment plants needs electricity to abstract water and treat water respectively. During dry spell, city citizen's electricity consumption increased due to hot weather, as they extensively use fan and air cooler. Excessive electricity consumption, many instances create load shading in many part of the city. Due to load shading, the water treatment plants cannot operate properly and deep tube well cannot abstract water.

On the other hand, out of 366 dupe tube wells, only 230 deep tube wells have the generator and rest of them do not have generator. About 17 mobile generators are using to run rest of the deep tube well when electricity failed to continue water supply in the city. It is really hard for the DWASA within these limited facilities to make sure water supply (DWASA, 2001).

It create major water crisis in some part of the city during dry season. City dwellers experienced severe problem as water shortage in some part of the city last over 2-3 days. At that time DWASA continue water rationing, but the volume of water rationing is not adequate enough to meet the demand of water users. City dwellers are experiencing waster supply problem during dry season almost every year. Though DWASA put lot of effort and continuing negotiation with Dhaka Electricity Supply Authority (DESA) to ensure continuous electricity supply in the city during dry season to make sure adequate water supply for city citizens..

3.6 Future Plan of DWASA to Ensure Water Supply in Dhaka Metropolitan Area

DWASA aim is to ensure 100 % water supply to city dwellers. With this aim DWASA take necessary steps for infrastructure development, as well as measures taken to improve the experienced problem and to raise the revenue with providing services to city citizens (DWASA, 2001). The major steps taken by DWASA are as follows:

i) *Installation of new surface water treatment plant in Pagla* and water intake of this plant will be on Padma River (which is known as Ganages River in the upstream of Bangladesh) that will generate 450 MLD to make sure water supply as per Dhaka City Water Supply Master Plan 2020.

ii) Necessary measures to *reduce system loss* and make organization profitable by leasing out billing system and collection of revenue. DWASA leased out billing and collection of revenue of Revenue zone 4 to Messers EPC Limited, as well as Revenue Zone–5 to DWASA Employees Essential Commodities Cooperatives Union with mutual understanding as per terms and condition of Messers EPC Limited. Leasing out of billing and revenue collection system was the precondition for funding of 4th phase development project by the donor. In future, DWASA will make sure to increase revenue, reduce system loss and provide better services by lease out billing and revenue collection.

iii) Increase the Tariff of water to reduce the loss of DWASA to bring the organization as a profitable organisation. So far every year, DWASA is facing the problem as earning is less then the expenditure. In July 2002, the tariff of water per unit (cubic meter) was 4.30 Taka. Due to increase of the rate of electricity DWASA water production cost rise. That's why DWASA raised the water tariff in January 2003, which was 4.75 per unit. Due to inflation, money devaluation and raise of electricity price, DWASA increased water tariff in January 2004, which was 5.00 Taka per unit for household connection. The rate of water tariff for industrial connection was 15.75 Taka in January 2003 and the increased rate is 16.50 Taka as of January 2004. Even then, DWASA have to face problem and continue to be losing concern as the production cost of water per unit is 6.00 Taka (Prothom Alo, 22 December 2003).

iv) The water supply network in Dhaka city is largely built in between 1943 to1973 and most of the pipelines are dilapidated. DWASA is facing serious problems in terms of maintaining and hindering the illegal connection along those lines (DWASA, 2001; The Daily Star, 2003). To improve those dilapidated pipelines DWASA has taken necessary measures of renovation and replaced water supply pipelines.

v) To overcome water shortage problem during dry season, DWASA has taken necessary measures based on experienced problem in previous years. Among the running 336 dupe tube well pump stations, only 230 deep tube wells have permanent generator and rest of them have to rely on 17 mobile generators. With this existing facilities DWASA facing enormous problem to ensure water supply in Dhaka City. To reduce the problem specially during dry season, DWASA took steps to buy 20 new Mobile generator for pump stations and 8 water carrying trucks and two tractors to stream line water supply for city dwellers by water rationing in the city (The Daily Star, 2003). Besides this, DWASA is continuing negotiation with DESA to make sure adequate electricity supply during dry season in all location where deep tube wells are located.

vi) DWASA sets target to reduce use of ground water as sources of water supply in Dhaka city from 82% to 60 % by 2005 (DWASA, 2001). In line with that DWASA is planning to install new surface water treatment plant and trying to construct water intake point on Padma River and Megna River. Moreover, had taken also project on rain water harvesting on the top floor of WASA Bhaban to encourage city citizen to practice rain water harvesting as an alternative sources of water besides ground water for water uses in the city.

vii) DWASA also took steps to renovate the deep tube well those were affected by floods

3.7 Problem and Challenges experiences by DWASA to ensure sustainable water supply and way forward

DWASA is experiencing enormous problem and challenges in supplying potable water in Dhaka Metropolitan area. In many incidences, they don't have adequate facilities including finance, physical facilities, as well as technically skill human resources. Moreover, the production cost is higher than water tariff, system loss is 42.86 %, illegal connection, corruption of employees of DWASA, inadequate maintenance of water supply infrastructure, pressure from the employees union, inefficient billing and revenue collection system, absence of access to information, inadequate information available for the city citizen, inadequate ICT infrastructure for proper management of DWASA's activities and above all poor management of infrastructure. There is no mechanism exist in the present system to involve city citizen or community to participate in the policy decision process. Moreover, there is no five year development plan for DWASA, based on existing administration wish or will, projects were taken, not based on over all interest of the city citizen or organisation. To resolve the problem and make sure water supply for city citizens following recommendations were made which are as follows:

1. Need to formulate five year development plan for the sustainable water supply in Dhaka City and vision of water supply need to be clarify and adopt the mission to achieve vision.
2. People participation in policy decision process and community involvement in protection of DWASA resources, physical infrastructure and to attain 100% water supply as well as reduce system loss.

3. Raise the awareness of city citizen about rationale use of water; to reduce miss-use of water and to assist DWASA to make it as a profitable organisation. To do this, several community groups can be formed in each ward with the assistance of DCC zonal office, and they will provide information to Ward Commissioner as well as Zonal Officer of DCC, and sub-zonal or zonal office of DWASA to resolve any problem they experienced in terms of water supply.

4. This community organization will inform or report to DWASA if any leakage occurred in any places and notice any illegal connection. City citizen will be eager to assist DWASA to perform their responsibility efficiently. Mass media can also play an important role to raise the awareness of the city citizen. Religious installation such as Mosque, Temple, Pagoda and Church could be useful in conveying any message to the society through the Imam or priest who conduct the weekly prayer.

5. Raising the awareness of the employees of DWASA, as well as clarify their responsibility and accountability through continuous training to change their mind set and perform their responsibility efficiently with honesty.Collate and make sure availability of information for employees as well as city dwellers
6. Strong political support and supportive legal framework needed for successful implementation of DWASA Water Supply Master Plan

7. Transparency in administration and legal provision to dismiss or hire and fire corrupt employees from the organisation.
8. Short, mid and long –term water supply management strategies must be entail with increasing water supply facilities and capacity development (physical infrastructure development and software development means continuous local and foreign training for skill development) to perform efficiently by the employees and to ensure adequate water supply.

9. Continuous technical training is essential to develop technically sound manpower for the sustainable management of water supply installation.

10. Adequate budget is essential for the successful implementation of plan

11. Transparency in management/administration and adequate information require to be provided and make available for public/city citizen

12. Privatization of billing and revenue collection system

13. Measures require to stop corruption and mechanism is need to be developed to dismiss the corrupt employee from DWASA

14. Interference of Employee union need to be stopped in DWASA and strong political commitment is required to prevent untoward intervention in DWASA management by Employee union.

15. DWASA should be provided more authority and autonomy to perform their own job. Ministry should not interfere in every days work except policy level decision could be communicated to Ministry for their approval.

16. DWASA can be privatized to make it as a profitable organization and providing better services to the city citizens. But, government should keep their control DWASA in terms of regulation and policy. Privatize DWASA could be guided by the instruction and directives of the government.

17. DWASA should encourage rainwater harvesting and use rainwater for toilet flush, watering plants and car wash. In line with that DWASA should develop locally affordable technology or mechanism for city citizen to storage rain water in a limited space and that water can be use various purposes. That will help to reduce water consumption load on DWASA, and DWASA will be able to meet the need of increased citizen in Dhaka City. To popularize adapted cheap technology, DWASA could take the assistance of NGOs working in Dhaka City.

18. System loss should be reduced by repairing or replacing dilapidated pipelines and also by stopping corruption of meter reader of DWASA in Dhaka City.

4. Conclusion

DWASA is experiencing enormous problem in providing interrupted water supply to city dwellers. The problem turns severe during dry spell when water table decline. Moreover, due to excess abstraction ground water table of the aquifer under Dhaka City is declining 1.5 m per year. Dhaka City also experiences land subsidence which also force to find alternative sources of water for Dhaka City. Due to rapid urbanization, Dhaka City will be the 2nd largest city in the world which will be populated by 25.2 million people by 2030 that will put enormous pressure on DWASA to find alternative sources of water supply. Surface water quality of different rivers around Dhaka City is also deteriorated in such a level that DWASA unable to treat water of those river during dry season. DWASA needs to shift to surface water treatment and reduce system loss to ensure sustainable water supply for city citizens of today and for population in 2030. DWASA has to establish several surface water treatment plants and several water intake points needs to be identified in the Padma and Meghna river and water could be abstracted through pipe line from far to make sure potable water supply in the Dhaka Metropolitan to meet the water supply demand of city citizens in 2030.

5. Acknowledgement

The work reported in this paper funded by the Japan Society for Promotion of Science (JSPS).

6. References

Ahmed M. F. (2001). Water Supply. In: *People's Report on Bangladesh Environment 2001*, Volume I & II, Main Report and Data Base, Rahman, A., Ali, A. and Choudhury, F. (eds). The Unnayan Shamannay. The University Press Limited, Red Cresent Building, 114 Motijheel Commercial Area, Dhaka.

Ahmed M. F. (2001a). *Water Supply. In: People's Report on Bangladesh Environment 2001*, Volume I, Main Report; Rahman, A., Ali, A. and Choudhury, F. (eds). The Unnayan Shamannay. The University Press Limited, Red Crescent Building, 114 Motijheel Commercial Area, Dhaka.

DWASA (2001). *Annual Report 2000-2001*. Dhaka Water Supply and Sewerage Authority, WASA Building, 98 Kazi Nazrul Islam Avenue, Dhaka-1215.

The Daily Star (2003). *Water to run smoothly*. The Daily Star 16 June 2003.

The Daily Star (2003a). *Cry for water peaks on summer blast*. The Daily Star 13 April 2003.

The Prothom Alo (2003). *Water Tariff Rise again by DWASA*. The Prothom Alo 22 December 2003.

Arsenic Removal by Iron Coated Sponge in Drinking Water Treatment

Tien Vinh Nguyen[1)], Saravanamuthu Vigneswaran[1*)], Huu Hao Ngo[1)], Damoda Pokhrel[2)] and Thiruvenkatachari Viraraghavan [2)]

[1] Faculty of Engineering, University of Technology, Sydney (UTS), PO Box 123, Broadway, Sydney, NSW 2007, Australia
[2] Faculty of Engineering, University of Regina, Regina, Saskatchewan S4S 0A2, Canada
*Correspondence: S.Vigneswaran (Tel:61-2-95142641; Fax: 61-2-9514-2633; Email:s.vigneswaran@uts.edu.au)

Abstract

Arsenic (As) contamination in drinking water is a serious problem in a number of countries in the world, especially in small communities and developing countries. Arsenic is related to many health diseases such as blackfood disease, skin and bladder cancer. Several technologies are used in removing arsenic from water namely coagulation, adsorption, ion exchange, and membrane processes, etc. Adsorption in fixed bed treatment system is getting increasingly popular for arsenic removal in small-scale treatment systems because of their simplicity, ease of operation and handling, regeneration capacity and sludge free operation. In this study, a new absorbent material was studied to evaluate the Arsenic removal capacity in water. It was found that iron coated sponge (IOCSp) has high removal efficiency in removing Arsenic from water. In term of weight, each gram of IOCSp can remove up to about 160μg/l of both As (III) and As (V). From the initial laboratory – scale results, IOCSp seems to be a promising absorbent material for arsenic removal and can be applied for arsenic removal in small communities.

Keywords: absorbent, Arsenic, drinking water, iron coated sponge (IOCSp), water contamination

1. Introduction

Arsenic (As) contamination in groundwater is a serious problem in many parts of the world. The As pollution is not only by human activities (mining operations, industrial waste discharges, fertilizers, pesticides, etc.) but also by natural activities (dissolution of or desorption from ores; volcanic emissions) (Thirunavukkarasu et al., (2001).

Arsenic is a very toxic and carcinogenic substance. Arsenic is a cause of severe health problems, even at very low concentration. Arsenic can lead to both cancer (skin, lung and urinary bladder) and non – cancer effects. In addition, inorganic As in drinking water may affect many organs: central and peripheral nervous systems, dermal, cardiovascular, gastrointestinal and respiratory systems. Non cancer health effects include gangrene, limb loss, cardiovascular and pulmonary disease, endocrine and hematological disorders, and reproductive/developmental problems. In addition, As is an "accumulative enabler," meaning people who are predisposed to various cancers, diabetes, high blood pressure and other ailments are more likely to fall ill. Because of the high toxicity of Arsenic, the standard is strict, from 7 to 50 parts per billion (μg/l).

Different treatment technologies such as coagulation/filtration, ion exchange, adsorption, membrane are studied to treat As in water. Among the above methods, adsorption are getting increasingly popular for arsenic removal in small-scale treatment systems because of their simplicity, easy of operation and handling, regeneration capacity and sludge operation. In adsorption technologies, a number of adsorbent materials (activated alumina, activated carbon, fly ash, iron oxide coated sand (IOCS), sulphur modified iron (SMI), granular ferric hydroxide (GFH) and iron filings (IF), manganese dioxide coated sand) have been tried to adsorb As to the surface of adsorbent. Nevertheless, to explore the suitable treatment technology, which can satisfy both drinking water guidelines or criteria and economical aspect, further studies are necessary.

Sponge is seen to be an ideal material for adsorption because of its characteristics. Sponge has high internal porosity and specific surface which promote rate of absorption. In addition, sponge can be easily compacted to an extremely small volume to facilitate disposal. Importantly, sponge is cost effective so it can be applied in small communities as well as in developing countries. To increase the efficiency of the adsorption, sponge should be coated with metal oxide. After adsorption of As onto a sponge matrix, the sponge can be directly disposed or regenerated with chemical solutions.

The objectives of this research were to develop the new metal oxide coated sponge (IOCSp) with the appropriate coating method. The capacity and effectiveness of the metal oxide coated sponge in removing arsenic in drinking water was evaluated.

2. Experiments, results and discussion

In this study, commercial polyurethane (PU) sponge was selected. The synthetic water was used as As contaminated influent in all experiments. Synthetic water was prepared by adding As in tap water to obtain an Arsenic concentration from 144.5μg to 530μg/l. The experiments were conducted at different pH values (6, 7 and 8). Solution 1M HNO_3 and NaOH was used to adjust the pH to a designed value. A graphite furnace atomic absorption spectrometry (SpectrAA-600) was used to determine the concentration of Arsenic.

Experiments with uncoated sponge: Haft-cylinder uncoated sponges with 3cm in diameter and 1.5cm in length were put into 100ml As solution (arsenic concentration was 530 μg/l) and shaked well from one to four days. The As concentrations of solution after shaking were analyzed for As residue. The experimental results showed that uncoated sponge had a very low capacity in arsenic removal, approximately 10% with Arsenic solution of 530μg/l (Figure 1).

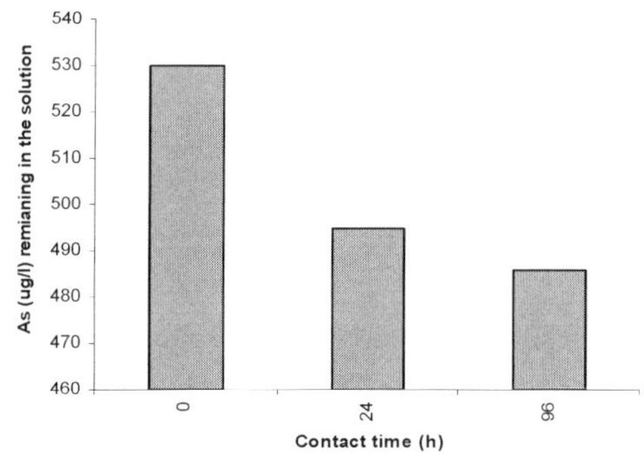

Figure 1: Arsenic removal by uncoated sponge

Optimization of iron coated sponge: The principal method for preparation of IOCSp is as follows: following wash and dry, uncoated sponges were soaked in 2M Ferric Nitrate solution at a different pH value (pH from 1, 2 ... to 12) and time (5, 10, 15, 20, 30 minutes and up to 10 hours). Then these sponges were put into the oven at a temperature (110^0C, 220^0C) in different time (8h, 12h, 14h, 18h, 20h). After cooling, the heated sponges were washed again and were used in batch experiments to find out the optimum condition for coating sponge with the iron oxide. The optimum coating conditions for preparing IOCSp were be found and applied in the succeeding batch and dynamic experiments. Experimental results showed that sponge could not be coated in an adequate manner when the soaking time was less than 1 hour. The sponge was damaged when it was coated at a temperature of 220^0C and above. Experimental results also showed that sponge could not be coated well at a pH above 5 and was damaged when drying time was more than 34 hours. From a series of experiments, the optimum condition for coating the sponge with iron oxide was found to be: pH of 4, soaking time of 10 hours, heating temperature of 110^0C and heating temperature of 20hours. These optimum conditions were used for the sponge preparation in the subsequent batch and dynamic experiments.

Batch kinetic studies: The IOCSp prepared at the optimum coating conditions was first cut into half-cylinder pieces each with 1.5 cm in length, 3 cm in diameter and 0.15g in weight. The IOCSp was then placed into 250ml Erlenmeyer flasks containing the synthetic water. The flasks were sealed with laboratory film. The samples were placed on a mechanical shaker and shaken at 175rpm. Samples from the flask were collected at regular time intervals and analyzed for residual As.

Figures 2 and 3 present the removal of As (III) and As (V) as a function of adsorption time of IOCSp respectively. The results showed that IOCSp could remove 47.3% to 65% of As (III) and As (V) after a 1 hour contact with IOCSp. A nine - hour adsorption led to a very high amount of As adsorption (less than 18μg/L of As remaining in the water). In these experiments, 0.15g of IOCSp was placed in a flask containing 100ml of As solution. Each gram of IOCSp adsorbed about 160μg of As (III) and As (V) upon a 9 hours contact of IOCSp with the As solution (Figures 2(b) and 3(b)).

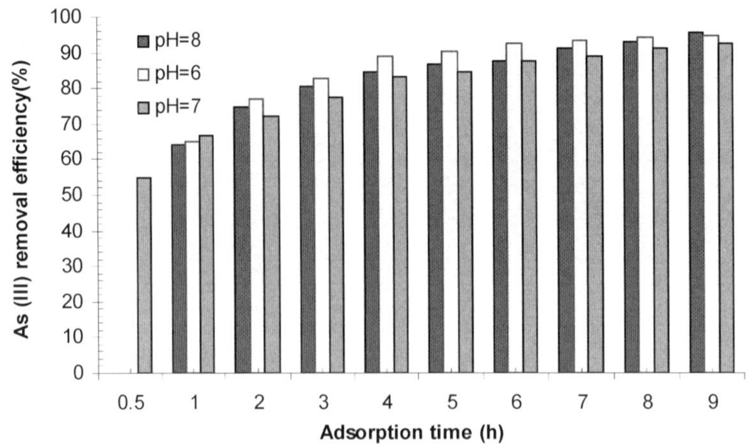

Figure 2(a). Removal efficiency of As (III) as a function of adsorption time (Initial As (III) concentration = 260 μg/L; IOCSp's weight = 0.15g, Temperature = 20^0C, volume of As solution = 100mL)

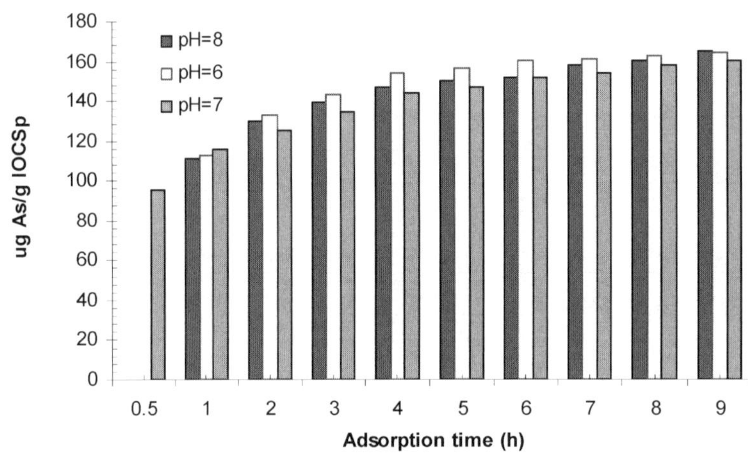

Figure 2(b). Amount of As (III) adsorbed onto a unit weight of IOCSp (Initial As(III) concentration=260 μg/L; IOCSp's weight=0.15g, volume of As solution=100mL)

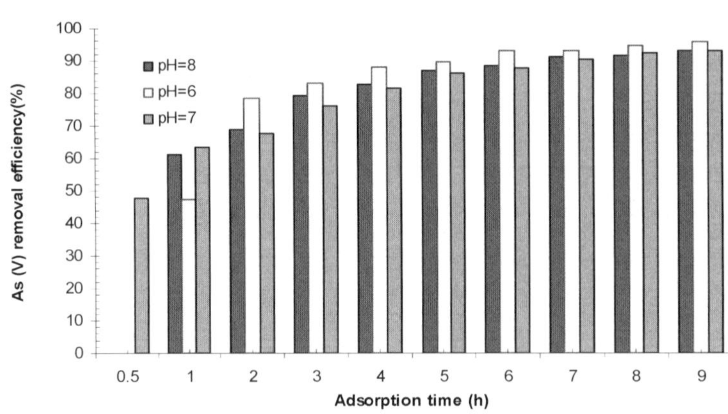

Figure 3(a). Removal efficiency of As (V) as a function of adsorption time (Initial As (V) concentration = 260 μg/L; IOCSp's weight = 0.15g, Temperature = 20^0C, volume of As solution = 100mL)

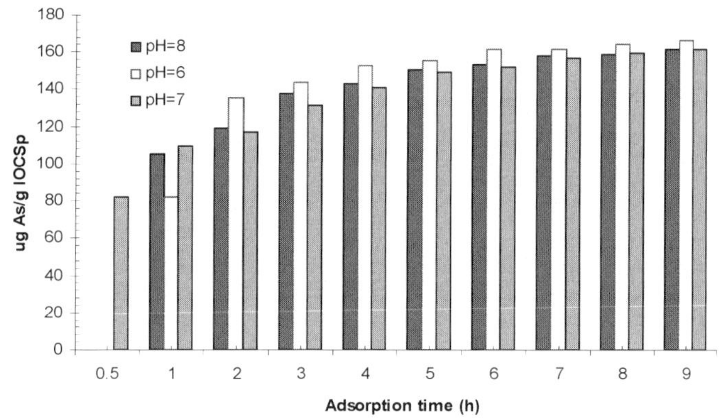

Figure 3(b). Amount of As (V) adsorbed onto a unit weight of IOCSp
(Initial As (V) concentration = 260 μg/L; IOCSp's weight = 0.15g, volume = 100mL)

IOCSp Sponge tray experiments: a semi – continous tray system consisting of IOCSp was then studied. The tray was designed to hold 10 rows of semi circular sponges each with a length of 19.5cm and diameter of 3cm. Each row of sponge weighs of 1.92g. The tray was placed at an angle of inclination of 10^0 to the horizontal to facilitate the water flow in gravitational mode. The tray had an influent chamber which helped to evenly distribute the water across the cross-section of the tray. In the first set of experiments, synthetic water (spiked with the As (III) concentration of 144.5 μg/l) was pumped through the tray using a peristaltic pump at a flow rate 15.6 mL/min. The water at the outlet was sampled for As (III). After that, the IOCSp was regenerated with 0.3M NaOH solution before running the second cycle with higher influent As (III) concentration (260 to 280μg/l) in order to study the effect of influent concentration. The operation time of experiment was 30 hours.

Similar experiments were conducted with As (V) solution with a determined As (V) concentration of 260-265μg/l. In these experiments, the IOCSp was also regenerated with 0.3M NaOH solution. In the experiments with As (V) solution, the tray contained only 7 rows (weight = 13.4g) of semi circular IOCSp. The angle of inclination of the tray was decreased to 6^0 to increase contact time between IOCSp and water when the water passes though each row of IOCSp.

Figure 4 Experimental set-up used in IOCSp tray experiments

The results on the removal of both As (III) and As (V) by the IOCSp tray system are presented in Figures 5 to 6.

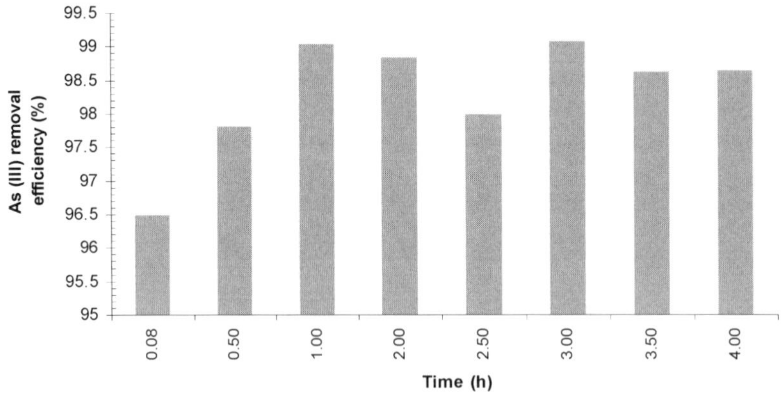

Figure 5(a). Experimental results with IOCSp tray system, cycle 1 (Initial As (III) concentration = 144.5 μg/l; IOCS'p weight = 19.2g)

The results of the first set of experiments showed that the tray can remove more than 96% of As(III). In the second series of experiments, after the IOCSp in the tray was regenerated by 0.3M of NaOH, the IOCSp could continue to adsorb nearly 10 L of As (III) (concentration = 260μg/l) to a value less then 50μg/l. After 30 hours running of operation, the efficiency decreased gradually to a value of 52.3% (123.9μg/l in the effluent) (Figure 5(b)).

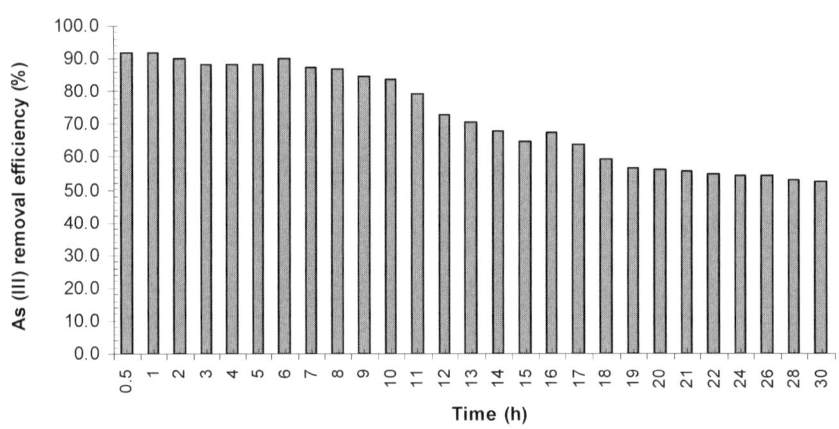

Figure 5(b): Arsenic (III) removal efficiency with the regenerated IOCSp tray system, cycle 2 (Initial As (III) concentration = 260 μg/l; IOCSp's weight = 19.2g)

In the study with As (V), after 6 hours of running, the As(V) concentration in the effluent reached to the value of 49.2μg/L, efficiency was 81.4%. This effluent As(V) was nearly 110μg/l (58.6% removal) after 28hours running (Figure 6(a)).

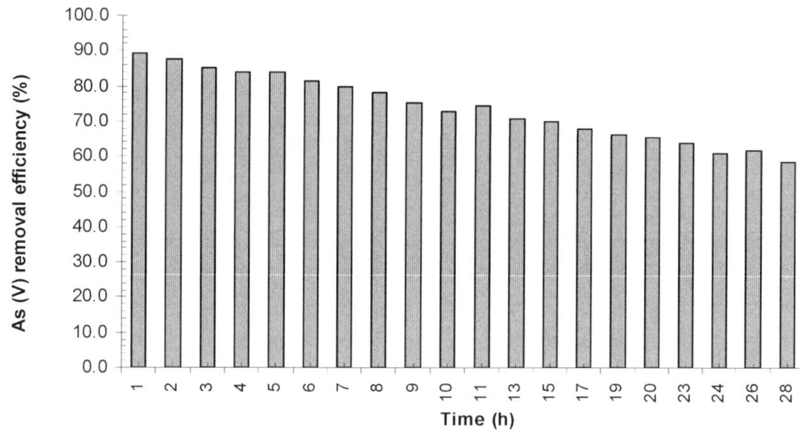

Figure 6(a): Experimental results with IOCSp tray system, cycle 1
(Initial As (V) concentration = 265 μg/l; IOCSp's weight = 13.4g)

After IOCSp in the tray was regenerated by 0.3M of NaOH, IOCSp continued to adsorb As(V) to a value less then 20μg/l (more than 93% removal) even after 6hours of running time (Figure 6(b)).

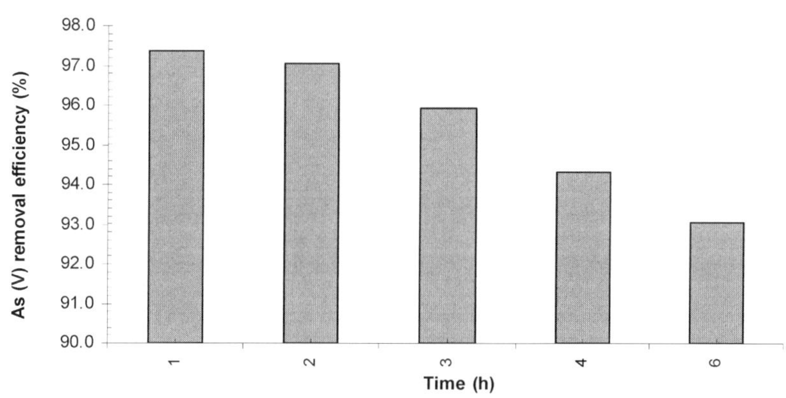

Figure 6(b): Arsenic (V) removal efficiency with the regenerated IOCSp tray system, cycle 2
(Initial As (V) concentration = 265 μg/l; IOCSp's weight = 13.4g)

Column studies:
Column experiments were carried out to study the removal of As in synthetic water. A glass column of 19 mm diameter and 400 mm height was used in the study. The synthetic water was fed through the packed column at a flow rate 8 mL/minute. The column tests were conducted in the down-flow mode and at the normal pH (7.4-7.6) of the synthetic water. The column was packed with 4g (57 mL) of the IOCSp.

The results on the removal of As using IOCSp column system are presented in Figures 7.

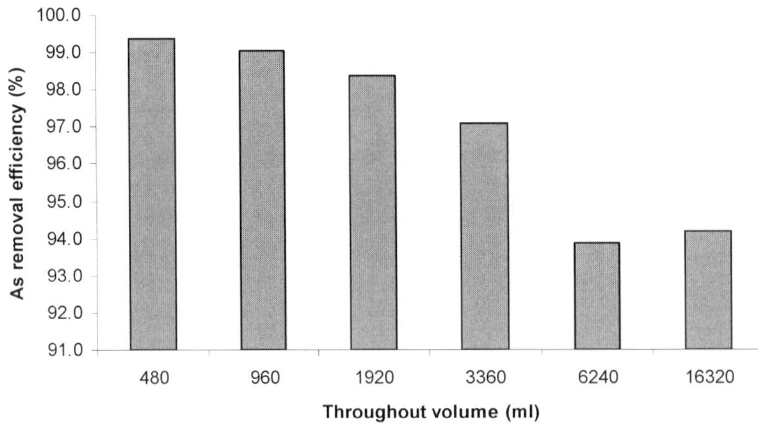

Figure 7. Arsenic removal efficiency with the IOCSp column
(Initial As = 310 μg/L; flow rate = 8mL/minute, Q = 1.7m^3/m^2.h, IOCSp's weight = 4g, volume of IOCSp = 57ml)

The initial results with IOCSp column showed that up to 287 bed volume, the As effluent concentration was lower than 20ug/l with removal efficiency is more than 94%.

3. Conclusions

The initial experimental results showed that iron coated sponge (IOCSp) has high removal efficiency in removing both As (III) and As (V) from water. However, the experiments is still in initial stage and it needs further long-term experimnets for evaluation.

4. Acknowledgement

This research is mainly funded by the Australian Research Council (ARC) Discovery Grant (2004-2006) and ARC International Linkage Grant (2003-2004). The first author would like to thank the Vietnamese Ministry of Education and Training for providing scholarship to study at University of Technology, Sydney.

5. References

Thirunavukkarasu, O.S., Viraraghavan, T., and Subramanian, K.S. and Tanjore, S. (2002). Organic arsenic removal from drinking water, Journal of Urban Water, 4, 415-421.

Section 3

Appropriate Technologies for Water Supply and Wastewater

Spatial Planning for Surface Water Protected Area: a Case Study of the Pong River, Northeast Thailand

Wanpen, Wirojanagud[1)] and Peter G. Jensen[2)]

1) Research Center for Environmental and Hazardous Substance Management, Khon Kaen University/ National Research Center for Environmental and Hazardous Waste Management, Thailand
wanpen@kku.ac.th
2) Carl Bro International, Denmark

Abstract

Spatial planning is a valuable tool to support sustainable development by basing the plan measures and interventions on key sustainability principles. Developing a spatial plan runs through three main work processes; baseline data analysis of the existing situation and problems and potentials, formulation of objectives and strategies, and integration across sectors and synthesis of all aspects into a coherent balanced spatial framework. Geographic Information System (GIS) is a very useful tool for spatial planning and management of surface water quality. Planning map for water quality protection of the Pong River, Northeast, Thailand was constructed according to the classification of water bodies related to their designated use. The Pong River is evidently designated as the surface water classification type 3 except the natural lagoons of Sue Ten and Tung Tieu as for class 2. Both lagoons are natural wetland to be conserved for ecological diversity as well as their natural beauty.

Keywords : spatial planning, the Pong River, surface water quality protected area, surface water classification

1. Introduction

Generally, the problems of water resources are concerned with degradation of water quality due to pollution from industries, communities wastewater, fishery and agriculture; water shortage in dry season; flooding in rainy season. This paper presents the water quality protection of the Pong River as the case study. The Pong River is an important water source for the upper northeastern provinces of Thailand, especially Khon Kaen Province that use this river as its sole water supply source for community, industry, agriculture and aquaculture. At the same time, the river is a recipient of wastewater from communities and industries and runoff from agriculture. Land use activities along the river, which relate to pollution sources, have an impact on the river water quality. In order to manage water quality of the water resource to conform with the water uses, it is strongly required to know the exact location or space and relevant data which are evidently presented on the map where the strategic measures can be implemented. This is known as a spatial planning map. The word spatial comes from space, it means that spatial planning has to do with our surroundings, the environment, nature, and everything that we build, construct or do to our surroundings.

2. Methodology

2.1 Baseline data analysis of the existing conditions
Water use and land use along the Pong River were reviewed from the previous studies (Ref. 1 and 2). Water quality of the Pong River was studied by collecting samples from various sampling points designate in according with the pollution sources relate to land use as shown in Figure 1.

Water samples were analyzed for physical parameters, chemical parameters, heavy metals and toxic substances, but emphasizing on DO, BOD, COD, NH_3-N, and Total Coliform.

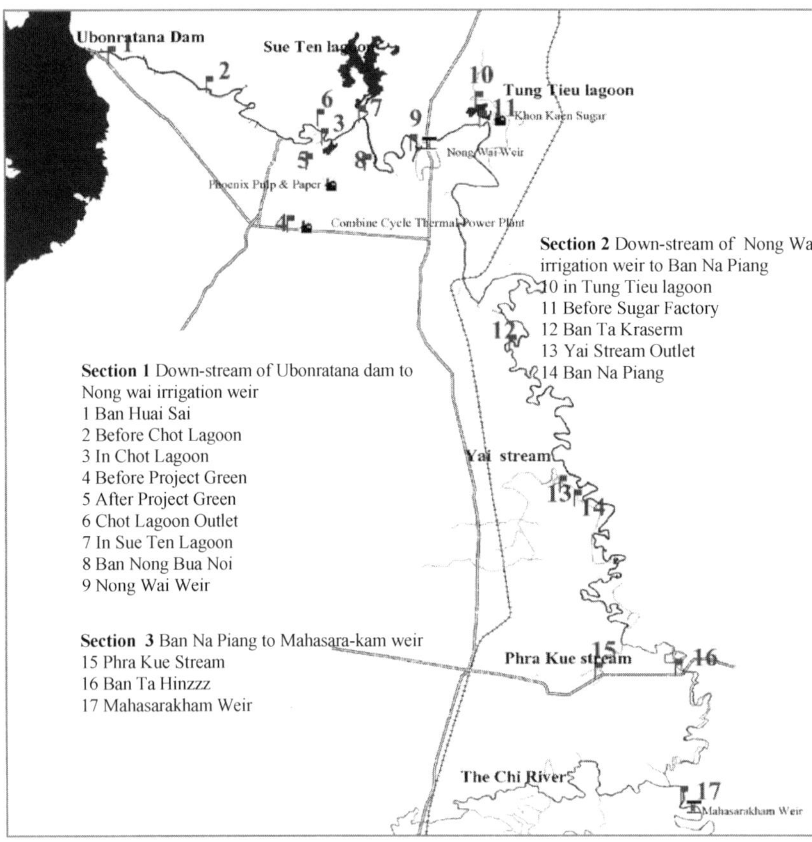

Figure 1. Sampling Station along the Pong River

2.2 Preparation of the map of surface water protected area
The surface water protected areas were constructed as follows:

1) A set of maps with baseline data; i.e. administrative boundaries, cities, towns, villages, infrastructure, industries, waterworks, land use, water bodies and rivers, topography were prepared.
2) The criteria of surface water protected area map is set up. The GIS layers used in this study are:

- Land use
- Watershed boundary
- Industry type 3 location
- Waste disposal site
- Community waterworks
- Mining industry
- Location of raw water pumping for waterworks
- Location of waste water treatment system

All of these GIS layers were overlain together for consideration of the protected area.

3) The stakeholders selected the main surface water resources as the protected area by considering the base line data.
4) Field studies and consultation with farmer networks, environmental volunteers and local administrators were conducted to collect the related data such as water quality data, water pumping stations, water consumption data, sources of pollution, existing protected water bodies. The surface water is classified by following the criteria of Pollution Control Department, Ministry of Natural Resources and Environment, as follows:

Class	Description	Water Use
1	Extra clean fresh surface water resources	❑ conservation not necessary pass through water treatment process require only ordinary process for pathogenic destruction ❑ ecosystem conservation where basic organisms can breed naturally
2	Very clean fresh surface water resources	❑ consumption which requires ordinary water treatment process before use ❑ aquatic organism of conservation ❑ fisheries ❑ recreation
3	Medium clean fresh surface water resources	❑ consumption, but passing through an ordinary treatment process before using ❑ agriculture
4	Fairly clean fresh surface water resources	❑ consumption, but requires special water treatment process before using ❑ industry
5	The sources which are not classification in type 1-4	❑ navigation

Figure 2. illustrates the criteria of surface water protected area. The objective of these criteria is to protect the quality of water bodies according to their designated use. The map of designated water use is prepared in this step.

5) The planning map of surface water protected area was constructed according to the classification of water bodies by combining the map of designated water use and all GIS layers. The measures of planning map were recommended to manage water quality.

3. Results and Discussion

3.1 Existing conditions

1) Land use

The interested section of the Pong River is downstream from the Ubonratana Dam to the Mahasarakam Dam which is divided into 2 sections, 34 km for the first section from Ubonratana Dam km to Nong Wai irrigation weir and 180 km for the second section from Nong Wai irrigation weir to Mahasarakam Dam. Land use along the first river section comprises a large pulp and paper mill, a thermal power plant, some rainfed agriculture areas and villages. At this river section, there is an intensive fish pen aquaculture production. Land use along the second river section consist of villages and large communities (e.g. municipality), a large irrigation area, and large industries including sugar production, distillery, paper board, and tapioca mills. These various land uses are generally considered as potential pollution sources that might cause water pollution problems to the river.

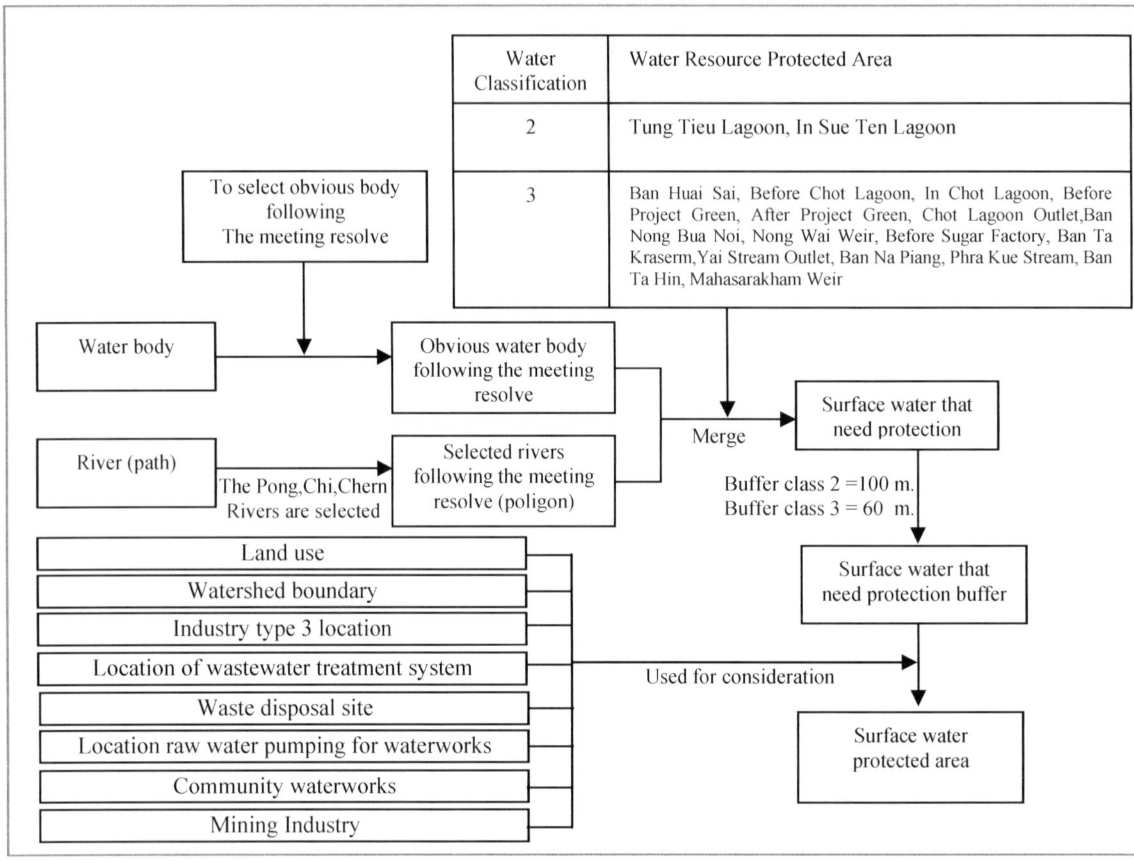

Figure 2 Criteria of surface water protected area

2) Water use

The Pong river is the source of water for communities, agriculture and industry. For the community water supply, there are many water treatment plants operated by the Provincial Water Work, Local Water Work as well as the government agencies have operated their own supply such as the military area, university, etc. Large industries also have their own water treatment plants. Regarding the water for agriculture, it is mainly for irrigation of about 1,056 million cubic meters (MCM) a year for the irrigation area of 523.35 km^2. Consideration for water to be protected in accordance with the water use, it is evidently that water quality has to be maintained as the surface water classification type 3.

3) Water quality

Water quality of the Pong River is presented in Figure 3. Generally, water quality of the Pong River (during 2003; January, April and November) was evaluated as surface water classification type 3; DO 4.3-7.2 mg/l and BOD 1.0-2.0 mg/l. Except sampling points 8, 9, and 13 were categorized in surface water classification type 4, and sampling point 15 as type 5 as indicated by DO 1-3.8 mg/l and BOD 1.0-6.0 mg/l.

When compared the existing surface water quality with the surface water classification designated to be protected as type 3, some sections of the Pong River in some seasons were in type 4-5, particularly the section downstream of the Project Green area, Tung Tieu lagoon and Pra Kue stream. In addition, Tung Tieu lagoon is the water body designated for aquatic life protection as it is the natural source of bird diversity and a recreation area for the nearby communities. It therefore should be protected as surface water classification type 2.

Existing land use related to pollution sources, water use and water quality are summarized in Table 1.

3.2 Planning map of the surface water quality protection area

According to the water uses of the Pong River are for agriculture, industry and raw water for tap water supply. Therefore, the Pong River has to be protected as for the surface water classification type 3, except the natural lagoons of Sue Ten and Tung Tieu where should be protected as type 2 of surface water classification (Figure 4). They are the local wetland where existed by natural beauty, biodiversity of bird species and the place for birds watching. Water classification of the Pong River should be protected are also summarized in Table 1.

3.3 Strategic measures

- Concerned agencies shall within their responsibilities and mandates control wastewater pollution from industries and communities to ensure that the specified effluent limit values are not exceeded, as well as monitor the water quality in all of the surface water protected areas.
- Regulation of land use around water bodies using buffer zones of 60-100 m.
- Concerned agencies should specify what kind of activities and projects that are not allowed in a certain water body and buffer zone around the water body.

4. Conclusions

Based on the physical environment conditions of the Pong River, the water quality protection is designated as for surface water classification type 3 except the natural lagoons of Tung Tieu and Sue Ten as for type 2. With the spatial planning map that indicates location of pollution sources as well as socio-economic data, the mitigation measures can be comprehensively implemented. Therefore, the spatial planning can be applied in the management of surface water quality, and web mapping is a very useful management tool, which gives easy access to all necessary data and helps planners to understand the needs and problems of the communities that live around the water bodies. The planning map can be used as a tool for planning in local level as well as in provincial and regional level.

5. Acknowledgement

The authors would like to thank to the Changwat Action Plans for Environmental Quality Management Project (CAPEQM-Project) for providing us the concept of spatial planning as well as the study team of the Integrated Pong River Management for all technical support.

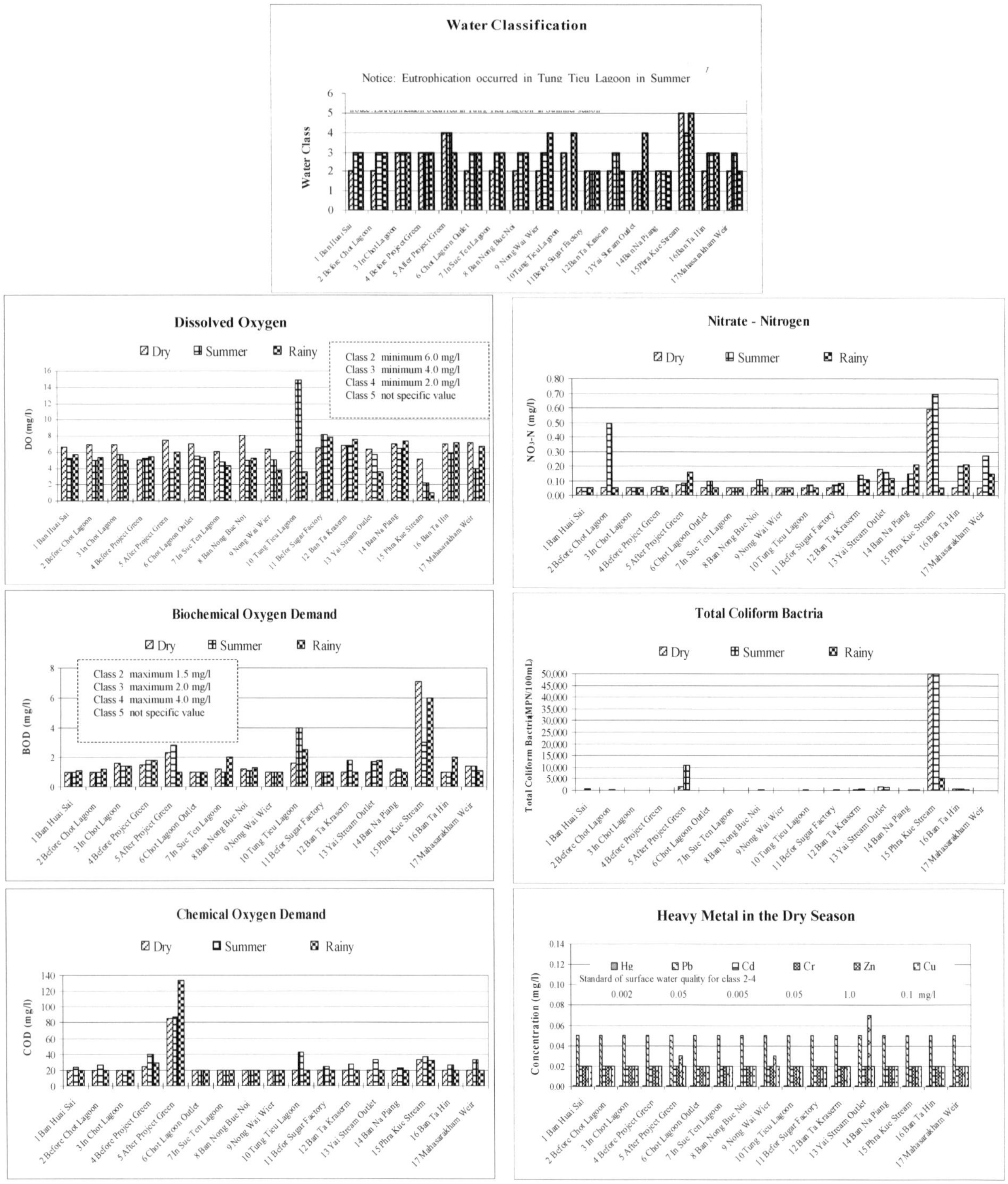

Figure 3. Existing Water Quality for the Pong River

Table 1. Land Use , Water Use and Water Quality of the Pong River and Tributary

Section	Land Use	Water Use	Station No.	Existing Water Class (Rainy season)	Target Water Class
Down-stream of Ubonratana dam to Nong wai irrigation weir	- Agriculture area of 200 km^2	- 1,056 MCM/year	1	3	3
			2	3	3
	- Industrial area of 2.30 km^2	- Pulp and Paper mill 1,075,200 m^3/month	3	3	3
		- Ice mill 1,800 m^3/month	4	3	3
		- Thermal power plant 437,760 m^3/month	5	3	3
	- Communities: Ubonratana Dam and Nampong Subdistrict Municipalities, 23 villages within 5 km by 9 villages within 1 km from both river bank with the area of 3.55 km^2	- Tap water supply 140,100 m^3/month	6	3	3
			7	3	2 (Tung Tieu)
			8	3	3
			9	4	3
Down-stream of Nong Wai irrigation weir to Ban Na Piang	- Agriculture area mostly is irrigation rice field area of 318.8 km^2		10	4	2 (Sue Ten)
			11	2	3
			12	2	3
	- Industrial area of 0.68 km^2	- Distillery mill 19,800 m^3/month	13	4	3
		- Sugar mill 117,540 m^3/month	14	2	3
		- Tapioca mill 37,500 m^3/month			
	- Communities: 55 villages in 5 km, and 23 villages in 1 km from both river banks with the area of 4.84 km^2	- Tap water supply 2,953,200 m^3/month			
Ban Na Piang to Mahasara-kam weir	- Agriculture area is mainly irrigated rice field area of 204.55 km^2		15	5	3
	- Industrial area of 0.05 km^2		16	3	3
	- Communities: 38 villages within 5 km and 21 villages within 1 km from both river bank with the area of 2.286 km^2		17	2	3

Source : Provincial Water Work
: Provincial Industrial Office

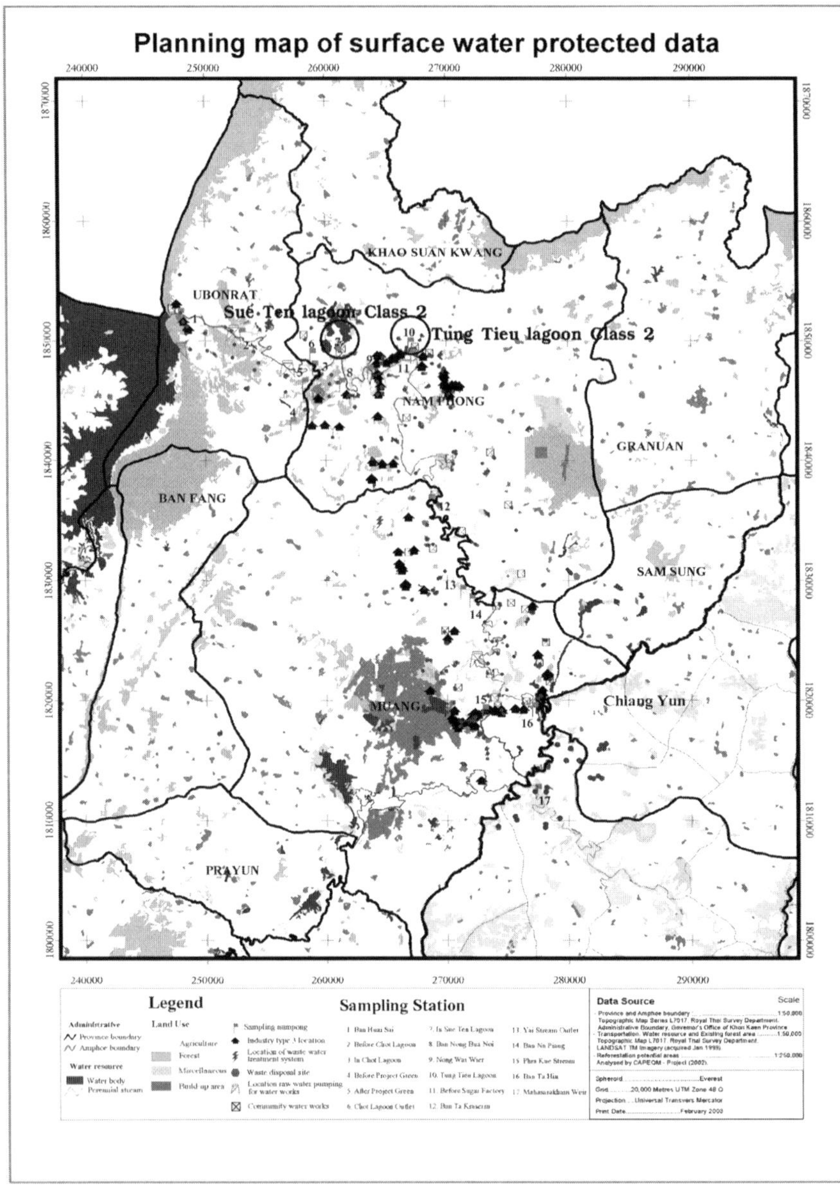

Figure 4. Water Quality Protected Area of the Pong River

6. References

Khon Kaen Province. Regional Environment Office 10, Ministry of Natural Resources and Environment, Thailand. and Danish International Development Assistance (2003). *Spatial Planning for Sustainable Development.*

Khon Kaen University (1995). Study Project for *Preparation of the Action Plan for the Pong River Water Quality Rehabilitation, Khon Kaen Province, Executive Summary.*

Office of Natural Resources and Environmental Policy and Planning, Ministry of Natural Resources and Environment, Thailand (1998). *Thailand State of Environmental Report.*

Khon Kaen University (2002). *Research on Environmental Management of the Pong Watershed.*

Reactivity and Sensitivity of DOM Fractions to Form THMs in Raw Water Supply and in Treated Wastewater Used for Reclaimed Water of the Northern-Region Industrial Estate, Thailand

Tansiporn Janhom[1], Charongpun Musikavong [2] Suraphong Wattanachira [3*] and Hiroaki Furumai [4]

1) National Research Center for Environmental and Hazardous Waste Management (NRC-EHWM), Chulalongkorn University, Bangkok, 10330, Thailand
e-mail:tansiphorn@yahoo.co.th
2) National Research Center for Environmental and Hazardous Waste Management (NRC-EHWM), Chulalongkorn University, Bangkok, 10330, Thailand
e-mail:jarongpun@yahoo.com
3) Department of Environmental Engineering /NRC-EHWM, Chiang Mai University, Chiang Mai, 50200, Thailand,
*Corresponding author: e-mail:suraphong@eng.cmu.ac.th
4) Department of Urban Engineering, the University of Tokyo, 7-3-1, Hongo, Bunkyo, 113-8656, Japan
e-mail:furumai@env.t.u-tokyo.ac.jp

Abstract

Fractionation of raw water supply and treated industrial estate wastewater for use as reclaimed water in the Northern-Region Industrial Estate located in northern Thailand, were performed so as to isolate dissolved organic matter (DOM) into six fractions including hydrophobic neutral (HPON), hydrophobic base (HPOB), hydrophobic acid (HPOA), hydrophilic base (HPIB), hydrophilic acid (HPIA) and hydrophilic neutral (HPIN) fractions. Reactivity and sensitivity of each DOM fraction to create THMs were also studied. The results show that HPOA and HPIA fractions in terms of dissolved organic carbon (DOC) in raw water supply were 41.6% and 20.5%, respectively, whereas HPOA of 30.6% and HPIA of 26.4% in treated wastewater were observed. It can be stated that HPOA and HPIA were the first and second largest DOM fractions in both waters in which the summation of their DOC were 62.1% of total DOC in raw water supply and 57.0% of total DOC in treated wastewater. As a direct result, HPOA and HPIA were also realized to be the major trihalomethanes (THMs) precursors in both waters. The summary of $THMFP_{HPOA}$ and $THMFP_{HPIA}$ in the respective waters were totally 59.8% and 58.1% of total THMFP. According to the reactivity of each DOM fraction, HPIB demonstrated the highest active source on the formation of THMs in raw water supply followed by HPOB, whereas HPOB and HPIB were the first and second highest active sources in treated wastewater. Regarding the sensitivity of each DOM fraction, HPIB and HPOB were determined as the first and second largest sensitive DOM fractions to form THMs in raw water supply whereas HPOB was the highest sensitive DOM fraction to create THMs in treated wastewater followed by HPIB.

Keywords: DOM fractions (HPON, HPOB, HPOA, HPIB, HPIA and HPIN), THMs, THMFP, reactivity, sensitivity, raw water supply and treated industrial estate wastewater.

1. Introduction

The Northern-Region Industrial Estate is the biggest industrial estate in northern Thailand and which consumes a water supply of about 14,000 m^3/day. Sometimes during the dry season the quantity of reservoir water is inadequate to produce enough water supplies. The industrial estate, therefore, has proposed the utilization of reclaimed water from treated industrial estate wastewater.

In general, dissolved organic matter (DOM) surrogates, termed trihalomethanes formation potential (THMFP) is one of the parameters used for representing organic contaminants in water. This is attributed to the fact that DOM in raw water supply, particularly in reclaimed water from treated industrial estate wastewater may react with chlorine during the disinfection process to form trihalomethanes (THMs), which are classified as potentially carcinogenic substances (Rook, 1974). At the time being, the resin fractionation process was widely used by many researchers (Leenheer, 1981; Imai *et al.* 2002; Marhaba *et al.* 2003 and Panyapinyopol *et al.* 2005) to isolate DOM in water into six fractions including hydrophobic neutral (HPON), hydrophobic base (HPOB), hydrophobic acid (HPOA), hydrophilic base (HPIB), hydrophilic acid (HPIA) and hydrophilic neutral (HPIN) fractions in order to characterize DOM more closely.

The aims of this research were to investigate the six fractions of DOM termed HPON, HPOB, HPOA, HPIB, HPIA and HPIN in raw water supply and in treated industrial estate wastewater. The reactivity and sensitivity to form THMs of each DOM fraction are useful for predicting and control the possible formation of disinfection by-product (DBPs) and these were also determined in this study.

2. Materials and methods

2.1 Study area and sample collection

As can be seen in Figure 1, water samples were collected from two sampling points in the polishing pond ⑦ of the central wastewater treatment plant and in the raw water supply reservoir ⑧. Sample collections were done in June, 2004 from the polishing pond and in October, 2004 from raw water supply reservoir. The raw water supply reservoir normally receives water from the Mae-Kuang River ⑩. About 600,000 m^3 of raw water is stored in the reservoir. Water from this reservoir is utilized as raw water to produce a water supply of approximately 14,000 m^3 per day and distributed to the Northern-Region Industrial Estate area. More or less 12,000 m^3 of wastewater per day is treated by the pond system of the central wastewater treatment plant. The polishing pond of approximately 85,000 m^3 capacity receives treated wastewater where it is planned to be retreated by the water reclamation process before being pumped into the raw water supply reservoir of the industrial estate.

Raw water supply was filtered through 0.7 μm GF/F filters and treated industrial estate wastewater was filtered through 0.45 cellulose acetate membranes. Filtered waters were then analyzed for DOC, UV-254, specific ultraviolet adsorptions (SUVA), and THMFP. A series of three resins (DAX-8, AG-MP-50, and WA-10) were used to isolate water samples into six fractions, including HPON, HPOB, HPOA, HPIB, HPIA, and HPIN in accordance with the resin fractionation procedure developed by Leenheer (1981) and Marhaba *et al.*, (2003). Then, each DOM fraction was analyzed through parameters including DOC, UV-254, SUVA, and THMFP.

2.2 Analytical method

The summary of analytical methods, standards and instruments used for analyzing the above-mentioned parameters are depicted in Table 1.

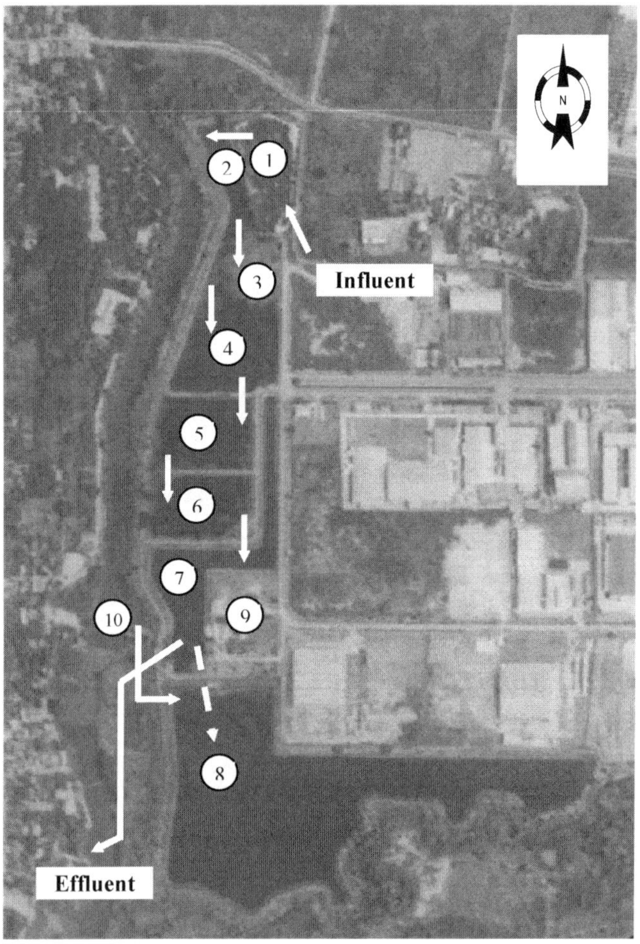

Symbol	Description	Appox. detention time
1	Aeration pond No1	4 days*
2	Aeration pond No.2	1day*
3	Stabilization pond No.1	3 days*
4	Stabilization pond No.2	3 days*
5	Stabilization pond No.3	2 days*
6	Stabilization pond No.4	2 days*
7	The polishing pond of treated wastewater (sampling point)	7 days*
8	Reservoir of raw water supply (sampling point)	42 days**
9	Water supply plant	
10	Mae-Kuang River	

* Based on wastewater treatment flow rate
** Based on the water supply production flow rate
——→ Water flow direction
– – → Proposed plan for the future

Figure 1 Map of sampling sites

Table1. Analytical methods

Parameters	Analytical Method	Standard	Instruments
UV-254	Ultraviolet Absorption Method	Standard method 5910 B (AWWA, 1995)	Jasco V-350 UV/VIS spectrophotometer
DOC	Wet - Oxidation Method	Standard method 5310 D (AWWA, 1995)	O.I. Analytical 1010 TOC Analyzer
Free chlorine residual	DPD Colorimetric Method	Standard method 4500-Cl G (AWWA, 1995)	Jasco V-350 UV/VIS Spectrophotometer
THMFP	Formation of Trihalomethanes and Other Disinfection By-Products and Liquid-Liquid Extraction Gas Chromatography Method	Standard method 5710 and 6232 B (AWWA, 1995)	Gas Chromatography with electron capture detector (GC/ECD)

3. Results and discussion

3.1 DOM fractions and its THMFP

DOC, UV-254, SUVA, and THMFP values of each DOM fraction in raw water supply and in treated industrial estate wastewater are reported in Table 2. The percent distribution of DOM fractions in raw water supply and in treated industrial estate wastewater is depicted in Figure 2. It was observed that in the case of raw water supply, hydrophobic (HPO) and hydrophilic (HPI) organics accounted for 57.1% and 43.9% of total DOC, respectively. In terms of the distribution of organic fractions, the HPOA fraction was found to be the major DOM fraction in raw water supply as DOC of HPOA accounted for 41.6 % of total DOC. The other organic fractions ordered according to their quantifiable presence in water were as follows: HPIA (20.5%), HPIN (17.7%), HPON (12.1 %), HPIB (4.8%), and HPOB (3.4%). In the case of treated industrial estate wastewater, the HPO and HPI organics accounted for 40.5 % and 59.5 % of total DOC, respectively. These levels were very close to the reported range of 33-40% for hydrophobic and 60-67% for hydrophilic in the effluent from sewage treatment plant receiving domestic sewage with treated industrial wastewater in Japan (Imai *et al*. 2002). Due to the percentage distribution of the organic fractions, the HPOA fraction was also found to be the major DOM fraction in treated industrial estate wastewater as DOC of HPOA accounted for 30.6% of total DOC. The other organic fractions ordered according to their quantifiable presence in this water are as follows: HPIA (26.4%), HPIN (22.7%), HPIB (10.4%), HPON (6.1%) and HPOB (3.8%).

With regard to the aforementioned results, it can be stated that HPO was the main organic group in raw water supply while HPI was the main organic group in treated industrial estate wastewater. With more detailed consideration, HPOA was the major DOM fraction in both raw water supply and treated industrial estate wastewater followed by HPIA. The summation of DOC of HPOA and HPIA accounted for 62.1% and 57.0 % of total DOC in raw water supply and treated industrial estate wastewater, respectively.

Table 2. DOC, UV-254, SUVA and THMFP of each DOM fraction in raw water supply and in treated industrial estate wastewater

	Parameters	Unfractionated water	Fractionated water							% Diff
			Hydrophobic fraction (HPO)			Hydrophilic fraction(HPI)				
		Total	HPON	HPOB	HPOA	HPIB	HPIA	HPIN	Total	
Reservoir water	DOC (mg/L)	5.37	0.71	0.20	2.45	0.28	1.21	1.04	5.89	+9.7
	UV-254(1/cm)	0.139	0.007	0.003	0.048	0.006	0.008	0.003	-	-
	SUVA(L/mg-m)	2.59	0.93	1.50	1.95	2.21	0.69	0.28	-	-
	THMFP(μg/L)	382.4	68.0	41.4	247.1	67.9	112.6	64.9	601.9	+57.4
Treated wastewater	DOC (mg/L)	6.83	0.43	0.27	2.16	0.73	1.86	1.60	7.05	+2.9
	UV-254(1/cm)	0.208	0.006	0.010	0.035	0.015	0.045	0.014	-	-
	SUVA(L/mg-m)	3.05	1.40	3.70	1.62	2.05	2.42	0.88	-	-
	THMFP(μg/L)	635.1	53.9	50.5	305.8	111.8	242.9	178.8	943.6	+33.8

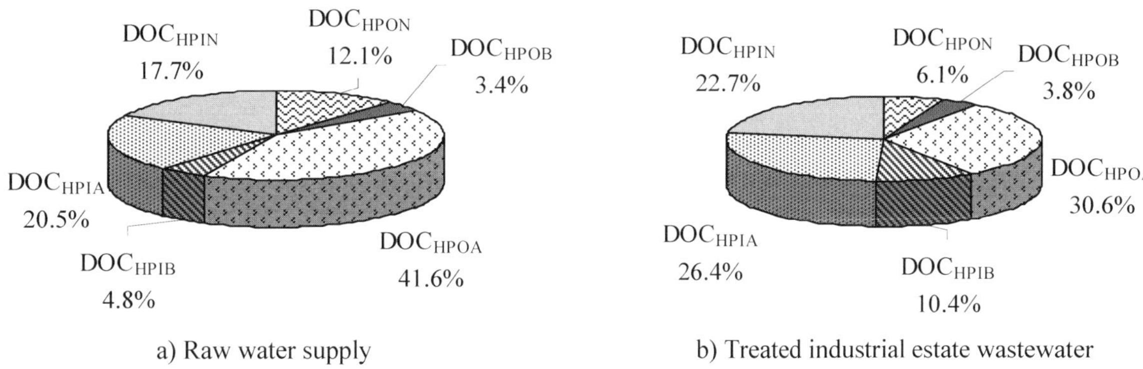

Figure 2 Percent distribution of each DOM fraction

THMFP distribution of each DOM fraction in raw water supply and treated industrial estate wastewater is illustrated in Figure 3. In the case of raw water supply, the main THMs precursors were HPOA and HPIA where $THMFP_{HPOA}$ and $THMFP_{HPIA}$ accounted for as much as 41.1 % and 18.7% by weight of the total THMFP from all fractions. The other THMs precursors ordered according to their THMFP are as follows: HPIB (11.3%), HPON (11.3%), HPIN (10.8%) and HPOB (6.9%). In the case of treated industrial estate wastewater, HPOA and HPIA were also the main THMs precursors where $THMFP_{HPOA}$ and $THMFP_{HPIA}$ accounted for as much as 32.4% and 25.7% by weight of the total THMFP from all fractions. This evaluation corresponds well with the report of Imai *et al.* (2003). The other THMs precursors in treated industrial estate wastewater ordered according to their THMFP are as follows: HPIN (18.9%), HPIB (11.8%), HPON (5.7%), and HPOB (5.4%).

From the obtained results, HPOA and HPIA were the main THMs precursor in both raw water supply and treated industrial estate wastewater. The summation of $THMFP_{HPOA}$ and $THMFP_{HPIA}$ accounted for 59.8% and 58.1% by weight of total THMFP in raw water supply and treated industrial estate wastewater, respectively. It should be mentioned however that the water treatment process could sufficiently reduce HPOA and HPIA prior to water being exposed to reactions with chlorine in the disinfection process.

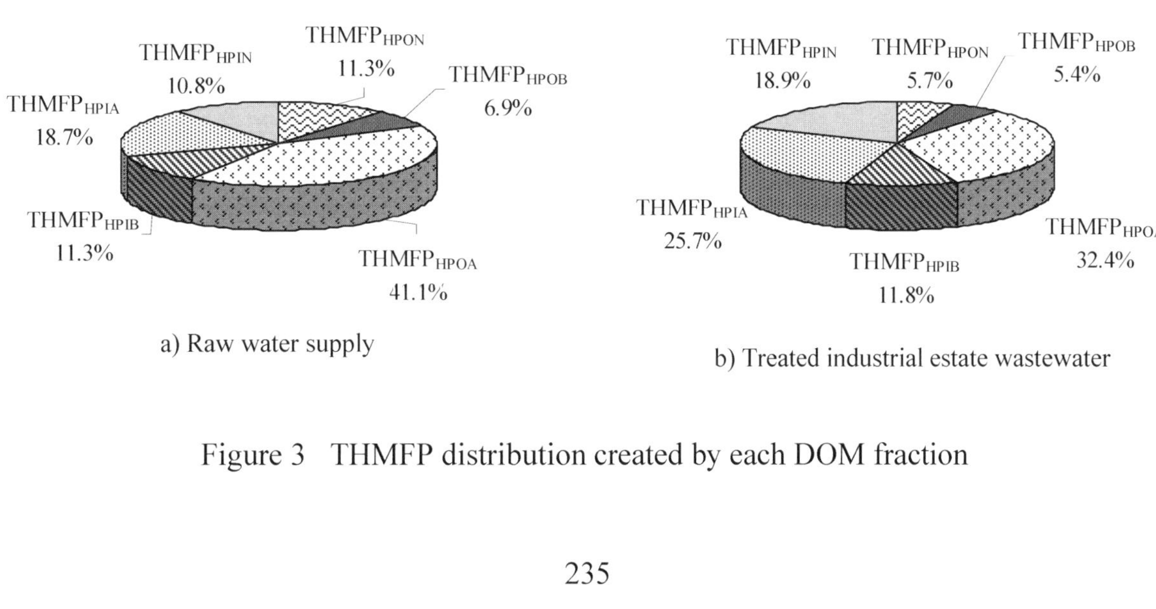

Figure 3 THMFP distribution created by each DOM fraction

3.2 Reactivity of each DOM fraction to form THMs

The term of reactivity of each DOM fraction is defined as follow:

$$\text{Reactivity of each DOM fraction} = \frac{\text{THMs created from each DOM fraction } (\mu\text{g THMFP/L})}{\text{DOC of each DOM fraction (mg DOC/L)}}$$

The reactivity of each DOM fraction facilitated the ability of each on the formation of THMs. As can be seen in Figure 4, when comparing the reactivity of each DOM fraction of raw water supply, the HPIB (reactivity of 240μgTHMFP/mgDOC) demonstrated the highest active source on the formation of THMs followed by HPOB (reactivity of 206ugTHMFP/mgDOC). In the case of treated industrial estate wastewater, HPOB (reactivity of 187μgTHMFP/mgDOC) demonstrated the highest active source followed by HPIB (reactivity of 153μgTHMFP/mgDOC). By these results, it is shown that HPOB and HPIB fractions in raw water supply and treated industrial estate wastewater, even when present in small quantities, were more highly reactive to form THMs than the major DOM fractions, HPOA and HPIA, which were observed as the first and second most prevalent DOM fractions in both waters. This corresponds with the report of Panyapinyopol *et al.* (2005) which notified that HPOB and HPIB fractions from a raw water supply source in Bangkok, Thailand were more highly reactive to form THMs than the major DOM fractions, HPIN and HPOA.

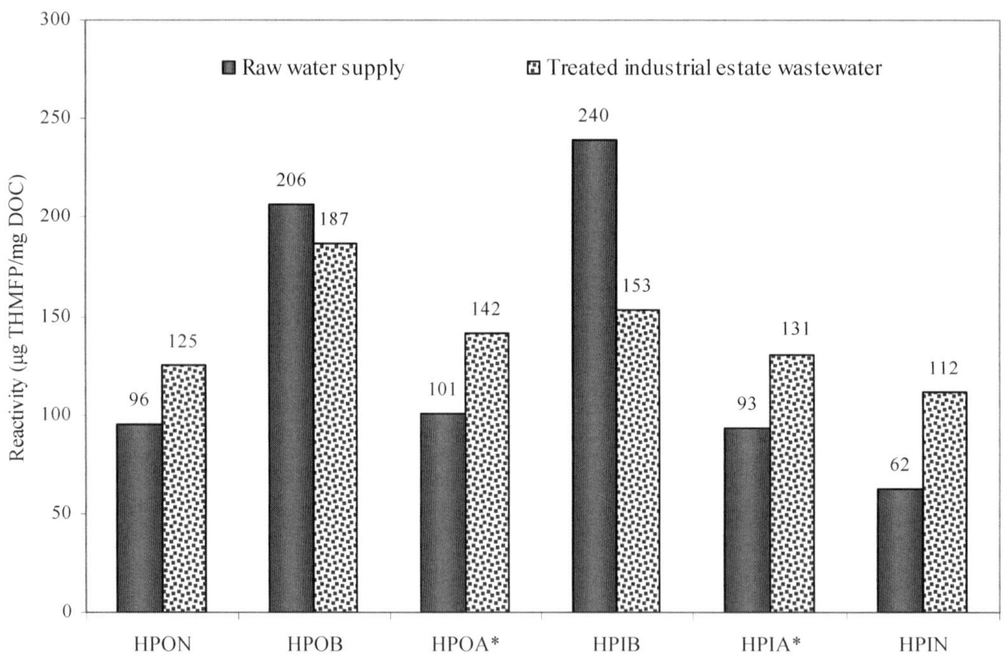

Remark: * HPOA and HPIA = Major DOM fractions in both waters.

Figure 4 Reactivity of each DOM fraction

3.3 Sensitivity of each organic fraction to form THMs

The definition of sensitivity of each DOM fraction is proposed as follow:

$$\text{Sensitivity of each DOM fraction} = \frac{\dfrac{\text{THMs created from each DOM fraction } (\mu\text{g THMFP/L})}{\text{DOC of each DOM fraction (mg DOC/L)}}}{\dfrac{\text{THMs created from total DOM in unfractionated water } (\mu\text{gTHMFP/L})}{\text{DOC of total DOM in unfractionated water (mg DOC/L)}}}$$

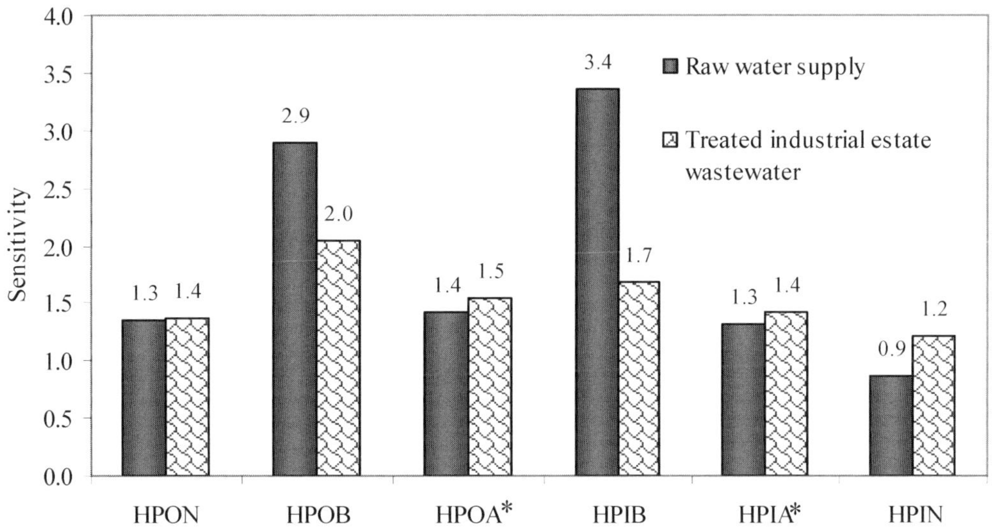

Remark: * HPOA and HPIA = Major DOM fractions in both waters.

Figure 5 Sensitivity of each DOM fraction

Figure 5 illustrates the sensitivity of each DOM fraction in raw water supply and in treated industrial estate wastewater. It can be seen that HPIB in raw water supply (sensitivity of 3.4) was the largest sensitive DOM fraction to form THMs followed by HPOB (sensitivity of 2.9). Regarding sensitivity of each DOM fraction in treated industrial estate wastewater, it was observed that sensitivities of 2.0 and 1.7 of HPOB and HPIB, respectively, were significantly determined as the first and second highest sensitive DOM fractions. These results were similar to the reactivities of HPOB and HPIB. By using the obtained sensitivity as an index for reducing THMs, it meant that the most effectively reduction of THMs created in the waters would be done by the reduction of HPOB and HPIB. However, HPOA and HPIA are still important DOM fractions to be considered for THMs removal since HPOA and HPIA were the most outstanding DOM fractions involved in forming THMs in both raw water supply and treated industrial estate wastewater.

4. Conclusion

HPOA and HPIA fractions in terms of dissolved organic carbon (DOC) in raw water supply were 41.6% and 20.5%, respectively, whereas HPOA of 30.6% and HPIA of 26.4% in treated wastewater were observed. It can be stated that HPOA and HPIA were the first and second most prevalent DOM fractions in both waters in which the summation of their DOC were 62.1% of total DOC in raw water supply and 57.0% of total DOC in treated wastewater. As a direct result, HPOA and HPIA were also realized to be the major trihalomethanes (THMs) precursors in both waters. The sum of $THMFP_{HPOA}$ and $THMFP_{HPIA}$ in the respective waters were 59.8% and 58.1% of total THMFP. According to the reactivity of each DOM fraction, HPIB demonstrated the highest active source on the formation of THMs in raw water supply followed by HPOB, whereas HPOB and HPIB were the first and second highest active source in treated wastewater. Regarding the sensitivity of each DOM fraction, HPIB and HPOB were determined as the first and second largest sensitive DOM fractions to form THMs in raw water supply whereas HPOB was the highest sensitive DOM fraction to create THMs in treated wastewater followed by HPIB.

5. Acknowledgements

This work was funded by the National Research Center for Environmental and Hazardous Waste Management (NRC-EHWM) of Thailand. Special thanks are extended to Associate Professor Dr. Pasert Pavasant for his valuable suggestions. Sincere gratitude is also given to the Northern-Region Industrial Estate, Thailand for assistance with sample collections.

6. References

AWWA and WPCF. (1995). Standard methods for the examination of water and wastewater. 19thed. APHA, Washington, D.C.

Imai A., Fukushima T., Matsushige K., Kim Y-H. and Choi K. 2002. Characterization of dissolved organic matter in effluents from wastewater treatment plants. *Wat. Res.,* 36:859-870.

Imai A., Matsushige K. and Nagai T. 2003. Trihalomethane formation potential of dissolved organic matter in a shallow eutrophic lake. *Wat. Res.,* 37: 4284-4294.

Leenheer J.A.,1981. Comprehensive approach to preparative isolation and fractionation of dissolved organic carbon from natural waters and wastewaters. *Environ. Sci. Technol.,* 15(5): 578-587.

Marhaba T.F., Pu, Y. and Bengraine K., 2003. Modified dissolved organic matter fractionation technique for natural water. *Wat. Res.,* 34(14):3543-3550.

Panyapinyopol B., Marhaba T.F. Kanokkantapong V. and Pavasant P. 2005. Characterization of precursors of trihalomethane formation potential in Bangkok source, *Journal of Hazardous Material,* 120 (1-3):229-236

Rook, J. J. 1974. Formation of haloforms during chlorination of natural water *Wtr.Trmt and Exam*., 23, 243.

Decentralized Wastewater Treatment Systems for Small-Scale Communities

Atitaya Panuvatvanich[1)], Narong Surinkul[2)] and Thammarat Koottatep[3)]

School of Environment, Resources and Development (SERD),
Asian Institute of Technology (AIT)
P.O. Box 4 Klong Laug, Pahtumthani 12120, Thailand
e-mail: 1) Atitaya.Panuvatvanich@ait.ac.th, 2) Narong.Surinkul@ait.ac.th, 3) thamarat@ait.ac.th

Abstract

In this study, the experiments were setup at AIT Environmental Research Station including three treatment units: (i) baffled septic tank with anaerobic filter, (ii) baffled septic tank with intermittent sand filter and (iii) baffled septic tank with constructed wetland, the ultimate goal is to develop and determine treatment performance of appropriate on-site treatment systems for small-scale communities. To imitate the characteristics of toilet wastewater, the experimental units were fed with a mixture of septage and sewage in order resulting in chemical oxygen demand (COD) and biochemical oxygen demand (BOD) concentrations of 1,000 and 300 mg/L, respectively. The baffled septic tank and anaerobic filter were operated at hydraulic retention times (HRT) of 2 days, whereas sand filter and constructed wetland units were operated at hydraulic loading rates (HLR) of 0.05 and 0.1 m^3/m^2.day, respectively. Base on the data obtained from 5-month operation, it was found that. The baffled septic tank with sand filter had average effluent BOD concentration of 7 mg/L, had the highest removal percentage (98%) when compare with the other treatment systems. The highest total nitrogen (TN) removal efficiencies were observed in the baffled septic tank with constructed wetland; the average TN removal and effluent concentration were 65% and 33 mg/L, respectively.

Keywords: anaerobic filter, baffled septic tank, constructed wetland, decentralized wastewater management, sand filter

1. Introduction

Due to the rapid growth of economic and urbanization in Thailand, it is very essential to treat domestic or municipal wastewater in order to avoid public health and environmental problems. The domestic wastewater treatment systems, typically used in small-scale communities, include individual or on-site treatment processes (or decentralized approach), while the centralized wastewater system is used in urban areas having dense settlements or large communities. It is apparent that the centralized system is inappropriate due to high investment and operation costs especially for small communities. The Royal Thai Government, for instance, has invested about 1.5 billion USD since 1992 for the construction of wastewater collection and treatment systems in 85 municipalities, out of which more than 60% are malfunction (Koottatep *et al.*, 2003).

Decentralized wastewater management could be an appropriate alternative to treat household or municipal wastewater for small-scale communities. However, limitations in applications of the decentralized approach in Thailand may include: absence of economic of scale and low per capita income, low level of technical expertise in operation and maintenance and limited access to the appropriate technologies. Therefore, the decentralized management and treatment system should be developed and experimented with regard to the applicability and economic viability for small-scale communities. Modifications of the conventional treatment systems should be based on the available processes such as anaerobic baffled reactor, anaerobic filter, sand filter, constructed wetland, etc. In addition to the low costs in investment and operation, the treatment systems should be able to treat the wastewater in the small communities. Objective of this study is to develop and determine treatment performance of appropriate on-site treatment systems for small-scale communities.

2. Materials and methods

The experiments were conducted at Environmental Engineering Research Station of AIT, Thailand. The influent wastewater was firstly fed into the baffled septic tank. Then, the effluent of baffled septic tank was fed into three different systems: anaerobic filter, sand filter, and constructed wetland. The samples were collected twice a week from five different points as shown in Figure 1. Samples were analyzed for pH, BOD, COD, total suspended solids (TSS), total kjedahl nitrogen (TKN), ammonia nitrogen (NH_4-N), nitrate nitrogen (NO_3-N) and total phosphorus (TP), according to Standard Methods for Examination of Water and Wastewater (APHA-AWWA-WPCF, 1998).

The experimental setups of the four units were fed with wastewater at intermittent basis having hydraulic retention time (HRT) and hydraulic loading rate (HLR) as explained in Table 1. Four pilot-scale treatment units are shown in Figure 1. Baffled septic tank with a total volume of 7.5 m^3 was constructed (unit dimensions: 5.25 m. length x 1.2 m. wide x 1.5 m. height) Anaerobic filter with a total volume of 5.4 m^3 (unit dimensions: 3.0 m. length x 1.5 m. wide x 1.5 m. height) had installed media in each chamber of the tank, anaerobic media used is a commercial media with surface area per volume of 105 m^2/m^3. Sand filter has a dimension of 3.0 m. length, 1.5 m. wide and 2.0 m. height. In the constructed wetland unit (unit dimensions: 3.0 m. length, x 1.5 m. wide x 2.0 m. height), the cattail plants were grown in the system (plant 8-12 Bulb/m^2).

Influent was a mixture of Bangkok septage and AIT wastewater at the ratio of 1:20, and fed into baffled septic tank. The characteristics of Bangkok septage AIT wastewater are shown in Table 2.

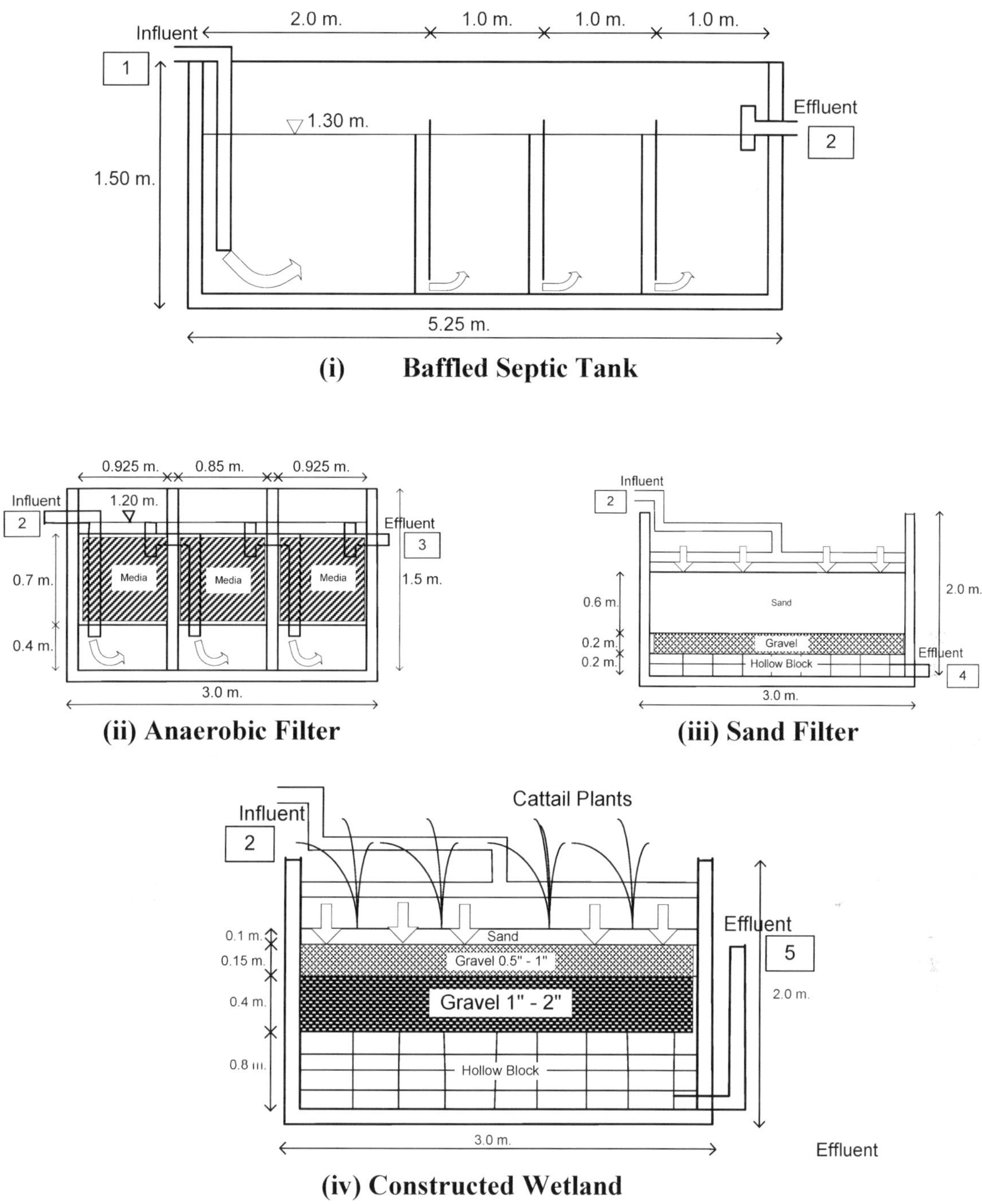

Figure 1. Schematic diagram of the experimental setups

Table 1. Operating conditions

No.	System	Selected loading
1.	Baffled Septic Tank	HRT = 2 day
2.	Anaerobic Filter	HRT = 2 day
3.	Sand Filter	HLR = 0.05 m^3/m^2.day
4.	Constructed Wetland	HLR = 0.1 m^3/m^2.day

Table 2. Characteristics of municipal wastewater, AIT wastewater and influent wastewater

Parameters	Bangkok Septage*	AIT wastewater	Influent wastewater		
	Range		Range	Average	Standard deviation
pH	7.4-8.25	7.0-8.15	6.4-7.9	7.1	0.5
TSS (mg/L)	4,800-10,320	10-80	96-2,620	1,010	570
COD (mg/L)	13,919-24,885	90-120	315-3,720	1,105	780
BOD (mg/L)	n/a	n/a	135-480	320	105
TKN (mg/L)	610-1,736	610-1,070	30-380	90	60
NH_4-N (mg/L)	246-1,081	15-40	20-55	40	10
NO_3-N (mg/L)	n/a	n/a	0-13	5	3
TN (mg/L)	n/a	n/a	35-380	95	60
TP (mg/L)	n/a	n/a	6-25	15	6

Note: Values are based on 40 samples, except BOD is averaged from 15 samples
*Data of Bangkok Septage and AIT wastewater were based on 40 samples

3. Results and discussion

The result obtained from pilot-scale experiments during December 2004 to April 2004 are summarized in Table 3.

Table 3. Overall treatment performance*

Parameters	Influent (1)**	Baffled Septic Tank (2)**	Baffled Septic Tank + Anaerobic Filter (3)**	Baffled Septic Tank + Sand Filter (4)**	Baffled Septic Tank + Constructed Wetland (5)**
pH	7.1	7.4	7.7	8.3	8.0
COD (mg/L)	1,105	150 (86%)	70 (94%)	20 (98%)	55 (95%)
BOD (mg/L)	320	55 (83%)	25 (92%)	7 (98%)	17 (95%)
TSS (mg/L)	1,010	65 (94%)	17 (98%)	7 (99%)	15 (98.5%)
TKN (mg/L)	90	45 (50%)	40 (56%)	8 (91%)	30 (67%)
NH_4-N (mg/L)	40	35 (13%)	34 (15%)	5 (88%)	25 (38%)
NO_3-N (mg/L)	5	4 (-)	3 (-)	30 (-)	3 (-)
TN (mg/L)	95	49 (48%)	43 (55%)	38 (60%)	33 (65%)
TP (mg/L)	15	10 (33%)	8 (47%)	5 (67%)	7 (53%)

Note: Values are based on 40 samples, except BOD is averaged from 15 samples.
*Removal efficiencies are shown in the brackets.
**Sampling point location

With regard to organic matter, it could be observed that the overall treatment efficiencies in terms of BOD and COD of the each unit: baffled septic tank, baffled septic tank with anaerobic filter, baffled septic tank with sand filter and baffled septic tank with constructed wetland ranged from 70-95%, 75-98%, 90-99% and 85-98%, respectively, as shown in Figure 2. Due to the average influent BOD and COD concentrations of was 320 and 1010 mg/L, respectively, the COD:BOD ratio of about 3:1, reveal that the influent contain organic matters which is probably non-biodegradable. The effluent BOD concentration of baffled septic tank with sand filter ranged from 0.6 to 15 mg/L, and having an average BOD concentration of 7 mg/L, and the average BOD removal about 98 %. However, it is difficult to determine which treatment system had the best performance because the baffled septic tank can remove about 85-95% of both organic matters whereas the individual anaerobic filter, sand filter and constructed wetland units have BOD removals of 55%, 87% and 69%, respectively.

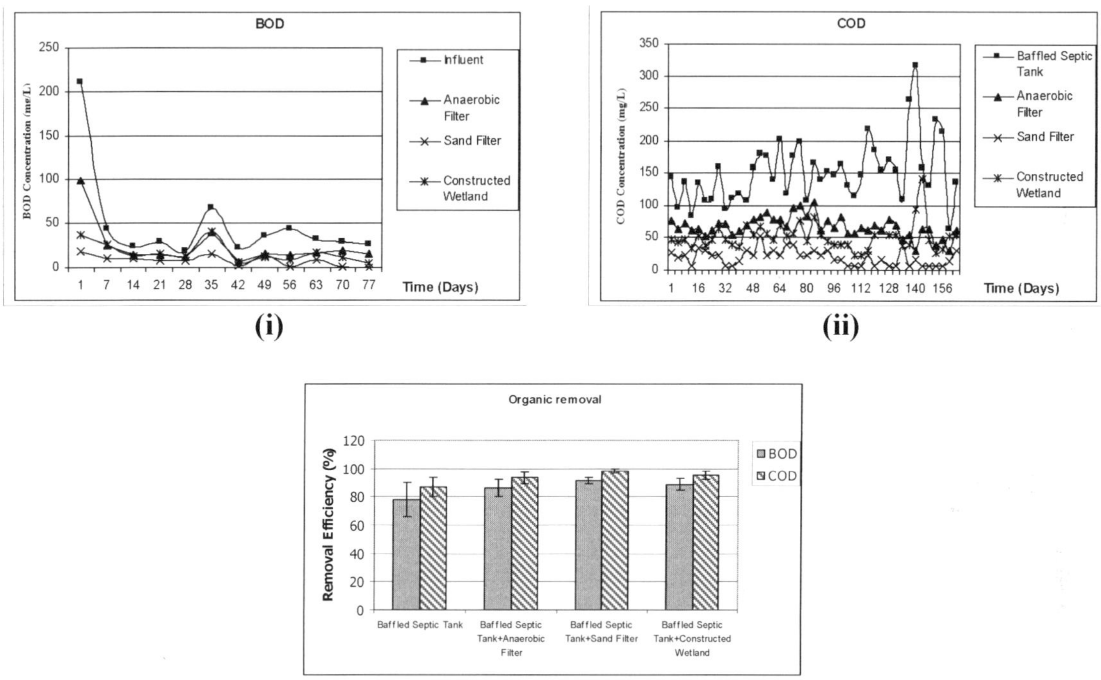

(i) (ii) (iii)

Figure 2. Effluent Concentration and Removal efficiencies in term of BOD and COD of each unit

Table 3 and Figure 3 illustrated that the average effluent TSS concentrations of baffled septic tank, baffled septic tank with anaerobic filter, baffled septic tank with sand filter and baffled septic tank with constructed wetland were 65, 17, 7 and 15 mg/L, respectively. Baffled septic tank with sand filter had the highest TSS of 99 %, whereas, baffled septic tank with anaerobic filter and baffled septic tank with constructed wetland had TSS removal of 98% and 98.5%, respectively.

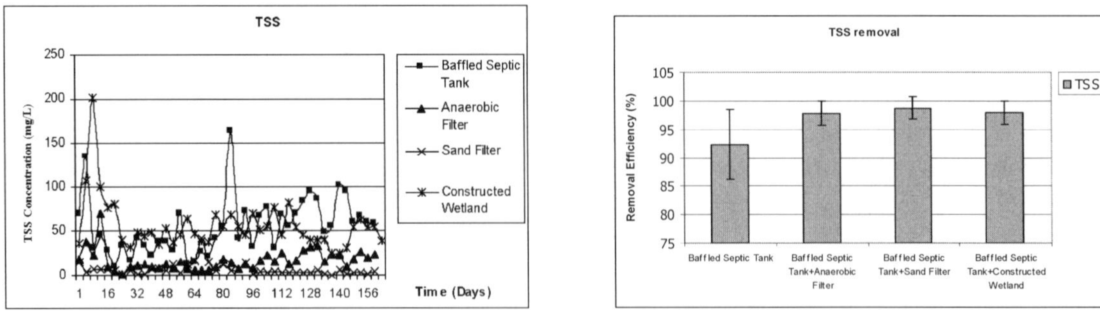

(i) (ii)

Figure 3. Effluent Concentration and Removal efficiencies in term of TSS of each unit

The average influent concentrations of TKN, NH_4, NO_3 and TN were found to be 90, 40, 5 and 95 mg/L, respectively (Figure 4). The experimental data reveals that baffled septic tank and sand filter can obtain the lowest effluent TKN and NH_4 concentrations (8 and 5 mg/L, respectively), but resulting in the highest NO_3 concentration of 30 mg/L. This occurrence could be due to the nitrification reaction taking place in sand filter unit. However, the lowest TN concentration was observed in the effluent of baffled septic tank and constructed wetland at the average concentration of 33 mg/L. It can be hypothesized that nitrification/ denitrification reactions are well taking place in the vertical-flow constructed wetland unit, similar to those reported by Koottatep *et al.* (2001).

Figure 5(ii) illustrated that influent wastewater consist of organic nitrogen 57%, NH_4 38% and NO_3 5% of TN concentration. After treatment, the N composition in wastewater has been changed as observed by the effluent of baffled septic tank consisting of organic nitrogen 70%, NH_4 21% and NO_3 9% of TN concentration. Effluent of sand filter consist of organic nitrogen 12.5%, NH_4 7.5% and NO_3 80% of TN concentration. According to Figure 5(i) and 5(ii), it can be noticed that the major N removal mechanism in baffled septic tank and anaerobic filter is the sedimentation of or-N particles, whereas the nitrification/denitrification reaction could not well perform in N removal, possibly due to absence in oxygen concentration. In the sand filter and constructed wetland units, it could support the nitrification/denitrification reactions, likely due to the presence of oxygen through wastewater percolating in the media and oxygen provision for N removal is the N uptake by cattail plants in the constructed wetland unit, as reported by Koottatep and Polprasart (1997).

Considering TP removal, the average influent TP concentration was 15 mg/L. Figure 6 illustrates that the average effluent TP concentrations of baffled septic tank, baffled septic tank with anaerobic filter, baffled septic tank with sand filter and baffled septic tank with constructed wetland were 10, 8, 5 and 7 mg/L, respectively. Baffled septic tank with sand filter had higher percent removal of TP concentration (67%) than those of baffled septic tank with anaerobic filter and baffled septic tank with constructed wetland, probably because adsorption process in sand filter could react with PO_4.

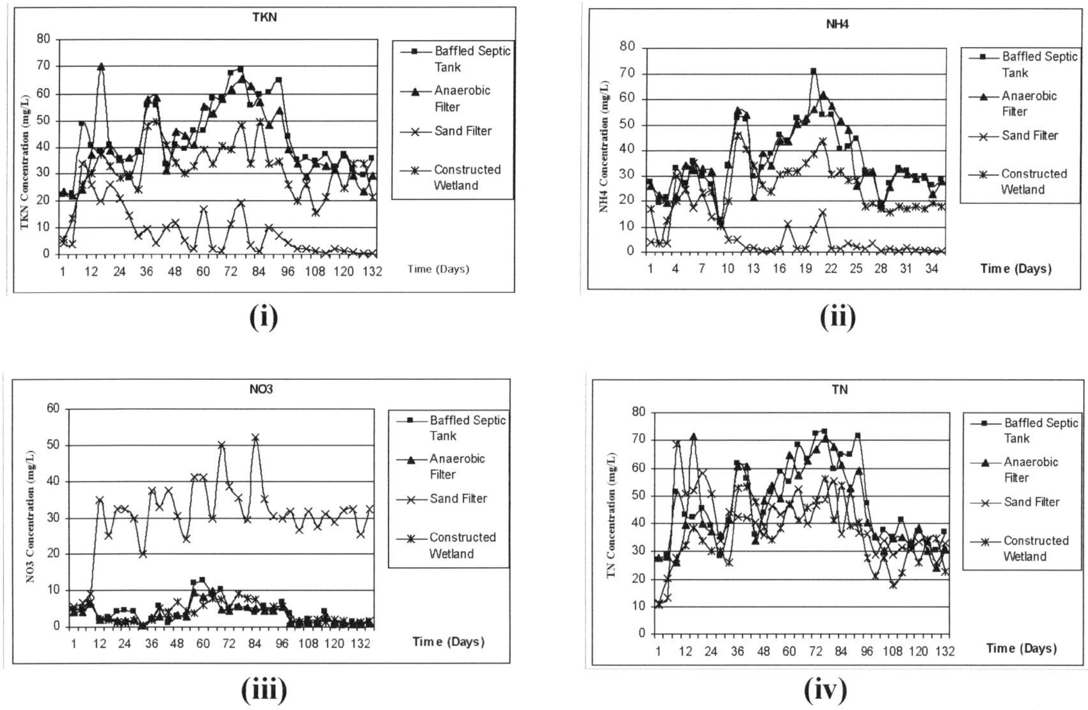

Figure 4. Effluent N Concentration of each unit

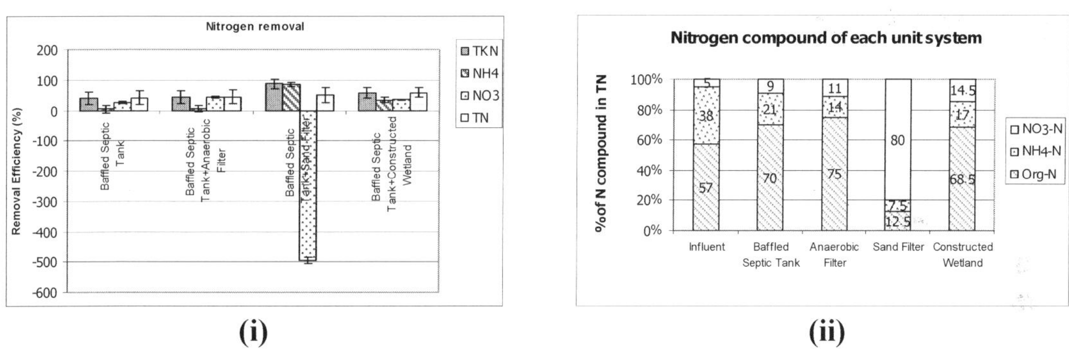

Figure 5. N Removal Efficiencies of each unit

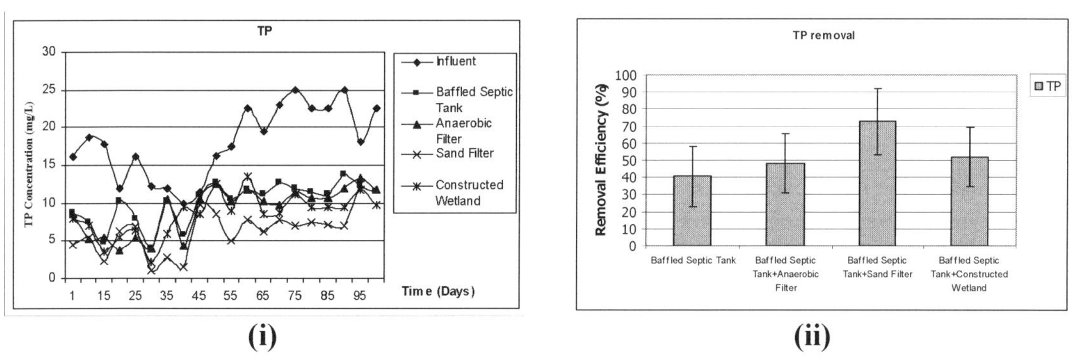

Figure 6. Effluent Concentration and Removal efficiencies of TP of each unit

4. Conclusions

With regard to the previous studies and recommend design criteria, the baffled septic tank and anaerobic filter were operated at hydraulic retention time (HRT) of 2 days, whereas sand filter and constructed wetland were operated at hydraulic loading rates (HLR) of 0.05 and 0.1 m^3/m^2.day, respectively.

The 5-month experiment results demonstrate that the baffled septic tank with sand filter had the highest treatment performance in term of BOD, COD, TSS and TP (98%, 98%, 99% and 67%, respectively). Whereas baffled septic tank with constructed wetland could achieve the highest TN removal.

Even the pilot-scale experiment results of this 5-month study reveal promising treatment performance of baffled septic tank, anaerobic filter, sand filter and constructed wetland, the long-term experiments with variation in operating conditions should be undertaken in order to reaffirm the appropriate criteria for design and operation.

5. Acknowledgement

The authors express their appreciation for the generous support of the National Center for Competence in Research North – South, Switzerland, which made this study possible.

6. References

APHA-AWWA-WEF, (1998), *Standrads Methods for the Examination of Water and Wastewater, 20th Edition.* American Public Health Association, Washington DC, USA

Koottatep, T. and Polprasert, C. (1997), Role of Plant Uptake on Nitrogen Removal in Constructed Wetland Located in the Tropics, *Water Science and Technology*, Vol. 36, No 12, pp.1-8

Koottatep, T., Kim Oanh, N.T., Polprasert, C., Heinss, U., Montangero A. and strauss, M. (2001), Setage Dewatering in Vertical-Flow Lacated in the Tropics, *Water Science and Technology*, Vol. 44, No 2-3, pp.181-188

Kootatep, T., Sitarnrassamee, B., and Polprasert, C., (2003), Water Environment and Remediation: Thailand's Way Forward, *Proc. of Alliance for Global Sustainability (AGS) Technical Workshop on Southeast Asian Water Environment*, the University of Tokyo, Tokyo, Japan

Selection of Appropriate Decentralized Sanitation Alternatives in Vietnam

Montangero Agnes[1)], Cau Le Ngoc[2)], Viet Anh Nguyen[3)] and Belevi Hasan[4)]

1) Department of Water and Sanitation in Developing Countries (SANDEC),
Swiss Federal Institute for Environmental Science and Technology (EAWAG)
P. O. Box 611, CH-8600 Duebendorf, Switzerland
e-mail: montangero@eawag.ch
2) Asian Institute of Technology Center in Vietnam (AITCV)
21 Le Thanh Torg, GPO Box 136, Hanoi, Vietnam
e-mail: lncau@aitcv.ac.vn
3) Center for Environmental Engineering of Towns and Industrial Areas (CEETIA),
Hanoi University of Civil Engineering
DHXD, 55 Giai Phong Road, Hanoi, Vietnam
e-mail: vietanhctn@hotmail.com
4) Department of Environmental Engineering, University Innsbruck
Technikerstrasse 13, A-6020 Innsbruck, Austria
e-mail: hasan.belevi@iubk.ac.at

Abstract

The rapid urbanization and industrialization process taking place in Vietnam is accompanied by an improvement of the socio-economic conditions but also by increasing environmental pollution and pressure on natural resources. The conventional centralized sanitation system is often considered as the most appropriate sanitation option. The authors state that several possible sanitation options exist and that depending on the context, one will be more appropriate than another. The development and evaluation of sanitation alternatives should be based on a deep understanding of stakeholders' perceptions and needs. Costs, socio-economic, health and environmental impacts of the different alternatives should be clearly demonstrated to all stakeholders as a basis for consensus finding. The method of material flow analysis (MFA) enables to evaluate the impacts of the current sanitation situation as well as to simulate the impacts of potential sanitation alternatives on environmental pollution and resource recovery. It can therefore be used as a tool for the evaluation of options. In the context of Vietnam where local authorities, local communities and increasingly also the private sector play an important role in public service provision, decentralized sanitation seems to be a promising alternative to conventional centralized sanitation.

Keywords: decentralised sanitation, developing countries, material flow analysis, nutrient recovery, Vietnam, wastewater reuse

1. Introduction

Vietnam is undergoing a process of rapid urbanization and industrialization. The urban population has increased from 19% of the total population in 1986 to 23.5% in 1999 and is expected to reach 40-45% by 2020 (Viet Anh, 2003). Urban centers attract people from rural areas in search of work, higher income and higher quality of life. The positive socio-economic development – the GDP doubled between 1990 and 2000 – is accompanied by increasing water demand as well as environmental pollution and health problems.

Only a very low ratio of the population is served by safe drinking water and adequate sanitation, especially in rural, peri-urban and poor urban areas. Septic tank is the most common on-site treatment facility in urban and peri-urban areas in Vietnam. In urban centers, 50-80% of the households are equipped with septic tanks (Viet Anh et al., 2002). The rest of the population have pit latrines, double vault latrines, bucket latrines or are not equipped with sanitation facilities. Septic tanks are rarely desludged. Urban drainage systems are poorly developed and under degradation. Most cities and towns are often inundated during the rainy season. Only a minority of industrial areas are equipped with wastewater treatment facilities. Septic tank effluent, untreated greywater and industrial wastewater are mainly discharged into the drainage network from where it flows to canals, lakes and ponds without any treatment. The increase of agrochemicals used and of livestock production in rural and peri-urban areas also contributes significantly to environmental pollution. The self purification capacity of receiving water bodies cannot cope with the high pollution loads discharged from urban areas. Lakes, ponds and canals in many cities in Vietnam are characterized by high organic and nutrient concentrations. Drainage water is widely used by farmers living on the edge of the city. The reuse of wastewater increases productivity in agricultural and aquacultural systems but also poses health risks (Viet Anh et al., 2004).

Developing appropriate solutions for the supply of safe drinking water, the management of wastewater, solid waste and storm water is one of the challenges faced by Vietnam as well as most developing or newly industrialized countries. Even though decentralization has a very long tradition in Vietnam, centralized sanitation is usually considered as an appropriate solution. However, conventional centralized sanitation also has its drawbacks and may not be appropriate in every context. It is characterized by large investment as well as operation and maintenance costs. Managing and financing large central sewerage and treatment facilities represent huge challenges for municipal authorities. Moreover, the management of resources (e.g. water, nutrients, energy, artificial fertilizers) within the conventional sanitation system is neither sustainable (linear flows) nor economical (Otterpohl et al., 1997; Berndtsson and Hyvönen, 2002).

2. Wastewater reuse in Vietnam: potential and risks

In Vietnam, the application of raw wastewater, including domestic sewage, industrial effluents and human excreta in agricultural and aquacultural production is widespread, especially in northern Vietnam. In Hanoi, more than 500,000 cubic meters of wastewater are discharged daily, and a large amount of this wastewater is pumped or diverted for irrigation and aquacultural production. It is evidenced that wastewater application in agriculture, particularly in fish farming, has brought benefits to the sector as well as surplus profits to farmers from better harvest yields (Hoan and Edwards, 2002; Viet Anh et al., 2004).

The efficiency gained through the reuse of wastewater also can be seen in the integration of fish farming and the rearing of domestic animals such as pigs, chickens and ducks. In the commune, the application of the garden-fish farming-livestock breeding model (the VAC model) is rather widespread, especially in those households with proximity to the fish ponds supplied with wastewater. Water from the ponds is often used for irrigation of the households' gardens as well as surrounding rice and vegetable fields. The rearing of pigs, chicken and ducks has contributed to the increase of fish productivity because of surplus feeds and manure used for aquaculture, at the same time, sub-standard fish could be supplemented as extra foods for the animals (Viet Anh, 2003).

Besides the undeniable benefits and efficiency that the reuse of wastewater has brought to agricultural and aquacultural production, possible negative impacts on human health of wastewater irrigation activities are of great concern to the public health sector. Little is known in Vietnam about the impact of wastewater on the health of farmers, especially those who are directly exposed to the wastewater. Awareness on the health impacts of wastewater and improvement of hygiene behaviors should be raised in the community so that farmers can protect themselves, as well as protect the society (Viet Anh, 2004).

There has recently been a strong tendency for agricultural production in peri-urban areas in southern Vietnam to move away from low input rice production to more intensive fruit and vegetable production, often in combination with pig production. The access to a market has led to increased availability of fertilizers and livestock feed in these areas, which has resulted in a dramatic increase in nutrient inputs. Some previous studies showed that even in rural and peri-urban areas in Vietnam, the different production units were poorly integrated as nutrient output from one unit seldom matched nutrient demand in the next unit. The crop units were characterized by large nutrient surpluses and low nutrient efficiencies, indicating a high risk for nutrient losses to the environment (Hedlund et al., 2003).

3. New approaches for the selection of appropriate sanitation options

New environmental sanitation approaches must be developed – both in industrialised and in developing countries – in response to the challenge caused by the large number of people without access to adequate environmental sanitation, the rapid population growth in urban areas, the water crisis, and the inability of the conventional sanitation system to manage water and nutrients in a sustainable way. The following principles (Bellagio principles) form the basis of a new approach – the household centred approach (Schertenleib and Morel, 2003):

- Human dignity, quality of life and environmental security at household level should be at the centre of the new approach, which should be responsive to needs and demands in the local and national setting.
- Decision-making should involve participation of all stakeholders, especially the consumers and providers of services.
- Waste should be considered as a resource, and its management should be holistic and form part of integrated water resources, nutrient flows and waste management processes.
- The domain in which environmental sanitation problems are resolved should be kept to the minimum practicable size (household, community, town, district, catchment, city) and wastes diluted as little as possible.

Based on these principles, a procedure for the development and evaluation of potential environmental sanitation options is proposed (Figure 1). Besides the knowledge of the infrastructural, managerial, financial, legal and institutional aspects of environmental sanitation in a specific context (e.g. a district or a city in Vietnam), a deep understanding of the perceptions and values related to environmental sanitation of the different stakeholders is of utmost importance for the development of potential environmental sanitation concepts (scenarios). Costs and socio-economic benefit, acceptance by users, health and environmental impacts as well as potential for resource recovery of the proposed scenarios are relevant criteria for their evaluation. The above mentioned principles such as considering waste as a resource and solving the problems as close as possible to where they are generated (decentralised options) should be taken into account in the development and the evaluation of scenarios.

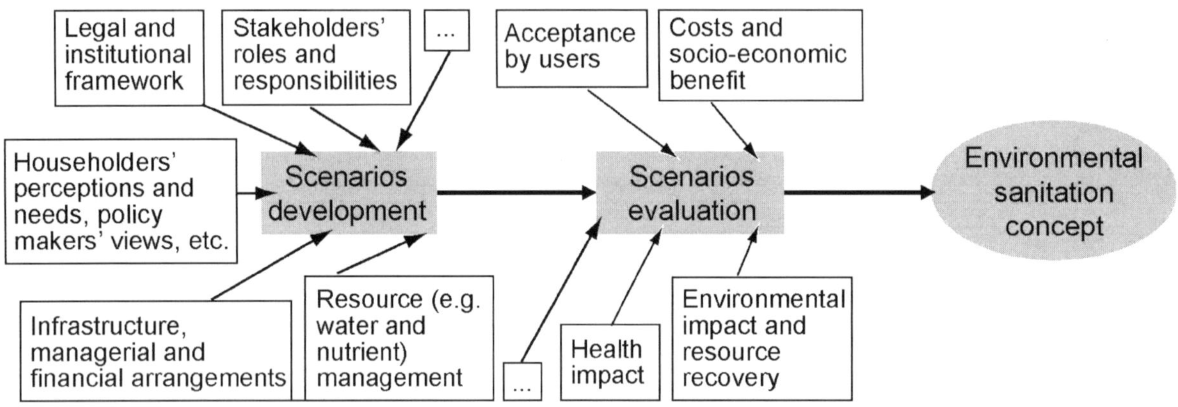

Figure 1. Proposed procedure for multi-stakeholder environmental sanitation planning

The decentralized approach is a new means of addressing wastewater management needs of sewered and unsewered areas in a comprehensive fashion. The basic idea of that is to treat the wastewater (possibly together with refuses) on-site by means of low-cost treatment systems, and make direct use of the treatment products (water, compost, biogas). This alternative can improve the sustainability of wastewater management and has a promising future, especially for developing countries like Vietnam, which are under a new period of infrastructure development and where water and sanitation issues are becoming more and more important. The elements included in decentralized wastewater management systems comprise: (1) wastewater pre-treatment, (2) wastewater collection, (3) wastewater treatment, (4) effluent reuse or disposal, and (5) biosolids and septage management. Although the components are the same as for large centralized systems, the difference is in the type of technology applied, and the form and mechanism of management. It should also be noted that not every decentralized wastewater management system should incorporate all the above mentioned elements.

Selection algorithms can assist planners and decision-makers in selecting appropriate environmental sanitation options. Figure 2 shows an algorithm being developed for the selection of appropriate sanitation options in the context of Vietnam. The user goes through a list of questions (criteria) and, depending on the answers, appropriate sanitation options are proposed.

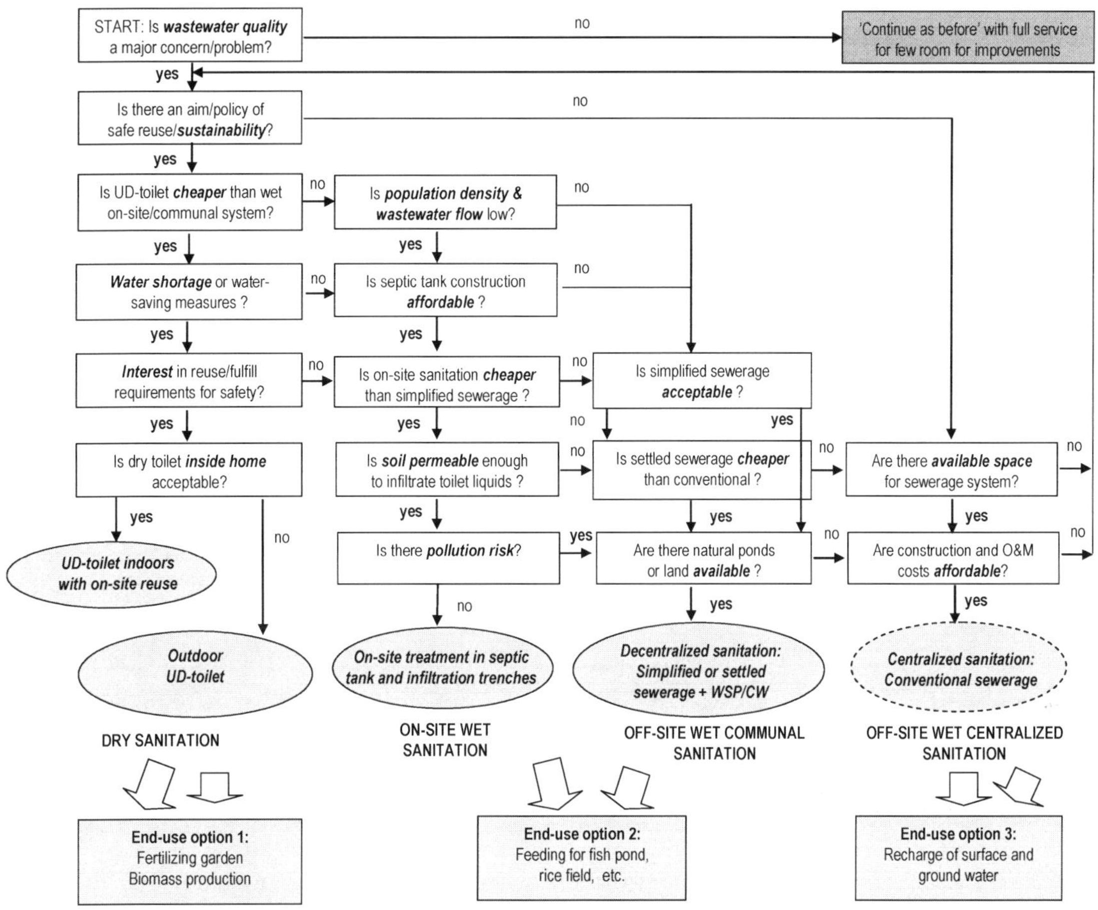

Figure 2. Algorithm for the selection of sanitation alternatives in the context of Vietnam (Viet Anh et al., 2004)

4. Material flow analysis as a tool for environmental sanitation planning

An appropriate environmental sanitation system should guarantee sustainable resource management (e.g. minimize untreated wastewater load in water bodies, maximize saving of water resources and nutrient recovery). The method of material flow analysis (MFA) studies the fluxes of resources used and transformed as they flow through a single process (e.g. a wastewater treatment option) or via a combination of processes (e.g. a region). In industrialized countries, MFA proved to be a suitable instrument for the early recognition of environmental problems and development of measures (Baccini and Brunner, 1991). One of the possible applications of the method is the quantification of urban waste and food flows and the corresponding water and nutrient flows in a given district or city. This quantification enables the identification of key processes and flows as well as the simulation of new environmental sanitation concepts or measures. These can be evaluated according to their impact on nutrient load into the environment as well as on nutrient recovery.

MFA has already been applied in urban areas of developing countries in the field of environmental sanitation (Binder, 1996; Binder et al., 1997). The conclusions were that MFA can be applied to regions in developing countries in spite of poor data availability and quality and that MFA and its mathematical description can be used for setting up monitoring concepts, the early recognition of resource demand and environmental impacts, and evaluating the effect of technical measures in mitigating these impacts. Leitzinger applied the methodology of MFA in the city (urban and peri-urban area) of Kumasi (Ghana) in order to estimate how much of the nitrogen and phosphorous demand in urban and peri-urban agriculture could be covered by compost produced with urban solid waste and excreta (Leitzinger, 2000; Belevi, 2002). Phosphorous flows in the urban water management system in Harare (Zimbabwe) are being investigated by Gumbo (2003) using MFA.

The first step of the method of MFA consists in analyzing the investigated system (e.g. a city and its surrounding peri-urban area). The system is defined by one or several processes (e.g. households, wastewater treatment, crop production, water bodies), the flows of goods (e.g. human excreta, greywater, solid waste, food, fertilizer) or substances (e.g. nitrogen, phosphorous) and the system boundary. Once the system is defined, mass flows of all goods and concentrations of the selected indicator substances are determined. Then indicator flows are calculated based on mass flows of goods and indicator concentrations. Mass flows or indicator flows are determined using information obtained through literature review, field measurements (surveys, sampling and laboratory analysis) or calculated by mass balances over processes. Once indicator flows have been determined, relevant processes and flows in the system can be identified and measures aiming at reducing pollution load into the environment and optimizing resource recovery can be developed and their impacts simulated. This information can then be used by planners and decision-makers to evaluate potential environmental sanitation improvement measures and select alternatives that are appropriate in their particular context.

Figure 3 illustrates how the method could be applied to identify problems in the current sanitation system and to develop potential measures. The example represents the situation in urban areas in Vietnam but is not based on real data. Flows are therefore not quantified, the thickness of the arrows only indicate the relative importance of the flows. In the current system, almost all nitrogen present in urban wastewater (human excreta, greywater and industrial wastewater) is discharged untreated in the drainage network from where it flows to ponds, lakes and rivers or is reused in agriculture. Valuable nutrients are mixed with pollutants originating from industries.

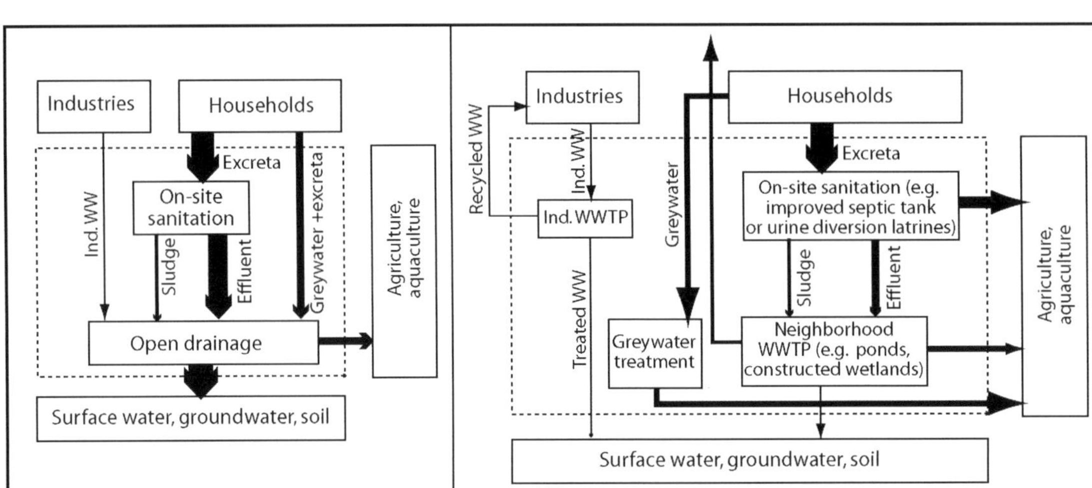

Figure 3. Simplified systems representing nitrogen flows in the current sanitation system (left) and in a potential future system (right)

A potential solution (Figure 3 right) could consist in improving the removal efficiencies of on-site sanitation facilities and introducing small-scale decentralized (e.g. community) wastewater treatment facilities such as constructed wetlands or waste stabilization pond systems (in case land is available) for the treatment of greywater as well as effluent and sludge from household treatment facilities. Treatment products (effluent and biosolids) would then be reused as irrigation water and organic fertilizer in agriculture and aquaculture. Treatment would aim at eliminating organic matter and nutrient in case reuse is not feasible and treated wastewater is to be discharged in surface water. Industrial wastewater would be treated separately and as far as possible reused internally. In this way, treatment products reused in agriculture would not be mixed with problematic substances such as heavy metals. The solution proposed leads to a reduction of the nitrogen discharged into the environment and to an increase in nitrogen recovery through treated wastewater reuse in agriculture. The quantification of the reduction of the nitrogen discharge and of the nitrogen reuse in agriculture/aquaculture could be carried out by MFA. For this purpose, the current nitrogen flows (Figure 3 left) should be quantified and a MFA model should be elaborated. Consequently, it is possible to simulate future scenarios by using this MFA model (Figure 3, right).

5. Conclusions

Decentralization does not only mean the physical division of a wastewater catchment area, but can also be understood as the devolution of responsibility to local stakeholders in all steps of decision-making related to construction and O&M of the sanitation system. In Vietnam, given the important role of local authority (ward People Committee – PC) and household initiatives as well as the growing role of private sector in public service provision, decentralized sanitation seems to be a promising alternative to conventional centralized sanitation.

The resolution of conflicting stakeholders' interests (householders, farmers, policy makers, sanitation experts, etc.) can be facilitated by clarifying stakeholders' positions and improving transparency in outcomes of alternatives. It is therefore important to involve all stakeholders in the development of potential scenarios, the underprivileged such as women and the urban poor in particular. Moreover, costs, socio-economic, health and environmental impact of the proposed scenarios should be assessed and clearly presented to the stakeholders.

MFA enable planners to evaluate the impacts of the current environmental sanitation system as well as simulate the impacts of new environmental sanitation concepts or measures on environmental pollution and resource recovery. It represents therefore a useful tool to increase knowledge on alternatives among the different stakeholders. However, there is a need for practical tools and recommendations to assist planners and decision-makers in applying the method in a context of poor data availability and reliability.

6. Acknowledgement

The authors gratefully acknowledge the Swiss Agency for Development and Cooperation (SDC) and the Swiss National Science Foundation (SNSF) for their financial support through the National Centre of Competence in Research (NCCR) North-South programme.

7. References

Baccini P. and Brunner P.H. (1991). *Metabolism of the Anthroposphere*. Springer, New York

Belevi H. (2002). Material flow analysis as a strategic planning tool for regional waste water and solid waste management. In: *Proceedings of the workshop "Globale Zukunft: Kreislaufwirtschaftskonzepte im kommunalen Abwasser- und Fäkalienmanagement"*. Munich, 13-15 May

Berndtsson J.C. and Hyvönen I. (2002). Are there sustainable alternatives to water-based sanitation system? Practical illustrations and policy issues. *Water Policy* 4 (2002) 515-530

Binder C. (1996). *The early recognition of environmental impacts of human activities in developing countries*. ETHZ PhD thesis Nr. 11748, Zürich, Switzerland

Binder C., Schertenleib R., Diaz J., Bader H.-P., Baccini P. (1997). Regional water balance as a tool for water management in developing countries. *Water Resources Development*, Vol.13, No. 1, 5-20

Gumbo B., Savenije H., Kelderman P. (2003). The phosphorus calculator: a planning tool for closing nutrient cycles in urban eco-systems. In: *Proceedings of the 2nd int. symposium on ecological sanitation*, Lübeck, 7-11 April

Hedlund A, Witter E., Bui Xuan An (2003). Assessment of N, P and K management by nutrient balances and flows on peri-urban smallholder farms in southern Vietnam. *Europ. J. Agronomy* 20 (2003) 71-87

Hoan V.Q and Edwards P. (2002). *Aquaculture and Wastewater reuse in Hanoi, Vietnam.* (Hand out from personel contact).

Leitzinger C. (2000). *"Ist eine Co-compostierung aus stofflicher Sicht in Kumasi/Ghana sinnvoll?"*, Diploma Thesis, ETH Zurich, Switzerland

Otterpohl R., Grottker M., Lange J. (1997). Sustainable water and waste management in urban areas. *Wat.Sci.Tech.* Vol. 35, No 9, pp. 121-133

Schertenleib R. and Morel A. (2003). The household-centred environmental sanitation approach (HCES). Paper presented at the 3rd world water forum in Kyoto, 16-23 March

Viet Anh N. (2003). Water environment and water pollution control in Vietnam: overview of status and measures for future. In: *Proceedings from the first international symposium on South-East Asia water environment - biodiversity and water environment*. Bangkok, Thailand, Oct. 2003

Viet Anh N. (2004). *Report from a second workshop on Capacity building for decentralized wastewater management (DWWM) in Vietnam,* CEETIA, Hanoi, 15th January, 2004. Prepared for GHK International. Working report. March 2004.

Viet Anh N., Hanh T.T.H, Thanh V.T.M, Parkinson J., Barreiro W. (2004). Decentralized wastewater management in Vietnam – a Hanoi case study. In: *Proceedings of the 30th WEDC Conference (in press)*, Vientiane, Laos PDR, October 2004.

Viet Anh N., Hung Long N., Ngoc Quang T., Thanh Son N., Tan Sinh B., Lan Huong T. (2003). *Towards application of Ecosan for urban areas in Vietnam.* Globconsult JSC. Report from a country study, prepared for SENSA/SIDA. Hanoi, December 2003.

Viet Anh N., Mara D., Drangert J., Gulyas H., Tonderski K., Tonderski A. *Low-cost and sustainable sanitation (LCST)*. Training material (under development). EU funded Asia-link program ASI/B7-301/98/679-024. 2003 – 2004.

Viet Anh N., Tran D.A, Tran H.N, Heinss U., Morel A., Moura M., Schertenleib R. (2002). Decentralized wastewater treatment – new concept and technologies for Vietnamese conditions. In: *Proceedings from the 5th IWA Conference on small wastewater treatment systems*, Istanbul, Turkey. 2002 Sept.

Life Cycle Assessments of Wastewater Treatment Systems in Bangkok City

Thammarat Koottatep[1] and Thitirat Chaosakul[2]

1) Environmental Engineering and Management, Asian Institute of Technology
PO. Box 4 Klong Laung, Pathumthani, 12120, Thailand
e-mail: thamarat@ait.ac.th
2) TEAM Consulting Engineering and Management Co., Ltd.
151 TEAM Building, Nual Chan Rd., Bangkok, 10230, Thailand
e-mail: thitirat@team.co.th

Abstract

Life Cycle Assessment (LCA) is as a tool for optimal resource utilization and anticipation of the environmental impacts, but only a few applications for wastewater treatment systems. In this study, the small- and large-scale conventional treatment system in Bangkok city were evaluated by comparing the eco-indicator 95 and 99 as well as appraising its modification scenarios for mitigation of its environmental impacts and damages. Based on the LCA analysis, the wastewater treatment systems could apparently cause the significant environmental impacts especially on eutrophication, whereas the environmental damages in terms of human health, ecosystem quality and resources were mainly due to the high energy consumption. Eco-indicator 95 reveals that the small-scale wastewater treatment plant at Si Phraya result in the slighter environmental impacts per 1 m^3 of wastewater than that of a large-scale system at Huay Kwang. On the contrary, the small-scale system would cause less environmental damage than those resulted from a large-scale treatment plant. Even the LCA could predict the overall environmental impact and damages, the in-depth considerations on economic, social-cultural aspects as well as institutional arrangements should be analyzed in order to determine the system sustainability.

Keywords: domestic wastewater treatment systems, life cycle assessment (LCA), sustainability

1. Introduction

During the past decades, there have been several millions dollars for investment in the conventional sewerage and domestic wastewater treatment systems in abating of the water pollution problems in Thailand, but their environmental sustainability are questionable. To systematically assess the wastewater treatment systems, the Life Cycle Assessment (LCA) could be applied in evaluating the environmental impacts over their entire life cycle, requiring a system analysis based on mass and energy balances providing an indication of material use, emissions, costs, and required land area (Belkema et al., 2002). The characterization step of LCA where different environmental burdens are added into environmental themes can be useful for the aggregation of environmental performance indicator and the valuation step where different environmental issues are weighted may be useful in the selection of different indicators.

In addition to the LCA approach, application of the environmental sustainability indicators (ESIs) could be the most effective approach to measure and describe different aspects of sustainability, with the focus often being on the environmental aspect (Lundin and Morrison, 2002). The majority of the ESI applications include the assessment of environmental performance of infrastructure, agriculture and production. The objective of this study is to appraise the conventional sewerage and domestic wastewater treatment systems in Thailand by using LCA together with ESI. The study compared LCA and ESI between small- and large-scale wastewater treatment plants of Bangkok metropolitan: Huay Kwang and Si Phraya with the design treatment capacity of 3,000 and 30,000 m^3/day, respectively.

2. Materials and methods

Because the long-term database and related information avail, this study selected the Huay Kwang and Si Phraya wastewater treatment plants, representing small- and large-scale systems, respectively, in Bangkok metropolitan. These treatment plants have been operated for more than a decade, treating wastewater from the combined sewerage that receive greywater, effluent of septic tanks (or cesspools) and storm water.

In LCA analysis, the environmental impact assessments were calculated per one $m^3$of treated wastewater by considering the impacts on global warming, acidification, nutrient enrichment, and discharge of heavy metals. System boundary of this analysis can be described in Figure 1.

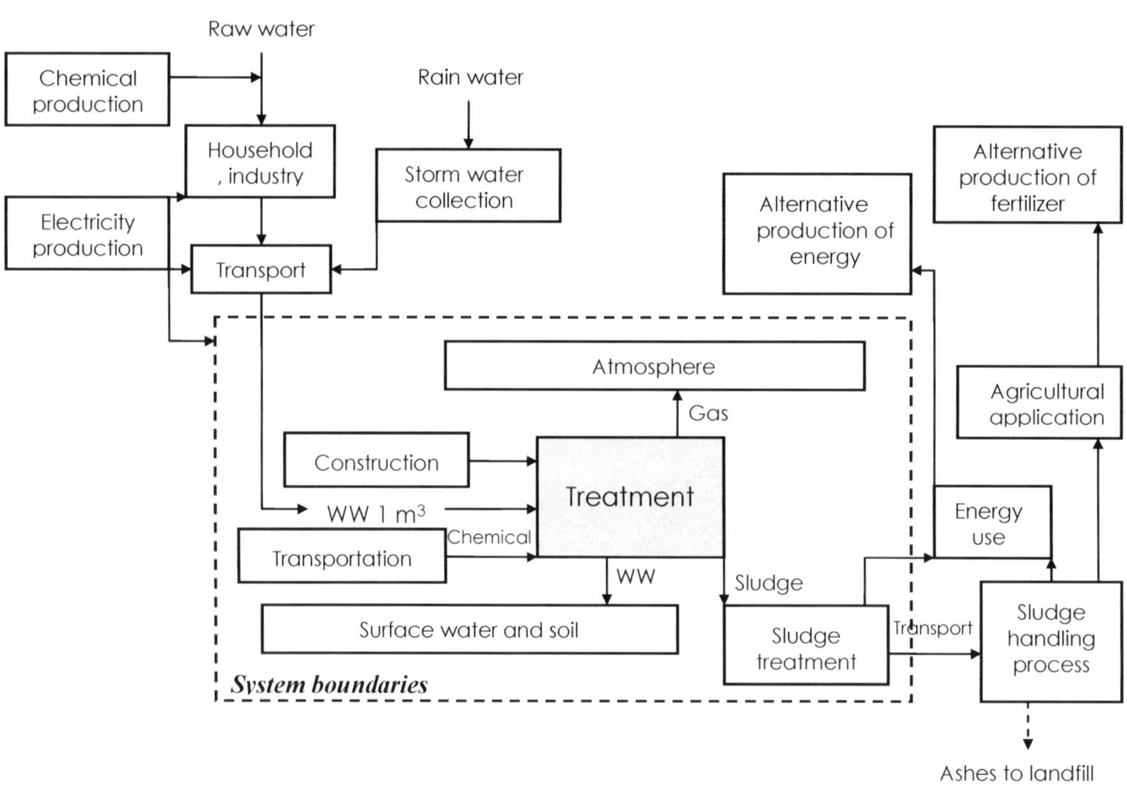

Figure 1. System boundaries for wastewater treatment systems (Modified from Balkema, 2000)

To systematically compare the systems, the required database of wastewater treatment plants were categorized as depicted in Figure 2. The impact assessment were then calculated according to the Environmental Design of Industrial Products (Wenzel et al., 1997), interpretation of the inventory were done by assessing the magnitude of the contribution using SimaPro 5.1 program.

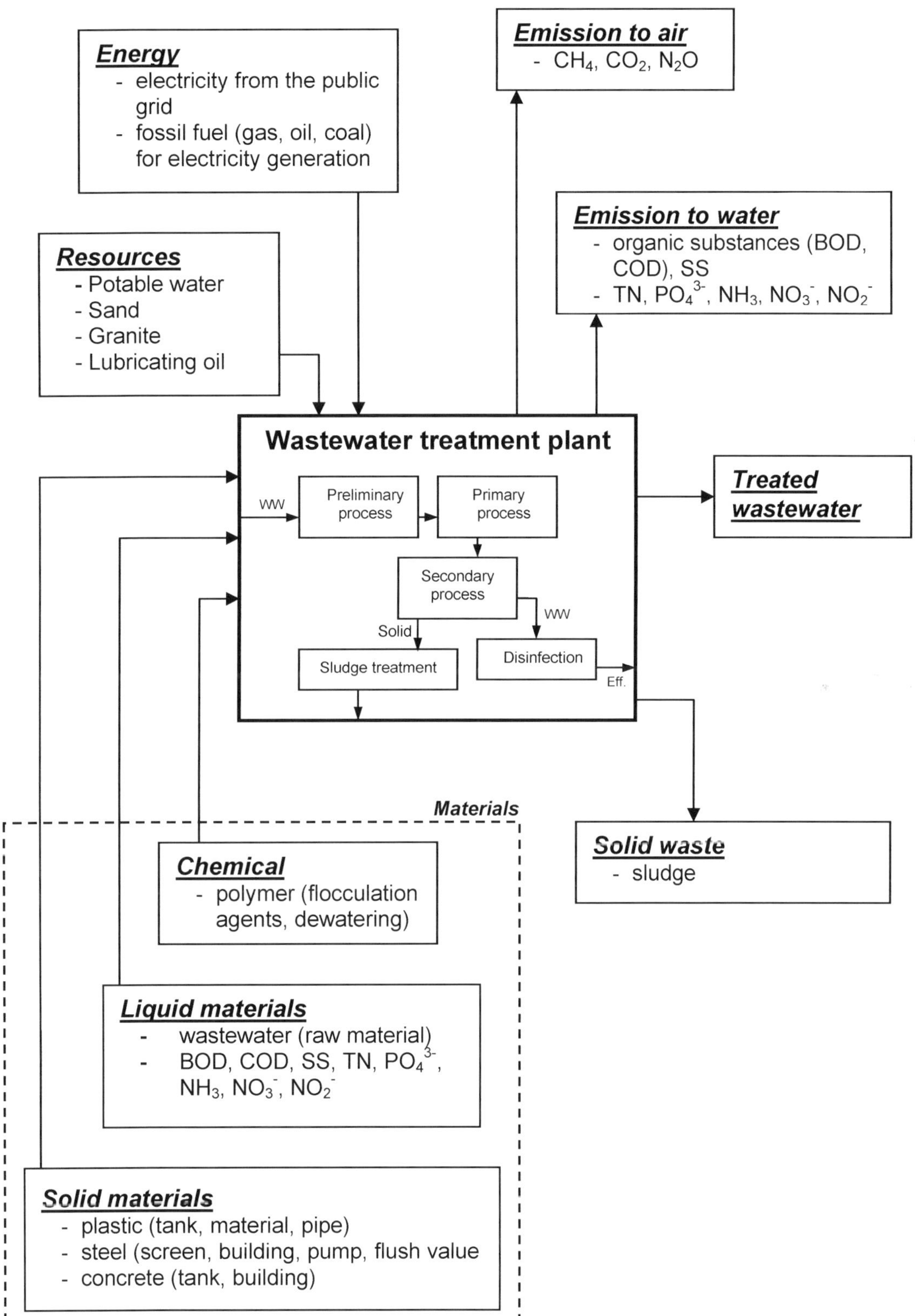

Figure 2. Typical inventory analysis of wastewater treatment plant

3. Results and discussion

Table 1 shows characteristics of the influent and effluent of the Huay Kwang and Si Phraya wastewater treatment plants, which were based on the monitoring reports during 2000 – 2005. It is apparent that the treated effluent of both treatment plants can meet the domestic effluent standard of Thailand, but still discharging TKN and fecal coliforms at the relatively high concentrations.

Table 1. Wastewater characteristics of treatment plants (Adapted from Chaosakul, 2005)

Parameters	Huay Kwang		Si Phraya	
	Influent	Effluent	Influent	Effluent
BOD, mg/L	213	16	53	4
COD, mg/L	374	51	108	16
SS, mg/L	173	18	58	6
TKN, mg/L	40	27	21	10
P , mg/L	0.2	0.25	0.15	0.2
Fecal coliform, MPN/100 mL	1,600,000	92,000	920,000	540,000

Source: Average from 60 data of the monthly records during year 2000 – 2005

According to the database and unit process diagrams of the wastewater treatment systems, the inventory analyses per 1 m^3 of wastewater including resources, material uses and emissions (air, water and soil) were conducted for assessment of the environmental impacts of treatment plants: Huay Kwang and Si Phraya, as depicted in Figures 3 and 4, respectively.

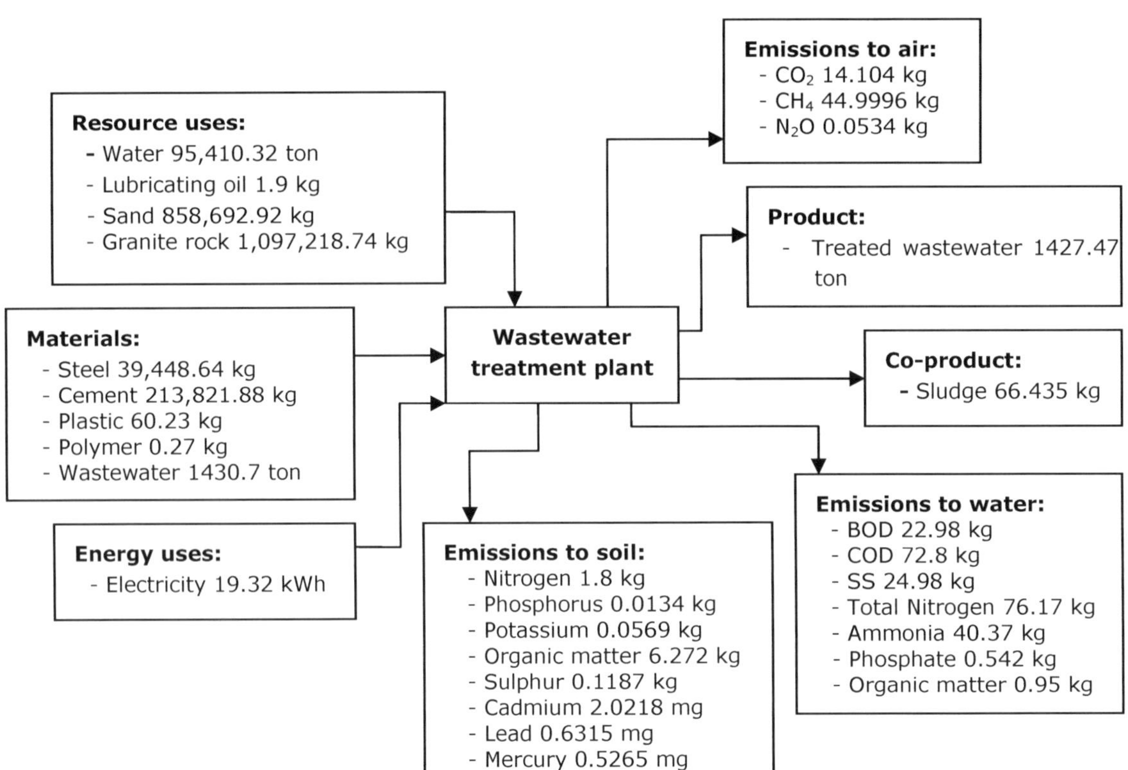

Figure 3. Inventory analysis of Huay Kwang wastewater treatment units.

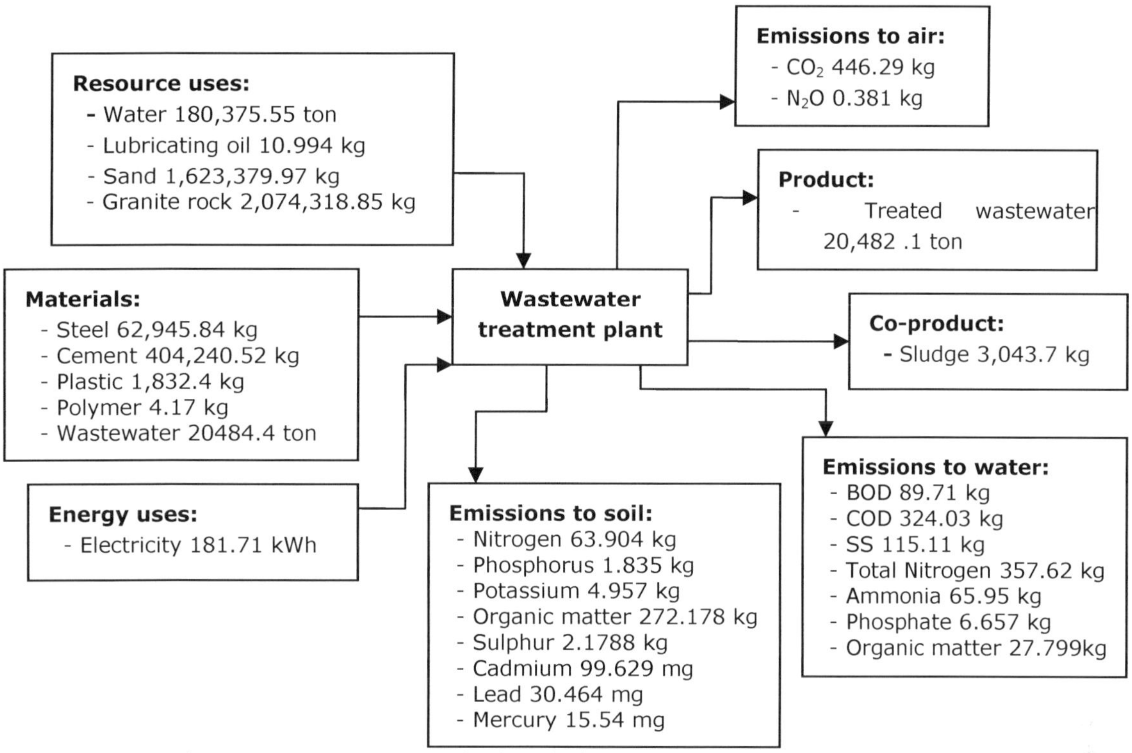

Figure 4. Inventory analysis of Si Phraya wastewater treatment units

In the wastewater treatment processes, the inputs consist of resources uses; materials; and energy using for plant operation. Apart from the treated wastewater, the outputs of wastewater treatment plant can be classified into four main groups; emission to air (CO_2, CH_4 and N_2O); emission to water (BOD, COD, SS, TN, P, NH_3 and organic matters); emission to soil (N, P, K, organic matters, S, Cd, Pb and Hg); and co-product (dried sludge). The data can reveal that Si Phraya wastewater treatment plant has used less resources and materials for treatment of 1 m^3 of wastewater, as compared with those at Huay Kwang treatment plant. In addition, the lower missions to air and water could be obtained at the Si Phraya treatment plant, but the higher emission to soil as compared with that of Huay Kwang system.

The abovementioned inventory of wastewater treatment plants were further analyzed for impact and damage assessments using Eco-indicators 95 and 99, respectively, as shown in Figures 5 and 6. The total eco-score for eco-indicator 95 include the environmental impacts due to summer smog, discharges of heavy metals, eutrophication, acidification, and greenhouse gas emission, whereas the Eco-indicator 99 considers the damages on resources ecosystem quality and human health. The results reveal that the total eco-score for eco-indicator 95 of Huay Kwang treatment plant is higher than that of Si Phraya treatment plant, but vice versa for the eco-indicator 99. Among the environmental impacts of the wastewater treatment plant, the highest eco-score of 2.44 and 0.72 milli-points at Huay Kwang and Si Phraya plants, respectively, are due to the potential to cause eutrophication, as the consequence of N and P discharges into water bodies. However, the environmental impacts have less significant to the eco-indicator 99 as compared with the damages on natural resources and human health. In a large-scale treatment plant, it could cause the damages to natural resources higher than a small-scale treatment plant.

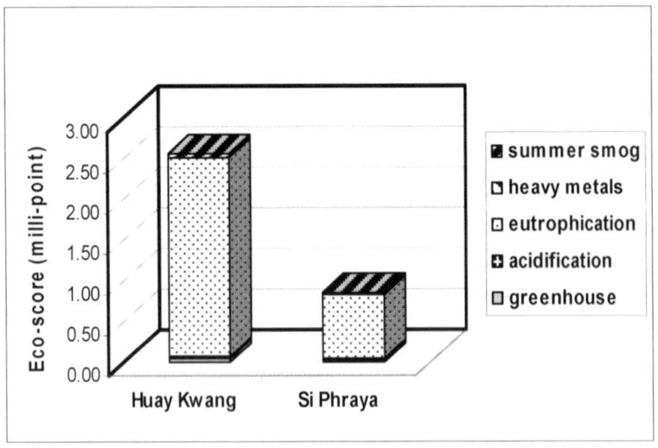

Figure 5. Eco-indicator 95

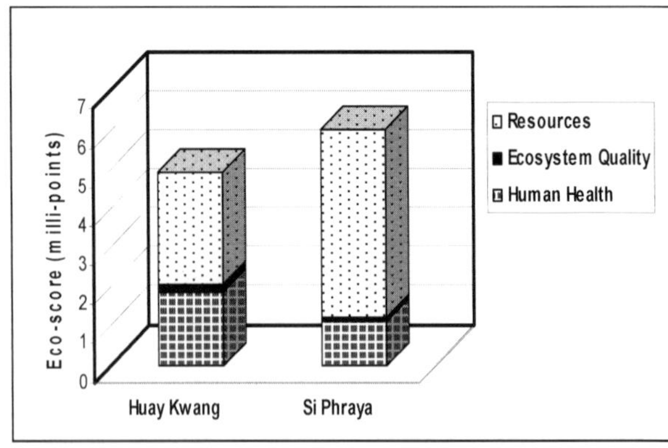

Figure 6. Eco-indicator 99

Apart from the results of SimaPro 5.1, the economic considerations of wastewater treatment plant was conducted in order to further assess the economic sustainability. The cost analysis refers to the construction, annual operation and maintenance, while the benefits include saving of medical expenses due to waterborne diseases: more earnings for not being ill; revenue from wastewater charges; and revenue from saving of potable for watering plants. Figure 7 shows the substantial reduction of annual cases of total waterborne diseases in the service areas of Si Phraya and Huay Kwang after the operation of treatment plants since 1991 and 1993, respectively.

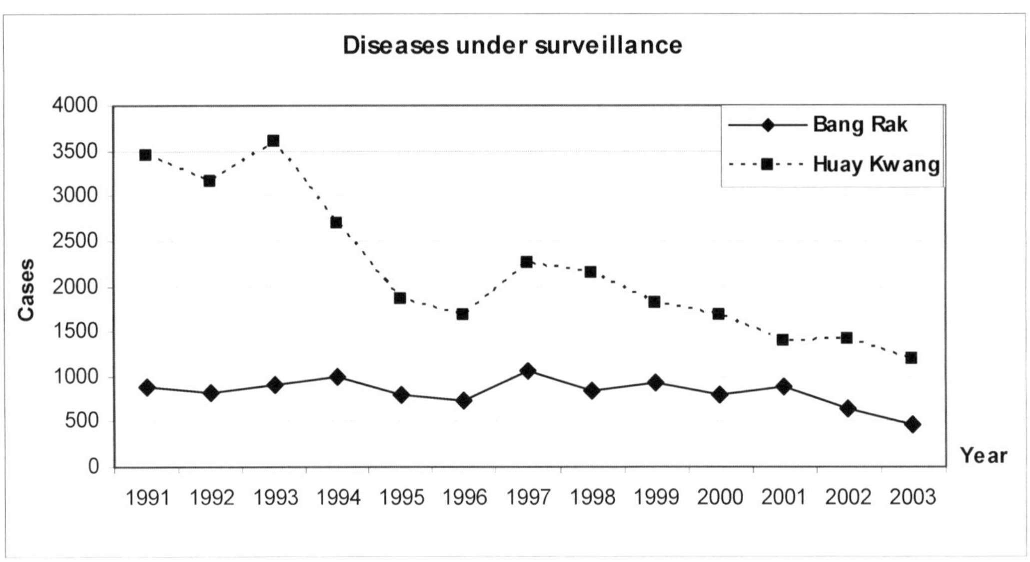

Figure 7. Annual cases of total waterborne diseases in the service areas of Si Pharaya (Bang Rak) and Huay Kwang treatment plants during 1991-2003

The costs and benefits per 1 m^3 wastewater of wastewater treatment plants, as shown in Table 2, reveals that Huay Kwang treatment plant has lower benefit to cost (B/C) ratio than that of Si Phraya treatment plant. This could be a result of economy of scale of the treatment plant, but regardless of wastewater collection systems. The negative net present values (NPV) of wastewater treatment could indicate that the investment was greater than the benefit, thereby rendering the non-profitable investment plan because of assuming the benefit of treatment plants are equal until the end of their life span.

Table 2. Costs and benefits per 1 m^3 wastewater of wastewater treatment plants (Adapted from Chaosakul, 2005)

Cost and benefit (baht)	Huay Kwang	Si Phraya
Cost		
- Construction cost	16,775	22,670
- Annual operation & maintenance cost	2,275	627
Benefit		
- Saving of medical expenses due to waterborne diseases	14	0.11
- Saving of earnings for not being ill	138	1.1
- Revenue from wastewater charges	-	1,000
- Revenue from saving of potable for watering plants	-	38
- Total benefit	152	1,040
Benefit/cost ratio (B/C)	0.04	0.19
Net present value (NPV)	-45,627	-30,419

To reduce the relatively high impacts and damages of both wastewater treatment plants, as mentioned earlier, the comparisons between modification scenarios were undertaken. First modification scenario is to increase the treatment capacity to its full design capacity because only 50 – 70% of the wastewater has been collected. The second scenario is to reduce energy consumption t the treatment plants by 50%, which should reduce the environmental impacts. Table 3 shows the reductions in impacts and damages due to the modification scenarios at Huay Kwang and Si Phraya treatment plants. It could be observed that the increase in treatment capacity can relatively reduce the impacts and damages per 1 m^3 of wastewater of 50% and 18% at Huay Kwang and Si Phraya treatment plants, whereas no decrease in eutrophication impact could be notified.

Table 3. Impacts and damage assessment of Huay Kwang and Si Phraya treatment plants

Category	Unit	Huay Kwang			Si Phraya		
		Reference	Scenario 1	Scenario 2	Reference	Scenario 1	Scenario 2
Impact category							
Greenhouse	kg CO_2-eq	0.21	0.105	0.164	0.126	0.103	9.55E-02
Acidification	kg SO_2-eq	3.56E-04	1.78E-04	3.37E-04	2.31E-04	1.88E-04	2.19E-04
Eutrophication	kg PO_4-eq	1.87E-02	1.87E-02	1.87E-02	5.99E-03	5.98E-03	5.98E-03
Heavy metals	kg Pb-eq	1.26E-07	6.28E-08	1.26E-07	5.62E-08	4.58E-08	5.62E-08
summer smog	kg C_2H_4-eq	3.13E-04	1.56E-04	2.19E-04	2.25E-04	1.83E-04	1.63E-04
Damage category							
Human Health	DALY	9.48E-08	4.74E-08	8.08E-08	5.67E-08	4.63E-08	4.75E-08
Ecosystem Quality	PDF*m2yr	2.54E-03	1.27E-03	2.44E-03	1.35E-03	1.10E-03	1.29E-03
Resources	MJ surplus	8.07E-02	4.03E-02	7.72E-02	1.35E-01	1.10E-01	1.33E-01

Adapted from Chaosakul (2005)

For the modification scenario 2, it could result in the reduction of 5 – 30% in global warming, acidification and summer smog potential, but no effects on eutrophication and disposal of heavy metals. The 50% reduction in energy consumption could also result in diminishing of the impacts on human health, ecosystem quality and resources of about 20 - 50% for both Huay Kwang and Si Phraya treatment plants.

4. Conclusions

Based on the LCA results in terms of Eco-Indicator 95 and 99, the evaluations of small- and large-scale wastewater treatment could be summarized as follows:

1. The treatment efficiencies of either small- or large-scale treatment systems could meet the domestic effluent standards, but still discharging N at high concentrations. However, a small-scale treatment plant could cause the environmental impacts per 1 m^3 of wastewater at the greater magnitude than that of a large-scale system, by which could the eutrophication would be a major incidence.
2. On the contrary, a large-scale treatment system would cause the environmental damage per 1 m^3 of wastewater at the greater magnitude than that of a small-scale treatment plant, by taken resources and human health into consideration.
3. No monetary profit from the operation of wastewater treatment plants could be obtained due to the installation of treatment plants was for environmental infrastructure services with the aims at improving quality of life and public welfare.
4. To mitigate the environmental impact potentials of wastewater treatment plants, the modification scenarios i.e. increasing of treatment capacity and reduction of electricity consumption could reduce the environmental impacts and damages at a range of 2 - 50%.

5. Acknowledgements

The authors express their grateful appreciations for the generous support of the National Centre for Competence in Research (NCCR) North – South, Switzerland, and the University of Tokyo through its Center for Sustainable Urban Regeneration (cSUR) which made this study possible.

6. References

Balkema, A. J. (2000). Multi-Criteria Analysis for Sustainable Wastewater Treatment. Eindhoven University of Technology, Netherlands. Available from www.iees.ch/Eco Eng001/ EcoEng001_R1.html

Balkema, A. J., Preisig, H. A., Otterpohl, R. and Lambert, F. J. D. (2002). Indicators for the sustainability assessment of wastewater treatment systems. *Urban water*, 4(2), 153-161.

Chaosakul, T. (2005). Appraisals of environmental sustainability indicators on the conventional sewerage and domestic wastewater treatment systems in Thailand: case study in bangkok city, *Master Thesis,* EV-05-30, Asian Institute of Technology, Bangkok, Thailand

Lundin, M. and Morrison, G. M. (2002). A life cycle assessment based procedure for development of environmental sustainability indicators for urban water systems. *Urban Water*, 4(2), 145-152

Wenzel, H., Hauschild, M. and Alting, L. (1997). Environmental assessment of products, Vol 1: Methodology, tools and case studies in product development. Chapman & Hall.

A Survey of Wastewater Management in the Concentrated Latex Industry in Southern Thailand

Somtip Danteravanich[1)], Udomphon Puetpaiboon[2)] and Nitas Proukaew[3)]

1)Faculty of Technology and Management, Prince of Songkla University, Surat Thani Campus, Muang, Surat Thani, 84100 Thailand.
e-mail: dsomtip@ratree.psu.ac.th
2) Faculty of Engineering, Prince of Songkla University, Hat Yai Campus, Hat Yai, Songkhla, 90112, Thailand.
e-mail: pudompho@ratree.psu.ac.th
3)Surat Thani Community College, Prince of Songkla University, Surat Thani Campus, Muang, Surat Thani, 84100 Thailand.
e-mail: pnitas@ratree.psu.ac.th

Abstract

Concentrated latex industry is an important part of the rubber industries in the South of Thailand. There are currently 55 factories distributed all over the South of Thailand. This article presents the survey results of wastewater management of concentrated latex factories in Southern Thailand conducted in 2002. The current performance of wastewater treatment, wastewater treatment technology, and wastewater management of the factories were also reported. It was estimated that concentrated latex factories altogether generate about 21,500 m^3/day of wastewater with a biological oxygen demand (BOD_5) load of approximately 75.25 tonnes/day. The present survey revealed that 28 factories directly discharge their effluent of 13,656 m^3/day to the watercourses. Most concentrated latex factories have not fully complied with the regulatory of Thai industrial effluent standard. The majority of factories using the pond treatment system, generally had problem complying with BOD_5, COD, SS and TKN standards. Poor operation and/or maintenance could be contributed to low percentage of compliance. In addition, suggestions and alternative approaches for better and sustainable wastewater management of concentrated latex industry were also addressed.

Keywords: concentrated latex; wastewater; Southern Thailand; wastewater management

1.Introduction

In Thailand, rubber industries play a very important role in the country economic development. At present, Thailand is recognized as the major rubber producer and exporter. During the year of 1999-2002, Thailand produced 2.1-2.5 million tonnes of natural rubber yearly with the production in terms of 50 % smoked sheet rubber, 32 % block rubber, 14 % concentrated latex and 4 % other rubber.

Concentrated latex industry is an essential part of the rubber industries in the South of Thailand. The investment trend of this rubber industry is currently positive. The factories are located predominantly in rural and suburb areas. In 2002, there are 55 concentrated latex factories distributed all over the South of Thailand. Although the concentrated latex factories are crucial for the rubber industry in Southern Thailand, they have also been identified as the major source of water pollution. The processing of concentrated latex consumes large volumes of water resulting in production of highly polluting effluents, which if discharged untreated into watercourses will have adverse impact on the water quality. Therefore, it should recognize and take urgently to control pollution of this industry.

Investigation of the concentrated latex wastewater management is important primarily to know the current wastewater treatment efficiency, to estimate the possible effects of the effluent upon the quality of receiving water and subsequently enable appropriate treatment technologies and to evaluate the feasibility of utilizing the effluents. Also, it is useful to verify the difficulties encountered in the control of pollution from concentrated latex industry. This paper presents the survey results of wastewater management of concentrated latex industry in Southern Thailand. The current status of chemicals and water usage, wastewater treatment technology, and wastewater treatment performance of the factories were reported. Discussion of various aspects of the survey results obtained is also included to learn the lessons about the present wastewater management of concentrated latex industry.

2. Methodology

In 2002, an investigation of concentrated latex wastewater management was conducted by using questionnaires, field survey and secondary data obtained from industrial and research agencies. The questionnaires were developed by the researchers and used for preliminary data collection. The 55 questionnaires were directly mailed to the factories and only 16 (29%) were replied. Before the questionnaires were analyzed, the data editing was undertaken. Phone calls to the factory in order to edit the data obtained was performed. Factory visiting and interview were also conducted from the factories replied the questionnaires. During factory visiting, in plant investigation, the interview with the factory managers and the persons working on wastewater treatment were conducted. Data cross-checked such as water and chemicals usage were carried out during the field survey. Moreover, the secondary data, such as the concentrated latex production, the characteristics and quantity of wastewater obtained from the Ministry of Industry and various research reports were reviewed.

Wastewater sampling of concentrated latex factory was undertaken. Samples collected during each visit were raw effluent, final discharge and intermediate samples from the treatment system. A total of 132 wastewater samples were determined for physical and chemical characteristics. The following characteristics of wastewater samples collected from concentrated latex factories were examined: pH, COD, BOD_5, suspended solids, TKN, TP, sulfate, and magnesium. The wastewater samples were determined according to the procedures described by APHA, AWWA & WEF (1992). The statistical analysis of data obtained in terms of ratio, percentage, mean, and median values were conducted.

3. Results and Discussion

3.1 General information of the concentrated latex factories

There are 55 concentrated latex factories in the South of Thailand in 2002. The factories are located in 11 provinces out of the total 14 provinces in the South, as shown in Table 1. Songkhla province has the largest number of concentrated latex factories. The average concentrated latex production of each concentrated latex factory was found to be 10-300 tonnes/day.

Latex as it comes out of the rubber tree (*Hevea brasiliensis*) contains about 30% DRC (dry rubber content), 64-65% water, and the rest is non-rubbers. Centrifugation has been employed to concentrate the field latex. In the centrifugation process, the rubber particles from the serum phase are separated by a centrifugal force and produce a standard latex concentrate with about 60% DRC and equal volume of skim latex with about 6-8% DRC. The concentrated latex is further treated with suitable preservatives, especially ammonia. After the addition of preservatives, the concentrated latex is blended and transferred to storage tanks to mature before to be transported or shipped. As for skim latex, it is further coagulated by acid treatment with H_2SO_4. During the concentrated latex processing, large amount of water is used for washing, cleaning and diluting. Water is also used in the skim latex process. The water and chemicals usage in concentrated latex factories were observed as shown in Table 2.

Table 1. General information of concentrated latex in Southern Thailand.

Provinces	Number of factories	Production capacity (tonnes/day)
Chumporn	1	15
Pattani	2	16-30
Krabi	2	62-100
Surat Thani	6	20-80
Phuket	2	25-60
Nakhon Sri Thammarat	4	35-98
Phattalung	1	50
Songkhla	18	15-300
Trang	11	10-300
Yala	7	15-68
Pung-nga	1	77

Table 2. Water and chemicals usage in concentrated latex factories.

Values	Water (m^3/tonnes of concentrated latex product)	NH_3 (kg/tonnes of concentrated latex product)	H_2SO_4 (kg/tonnes of skim rubber)
Mean	5.25	15.22	114.7
Maximum	12.4	24.0	300
Minimum	2.6	2.7	20.8
Median	3.8	15.4	92.0

3.2 Raw effluent characteristics and quantity

As mentioned above, large volume of water was used in the concentrated latex factory and generated as wastewater from the concentrated latex and skimming processing. From in plant investigation, it was found that an average wastewater generated from concentrated latex process is 4.01 m^3/tonnes concentrated latex product, whereas wastewater from skimming process is determined as an average value of 20.82 m^3/tonnes skim rubber.

The effluent consists of process water, small amounts of uncoagulated latex and substantial quantities of proteins, sugars, lipids, inorganic and organic salts that originate from natural rubber latex. Table 3 shows the average properties of raw effluents from the investigated concentrated latex factories. Effluent from the concentrated latex factories are acidic as indicated by the pH values ranging from 2.33 to 7.23 and as an average value of 4.6.The acidic nature of the effluents is attributed to the use of H_2SO_4 in the process line for coagulation of skim latex. The effluents also contain fairly large amounts of suspended solids as of an average value of 914 mg/l. The average BOD_5 value of the concentrated latex effluent was 4,854 mg/l while the corresponding average COD value was 11,568 mg/l. The high BOD_5 and COD values of concentrated latex factories are mainly organic in origin with high oxygen requirement for their oxidation. The ratio of BOD_5 and COD is determined ranging from 0.23-0.84 or is equivalent to the average value of 0.52. This reflects that concentrated latex effluent is fairly biodegradable and thus amenable to biological treatment. In addition, concentrated latex effluent contains high levels of nitrogen and sulfate with the average values of 992 mg/l of TKN and 448 mg/l of sulfate, respectively. Total nitrogen includes ammonia-nitrogen combined with organic nitrogen. The main contribution to total nitrogen is ammonia-nitrogen due to the use of substantial quantities of ammonia in the latex preservation. During the centrifugation of field latex, a significant part of the ammonia remains in the skim latex. On the subsequent coagulation of the skim latex, the ammonia is retained in the resulting skim serum that is normally discharged as wastewater. Phosphorus and magnesium were also determined in concentrated latex effluent with the values ranging from 2.2 to 20.5mg/l and 4.8 to136 mg/l, respectively. These were due to the chemicals used in the process line, in particular DAP (Diammonium phosphate). This concentrated latex effluent characteristic is consistent with Maheswaran and John's study in Malaysia in 1991.They reported that concentration of COD, BOD, TN, SS and pH of concentrated latex effluent were 4,849 mg/l, 3,524 mg/l, 602 mg/l, 818 mg/l and 4.8, respectively. Other studies in Thailand have reported that the wastewater of concentrated latex factories was acidic and contained high organic matter, nitrogen, and sulfate. For example, Prabnakorn (2000) revealed that the concentrated latex effluent contained of 510 ± 624 mg/l, 4,440 ± 436 mg/l, 10,123 ± 620 mg/l and 6,650 ± 229 mg/l of suspended solids, sulfate, COD and BOD, respectively. In 1999, Rakkoed reported that TKN in concentrated latex effluent was ranging from 543-1,268 mg/l.

3.3 Wastewater treatment technology and performance of the treatment

The physical and biological treatment was combined as wastewater treatment system in concentrated latex factory. The physical treatment process is rubber trap and is used as the first unit of treatment system to remove suspended rubber. The biological wastewater

Table 3. Raw effluent and treated effluent characteristics of concentrated latex factories monitored.

Parameters	Raw effluent		Treated effluent	
	Range	Mean	Range	Mean
COD	672-17,000	11,568	24-1,272	489
BOD_5	571-10,592	4,854	18-485	141
SS	54-2,300	914	48-600	258
TKN	70-2,290	992	5.04-1,033	328
TP	2.2-20.5	5.42	1.1-36	5.3
SO_4^{2-}	51-1,118	448	16-255	67
Mg	4.8-136	64	1.5-36.6	8.65
pH	2.33-7.23	4.6	6.77-9.15	8.03

Remark: All units are mg/l, except pH.

treatment technology used in 22 concentrated latex factories is stabilization pond, included with anaerobic, facultative and aerobic ponds. 24 factories use anaerobic pond together with aerated lagoon for treatment their wastewater and 5 factories use activated sludge process. One factory uses UASB (upflow anaerobic sludge blanket) and land application of wastewater was adopted in 3 factories. It was observed that if land is a constraint as in the case of rubber factories sited in the near urban areas, the aeration system is employed. Also, the enclosed anaerobic digestion system together with a facultative pond was found in the areas where land is a constraint.

27 factories do not discharge their treated wastewater to the natural water sources. The rest of 28 factories discharge treated wastewater with total volume of 13,656 m^3/day to the watercourses nearby the factories. The organic and nitrogen loading produced from 55 concentrated latex factories in the South of Thailand were determined in the range of 5.6-236 kg BOD_5/day and 2.65-764 kg TKN/day of each factory. Figure 1 shows the wastewater quantity, BOD_5 and nitrogen loading discharged to the watercourses of each province in the South. It was estimated that the total volume of 21,500 m^3 of wastewater with organic load of 75.25 tonnes of BOD_5 was daily generated from the concentrated latex factories in the South.

Table 3 shows the quality of treated wastewater from the treatment plants monitored. It was observed that the treatment systems used are almost unable to produce effluents that can comply with Thai industrial effluent standard. Thai standard requires the concentrations of industrial effluent to be not more than 120 mg/l, and 20 mg/l for COD and BOD_5, respectively.

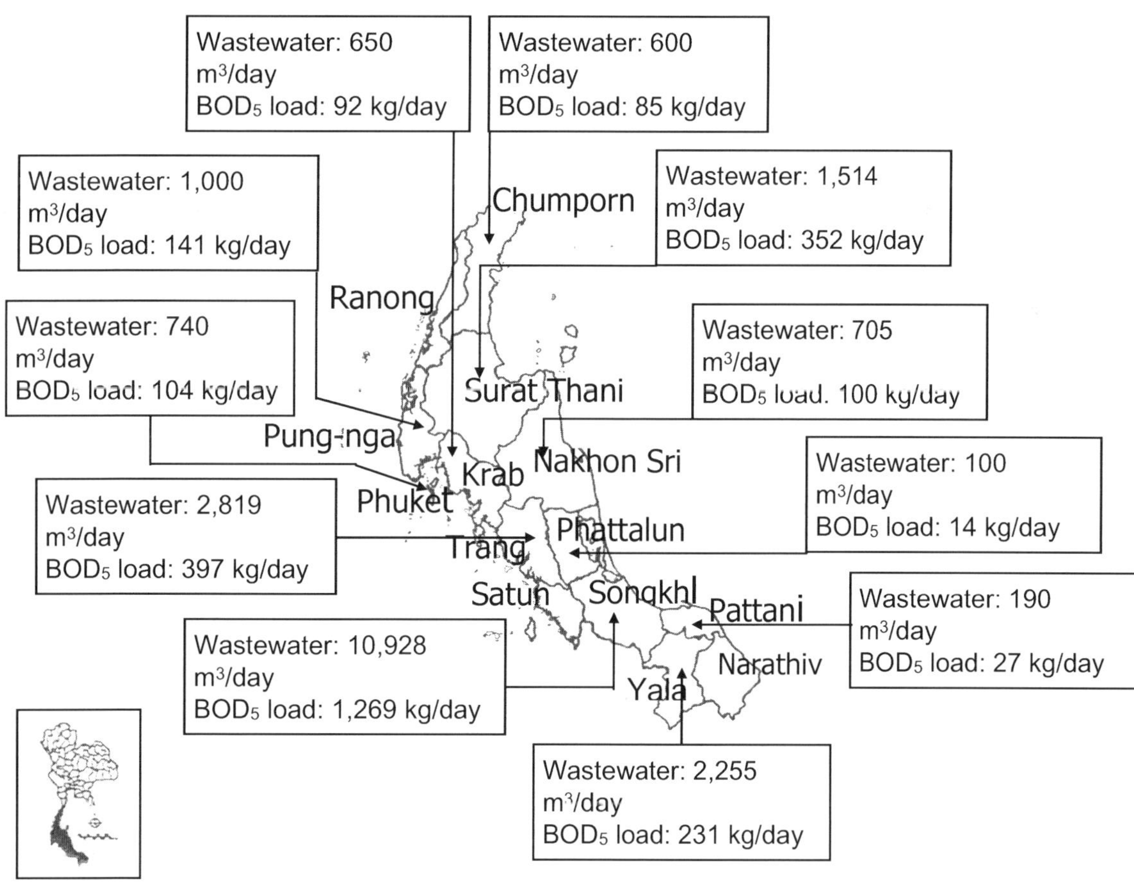

Figure 1. Wastewater generation, BOD_5 and nitrogen loading discharged to the watercourses of concentrated latex factories in Southern Thailand.

TKN and SS concentrations must not exceed values of 100 mg/l and 50 mg/l, respectively. Compared with these standards, it can be seen that the concentrated latex effluent does not meet the effluent standard criteria. Although most treatment plants were able to remove BOD, COD, Mg and sulfate efficiently as shown in Table 4, they were able to hardly remove nitrogen and phosphorus. Low nitrogen removal from ponding system could be due to insufficient retention time in the facultative pond. The treatment systems in some cases were observed to perform below expectation. This was because of poor design, operation and /or maintenance. Moreover, the quality of rubber factory effluent depended on the type of process employed, the quality of supervision and the control of emission.

Table 4. Treatment performance of concentrated latex factories.

Values	COD	BOD_5	SS	TKN	TP	SO_4^{2-}	Mg
Mean	90.4	93.1	69.6	68.8	31.6	81.9	77.7
Maximum	98.8	99.4	87.8	99.1	88.8	96.4	98.2
Minimum	65.7	46.9	11.1	2.4	1.82	39.2	36.0

Note: unit is % removal.

As mentioned above, only 22 factories did not discharge their treated wastewater to the watercourse. Wastewater just was contained in the pond and was evaporated by sunshine. However, there are 3 factories which employed the land application system for their wastewater by disposal to the oil palm. Land application of rubber factory effluent to rubber plants is rather limited. This is because the fertilizer demand for rubber is low compared to that of oil palm which requires more fertilizer. In this respect the interest has been to apply the concentrated latex effluent to oil palm. However, the application of concentrated latex effluent to oil palm can only be economical if the oil palm estates are near the concentrated latex factory.

3.4 Problems from the wastewater treatment.

Since concentrated latex wastewater contained high organic matter, and sulfate, under anaerobic treatment, the foul smell could be generated as noxious gases including hydrogen sulfide. If large amount of sulfide, a very toxic compound, is produced, it can cause unfavorable condition to the communities living nearby the factory. Thus, many concentrated latex factories in the South those employed the anaerobic treatment pond usually faced this problem. Besides this, some concentrated latex factories faced the precipitation of magnesium salts in the wastewater pipelines as well as at the paddles of aerators in aerated lagoon. Such precipitation caused the clogging of the pipelines and damaged the aerators and costed a lot to maintain the wastewater treatment machines and solve such problem. Observations from the survey reveal the following factors contributed to non-compliance of wastewater treatment of concentrated latex factories in the South of Thailand. Poor monitoring and maintenance of the wastewater treatment plant was performed. The treatment plant being under modified and not stable in treatment function results in ineffective treatment. Increase in production capacity causes overloading of the wastewater treatment plant.

3.5 Treatment cost and wastewater treatment operators

It can be said that most of concentrated latex factories did not have professional operators as a full time working. So, wastewater treatment system was inconsistently controlled and operated, and the treatment technology could not be accomplished in high potential employment. In terms of the costs of wastewater treatment, it was observed that each factory monitored has monthly operating cost of wastewater treatment varying from 15,300 to

173,500 Baht (equivalent to382-4,337 US$, when currency exchange is 40 Baht to 1 US$). The 51.5 % of this expenditure was used for electricity. The 37.8% of the expenditure was for chemicals for treatment and the rest was for labor and operators, maintenance, and wastewater analysis. It was found that the treatment cost of ponding system is very much lower than activated sludge process and/or aerated lagoon. This was because ponds do not require mechanical aerators and use less electricity.

3.6 Recommendation of wastewater management

Concentrated latex industry has a great economic importance to the South of Thailand. Therefore, any wastewater technology adopted for concentrated latex effluent should be within the economics, high efficient treatment achievement and has no further effect. In this regards some recommendation on the wastewater management for concentrated latex industry in order to adopt the pollution control measure in the factories can be drawn as follows:

Monitoring the performance of the wastewater treatment plant of each factory should be consistently conducted,

Professional operators should be employed and work as full time,

Promote more good housekeeping, cleaner technology, effective recovery, reuse and non-waste technology,

Strengthen the relationships among the experts and researchers from academic institutes located in the area to couple with wastewater management improvement, and ect.

Monitoring by rubber wastewater property analyses is important. This could help to evaluate the strength and characteristics of the effluent in order to apply proper treatment methods; to determine and ensure that the effluent complies with the standards for wastewater treatment, and production control; and to ascertain the by products which may possibly be recovered from the effluent.

The high potential use of wastewater treatment could be achieved if the professional operators could control the treatment system. Operator training of wastewater treatment plants together with start-up procedure and monitoring and surveillance of plant performance should be carried out. Wastewater sampling, analytical procedures and data interpretation courses should be conducted for personnel involved in treatment plant operation on theory. A practice of wastewater treatment system followed by practical training through concentrated latex factories which have full-fledged treatment for two to three weeks hands-on training is also recommended. Training on analytical techniques, sampling procedure and data interpretation for concentrated latex industry personnel could achieved through attachment to the universities located in the South. However, the new regulation on the operators for environmental control announced by the Ministry of Industry in 2002 was that the operators must be full time working for controlling and operating the wastewater treatment system. This regulation will enforce from the 5th April 2005. Regarding this regulation, the factories need to have qualified operators for full time working. Qualified and professional operators who can operate and control the concentrated latex wastewater contained with high organic, nitrogen, sulfate and acidic is rarely obtained. Thus, it should be urgently be arranged the special operators training for the concentrated latex factories in the South.

It is accepted that good housekeeping is a sign of good management. The basic actions in achieving good housekeeping in concentrated latex factories included regular maintenance of drains, sumps and pits in order to prevent any stagnation of wastewater and to remove the solids. Rubber recovered from factory effluent by rubber traps should be carefully preformed. These will promote the efficiency of treatment due to loading reduction before treating in the secondary treatment. Since the substrates remained in treated wastewater in particular nutrients are useful for waste recovery, the waste recovery technology is one important issue to be further developed. Hence, research cooperative activities between the research agencies and concentrated latex industries are recommended to strongly link. Likewise, the developments on the area of land disposal or effluent irrigation system, and enclosed anaerobic system are recommended to further develop and adopt such technologies to the practical use. Therefore, those technologies can be applied to solve the nuisance problem of bad smell from wastewater treatment plant as well as recovery the valuable substrates remained in the concentrated latex wastewater should be developed.

4. Conclusion

A survey of the wastewater management of concentrated latex factories in Southern Thailand was undertaken in 2002 in order to assess the state of employment of the treatment technology by the concentrated latex industry. Also, the study aimed to determine the performance of the concentrated latex factories in complying with the effluent regulation standards. The survey revealed that all concentrated latex factories had wastewater treatment system. The wastewater treatment technologies which have been widely used for concentrated latex factories in Southern Thailand are stabilization ponds, and aerated lagoons. The concentrated latex factories using such treatment system had difficulty in complying with the BOD, COD,TKN and SS of industrial effluent standard. New approach for the treatment of concentrated latex wastewater should be envisaged to improve the removal of suspended solids, organic, nitrogen and sulfate with low cost technology. In addition, increasing technical expertise of the operators is necessary to be promoted by different training course/ programs.

5. Acknowledgement

The authors would like to thank National Research Council of Thailand for financial support to conduct this study, and to all concentrated latex factories for providing information and full co-operation.

6. References

APHA, AWWA and WEF. (1992). *Standard Methods for the Examination of Water and Wastewater*, 18th ed., New York: American Public Health Association.

Danteravanich, S. (2002). *Waste Management of Concentrated Latex Industry*, Research report of the research project No. 05011504.

Maheswaran, A. and John, C.K. (1991). *Manual on Pollution Control in the Rubber and Palm Oil Industries*, Department of Environment, Ministry of Science, Technology and the Environment.

Prabnakorn, P. (2000). *Primary Treatment by Air Floatation in Rubber Trap Tank for Latex Rubber Plant Wastewater Treatment*, Master of Science Thesis in Environmental Management, Prince of Songkla University.

Rakkoed, A., (1999). *Assessment of Nitrogen Problem of Rubber Plant Wastewater and Nitrogen Removal with the Fixed Film in Stabilization Pond*, Master of Science Thesis in Environmental Management, Prince of Songkla University.

Development of Integrated Decentralized Sanitation System using Aerobic Membrane Bioreactor

Melissa Montalbo [1], S.M.S.M.K. Samarakoon [1], *C. Visvanathan[1], Kensuke Fukushi,[2] and KazuoYamamoto[2]

[1] Environmental Engineering and Management, Asian Institute of Technology, P.O.Box 4, Klong Luang, Pathumthani 12120, Thailand email: visu@ait.ac.th
[2] Environmental Science Center, University of Tokyo, 7-3-1 Hongo, Bunkyo-ku, Tokyo 113-0033, Japan
*Corresponding author

Abstract

Inspired from the concept of ecosanitation, an integrated decentralized sanitation system using aerobic membrane bioreactor is developed. This system is a holistic management approach that aims to develop a method to treat domestic wastewater and solid waste and reclaim wastewater effluent in a decentralized manner. The method includes an incorporation of the domestic wastewater and the liquid fraction of kitchen waste that contains the organic components, directing it in an aerobic membrane bioreactor (MBR) treatment module. As for the solid portion of the waste, it will be stored in a holding tank which is to be collected and pumped into an anaerobic digester within the village or the community. To further investigate its feasibility, a study was conducted in an aerobic membrane bioreactor with the objective of determining the performance in terms of effluent quality at HRT of 2, 4 and 6 hours. The MBR obtained a 98% COD removal and 95% TKN removal efficiency, throughout the experiment and in HRT of 2 hours, respectively. Taken as a whole, the proposed scheme is considered a good option for water sanitation and environmental conservation.

Keywords: decentralized sanitation, domestic wastewater, membrane bioreactor, water reuse

1. Introduction

Most rural areas in Asian countries are suffering from lack of proper sanitation. As a result of this poor management, domestic wastewater becomes a threat to public health and the environment. In order to facilitate an effective wastewater treatment design, it is necessary to consider the characteristics and sources of the wastewater. The wastewater generated from toilet is termed as black water and contributes a significant amount of organics, nitrogen and phosphorus. Black water can be further classified as yellow water (urine) and brown water (human feces). Grey water consists of water from clothes washing, bathing and dishwashing, which is relatively less polluted.

Principal technologies for domestic wastewater treatment can be classified as either on-site sanitation (decentralized) or off-site (centralized) sanitation systems. The off-site sanitation approach is usually found in rich countries that can afford constructions of huge pipelines, organized operation and maintenance, lots of manpower, and effectively vast wastewater treatment plants. However, this system is not possible in low-income or developing countries.

Thus, on-site treatment sanitations have been the alternative. Decentralized systems have been applied in rural and urban areas for several decades both in developed and developing countries because of its low investment costs and simplicity in operation and maintenance. The most common form of on-site wastewater treatment is the septic tank.

Solid Waste Scenario

The solid waste produced per capita often varies in different countries depending on its geographical location, climatic factor, economic status, and consumption pattern. In general, the primary fraction of the waste generated is dominated by biodegradable portion such as the food waste. Garbage collectors collect these wastes and it either ends up in an unsanitary dumpsite or in some other non-engineered landfill. This practice does not only pollute the environment due to biogas and leachate production in the landfills and dumpsites, but also affect the health of the people living near the area.

Integrated Decentralized Sanitation – Aerobic MBR System

Ecological sanitation is a conceptual shift in the relationship between people and the environment with its objective to protect human health and the environment, while reducing the use of water in sanitation systems and recycling nutrients to help reduce the need for artificial fertilizers in agriculture (EcoSanRes, 2004). Motivated from the success of ecosanitation, an integrated decentralized sanitation system using aerobic membrane bioreactor is developed. This system is a unique management plan that intends to not only sanitize the household, but also reclaim the wastewater for secondary reuse and treat the solid waste in a decentralized manner, rather than dumping in some barren land or non-engineered landfill. The scheme described in figure 1 incorporates the domestic wastewater and liquid portion of the kitchen waste, which was extracted, using a grinder, mixed together to make a combined wastewater. This combined wastewater will be fed into the proposed aerobic membrane bioreactor treatment module (Fig. 2). MBR was chosen as the proposed treatment as it has coupled the 3 major processes of wastewater treatment: sedimentation, biological reaction, and filtration. It also provides higher wastewater degradation performance in terms of quality effluent compared to conventional septic tanks. Since it is a compact system, therefore it is very convenient to be applied in residential households.

Generally, 70% of kitchen waste can be considered as organic in nature. If these kitchen wastes are sorted within the homes, rather than mixed with other household solid waste like plastic and paper, it may become a valuable component in the biological process of degradation of wastewater. Thus, the integrated decentralized sanitation system incorporated the liquid portion of the kitchen waste into the domestic wastewater to increase its organic content and thereby increase the MBR degradation performance. In this way, about 40-50% of the municipal solid waste stream gets reduced, enabling better options for recycle and disposal.

Anaerobic digestion of solid waste has been found to become a major focus in waste management throughout the world. It does not only eliminate the waste but also produce biogas, which could be a source of energy when converted to power. In this study, the residue of the kitchen waste after grinding will be mixed with the solid part of the black water in the holding tank. In addition, sludge withdrawn from the membrane bioreactor will also be disposed into the holding tank. The holding tank, designed to create a temporary storage for the biodegradable solid waste in the household, will later be diverted into the anaerobic digester available in the community, as clearly expressed in Figure 1. As installing an individual anaerobic digester per household will not be economical, thus it is practical to operate in a centralized manner or in clusters of households.

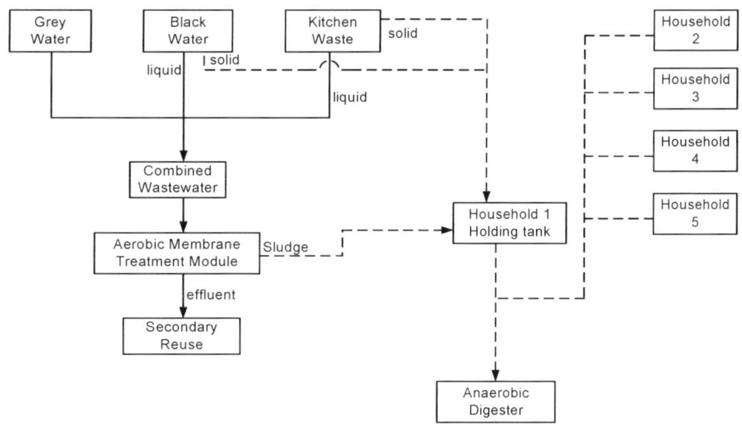

Figure 1. Integrated decentralized sanitation using aerobic MBR system

2. Experimental Study

In order to further the applicability of the integrated decentralized sanitation concept, an experimental study was conducted in laboratory scale. The objective of the study is to investigate the performance of MBR in terms of effluent quality at different operating conditions. Biofouling, which is the main constraint in most MBR studies, was also considered. The membrane used in the study was supplied was Mitsubishi Rayon, Japan with a pore size of 0.4 μm and surface area of 0.2 m^2.

The daily domestic wastewater and kitchen waste generation were estimated in a real household prior to the experiment. From these data, synthetic wastewater was prepared to represent domestic wastewater with partial urine separation. The synthetic wastewater compositions were as follows: glucose, 272 mg/L; soya protein, 290 mg/L; NH_4Cl, 97.1 mg/L; KH_2PO_4, 26.1 mg/L; $CaCl_2$, 10 mg/L; $MgSO_4 \cdot 7H_2O$, 10 mg/L; $FeCl_3$, 3 mg/L; $NaHCO_3$, 200 mg/L. Kitchen waste was collected from the cafeteria of Asian Institute of Technology (AIT), Thailand, then grinded to extract the liquid portion and mixed into the synthetic wastewater. The seed sludge for the MBR was collected from Thammasat University wastewater treatment plant, Pathumthani, Thailand. It was then acclimatize to the lab scale MBR treatment system, which is presented in Figure 2. The design is composed of 2 compartments: anoxic tank and membrane bioreactor. The anoxic compartment is included in the system to achieve denitrification of the wastewater, while the membrane bioreactor compartment will complete the degradation of the organic compounds.

Investigations were done in 3 hydraulic retention times (HRT) of 6, 4 and 2 hours at permeate flux of 0.125, 0.187, 0.360 m^3/m^2-day, respectively. In the 3 HRT conditions, mixed liquor suspended solids (MLSS) and COD were maintained between 9,000-10,000 mg/L and 650 mg/L, respectively.

3. Results and Discussion

3.1 *Performance of aerobic MBR*
COD Removal
Figure 3 reports the variation of influent and effluent COD and its percentage removal efficiency. It could be observed that the influent COD varies largely because of the different compositions of kitchen waste per batch fed into the reactor. COD removal during the first run was fluctuating due to the changes made in the operating condition. In the middle of the first run, partial urine separation was done to increase the nitrogen content of the influent and supply enough nitrogen for the denitrifier's growth. After the stabilization of the microorganism during the first run, the COD removal was maintained at 98% with MLSS concentration maintained at 10,000 mg/L.

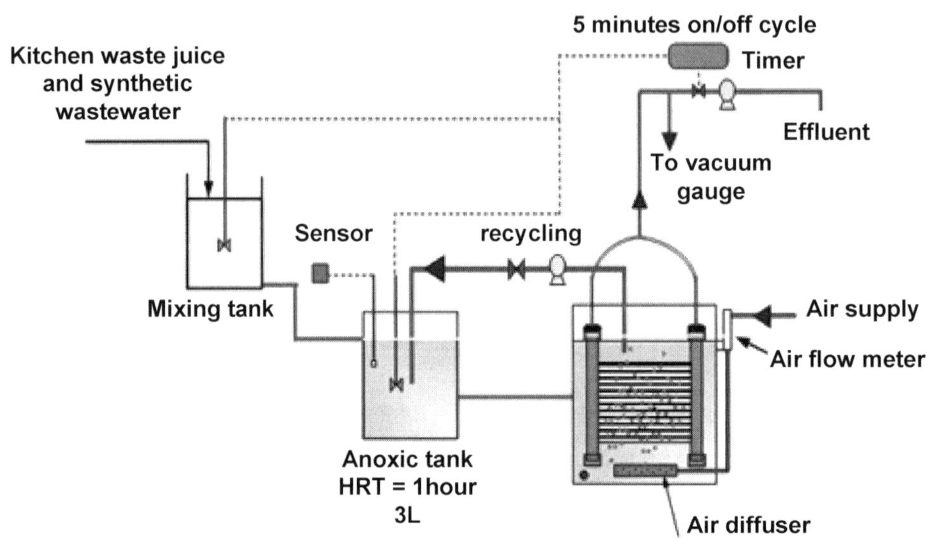

Figure 2. Continuous aerobic MBR experimental set up

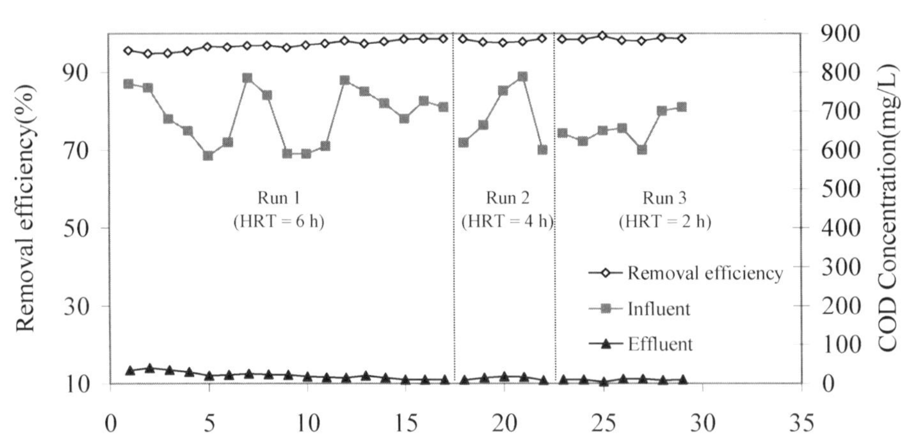

Figure 3. Variation of influent and effluent COD and its removal efficiency

Nitrogen Removal
The nitrogen removal was measured through the TKN concentrations in the influent and effluent. It is once again noticed that the influent concentration varies greatly due to the varying concentration of the kitchen waste juice. As shown in Figure 4, the TKN removal

efficiency was at 91%, 95% and 86% in run 1, 2, and 3, respectively. With run 3 showing the lowest removal efficiency, it could be stated that the operating condition at HRT of 2 hours was not favorable for the nitrogen removal.

Phosphorus Removal

The phosphorus removal efficiency was highest in run 2 at 92.6%, closely followed by run 3 at 90.6% and finally by run 1 at 87.4%. Due to the stable removal of phosphorus, results obtained at various HRTs could be concluded that the major pathway of phosphorus removal is the microbial assimilation. However, it might be necessary to investigate further by doing detailed phosphorus analysis of the sludge.

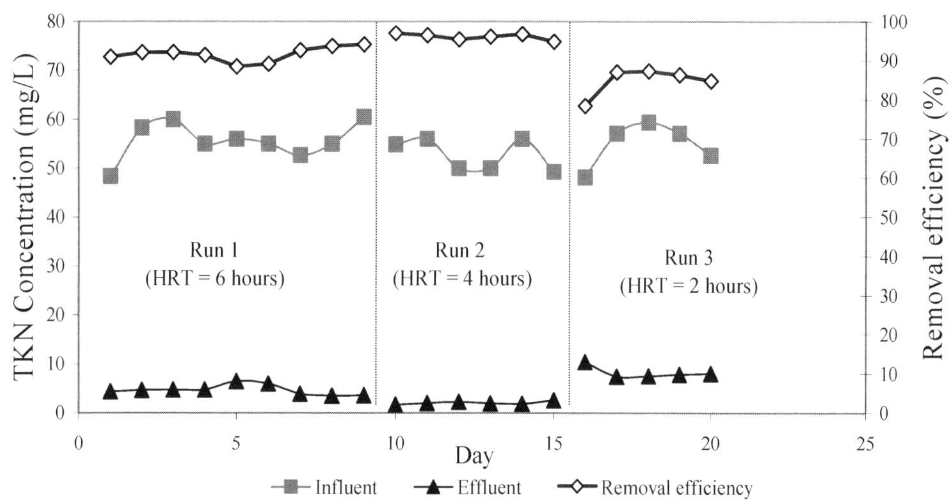

Figure 4. Variation of influent and effluent TKN and its removal efficiency

3.2 Comparison with Reuse Standards

Reclaimed water applications range from pasture irrigation to augmentation of potable water supplies. Water reclamation and reuse criteria are principally directed to health protection and may vary from country to country. In this study, various text and publications have been the sources of effluent criteria as enlisted in Table 1. In terms of COD removal, MBR has performed well, giving results at different experimental runs of below 20 mg/L. As of phosphorus, nitrate and nitrite level, only at HRT of 4 hours, is when the effluent became suitable for secondary reuse.

Table 1 Comparison of effluent characteristic for secondary reuse

Parameter	**Operating condition**			**Reuse Options**	
	Run 1 (HRT= 6 hours)	Run 2 (HRT= 4 hours)	Run 3 (HRT= 2 hours)	Range in secondary effluent	Treatment goal in reclaimed waster
COD/(mg/L)	13.5	10.33	10	50-150	<20-90
TKN(mg/L)	4.67	3.0	8.0	-	-
TP/(mg/L)	0.77	0.55	0.72	0.1-3.0	<1-20
NO_3^-(mg/L)	10.4	7.4	3.72	<10	<10
NO_2^-(mg/L)	0.12	0.143	-	<1	<1
Turbidity(NTU)	0.21	0.3	0.31	1-30	<0.1-30

Sources: *Metcalf and Eddy, 2003; Lazanova, et al., 2001; Crites and Tchobanoglous, 1998;*

3.3 Membrane Fouling

Figure 5 presented the variation in transmembrane pressure (TMP) during the entire duration of the experiment. At HRT of 6 hours, the membrane reached 27 days before it clogged (>60 KPa) due to unstable growth of microorganism, which was on its acclimatization stage. When the MLSS concentration was maintained at 10,000 mg/L, TMP also stabilized at around 17-19 days in run 2 and 3 with HRT of 4 and 6 hours, respectively. As the HRT was decreased to 2 hours, a sudden drop of TMP at 4-6 days was observed. Parallel to this result, a significant decrease in dissolved oxygen (DO) concentration in figure 7 was also noticed. Therefore, airflow rate was increased to 8.3 L/m^2-min that recovered the TMP to about 11 days.

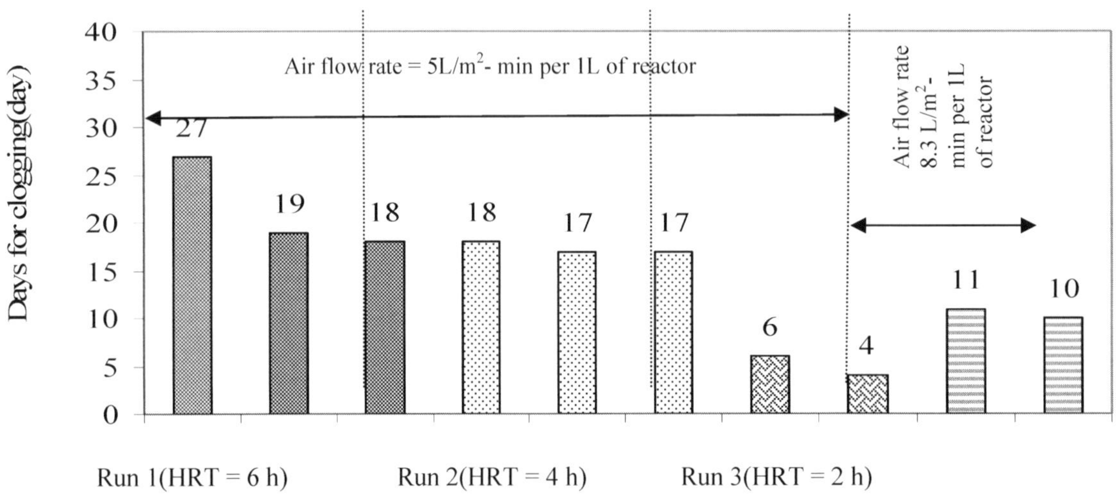

Figure 5. Rate of clogging at different experimental runs

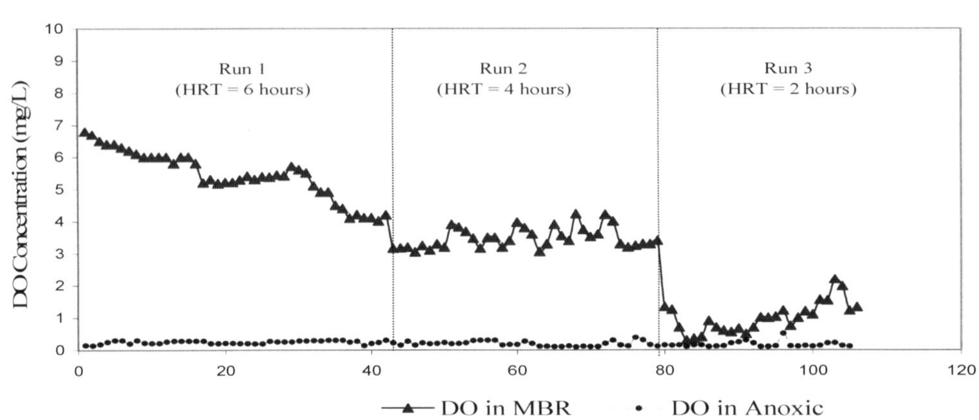

Figure 6. Dissolved oxygen concentrations at different experimental runs

A significant gradual decrease of DO concentration was observed, as the HRT was decreased from 6 hours to 2 hours (Figure 6). This rapid consumption of oxygen could be due to the consequent increase of organic loading rate in the system. As the microorganism was adapted to the new HRT condition, it tends to consume more organic compounds in a shorter period of time. It is logical that the tendency of the microorganism is also to consume more dissolved oxygen for respiration. It is therefore necessary to increase the airflow rate upon decrease of HRT to supply sufficient amount of oxygen into the system and maintain the degradation performance of the microorganism. Also, it could be suggested that this interpretation may be investigated further by taking microbial biokinetic test.

EPS Effects

In most studies, EPS is hypothesized to be the cause of clogging in membrane bioreactors. In this study, EPS was determined in terms of carbohydrate and protein bounded in the sludge. Figure 7 depicts the concentration of EPS in the system at varying experimental conditions. It could be viewed that EPS at HRT 6 and 4 hours is almost in the same range while a steep increase was observed at HRT of 2 hours. It could be recalled that in Figure 6, the clogging of membrane was rapidly observed during run 3, which operated under HRT of 2 hours. It was also in this operating condition where the EPS was observed to be in its peak. In addition, as commonly known, microorganism also evolves according to the demands of its changing environment. It might also be possible that due to microbial metabolic modifications that EPS has been produced extensively. Thus, it is proposed that EPS may be one of the causes of membrane clogging.

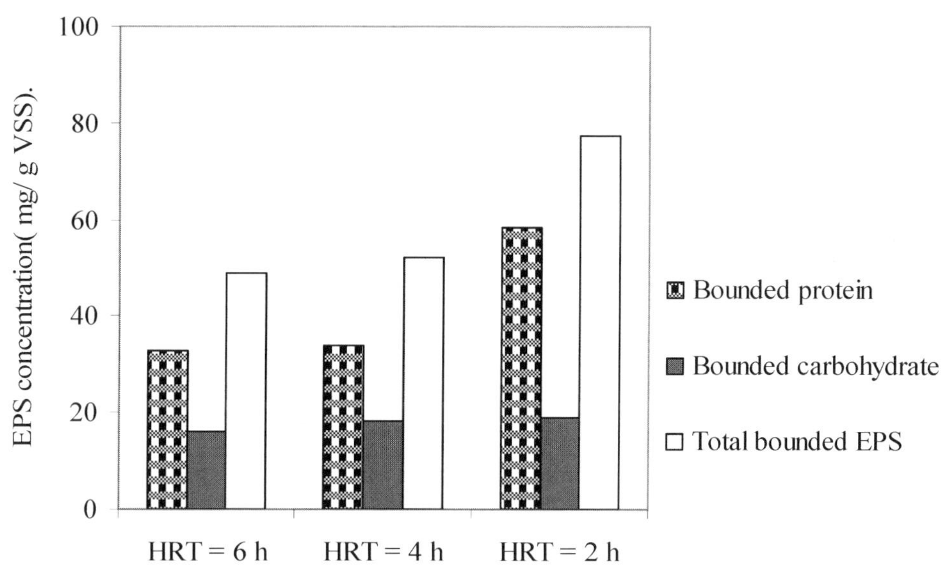

Figure 7. EPS concentration as different experimental conditions

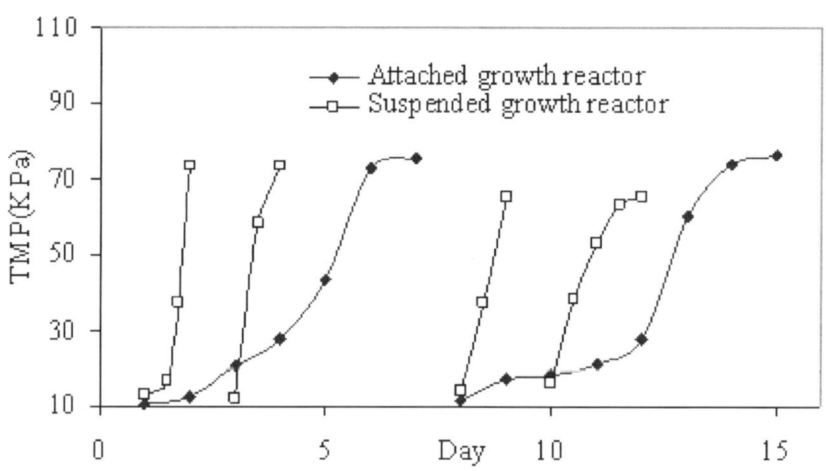

Figure 8. Variation of TMP in attached and suspended growth systems

Minimizing Membrane Fouling
In most membrane bioreactor studies, the main constraint is biofouling of the membrane. In order to achieve a maximum output in MBR, minimizing membrane fouling is very important. Hence, this study has tried to incorporate moving media into the reactor to investigate the performance of an attached growth system. Moving media was chosen, as an alternative option for it is cost effective and commonly available in the market. Experiments were run at MLSS concentration maintained at range of 9,000-10,000 mg/L and HRT of 2 hours. It is evident in Figure 8 that the suspended growth reactor experienced faster clogging than the attached growth reactor basing from the variations in the TMP. Since the study was only limited at investigating the HRT of 2 hours, it is therefore recommended to further carry out experiments to obtain the optimum operational conditions for the attached growth system.

4. Conclusion

Basing from the experimental results, the following conclusions were derived. There was more than 98% COD removal efficiency regardless of HRT. Effluent quality in terms of COD, TKN and Phosphorus removal efficiency obtained from 4 hours HRT was considered the optimum operating condition. Consequently, effluent at this stage also gives the quality that passes the criteria for secondary reuse. Attached growth MBR obtained more promising results than suspended growth MBR. Overall, the proposed scheme is considered a good option for water sanitation and environmental conservation.

5. Acknowledgement

The authors would like to thank the Southeast Asian Center for Water Environmental Technology (SACWET), a project under Tokyo University, Japan for the financial support extended to this research.

6. References

Crites, R. and Tchobanoglous, G. (1998). *Small and decentralized wastewater management systems.* McGraw Hill series in water resources and environmental engineering, Boston, MA. ISBN: 0-07-289087-8.

EcoSanRes (2004). *The Main Features of Ecological Sanitation.* Downloaded August 2004, Website: http://www.ecosanres.org/

Lazanova, V., Levine, B., Sack, J., Cirelli, G., Muntau, H., Salgot, M. and Brissaud, F. (2001). Role of water reuse for enhancing integrated water management in Europe and Mediterranean countries, *Wat. Sci. Te*ch. 43(10): 25-33.

Metcalf and Eddy (2003). *Wastewater engineering: treatment, disposal and reuse*, 4th ed., McGraw Hill Inc., Singapore, ISBN: 0-07-041690-7.

Performance Evaluation of Free Water Surface Constructed Wetlands at High Cadmium Loading

P. Pimpan[1)] and R. Jindal[2)]

1) School of Environmental Engineering, Suranaree University of Technology
111 University Avenue, Nakhon Ratchasima 30000, Thailand.
e-mail: d4440512@sut.ac.th
2) Department of Civil Engineering Faculty of Engineering, Mahidol University
25/25 Phuttamonthon 4 Rd., Salaya,Nakornprathom 73170, Thailand
e-mail: egrjd@mahidol.ac.th

Abstract

This study was conducted to investigate the performance of free water surface constructed wetlands at high cadmium loading under different environmental conditions. Influents were prepared by mixing the synthetic wastewater with $CdCl_2.H_2O$ at concentrations of 5, 10, 25, and 50 mg/L for each of the four simultaneous experiments during three runs. The performance of the wetland system was evaluated for three hydraulic retention times (HRT) 5, 7 and 10 days. Removal efficiencies of the soluble chemical oxygen demand (S-COD) were found to be in the range of 72-91% during the experimental period. The mean effluent cadmium concentrations varied between 0.17-12.73 mg/L. The overall average cadmium removal efficiency during the three runs ranged between 75-97%. Most of the cadmium was accumulated in soils (56-76% of total influent cadmium). Of the total cadmium uptakes, about 14 to 23% of the total cadmium intake was accumulated in bulrush plants. The accumulation in soil and in plants increased with influent loading during each run. The optimum HRT for efficient Cd removal from wastewater appears to be 10 days.

Keywords: cadmium, wetlands, free water surface, bulrush plants, plant uptake, soil adsorption

1. Introduction

Wetlands are described as nature's kidneys because they naturally function as filtering systems in the landscape. Principal processes in wetland systems include sedimentation, filtration, adsorption, chemical precipitation, decomposition and degradation of material by microorganisms. These ecosystems are being used as templates to design and construct wetlands that have enhanced treatment capabilities for managing domestic, municipal and industrial wastewater (Annonymous, 1994). Constructed wetlands can mimic the filtration processes of natural wetlands, effectively removing contaminants from wastewater (Lorion, 2001). Although constructed wetlands are being developed in many parts of the world for various functions, their wastewater treatment capabilities have attracted research efforts for a wide range of applications (Vigneswaran and Sundaravadivel, 2001).

Industry, being the largest water consumer is also the largest source of water pollution. All industrialized and nonindustrialized countries have used and are deliberately using heavy metals for various purposes. The wastewater discharged by industrial plants are thought to be the major source of heavy metals discharged to surface water. Heavy metals are being deposited widespread in the environment (Ernst, 1995; Suschka and Zielonka, 1995). The adverse effects of heavy metals on human health are well documented. At extreme levels, cadmium causes an illness called "Itai-Itai" disease, characterized by brittle bones and intense pain. At low levels of exposure over prolonged periods, it causes high blood pressure, sterility among males, kidney damage, and flu-like disorders (Sawyer *et al.*, 1994).

The mechanisms of metal removal in wetlands may include adsorption, complexation, precipitation, and plant uptake (Crites *et al.*, 1997). Many varieties of wetland plant species are tolerant to high concentrations of heavy metals, perhaps because of the protective effect of the iron plaque, which can develop around the roots (Vigneswaran and Sundaravadivel, 2001). Hence, wetlands can be designed and built for heavy metals removals.

The objective of this study was to investigate the performance evaluation of free water surface (FWS) constructed wetlands at high cadmium loading.

2. Materials and Methods

Laboratory-scale units of constructed wetlands were setup at the Suranaree University of Technology Campus in Thailand. Experimental setup (shown in Figure 1) consisted five laboratory-scale FWS constructed wetland units (designated as R1, R2, R3, R4, and R5) made of zinc plate, with the dimensions (length × width × depth = 2.5 × 0.25 × 0.85 m) and length-to-width ratio of 10:1. Each unit had a 20 cm thick layer of sands with 0.1 cm particle diameter at the bottom and a 40 cm layer of clay loam soils and sands mixture in 3:1 ratio at the top. Bulrush (*Cyperus corymbosus* Rottb.) plants were cultivated in the wetland units at approximately 0.15 m interval, corresponding to 23 rhizomes/m^2, with planting depth of 0.10 m. A water level of 0.15 m above the substratum surface was maintained in the FWS constructed wetlands.

Figure 1. Experimental setup of constructed wetlands

A synthetic wastewater resembling some industrial effluents was prepared for this study in the laboratory having 600-900 mg/L of soluble chemical oxygen demand (S-COD). Influents were prepared by mixing the synthetic wastewater with $CdCl_2.H_2O$ at concentrations of 5, 10, 25, and 50 mg/L for each of the four simultaneous experiments during three runs designated as Runs I-III, respectively. The performance of the wetland system was evaluated for three hydraulic retention times (HRT) 5, 7 and 10 days. The recommended HRT for FWS constructed wetlands has been reported to be 5-14 days (USEPA, 1988; Crites, 1994).

The three runs were conducted consecutively by changing the HRT 5, 7 and 10 days, respectively. The experimental period was from December 2003 to May 2005. Each experiment was run for 2 months and 22 days. Wastewater samples were collected daily from the influent and effluent points and analyzed for S-COD concentration until quasi-steady-state conditions were reached at 46-53 days. When the S-COD concentration was steady, four parameters were analyzed in the influent and effluent streams including: pH, dissolved oxygen (DO), S-COD, and cadmium (Cd). Samples from soils and plants were also analyzed for cadmium accumulations at 2 weeks intervals and at the end of each run.

3. Results and discussion

3.1 Influent and Effluent Concentrations of Monitored Parameters

During the experimental period (1 year and 6 months), the DO in the influent was in the range of 0-0.3 mg/L. The effluent DO concentrations in the five reactors during the three experimental runs varied between 1.6-6.1 mg/L. The ambient, as well as influent and effluent temperatures during the experimental period, varied between 20 and 32 $^{\circ}$C. The pH of the prepared synthetic wastewater mixed with different Cd concentrations varied between 4.04-5.72. The mean effluent pH during three experimental runs varied between 5.76 and 7.58. Bulrush plants are found to be growing well in a pH range of 4-9 (Reed *et al.*, 1988; USEPA, 1988).

Figure 2 shows average removal efficiency of S-COD during three experimental runs. Mean S-COD removal efficiencies in the FWS wetland systems were in the range of 71.58 90.89%.

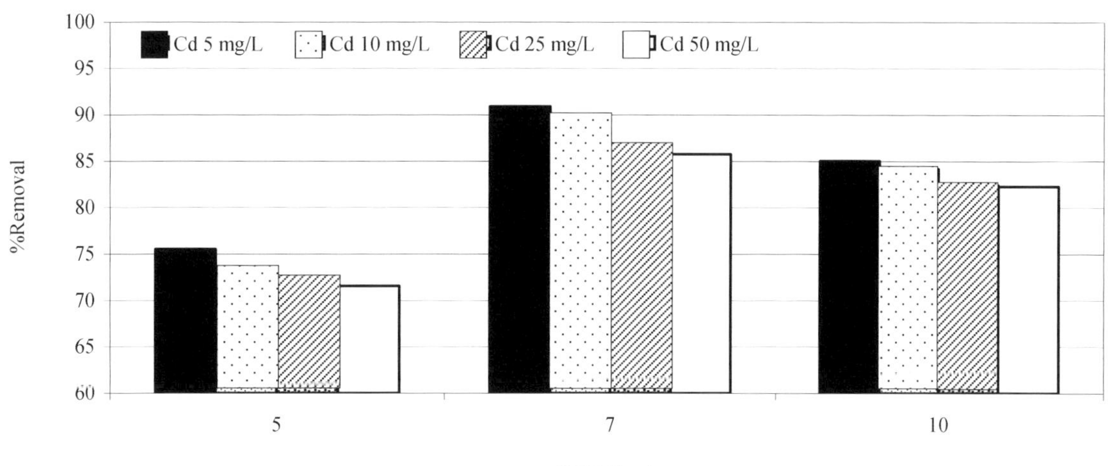

Figure 2. Average removal efficiency of S-COD during three experimental runs

It can be seen in Fig. 2 that S-COD removal in the wetland units was only slightly affected by the influent Cd concentrations. There was a reduction in S-COD removal with increased influent loading during each run (3.92% in Run I, 5.12% in Run II, and 2.76% in Run III, reapectively). With respect to HRT, the removal increased (15.24% on average) in Run II (HRT = 7 days) compared to in Run I (HRT = 5 days), but slightly decreased (4.95% on average) in Run III (HRT = 10 days). Therefore, 7 days seems to be the optimum HRT for S-COD removal in the FWS wetland systems with bulrush plants in this study.

3.2 Cadmium Removal in Wastewater

Figure 3 shows cadmium removal efficiency of wetland units with four different cadmium concentrations in influent during the three experimental runs. The overall average cadmium removal efficiency during the three runs ranged between 75-97% (74.6-81.9% in Run I at HRT = 5 days, 82.6-90.8% in Run II at HRT = 7 days, and 92.7-96.5% in Run III at HRT = 10 days, respectively).

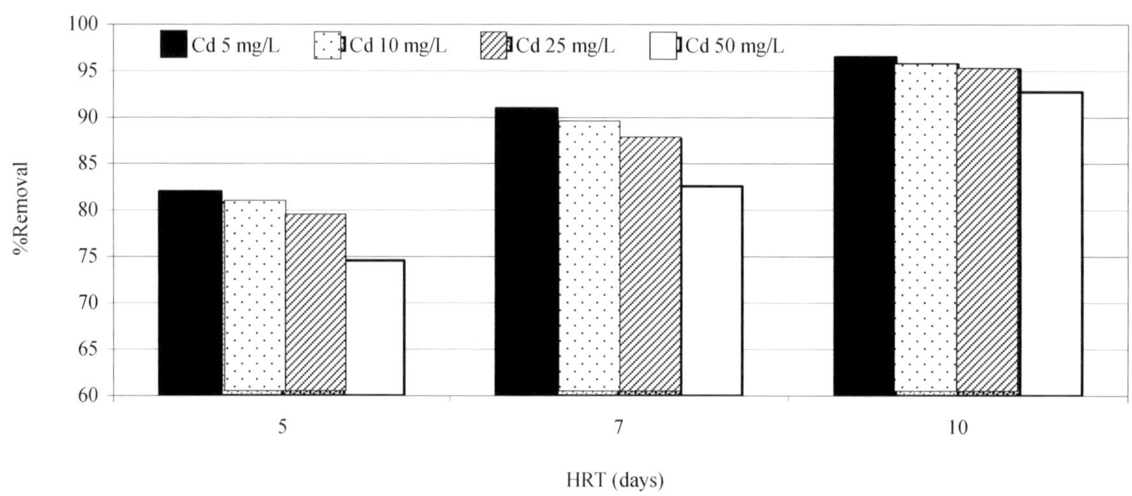

Figure 3. Average cadmium removal efficiencies of wetland units during three experimental runs

For HRT 5 days, influent and effluent cadmium loadings were 258.8, 517.5, 1,293.8, 2,587.5 mg/d, and 46.6, 99.4, 267.9, 658.5 mg/d, for reactors R2, R3, R4, and R5, respectively. In Run II, influent and effluent cadmium loadings were 184.8, 369.6, 924, 1,848 mg/d, and 16.7, 38.9, 113.4, 322.3 mg/d, for R2, R3, R4, and R5, respectively. For HRT 10 days, cadmium loadings were 129.4, 258.8, 647, 1,294 mg/d in the influent and 4.5, 10.9, 37.3, 93.9 mg/d in the effluent for R2, R3, R4, and R5, respectively.

It can be seen in Figure 3, that cadmium removal was affected by higher influent loading in each run. Removal decreased with increased loading. However, the removal efficiency for cadmium increased with HRT. Maximum removal occurred at HRT 10 days. Thus, in terms of Cd removal from wastewater in FWS constructed wetlands with bulrush plants, HRT of 10 days seems to be optimum for this study.

3.3 Cadmium Removal by Soil

Soil samples were collected at the end of each run along reactor lengths at distances of 0.05, 0.65, 1.25, 1.9, and 2.45 m from the inlet, and at depths of 0, 15, 30, and 45 cm from the top of the soil bed. The total accumulation of cadmium in the soils was determined at the end of each run. The results showed a high concentration of cadmium at the inlet that decreased along the reactor length in three experimental runs. Concentrations at the top were higher than at the depth of 15, 30 and 45 cm. The porosity of sand (0.49) being higher than that of the mixture of clay loam soils and sands (0.41) could have been the cause of a higher adsorption of cadmium at the top of soil bed. Clays are soils with very fine particles packed closely together. Clays typically have highest adsorption potential of any soils because of their high surface area to volume ratio resulting from their small particle size distribution. Loamy soils have excellent plant growth characteristics because of adequate nutrient holding capacity and higher hydraulic conductivities and gaseous diffusion rates (Kadlec and Knight, 1996). The total cadmium accumulation in soils during three experimental runs is shown in Figure 4.

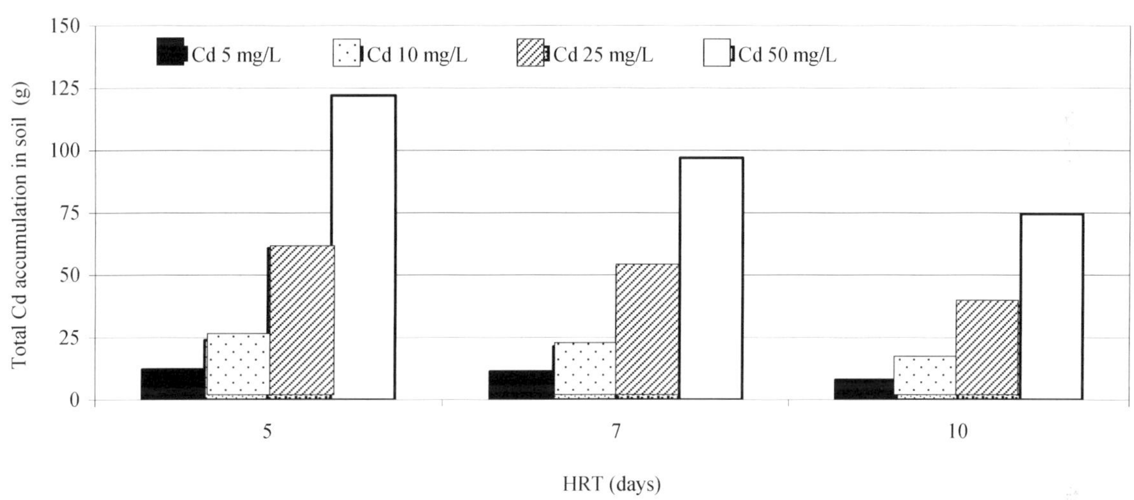

Figure 4. Total cadmium accumulation in soil bed in the FWS wetland system during three experimental runs

It can be seen in Figure 4, that total cadmium accumulation in soil bed increased with influent loading during each run. Also, the accumulation decreased with HRT for all influent concentration.

3.4 Cadmium Removal by Plant Uptake

The factors affecting the amount of metal absorbed by a plant are those controlling: (i) the concentrations and speciation of the metal in the soil solution, (ii) the movement of the metal from the bulk soil to the root surface, (iii) the transport of the metal from the root surface into the root, and (iv) its translocation from the root to the shoot. Plant uptake of mobile ions present in the soil solution is largely determined by the total quantity of this ion in the soil but, in the case of stroing adsorbed ions, absorptioni is more dependent upon the amount of root produced (Alloway, 1990).

Cadmium accumulations in roots and stems of the bulrush plants until the end of each experimental run are shown in Figure 5. It can be seen that total Cd accumulation in the stems was higher than in the roots for all four wetland units. It should be also be noted that the cadmium concentration in the roots (mg/g) was found to be higher than in the stems. However, as the dry weight of the roots of bulrushes was only 25% of the stems, the actual mass accumulation of Cd in stems was higher than in the roots.

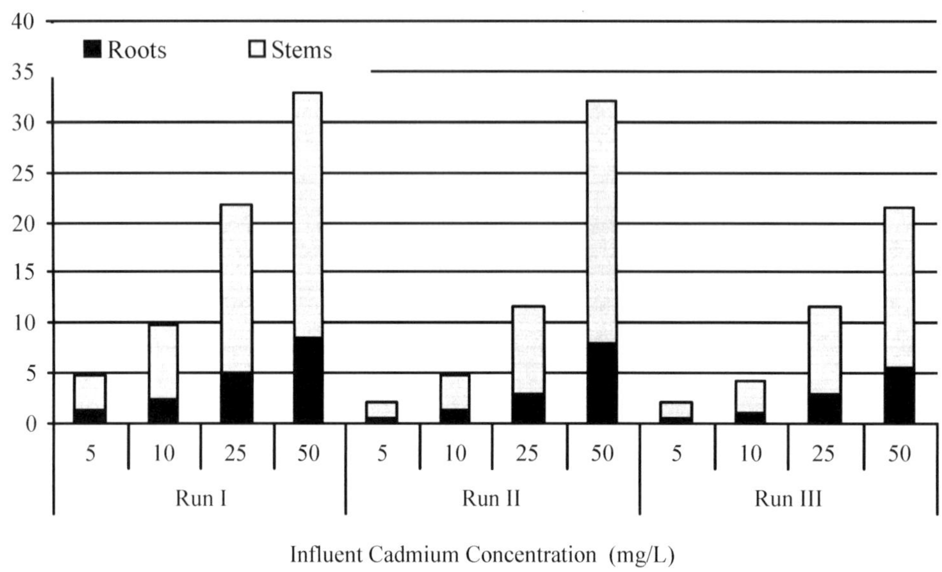

Figure 5. Total cadmium accumulation in roots and stems of plants in FWS wetland system at the end of each experimental run

It can be seen in Figure 5 that, during all the runs cadmium accumulations in stems and roots were higher at high influent loadings. However, the cadmium accumulations in stems and roots at higher HRT (lower loading for each concentration) decreased as compared to those at lower HRT (higher loadings).

3.5 Cadmium Mass Balance

The overall performance of the FWS wetlands in terms of Cd removal was very good. Both, the soils and plants were the sinks for the accumulation of cadmium. Cadmium was sorbed by the soil bed and plant uptake. Chemical precipitation of cadmium could also have occurred in wetland systems.

Figure 6 shows the mass balance for cadmium during each experimental runs. Relative to the total intake, the cadmium present in the effluents were in the range of 18.1-25.5, 9.1-13.2, and 3.5-7.4% in Run I, II, and III, respectively. The maximum cadmium removal occured through the accumulation in the soils with mass fraction values of 56.3-57.9, 63.9-74.8, and 69.9-75.5% for HRT = 5 days, 7 days, and 10 days, respectively. The lost Cd via other sinks ranged between 1.5-1.6, 1.7-1.9, and 2.1-2.3% of total cadmium uptake in Run I, II, and III, respectively. Mass fraction of cadmium accumulated in plants ranged between 15.4 and 22.9, 14.4 and 21.1, and 18.9 and 21.9, during Run I, Run II, and Run III, respectively.

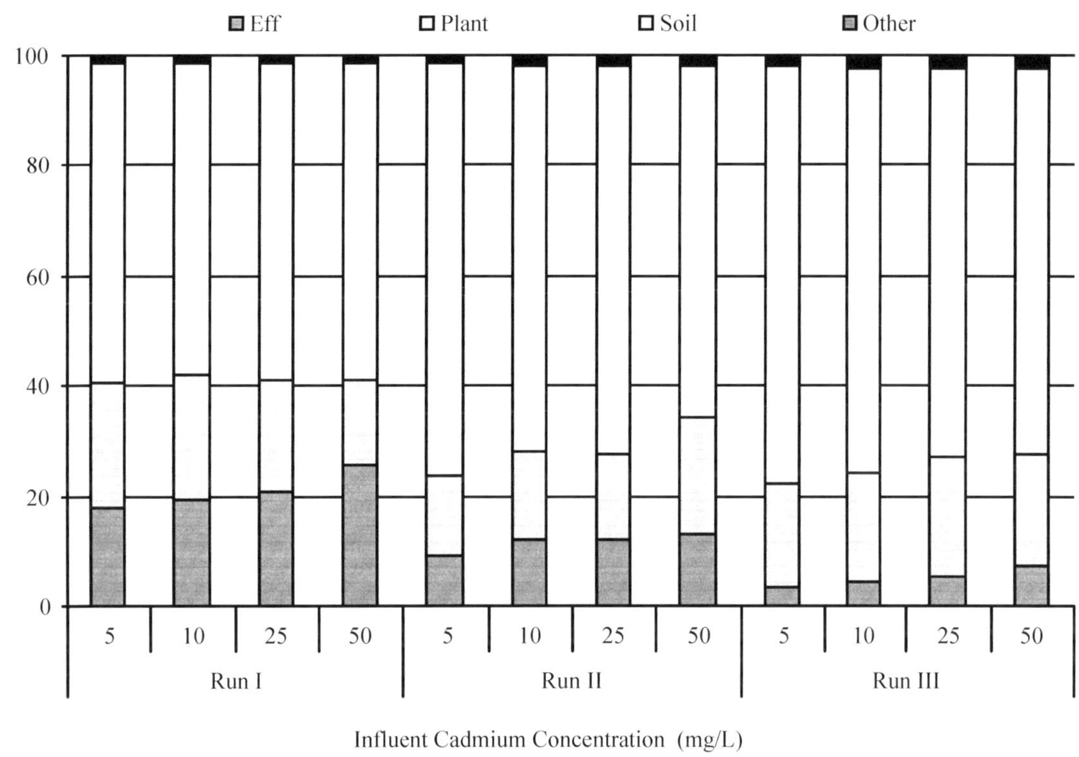

Figure 6. Cadmium mass balance in three experimental runs

4. Conclusions

The three FWS constructed wetlands with bulrush plants were used to treat synthetic industrial wastewater with 600-900 mg/L of S-COD. The performance of the wetland system was evaluated for three hydraulic retention times (HRT) 5, 7 and 10 days. The overall average cadmium removal efficiency during the three runs ranged between 75-97% (74.6-81.9% in Run I at HRT = 5 days, 82.6-90.8% in Run II at HRT = 7 days, and 92.7-96.5% in Run III at HRT = 10 days, respectively). Most of the cadmium was accumulated in soils (56-76% of total influent cadmium). Total accumulation of Cd in soil bed was found to be higher in Run I (HRT 5 days) and decreased with HRT for all influent concentration. Also, the accumulation increased with influent loading during each run. The cadmium accumulation in soil decreased along the reactor lengths. Cadmium accumulations at the top of soil bed were higher than at the depth of 15, 30 and 45 cm. The porosity of sand (0.49) being higher than that of the mixture of clay loam soils and sands (0.41) could have been the cause of a higher adsorption of cadmium at the top of soil bed. Amount of total Cd uptake by the stems of bulrush plants were higher than roots for all four wetland units fed with synthetic wastewater at different influent loading during the three runs. Of the total cadmium uptakes, about 14 to 23% of the total cadmium intake was accumulated in bulrush plants. During all the runs, cadmium accumulations in stems and roots were higher at high influent loadings. However, the cadmium accumulations in stems and roots at higher HRT (lower loading for each concentration) decreased as compared to those at lower HRT (higher loadings).

Based on the results of this study, it could be concluded that FWS constructed wetland system with bulrush plants may be effective in cadmium removal from industrial wastewater even at high influent loadings. The optimum HRT for efficient Cd removal from wastewater appears to be 10 days. Thus wetland systems can be used to improve the quality of final effluents from industrial wastewater treatment plants before their disposal into receiving water bodies.

5. Acknowledgements

This study was supported by the funding from the Royal Golden Jubilee Ph.D. Scholarship Grant (RGJ-Ph.D.) of Thailand Research Fund (TRF), Thailand. The authors acknowledge the support provided by the Suranaree University of Technology (SUT), Thailand.

6. References

Alloway, B. J. (1990). *Heavy metals in soils,* John Wiley & Sons, New York.

Annonymous (1994). Cobalt constructed wetland. *The Corporation of the Town of Cobalt Ontario*, <http://cobalt.ca/wetland_information.htm>.

Crites, R. W. (1994). Design criteria and practice for constructed wetland. *Water Sci. Technol.*, **29**(4), 1-6.

Crites, R. W., Dombeck, G. D., Watson, R. C., and Williams, C. R. (1997). Removal of metals and ammonia in constructed wetlands. *Water Environ. Res.*, **69**, 132-135.

Ernst, W. H. O. (1995). Decontamination or consolidation of metal-contaminated soils by biological means. In: *Heavy Metals Problems and Solutions*, W. Salomons, U. Forstner, and P. Marder (Eds), Springer-Verlag Berlin Heidelberg, Germany, pp. 141-149.

Kadlec, R. H., and Knight, R. L. (1996). *Treatment wetlands*, CRC, Boca Raton, Fla.

Lorion, R. (2001). Constructed wetlands: Passive systems for wastewater treatment. *Technology status report prepared for the USEPA technology innovation office under a national network of environmental management studies fellowship*, <http://ag.arizona.edu/AZWATER/arroyo/assets/wetlnd1.gif> (Aug., 2001).

Reed, S. C., Middlebrooks, E. J., and Crites, R. W. (1988). *Natural systems for waste management and treatment*, McGraw-Hill, New York.

Sawyer, C. N., McCarty, P. L., and Parkin, G. F. (1994). *Chemistry for Environmental Engineering*, McGraw-Hill.

Suschka, J., and Zielonka, U. (1995). Some remarks on the contamination of the environment with heavy metals in a part of Poland. In: *Heavy Metals Problems and Solutions*, W. Salomons, U. Forstner, and P. Marder (Eds), Springer-Verlag Berlin Heidelberg, Germany, pp. 393-401.

USEPA (1988). *Design manual: Constructed wetlands and aquatic plants systems for municipal wastewater treatment*, United States Environmental Protection Agency, Office of Research and Development, EPA 625/1-88/022.

Vigneswaran, S., and Sundaravadivel, M. (2001). Constructed wetlands for wastewater treatment. *Critical Reviews in Environmental Science and Technology*, **31**(4), 351-409.

Constructed Wetlands for Reactive Dye and Nitrogen Removal

Flora Wei-Ni Tan[1], Md Sani Ibrahim[2] and Poh-Eng Lim[3]

1) School of Chemical Sciences, Universiti Sains Malaysia, 11800 Penang, Malaysia
2) School of Chemical Sciences, Universiti Sains Malaysia, 11800 Penang, Malaysia email: sani@usm.my
3) School of Chemical Sciences, Universiti Sains Malaysia, 11800 Penang, Malaysia email: pelim@usm.my

Abstract

The objectives of this study are to evaluate the performance of planted and unplanted constructed wetlands in the removal of a reactive diazo dye and to investigate the effect of the dye on nitrogen removal in the wetland system. Two laboratory-scale gravel-filled horizontal subsurface-flow constructed wetland units, one planted with cattails (*Typha latifolia*) and the other unplanted, were operated outdoors at a nominal hydraulic retention time of 7 days. The wetland units were initially fed with a primary-treated domestic wastewater followed by the same domestic wastewater spiked with 35 mg/L Reactive Black 5 dye. The ammoniacal nitrogen (AN) and reactive dye concentrations were monitored at the inlet, outlet and 6 intermediate points along the centerline at the depths of 0.15 and 0.45 m. The results showed that the planted and unplanted wetland units were able to achieve the percentage of dye removal at above 68% for a monitoring period of 107 days and that wetland plants did not play an important role in dye removal. The inhibitory effect of dye on AN removal was observed in both types of wetland units but the effect was greater in the unplanted unit indicating that wetland vegetations play some role in reducing the inhibitory effect of dye.

Keywords: Constructed Wetlands; Nitrogen; Reactive Dye; Removal

1.Introduction

Large quantity of dyes are produced annually and used predominantly in the textile industry. The industry consumes a large proportion of reactive azo dyes due to the high demand for cotton fabrics with brilliant colours. Once discharged into receiving watercourses, the dyes not only constitute an aesthetic problem thus restricting the beneficial uses of receiving waters downstream but may also have acute or chronic effects on exposed aquatic organisms. It is therefore imperative to treat dye-containing wastewaters to minimize adverse environmental impacts. Treatment of dye-containing wastewaters by chemical, physical or biological process or a combination of them has been investigated and reported by many researchers (Shelley et al., 1976; Nicolaou and Hadjivassilis, 1992; Davies and Cottingham, 1994; Slokar and Marechal, 1998; Lim and Er, 2000; Lourenco et al., 2001; Fu et al., 2002).

Constructed wetland provides a good example of a wastewater treatment system utilizing environmental friendly technology. It has been proven to be an efficient treatment system for municipal wastewater as well as urban runoff and identified as a good candidate for river water quality improvement in removing oxygen demand, nutrients and solids. In an effort to broaden the scope of application for constructed wetlands, recent research interest have been focused on the potential of the system in treating toxic pollutants such as heavy metals (Lim et al., 2001; Scholz and Jing, 2002; Scholz, 2003; Lim et al., 2003a; Lim et al., 2003b) and organic priority compounds (Polprasert and Dan, 1994). Nonetheless, little has been reported on the capability of constructed wetland system in treating dyes. The objectives of this study, therefore, are to evaluate the performance of planted and unplanted constructed wetlands in the removal of a reactive diazo dye and to investigate the effect of the dye on nitrogen removal in the wetland system. The latter objective is particularly pertinent to situations in which dye-containing wastewater may be combined with domestic wastewater for treatment by constructed wetlands.

2.Materials and Methods

2.1Treatment Studies

Two laboratory-scale horizontal subsurface-flow wetland units made of acrylic plates with dimensions of 1.84 x 0.44 x 0.67 m were constructed. The support media of 0.6 m in depth consisted of graded gravel of size 2.4-4.8 mm diameter. One of the wetland units were planted with cattails (*Typha latifolia*) at a density of 44 rhizomes/m^2 whereas the other was unplanted. Clusters of perforated tubes (1.25 diameter) of depths 0.15 and 0.45 m were placed at locations A, B, C, D, E and F along the center line in the units at distances 0.26, 0.52, 0.78, 1.04, 1.30 and 1.56 m, respectively, from the inlet. Both the units were placed outdoors and fed with a primary-treated domestic wastewater at a nominal retention time of 7 days. Once the cattail plants had reached maturity, the ammoniacal nitrogen (AN) concentration was monitored at the inlet, outlet and 6 intermediate sampling points at the depths of 0.15 and 0.45 m for about 4 months to assess the performance of the system. Sampling at intermediate points in addition to inlet and outlet provides a better understanding of the pollutant removal pattern in the wetland system. At maturity, the plant roots were observed to have penetrated to a depth of 0.3 m. This means that sampling at 0.15 and 0.45 m covers the root and non-root zones, respectively. Subsequently, the wetland units were fed with the same domestic wastewater spiked with 35 mg/L Reactive Black 5 dye for a further 4-month period during which monitoring of dye and AN concentrations were conducted at those sampling points mentioned earlier.

Sampling time was always confined to 0800 – 1000 h to avoid significant loss of water due to evapotranspiration. The AN concentration was determined following the procedure described in the Standard Methods (APHA, 1992). The dye concentration was determined using an UV-VIS spectrophotometer (Hitachi Model U-2000) with λ_{max} at 602 nm.

2.2Kinetic Studies

About 1000 g of graded gravel media were excavated, respectively, from two unplanted wetland units, one of which was fed with dye-containing domestic wastewater for about a year whilst the other was treating only domestic wastewater. The media, bound in a nylon net, were drained of wetland water and immersed completely in 600 mL of 5 mg/L Reactive Black 5 dye solution with continuous stirring. At fixed interval, an aliquot of the dye solution was withdrawn, filtered and measured for the dye concentration at λ_{max} 602 nm. The sample was returned to the bulk solution after measurement.

To ascertain the role of the suspended biomass in dye removal, grab liquid samples were collected from locations A, B and C and monitored for the change of dye concentrations with time.

3.Results and Discussion

The percentages of dye and AN removal for the planted and unplanted wetland units are summarized in Table 1.

Table 1. Percentages of dye and AN removal for the planted and unplanted wetland units at increasing duration of dye addition

Duration of dye addition, days	Percentage of removal, %			
	Dye		AN	
	Unplanted	Planted	Unplanted	Planted
0	-	-	63	91
26	76	74	56	67
35	80	80	59	91
42	76	76	39	96
49	73	78	35	85
54	68	71	27	79
58	69	71	8	70
107	79	84	17	69

3.1 Removal of Dye

The data in Table 1 showed that the planted and unplanted wetland units were able to achieve the percentage of dye removal at above 71 and 68%, respectively, for a monitoring period of 107 days. There was no significant difference ($P<0.05$) in the percentages of dye removal between the planted and unplanted wetland units at increasing duration of 35 mg/L of dye addition. This suggests that wetland plants did not play an important role in dye removal. This is further supported by the removal profiles of dye after 35, 54 and 107 days of dye addition for planted and unplanted wetland units at 0.15 and 0.45 m depths as shown in Figure 1(a) and (b), respectively. For the same duration of dye addition, there was no significant difference ($P<0.05$) in the percentages of dye removal along the treatment path between the root and non-root zones for both planted and unplanted wetlands.

The shape of all the removal profiles indicates a rapid dye concentration reduction in the initial part of the wetland unit, namely from the inlet to location A, followed by a slow removal for the remaining part of the unit. This may be explained by the fact that many solids were trapped in the initial part of the wetland media. As such, more dyes were removed by these settled solids through adsorption and subsequent biodegradation processes. The removal of dye for the latter part of the unit was primarily carried out by the biofilms on the media surfaces. The results of the comparative kinetic studies expressed as the time course of dye concentrations are shown in Figure 2. Since studies (Aksu and Tezer, 2000; Aksu and Donmez, 200 3) have shown that biosorption of reactive dyes attained equilibrium within 2 hours, the time scale of the kinetic studies immediately rules out the importance of biosorption in dye removal. It is therefore evident that dye was primarily biodegraded by biofilms on the media surfaces and the data fitted the zero-order kinetics very well (R^2 > 0.970). Asexpected,

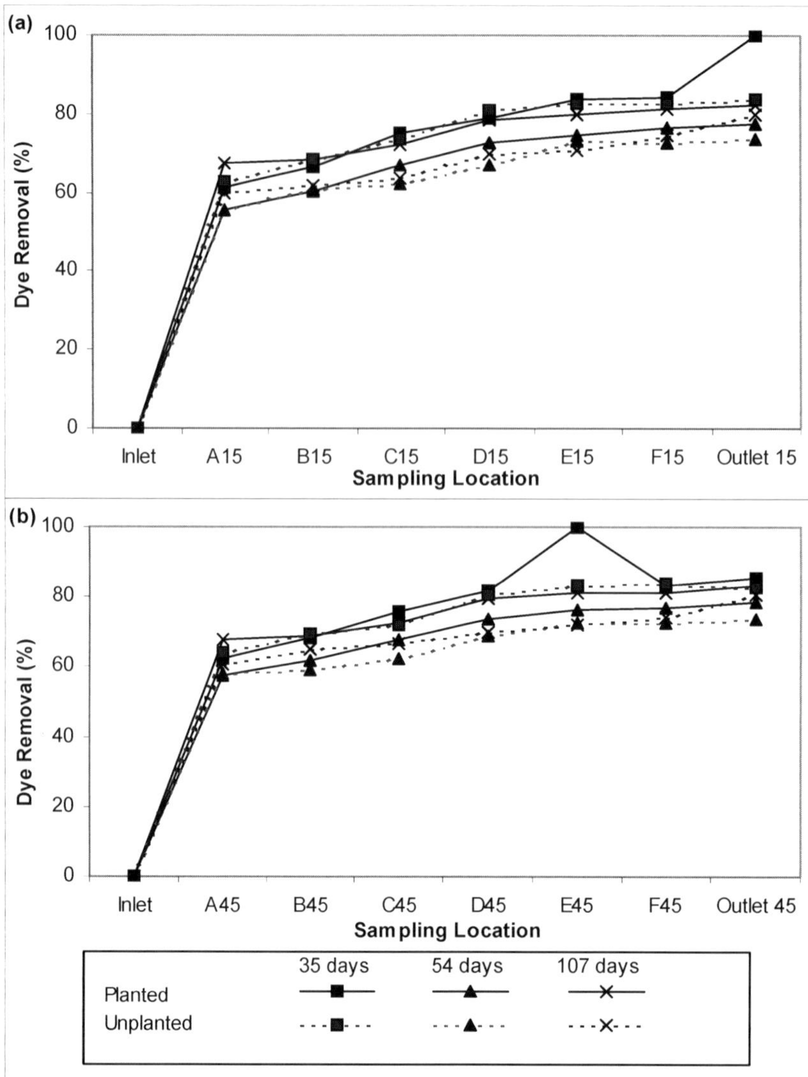

Figure1. Percentages of dye removal in planted and unplanted wetland units at (a) 0.15 m and (b) 0.45 m depths after 35, 54 and 107 days of dye addition.

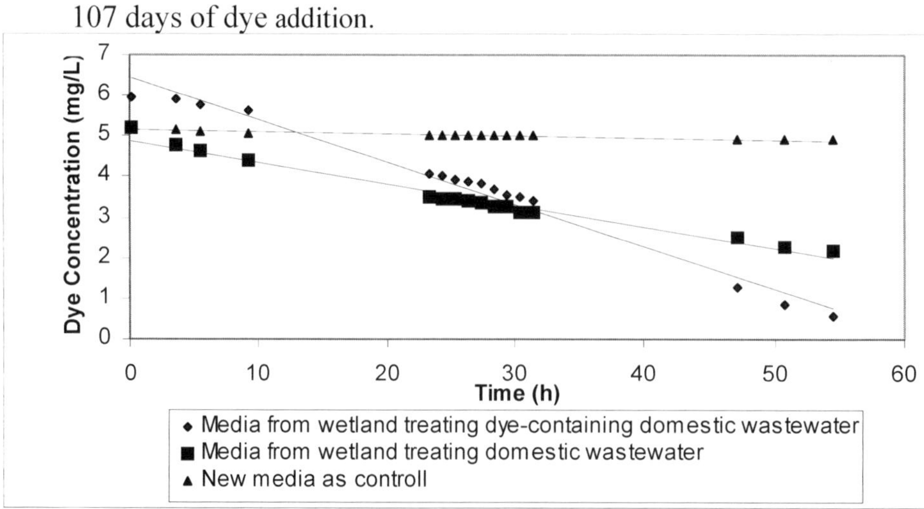

Figure 2. Time course of dye concentrations using three different types of gravel media of comparable weight.

the dye removal rate constant for acclimatized biofilms, at 0.102 mg/L h, was higher than the value of 0.054 mg/L h for unacclimatized biofilms. There was little evidence of inhibition, as exemplified by a lag phase, shown by the unacclimatized biofilms in the degradation of Reactive Black 5 dye. It should also be pointed out the rate of biodegradation of the dye is a function of the characteristics of the biofilms and is therefore location specific.

With regard to the liquid samples, there was little change in the dye concentration with respect to time indicating that the role of suspended biomass in the removal of dye was insignificant.

3.1 Effect of Dye on Ammoniacal Nitrogen (AN) Removal

Table 1 shows the performance of the planted and unplanted wetlands in AN removal with the duration of dye addition. Figures 3(a) and (b) show the AN removal profiles along the treatment path after 35, 54 and 107 days of dye addition at 0.15 and 0.45 m depths, respectively, for the planted wetland unit whereas those for the unplanted wetland are shown in Fig. 4(a) and (b), respectively. The results show that there was a progressive deterioration in the performance of both planted and unplanted wetland units as the duration of dye addition increased.

In order to assess the effect of dye on AN removal quantitatively, the first-order plug-flow model was employed to describe AN removal in the wetland (Reed et al., 1995; Crites and Tchobanoglous, 1998) as follows:

$$N_e/N_o = \exp(-kt) \tag{1}$$

where N_o and N_e are the AN concentrations before and after treatment, respectively, k the apparent AN removal rate constant and t the reaction time. By using the AN concentration data obtained from the intermediate sampling locations at 0.15 m depth, the value of k was estimated from the slope of the straight-line plot of $\ell n\ (N_e/N_o)$ versus t. This model gave a relatively good fit to the data with $R^2 \geq 0.67$. The values of k were calculated for AN removal before the addition of dye and at various durations after dye addition for the planted and unplanted units. Figure 5 shows that the planted and unplanted wetland units exhibited different patterns with regards to the dye effect on k values. For the unplanted wetland, AN removal was essentially halted after 107 days of 35 mg/L dye addition indicating that the bioactivity of the nitrifying microorganisms was effectively inhibited. In contrast, a rapid decrease in the k value was observed for the planted wetland during the initial stage of dye addition but the AN removal seemed to be able to sustain a fairly constant rate thereafter. This points to the important role of wetland plants in sustaining some AN removal even after the microbial activity has been inhibited.

4. Conclusions

Based on the findings of this study, the following conclusions can be drawn:

a) The planted and unplanted wetland units were able to achieve the percentage of dye removal at above 71 and 68%, respectively, for the duration of this study.
b) There was no significant difference in the percentages of dye removal between the planted and unplanted wetlands indicating that the cattail plants do not play an important role in dye removal.

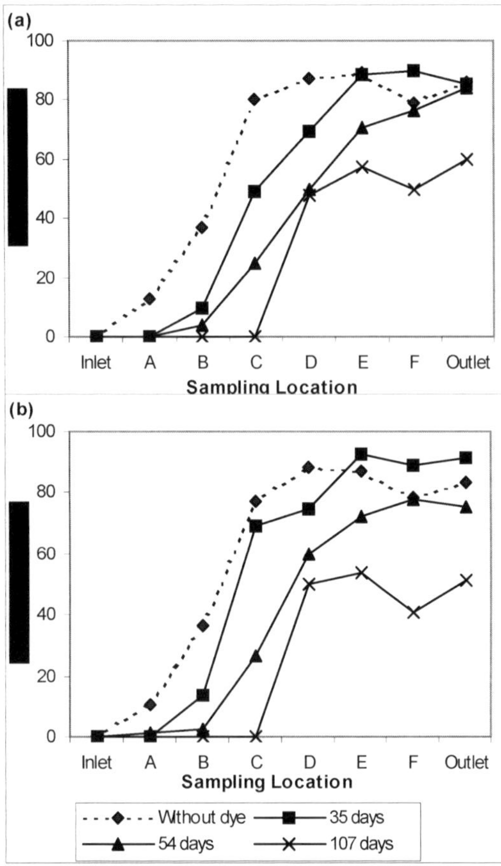

Figure 3. Percentage of AN removal in planted wetland at (a) 0.15 m and (b) 0.45 m depths before dye addition and after 35, 54 and 107 days of dye addition.

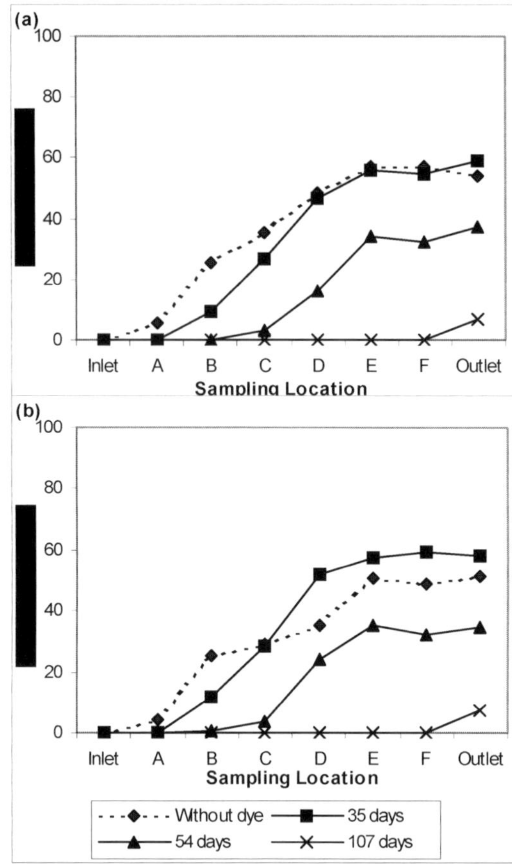

Figure 4. Percentage of AN removal in unplanted wetland at (a) 0.15 m and (b) 0.45 m depths before dye addition and after 35, 54 and 107 days of dye addition.

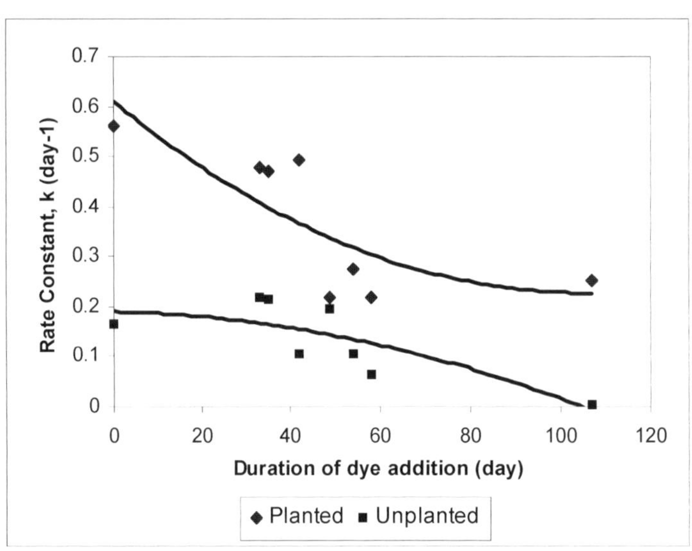

Figure 5. Effect of dye on the first-order AN removal rate constant estimated at 0.15 m depth for planted and unplanted wetland units.

c) After an initial rapid dye removal due to adsorption and biodegradation by settled solids, dye was biodegraded by biofilms on media surfaces following zero-order kinetics in the latter part of the wetlands. The role of suspended biomass in the liquid phase in dye removal was found to be insignificant.
d) The inhibition of AN removal by dye was assessed quantitatively using the apparent AN removal rate constant k based on the first-order plug-flow model. It was observed that AN removal was essentially halted in the unplanted wetland after 107 days of dye addition. In contrast, AN removal in the planted wetland seemed to be able to sustain a fairly constant rate after an initial rapid decrease indicating the important role of wetland plants in reducing the inhibitory effect of dye.

5. Ackowledgements

Financial support from the Ministry of Science, Technology and Innovation, Malaysia under the IRPA Grant No. 08-02-05-3221 EA 012 is gratefully acknowledged.

6. References

Aksu Z. and Donmez G. (2003). A comparative study on the biosorption characteristics of some yeasts for Remazol Blue reactive dye. *Chemosphere* 50, 1075-1083.

Aksu Z. and Tezer S. (2000). Equilibrium and kinetic modeling of biosorption of Remazol Black B by *Rhizopus arrhizus* in a batch system: effect of temperature. *Process Biochemistry* 36, 431-439.

APHA. (1992). *Standard Methods for the Examination of Water and Wastewater.* 18th ed. American Public Health Association, Washington DC.

Crites R.W. and Tchobanoglous G. (1998). *Small and Decentralized Wastewater Management Systems.* McGraw-Hill, Singapore

Davies T.H. and Cottingham P.D. (1994). The use of constructed wetlands for treating industrial effluent (textile dyes). *Wat. Sci. Technol.* 29, 227-232.

Fu L., Wen X., Xu L. and Qian Y. (2002). Removal of a copper-phthalocyanine dye from wastewater by acclimated sludge under anaerobic or aerobic conditions. *Process Biochemistry* 37, 1151-1156.

Lim P.E. and Er C.C. (2000). Treatment of dye-containing wastewater by sequencing batch reactor with powdered activated carbon addition. *Toxicol. Environ. Chem.*, 75, 75-87.

Lim P.E., Mak K.Y., Mohamed N. and Md. Noor A. (2003a). Removal and speciation of heavy metals along the treatment path of wastewater in subsurface-flow constructed wetlands. *Wat. Sci. Technol.* 48, 307-313.

Lim P.E.,Tay M.G., Mak K.Y. and Mohamed N, (2003b). The effect of heavy metals on nitrogen oxygen demand removal in constructed wetlands. *Sci. Tot. Environ.*, 301, 13-21.

Lim P.E., Wong T.F. and Lim D.V. (2001). Oxygen demand, nitrogen and copper removal by free-water-surface and subsurface-flow constructed wetlands under tropical conditions. *Environ. International* 26, 1-7.

Lourenco N.D., Novais J.M. and Pinheiro H.M. (2001). Effect of some operational parameters on textile dye biodegradation in a sequential batch reactor. *J. Biotechnol.*, 89, 163-174.
Nicolaou M. and Hadjivassilis I. (1992). Treatment of wastewater from textile industry. *Wat. Sci. Technol.* 25, 31-35.

Polprasert C. and Dan N.P. (1994). Phenol removal in model constructed wetlands located in the tropics. Paper #AC 941505. Chicago, Illinois: *Water Environment Federation 67th Annual Conference & Exposition.*

Reed S.C., Crites R.W. and Middlebrooks E.J. (1995). *Natural Systems for Waste Management & Treatment.* 2nd ed. McGrae-Hill.

Scholz M. and Xu J. (2002). Performance comparison of experimental constructed wetlands with different filter media and macrophytes treating industrial wastewater contaminated with lead and copper. *Bioresource Technol.*, 83, 71-79.

Scholz M. (2003). Performance predictions of mature experimental constructed wetlands which treat urban water receiving high loads of lead and copper. *Water Res.*, 37, 1270-1277.
Shelly M.L., Randall C.W. and King P.H. (1976). Evaluation of chemical-biological and chemical-physical treatment for textile dyeing and finishing waste. *J. Water Pollut. Control Fed.* 48, 753-761.

Slokar Y.M. and Majcen Le Marechal A. (1998). Methods of decoloration of textile wastewaters. *Dyes and Pigments* 37, 335-356.

Microbial Community Structures of a Constructed Wetland: a Biogeochemical Approach Based on the Fluorescence *In Situ* Hybridization Technique

Nguyen Tran Thien Khanh[1)] and O.V. Shipin[2)]

1) Former address: Environmental Engineering and Management Program, School of Environment, Resources and Management, Asian Institute of Technology, P.O. Box 4, Pathumthani, 12120, Thailand
Present address: 26/6 Dinh Tien Hoang St, My Binh – Long Xuyen , An Giang Province, Viet Nam
e-mail: thienkhanhag@yahoo.com
2) Environmental Engineering and Management Program, School of Environment, Resources and Management, Asian Institute of Technology, P.O. Box 4, Pathumthani, 12120, Thailand
e-mail: oshipin@ait.ac.th

Abstract

The study presents a pioneering effort in molecular microbiology of the Natural (wastewater) Treatment Systems (NTS). The microbial mechanisms were partially elucidated with a view to understanding N-removing performance of an integrated Free Surface Flow (FSF) constructed wetland of the AIT WWT plant. Fluorescence *In Situ* Hybridisation (FISH) was the main technique used as a tool in the present study. Though employed extensively in the activated sludge and anaerobic digestion research, it is still to receive its dues in the field of NTS. Analysis of the N species dynamics in two wetland sections revealed an erratic pattern of the N-removal in the FSF wetland, the most biologically complex of all the wastewater treatment systems. The FISH analysis (12 probes) provided insights into microbial community structures and revealed profound changes occurring microbiologically when one moves downstream in the wetland as well as seasonally. The profiling of the wetland appears to corroborate biogeochemical monitoring data and indicate that the FSF wetland is in fact characterized by the unfavourable anaerobic/anoxic conditions. It was concluded that one of the main factors contributing to its malfunction is improper harvesting of the vegetation in the wetland channels. Hence FISH technique linking biogeochemistry and microbiology can be considered as a useful tool for monitoring and trouble-shooting of constructed wetlands.

Keywords: Integrated natural treatment systems; constructed wetland; FISH technique; N-removal

1. Introduction

Integration of Natural (wastewater) Treatment Systems, NTS (waste stabilization ponds, wetlands *etc.*) with each other as well as with advanced unit processes (biofiltration) offers NTS a second lease of life. Long-term full and pilot scale experience in South Africa and Thailand have convincingly shown that, contrary to a common view, a low tech N-removal from municipal and light industrial wastewater is a reality for a developing community (Polprasert and Koottatep, 2004). The high treatment efficiency of the integrated systems was ascribed to interplay of N-related redox processes complementing each other seasonally

(Shipin et al., 2004, Baptista et al., 2003). The presented FISH-based approach to microbial community structures is a pioneering effort in the field of NTS. It endeavours to establish interrelationships between groups (nitrite- and ammonia-oxidizers, denitrifiers, *etc*) within various eco-niches of the integrated systems and link them to the treatment performance.

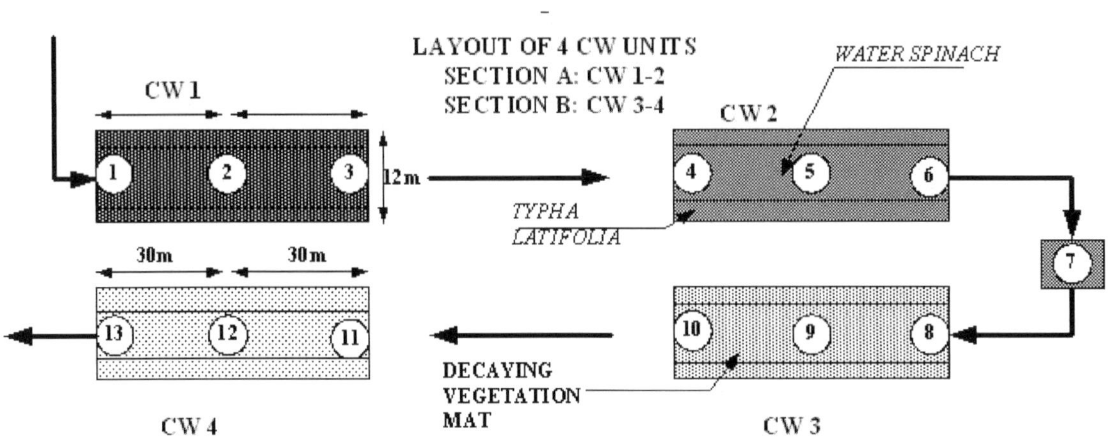

Figure 1. Layout of sampling points (1-13) at the AIT Free Surface Flow constructed wetland (CW) system. Wetland A (CW1-2); Wetland B (CW 3-4)

The main aim of the study was to gain understanding of the microbiological mechanisms underlying erratic and irregular N-removing performance of the FSF constructed wetland. Biogeochemical investigation of the 4 unit FSF wetland downstream of 4 pond series (integrated AIT WWT system) provided insights into its microbial phenomena. Culture-independent techniques, FISH- and PCR-based detection (not used in the study), developed over the last two decades have revolutionized our understanding of microbial processes *in vivo*. However, these powerful techniques are yet to be utilized by the researchers investigating Natural Treatment Systems. In reality these are being extensively used for studies of by far more microbiologically uniform and straightforward processes (e.g. activated sludge, anaerobic digestion, *etc*).

2. Materials and methods

Designed as a Free Surface Flow *Typha*-vegetated constructed wetland (FSF CW) the AIT CW is integrated with 4 upstream waste stabilization ponds. CW ADWF 410 m^3/day and AWWF = 800 m^3/day (rainy season: April-September), average $T_{CWwater}$= 28^0C, N-loading rate: 0.052 kg TN/m^3.day; dimensions of each unit: 60 x 12 x 0.7 m; V_{total} = 2000 m^3 (4 CW units), $HRT_{dry\ season}$ = 5.0 days ($HRT_{rainy\ season}$ = 2.5 days). To simplify analysis and interpretation of the monitoring results the wetland system was treated as consisting of two sections (section A: CW units 1-2; section B: CW units 3-4) (Figure 1). All biogeochemical analyses were performed in accordance with the Standard Methods (APHA, 1995). Sampling for the FISH analysis of the microbial community structures was undertaken monthly. Due to a heterogeneous nature of the free surface flow wetland in which microorganisms are attached to decaying vegetative mats, silt, sludge, plant roots, *etc* samples from the depths of 10-60 cm were proportionately averaged by homogenization to ensure that the representative microbial communities are reported. FISH technique data was generated according to the protocols described by Neef *et al.* (1996). Conventional Olympus microscope (Japan) with a reflected light fluorescence attachment was used. All FISH counts for 11 specific probes were expressed as percentages of EUB338 (probe-related counts (all *Eubacteria*). *Archaea* (probe Arch915), strictly anaerobic methanogens, were counted as % of DAPI counts.

Table 1. Major parameter monitoring data (averaged) of the AIT wastewater treatment (WWT) plant during the dry season and rainy season (in brackets).

PARAMETER	AIT WWT plant influent	Facultative pond influent	Wetland influent	Wetland effluent
DO (mg /L)	1.2	0.5	6.2	0.2
COD (mg /L)	160.0 (150.7)	112.0 (92.5)	80.0 (61.8)	56.0 (33.0)
NH_4^+ - N (mg/L)	43.0 (30.0)	26.0 (24.0)	13.3 (15.8)	3.1 (20.4)
Organic N (mg /L)	19.2 (20.6)	14.3 (12.5)	5.1 (6.1)	6.0 (8.9)
NO_3^- - N (mg/L)	4.3 (2.8)	1.8 (1.0)	3.6 (3.0)	9.4 (0.5)
S^{2-} (mg/L)	4.3	8.7	1.2	12.5

3. Results and discussion

Biogeochemical monitoring of the constructed wetland demonstrated that, on one hand, over the period from April to June 2004 the wetland system did not perform satisfactorily in terms of the removal of ammonia released from the upstream pond system (Figure 2 A,B). An average NH_4^+-N value in the wetland influent was 15.8 mg/L, while the wetland effluent value was 20.4 mg/L. No nitrification was observed with organic N even increasing in wastewater from 6.1 to 8.9 mg/L while passing through the wetland. The period loosely coincided with the rainy season ($HRT_{wetland}$ = 2.5 days). Overall, the sum of the main N species (ammonia, nitrate and organic N) entering and leaving the system was 24.9 and 29.8 mg/L, respectively (net increase: 4.9 mg/L). Amount of nitrite was negligible (and therefore not shown).

On the other hand, it was observed that at least during the period, which appears to roughly coincide with the dry season (January-April 2004; $HRT_{wetland}$ = 5.0 days), the wetland system showed an improved removal of N. The sum of the main N species (ammonia, nitrate and organic N) entering and leaving the system was 25.0 and 18.5 mg/L, respectively (net removal: 6.5 mg/L). As a possible explanation of the phenomenon one could suggest a longer retention time allowing more substantial microbial nitrifying action (as compared to a shorter rainy season $HRT_{wetland}$ = 2.5 days). DO profiling (not shown) indicated that aerobic conditions prevailed only at the very surface where DO concentration never exceeded 1.0 mg/L. Oxygen concentration sharply decreased to anoxic conditions below several centimetres of the water. Strata below 50 cm were characterized by the strongly anaerobic conditions. Low concentration of sulphide (as S) in the microalgae-rich wetland influent (pond effluent) and the high sulphide concentration in the final wetland effluent confirm it (Table 1). Extensive anaerobiosis appeared to prevent efficient nitrification or nitrite formation, the processes based on the production of molecular N from ammonia and critical for the N removal.

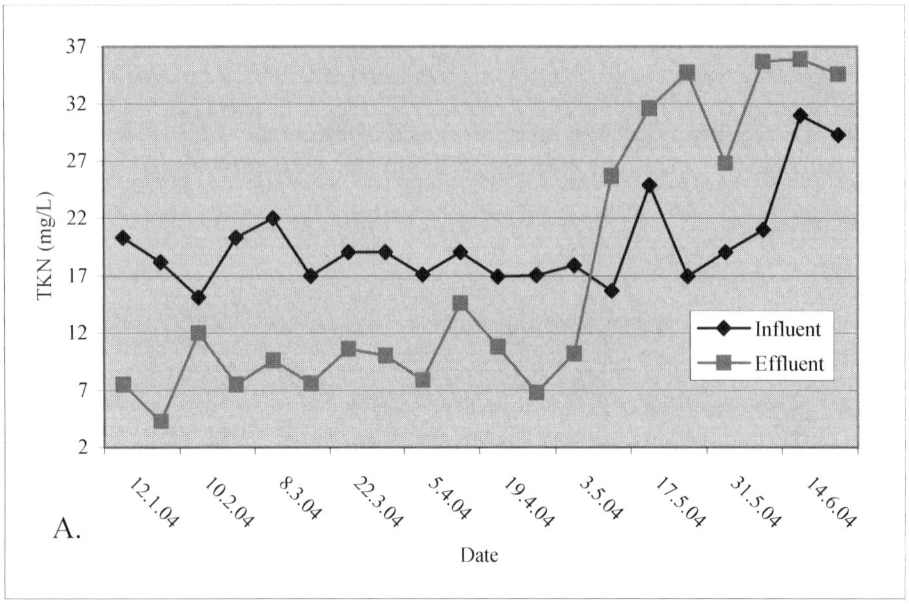

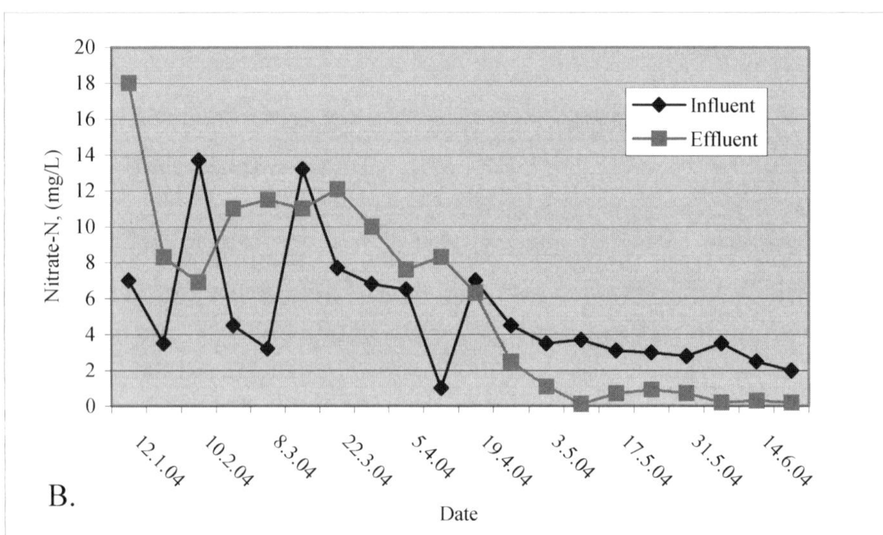

Figure 2A, B. Nitrogen species dynamics in the constructed wetland. Wetland effluent and influent levels of TKN (A) and nitrate-N (B) during the dry and rainy seasons.

FISH analysis employing 11 different probes (Table 2, Figure 3) have revealed profound changes occurring microbiologically when one moves downstream in the wetland sections, namely, from section A to B. It was observed that during the period coinciding with the dry season numbers of bacteria belonging to the major taxa, α- and β-proteobacteria (comprising critical nitrification-related bacteria: ammonia-oxidizers, specific to the Nso1225 probe and nitrite-oxidizers, the NIT3 probe) decrease. This indicated the progressively increasing environmental pressure on nitrification. The trend is by far less pronounced during the period coinciding with the rainy season. The much less significant negative shift in the percentages in the rainy season may indicate a low nitrifying activity (both ammonia and nitrite oxidation), which is incidentally reflected in a drastically lower nitrate concentration in the wetland effluent (Figure 2 B). The rRNA-targeted oligonucleotide probe, Ntspa662, specific to *Nitrospira spp.*, a bacterial group not phylogenetically related to *Proteobacteria* follows the trend, though in a less pronounced manner.

Table 2. Microbial community structure of the wetland bacterial consortia as revealed by the FISH analysis. Averaged samples of the 1st wetland section CW A (units 1-2) and 2nd section CW B (units 3-4).

rRNA-targeted oligonucleotide probes and targeted bacteria (FA, formamide % in hybridization buffer)	Percentage of EUB 338 probe-related counts (%)			
	Dry season period*		Rainy season period**	
	CW A (units 1-2)	CW B (units 3-4)	CW A (units 1-2)	CW B (units 3-4)
ALF1b (20,FA) α-proteobacteria (incl. common denitrifiers and nitrite-oxidisers)	36.3	8.8	32.3	19.0
NIT3 (40, FA) α-proteobacterial nitrite-oxidizers *Nitrobacter spp*	4.3	1.8	2.4	2.9
Par651 (35, FA) *Paracoccus spp.* common α-proteobacterial denitrifier	5.6	3.2	4.7	4.3
Rrp1088 (40, FA) common α-proteobacterial denitrifiers***	7.0	2.6	4.9	3.5
Bet42a (35, FA) β-proteobacteria	18.0	10.5	21.5	21.9
Nso1225 (35, FA) β-proteobacterial ammonia-oxidizers (*Nitrosomonas spp. etc*)	3.2	1.8	1.7	1.5
Gam42a (40, FA) γ-proteobacteria	7.1	3.0	5.1	4.0
Ntspa662 (35, FA) unrelated nitrite-oxidizers *Nitrospira spp.*	1.6	1.8	1.1	1.5
HGC69a (35, FA) High G+C *Actinomycetes*	5.1	1.8	7.5	2.9
CF319a (35, FA) *Cytophaga/Flavobacterium* cluster	5.2	3.5	6.5	4.4
Not identified bacteria	6.6	61.2	12.3	33.1

*January-April 2004, ** May-June. *** *Rhodovulum, Roseobacter, Paracoccus*, etc.

Prominent role of *β*-proteobacterial (possibly) aerobic ammonia-oxidizers and α-proteobacterial nitrite-oxidizers testifies to a substantial level of nitrification occurring in the channels. If it is aerobic or anaerobic remains to be seen and is a subject of the further investigation. However, common denitrifiers (*Rhodovulum sp, Roseobacter sp, Paracoccus sp.*) tend to decrease in numbers (Table 2) both in the 2nd (B) part of the wetland (coinciding with a decreased denitrification, Fig. 2, Table 1) and in the rainy season (compared to the dry season with a higher denitrification. Counts of γ-proteobacteria (comprising important heterotrophs and denitrifiers *Pseudomonas spp.* appear to decrease like-wise. The rRNA-targeted oligonucleotide probe CF319a indicating phylogenetic affiliation to the *Cytophaga-*

Flavobacterium phylum and therefore specific to such physiologically diverse heterotrophs and environmentally important gram-negative aerobic taxa as *Cytophaga spp.*, *Flavobacterium spp.* and *Bacteroides spp.* have been used to follow the population dynamics of these bacteria primarily common in the oxic water ecosystems. *Cytophaga spp.* is one of the most prominent obligately aerobic cellulose degraders in nature, while *Flavobacterium spp.* is a common heterotrophic obligate aerobe. The FISH profiles of both sections of the wetland (A and B) show that numbers of these primarily aerobic bacteria decrease in the section B compared to the section A which testifies to the increasing anaerobiosis in the downsream (2nd) part of the wetland.

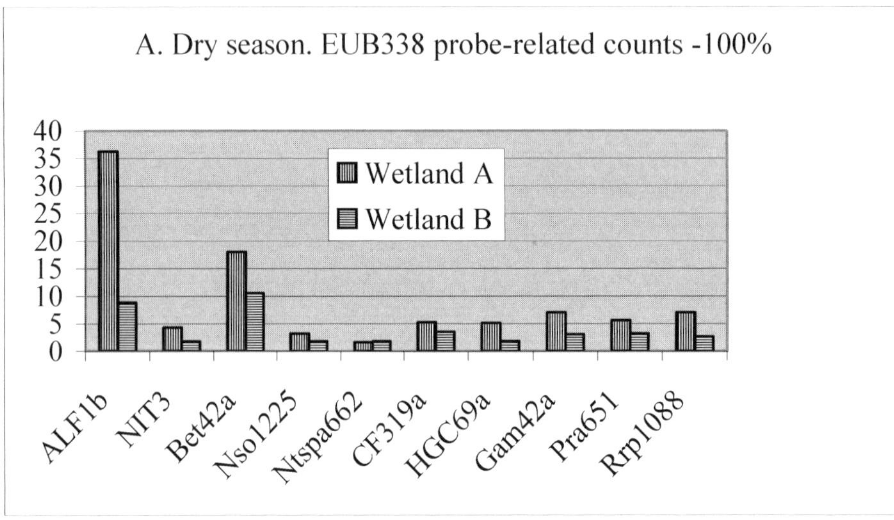

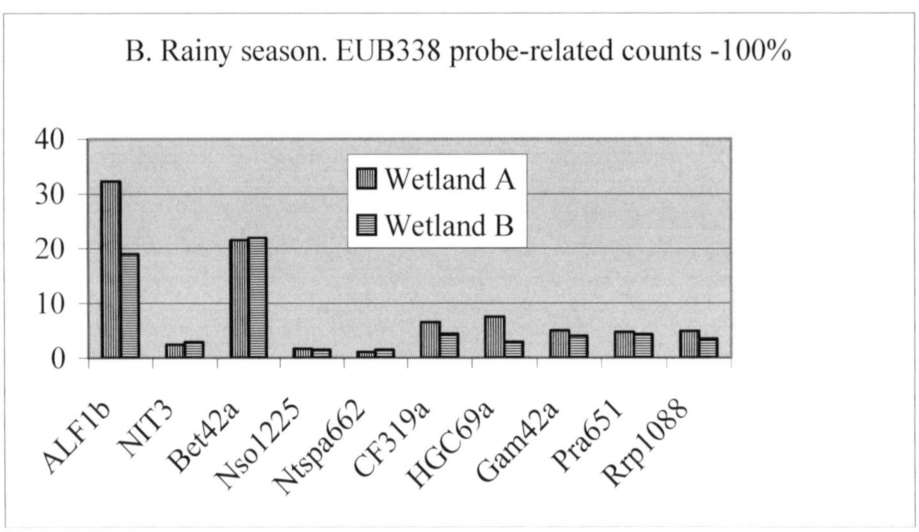

Figure 3A, B. Graphic representation of Table 2. Microbial community structures based on averaged percentages of probe-related bacterial counts. Section A (wetlands CW1 and CW2) versus section B (wetlands CW3 and CW4). Dry season period: January-April 2004. Rainy season period: May-June.

Probe HGC69a was used to profile the oxic/anoxic strata of both sections of the wetland. The probe indicates phylogenetic affiliation to the *Actinomycetes* which are high G+C gram-positive heterotrophs common mostly in the oxic environments similar to soil. The FISH profiles of both sections of the wetland (A and B) show that numbers of these primarily aerobic microorganisms decrease in the section B compared to the section A which further testifies to the increasing anaerobiosis in the downstream part of the wetland (Table 2, Figure 3 A,B).

Archaea counts (probe Arch915, strictly anaerobic methanogens, not shown) tend to increase in the rainy season in comparison to the dry season, as well as in the wetland A in comparison to the wetland B, further confirming onset of anaerobiosis.

Judging by the numbers of unidentified bacteria (i.e. those not belonging to the taxa specific to the 12 probes used) one may conclude that these yet unidentified taxa appear in the wetland section B in greater numbers in comparison to the section A. These groups are yet to be studied in future through the use of additional oligonucleotide probes.

Overall, it can be concluded that molecular microbiological profiling of the wetland (Table 2, Figure 3 A,B) appears to corroborate biogeochemical monitoring data (Table 1, Figure 2) and indicate that the FSF wetland, which due to its nature should be characterized by aerobic/anoxic conditions, is in fact currently characterized by the anaerobic/anoxic conditions. Since the organic (both carbonaceous and nitrogenous) load on the system is low, the reason for the anaerobic shift and ensuing environmental stress should be sought elsewhere: different hydraulic regime during the period of inferior N-removal (rainy season, $HRT_{wetland}$ = 2.5 days) may offer only a partial explanation.

To address the issue an investigation of the stratification of the wetland channels was undertaken. It allowed appraising of the physical state of the FSF wetland and it was found as a result that due to the accumulation of decaying vegetation, extensive underwater mats have been formed. Presence of the mats providing attachment medium as well as nutrient substrate for anoxic and anaerobic microbial consortia changed the nature of the processes in comparison to those, which were to occur in well-maintained free surface flow wetland channels. Therefore it would be more appropriate to refer to the AIT wetland in its present state as operating not in the FSF mode but rather in the mixed horizontal subsurface/free surface flow mode.

It is important to note that the latest improper vegetation harvesting has taken place in between dry and rainy seasons which made the microbial community shift appear as being caused by the shift in HRT.

Due to the improper management, the cattail *(Typha sp.)* plant biomass harvested irregularly at the wetland was left in the wetland channels where it has inevitably rotten which led to a significant anaerobiosis. As a result plant organic matter (protein) decomposed with a concomitant sulfate reduction. If $[SO_4^{2-}]$ above 40 mg/L was measured in the wetland influent the high sulfide concentration, $[S^{2-}]$ = 12 mg/L was routinely measured in the final CW effluent (Table 1). Since the wetland effluent is actually the final effluent of the entire WWT plant discharging into the nearby large canal (Khlong Neung), the environmental soundness of the existing situation is questionable.

Nevertheless, irregular but sometimes fairly reasonable (under the circumstances) N-removal takes place with overall 6.5 mg N/L removed from the influent (approx. 0.014 kg TN/m^3/day). Analysis of the biogeochemical monitoring in conjunction with the molecular microbiological data leads us to a preliminary conclusion that one of the main factors contributing to the malfunction of the AIT wetland performance is the improper harvesting of the vegetation in the wetland channels.

4. Conclusions

To the best of our knowledge there exists no available published information on the FISH-generated microbial structures of the constructed wetlands (as well as other NTS) and therefore the study was the first such approach to the microbial mechanisms underlying biogeochemistry of these ecosystems. The data obtained allowed to link the shifts in microbial populations to the changes in performance (rather non-performance, in this case) in terms of the removal of nitrogenous compounds from wastewater. The findings confirm that such molecular microbiological technique as Fluorescence In Situ Hybridization can be a useful tool for the monitoring and trouble-shooting of the performance of constructed wetlands and, potentially, other Natural Treatment Systems. However, additional research is needed to realize its full potential.

5. Acknowledgements

Authors gratefully acknowledge support provided by the University of Tokyo, Japan as well as help and advice of Dr H. Satoh and Dr X. Wang.

6. References

APHA (1995). *Standard Methods for the Examination of Water and Wastewater.* 19th edn. American Public Health Association/AWWA/Water Environment Federation, Washington DC, USA.

Baptista J.D.C., Donnelly T., Rayne D. and Davenport R.J. (2003) Microbial mechanisms of carbon removal in subsurface flow wetlands. *Water Sci. Techn.*, **48**(5), 127–134.

Neef A., Zaglauer A., Meier H., Amann R., Lemmer H. and Schleifer K.-H. (1996). Population analysis in a denitrifying sand filter: conventional and *in situ* identification of *Paracoccus spp.* in methanol-fed biofilms. *Appl. Env. Microbiol.,* **67**(12), 5810-5818.

Polprasert C. and Koottatep T. (2005). Integrated pond/wetland and pond/aquaulture systems. Chapter in: *Pond treatment technology* (ed. A. Shilton) Integrated Environmental Technology series, International Water Association, London.

Shipin O., Koottatep T., Nguyen Tran Thien Khanh and Polprasert C. (2004). Integrated Natural Treatment Systems for developing communities: low-tech N-removal through the fluctuating microbial pathways. *International Water Association Specialist Conference on Waste Stabilization ponds and Wetlands.* 28 Sept-1 Oct 2004, Avignon, FRANCE.

Multiple Evaluation of Soil Aquifer Treatment for Water Reclamation using Instrumental Analysis and Bioassay

Norihide Nakada[1)], Naoyuki Yamashita[1)], Kiyoshi Miyajima[1)], Yutaka Suzuki[1)], Hiroaki Tanaka[2)], Hiroyuki Shinohara[3)], Hideshige Takada[3)], Nobuyuki Sato[4)], Mikio Suzuki[4)], Mitsuaki Ito[4)], Fumiyuki Nakajima[5)] and Hiroaki Furumai[6)]

1) Water Environment Research Group, Public Works Research Institute
1-6, Minamihara, Tsukuba, Ibaraki, 305-8516, Japan e-mail: nakada55@pwri.go.jp
2) Research Center for Environmental Quality Control, Kyoto University
1-2, Yumihama, Otsu, Kyoto, 520-6222, Japan e-mail: htanaka@biwa.eqc.kyoto-u.ac.jp
3) Institute of Symbiotic Science and Technology, Tokyo University of Agriculture and Technology
Fuchu, Tokyo 183-8509, Japan e-mail: shige@cc.tuat.ac.jp
4) METOCEAN Environment Inc.
1334-5, Riemon, Ooigawa, Sida, Sizuoka, Japan e-mail: nobu@notes.metocean.co.jp
5) Research Center for Advanced Science and Technology (RCAST), the University of Tokyo
4-6-1 Komaba, Meguro, Tokyo 153-8904, Japan e-mail: nakajima@env.t.u-tokyo.ac.jp
6) Department of Urban Engineering, the University of Tokyo
7-3-1, Hongo, Bunkyou, Tokyo, 113-8656, Japan e-mail: furumai@env.t.u-tokyo.ac.jp

Abstract

The use of rainwater, groundwater and treated wastewater is helpful for sustainable utilization of water resources, especially in urban area. In this study, we performed soil column experiments in consideration of soil aquifer treatment of treated wastewater. The column was packed with surface soil and treated wastewater was fed into the column in a down-flow mode at a constant flow rate for 80 days. Water quality of the influent and effluent was evaluated by instrumental analysis and three types of bioassay. As a result, heavy metals were sufficiently removed from the column influent. The removal efficiency of nonylphenol, which has low biodegradability, showed a slowly declining tendency, whereas, that of estrogens and estrogenic activity was stably high. Total phosphorus and phosphate-phosphorus were effectively removed. Beside this, dramatically reduction of the concentration of ammonium-nitrogen was observed and subsequent increase in the concentration of nitrate-nitrogen was observed, thus suggesting groundwater pollution by nitrite may be concerned. In the results from bioassay, obvious inhibition for algal growth was observed for the effluent, not for marine bacteria. The results of our study suggested the soil aquifer treatment for the treated wastewater was effective for removal of inorganic phosphorus, heavy metals and estrogens.

Keywords: bioassay, instrumental analysis, reclamation, soil-aquifer penetration treatment, soil column experiment, wastewater treatment

1. Introduction

Anticipating shortage of water resources in the future of rapidly populated cities in Southeast Asian countries, a water recycle and reuse system in urban areas should be established. For this purpose, use of rainwater or groundwater and reuse of wastewater, including treated wastewater, are very prospecting. It is, however, very important that considering reuse of wastewater from the viewpoint of not only quantity but also quality in terms of risk that could rise under difference conditions of water use. Our research focused on soil aquifer treatment (SAT) of treated wastewater (secondary effluent from municipal sewage treatment plant) in order to recharge groundwater and establish a new water recycling system in urban areas.

In this research we conducted an experiments with columns packed with soil to simulate the SAT. Secondarily treated wastewater was used as influent of the column experiment, and influent and effluent quality was evaluated by instrumental analysis and bioassay. Conventionally, the quality of river water or groundwater is evaluated in terms of nutrients or conventional water quality parameters, such as biochemical oxygen demand (BOD) or chemical oxygen demand (COD_{Mn}). In our research, the water samples were measured for organic micro-pollutants, such as endocrine disrupters and estrogens, which recently draw much attention internationally, in addition to the conventional water quality parameters. The total estrogenic activity was also measured by DNA recombinant yeast. Furthermore, bioassay was conducted using bacteria and algae that are important constituents of aquatic ecosystem in order to evaluate the influence of the treated water on the ecosystem.

The main purpose of our research is to understand and evaluate changes of water quality in the process of soil penetration of treated wastewater for recharge of groundwater, one of water resources in the urban area, by instrumental analysis and bioassay.

2. Materials and methods

2.1 Soil Column Experiment

The columns used for the experiment are of stainless steel, measuring 200 mm in internal diameter and 600 mm in length. A stainless steel filter is attached to the bottom of each column. Water is fed to the column by a roller pump via a TYGON tube. The column influent and effluent are stored in a stainless steel tank.

Soil used in the experiment was sampled at a park in Tokyo Metropolis from depths of 90 to 120 cm (Ando sols) from the surface. The sample was taken from loam layer. The sampled soil was dried in air and sieved with 2 mm and 7 μm mesh screens to obtain soil particles of 75 μm to 2 mm in diameter. Then, the sieved soil was packed in two columns to a depth of 20 cm and 50 cm, respectively (each weighing 5 kg and 12.5 kg, respectively).

Prior to the soil column experiment using treated wastewater, groundwater was intermittently fed to the two columns at a flow rate of 1 to 5 mL/min to a total of about 60 L in order to review conditioning of the columns and optimization of the flow rate (48 to 240 mm/day).

The columns used for soil aquifer treatment experiment were those used in the preliminary experiment (groundwater-saturated soil columns). Secondary effluent from a municipal sewage treatment plant having activated sludge process was collected as influent every one to seven days from January to March, 2003. The column influent was sampled one day before it was fed to the columns and was chlorinated (in 24 hours) to have a chlorine concentration of 3 mg/L by adding sodium hypochlorite. The chlorinated column influent was then dropped to each column at a rate of 5 mL/min (240 mm/day) by a roller pump. All the experiments were conducted at a constant temperature room (20 °C). The column influent was continuously fed to the columns for a total of 80 days. About 15 L of column effluent is necessary for analysis of many items, including bioassays, in our research. Thus, the effluent water was stored for two days to gather the sufficient amount for sample and then periodically sampled for analysis.

2.2 Instrumental Analysis

The temperature, pH, electric conductivity, and turbidity of the column influent and effluent were measured at the experiment site. COD_{Mn}, BOD, total organic carbon (TOC), suspended solids (SS), dissolved organic carbon (DOC), nitrogens (NH_4^+-N, NO_2^--N, and NO_3^--N), and phosphorus were measured according to the standard method, JIS K0102 (1998). Heavy metals were measured by inductively coupled plasma / mass spectrometer (ICP/MS) according to the standard method JIS K 0102 (1998). These items were analyzed every other week.

Endocrine disrupters, including nonylphenol (NP), octylphenol (OP) and bisphenol A (BPA), and estrogens, including estrone (E1) and 17β-estradiol (E2), in the water samples were measured based on the previous report (Nakada *et al.*, 2004). In brief, the samples were filtered by GF/F glass fiber filter and the filtrate was extracted by Sep-Pak tC18 cartridge. The components adsorbed to the cartridge were eluted by methanol and the elution was fractionated and purified by silica gel chromatography for analysis by gas chromatograph / mass spectrometer (GC/MS). This analysis was made every two weeks.

2.3 Bioassay

To keep samples for bioassay stable until use, 10 mg/L and 1 g/L of ascorbic acid was added to the sample prepared for bacteria and algae assay and that for DNA recombinant yeast assay, respectively. Sampling and testing were conducted every two weeks in line with the measurements of endocrine disrupters and estrogens. The measurement methods for each test item are described below.

Bacteria Test Procedure (Microtox)

Samples for Microtox were filtered by a membrane filter (Advantec, 0.2 μm mesh) and the filtrate was stored in a frozen state at -30 °C until put to test. After unfreezing of the sample, a dilution series was prepared with sodium chloride solution. Luminescent marine bacterium *Vibro fischeri* was added to the dilution series and exposed for 5 minutes and 15 minutes. Light production after 5 min. and 15 min. exposure was measured with a photometer for Microtox test (AZUR Environmental, Microtox Model 500). EC50 value (concentration to inhibit 50% of light production) was calculated by comparing the result with the control.

Algal Growth Potential (AGP) and Algal Growth Inhibition (AGI) Test

Samples for AGP test were filtered by a glass fiber filter (Whatman, GF/B) and a membrane filter (Advantec, 0.2 μm mesh) and stored in a frozen state at -30 °C before put to test. After unfreezing, a dilution series was prepared. Then, the samples were exposed to algae (*Pseudokirchneriella subcapitata* (NIES-35)) in a 96-hole micro plate. *P. subcapitata* were subcultured in an AAP medium. The algae in a logarithmic growth phase, a few days after subculture, were put to use. Both exposure tests were conducted at a culture temperature of 24 °C and a shaking rate of 120 rpm and the initial cell concentration of algae was set at 1×10^4 cells/mL.

In the case of AGP test, a sample of 200 μL and a *P. subcapitata* cell suspension of 40 μL were put in each well on the micro plate. The light conditions were 4000 Lux of illuminance and a light-on/off cycle of 12 hours. During cultivation, absorbency (450 nm wavelength) was measured with a micro plate reader (ARVO SX-1420, Wallac) to monitor algal growth. The algae growth at a point in time when the amount of algae reached almost constant (approximately 2 weeks) was taken as AGP value.

In the case of AGI test, a sample of 200 μL, a *P. subcapitata* cell suspension of 20 μL, and an AAP medium of 20 μL were added to each on the plate. Light of 4000 Lux in illuminance was continuously radiated. The exposure time was 96 hours. Absorbency (450 nm wavelength) was measured with the micro plate reader during the cultivation to monitor algal growth. After the test, the algal growth inhibition rate was calculated from the area under the algal growth curve, while the EC50 value (concentration to inhibit 50 % of light production) was calculated based on the relationship between the sample concentration and the algal growth inhibition.

Yeast Estrogen Screen Assay
DNA recombinant yeast used for the assay was taken from the strain given by Prof. Sumpter, Brunel University, UK. The test was conducted according to the procedure used by Yakou *et al.* (1998). In brief, the sample was filtered with a GF/B glass fiber filter and the filtrate was extracted by Sep-Pak C18 cartridge. Components adsorbed by the cartridge were eluted by methanol. The elution components were condensed and resolved in dimethyl sulfoxide before assay. The EC50 value was calculated from the results, and estrogenic activity was calculated as a relative value for activity of 17β-estradiol, which is a positive control.

Table 1. Average value of each parameter of column influent during the soil column

Constituent	Unit	Concentration ± S.D.
Water quality parameter (n=11)		
pH	-	7.42 ± 0.2
EC	mS/m	59.1 ± 1.2
SS	mg/L	3.12 ± 0.9
COD_{Mn}	mg/L	8.93 ± 0.7
BOD	mg/L	2.74 ± 0.6
TOC	mg/L	5.73 ± 0.4
DOC	mg/L	5.42 ± 0.4
T-N	mg-N/L	35.1 ± 3
NH_4-N	mg-N/L	33.2 ± 4
NO_2-N	mg-N/L	0.02 ± 0.02
NO_3-N	mg-N/L	0.02 ± 0.01
T-P	mg-P/L	0.20 ± 0.05
PO_4-P	mg-P/L	0.11 ± 0.04
Heavy Metal (n=11)		
Cu	mg/L	0.006 ± 0.002
Zn	mg/L	0.021 ± 0.004
Ni	mg/L	0.003 ± 0.001
Mn	mg/L	0.022 ± 0.003
Endocrine disrupter (n=6)		
Nonylphenol	ng/L	133 ± 30
Octylphenol	ng/L	20.2 ± 3
Bisphenol A	ng/L	83.9 ± 20
Endogenous estrogen (n=6)		
Estrone	ng/L	16.1 ± 2
17ß-Estradiol	ng/L	3.10 ± 0.6
Estriol	ng/L	6.47 ± 2

3. Results and discussion

3.1 Instrumental Analysis
Table 1 shows the average values of the analysis results for the column influent measured during the experiment. As shown in the table, the results of influent quality analysis were stable.

Figure 1 shows the removal rate in the columns for all analysis items. Comparing with the changes of influent quality, a wide range of changes in removal rate was observed for many components. Among conventional water quality parameters, BOD and COD_{Mn} had the largest changes (-430 to 88 % and -110 to 100 %, respectively). Total

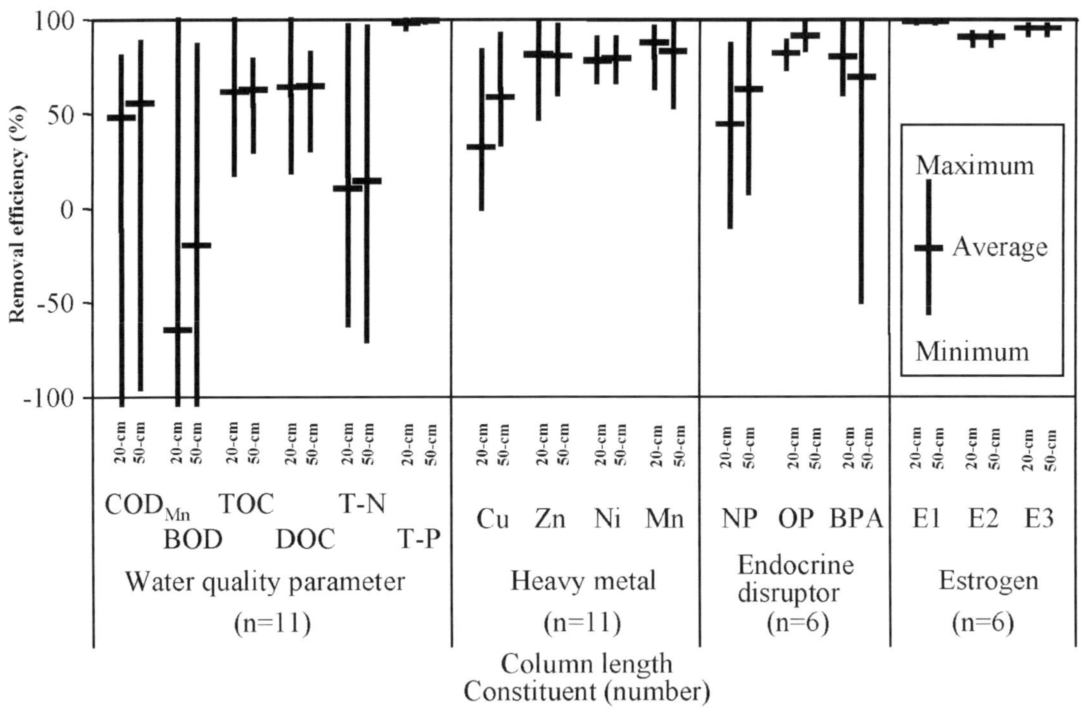

Figure 1. Averaged removal efficiency of each constituent with minimum and maximum value on the soil column experiment.

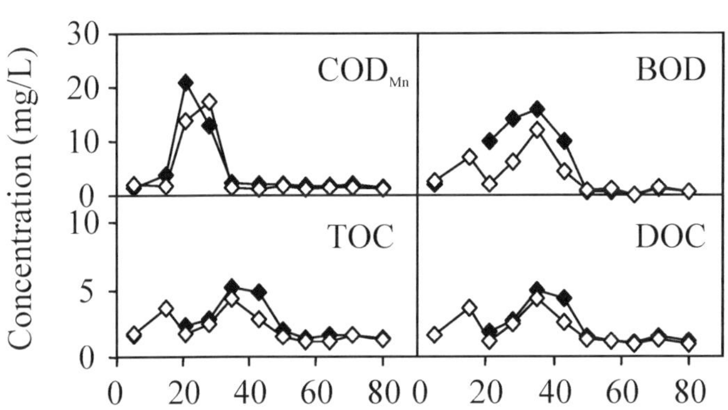

Figure 2. Breakthrough curves for COD_{Mn}, BOD, TOC, and DOC on the 20-cm column (closed symbol) and on the 50-cm column (open symbol). The results of BOD, TOC and DOC for 20 –cm column were not available.

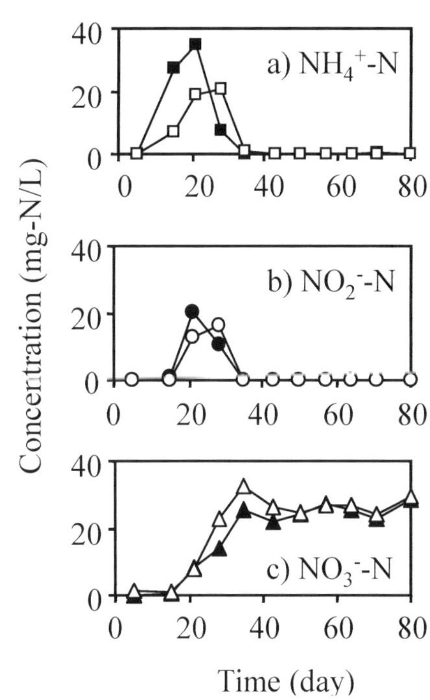

Figure 3. Breakthrough curves for NH_4^+-N (a), No_2^--N (b), and No_3^--N (c) on the 20-cm column (closed symbol) and on the 50-cm column (open symbol).

nitrogen removal rate also showed large changes. For total phosphorus, almost 100% removal was observed during the experiment. Significant removals of heavy metals (50 to 98 %) except copper and endocrine disrupters (44 to 98 %) were observed during the experiment. As with phosphorus, almost 100% removal of estrogens was achieved.

For BOD, COD_{Mn}, TOC, and DOC, their concentrations in the column effluent increased from both columns from the 20th to 40th day after the start of feeding (Figure 2). The concentration of COD_{Mn} and BOD in the effluent exceeded that of the influent at this period. These concentrations in the effluent decreased after the 50th day (Figure 2), and stable removal rates of COD_{Mn} and BOD were achieved (81 % and 69 %, respectively). The similar complicated behavior was observed for nitrogens (Fig. 3). From the 15th to 30th day of feeding, the concentrations of NH_4^+-N and NO_2^--N increased. A high correlation between the NO_2^--N concentration and COD_{Mn} changes was identified. Both ions began to decrease thereafter, while NO_3^--N increased, and a high concentration of NO_3^--N was kept until the end of the experiment (Figure 3). Eventually, the total nitrogens in the column effluent in the latter half of the experiment were almost all NO_3^--N. NH_4^+-N changed to NO_2^--N and then NO_3^--N. This transformation indicated the progress of nitrification in the columns during the experiment and furthermore suggests a transition of nitrifying microbial population.

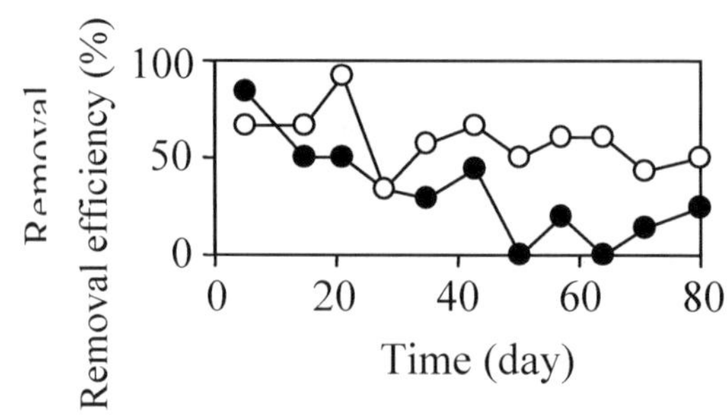

Figure 4. Removal efficiency of Copper on the 20-cm column trial (closed symbol) and on the 50 –cm column trial (open symbol).

A relatively high removal rates were observed for heavy metals (78 – 87 %) except for copper (Fig. 1). The removal efficiency of copper was relatively low and seemed to decrease with increasing amount of fed volume, especially for the 20-cm column trial (Figure 4). The faster decrease reflects that the shorter soil column had less sorption capacity.

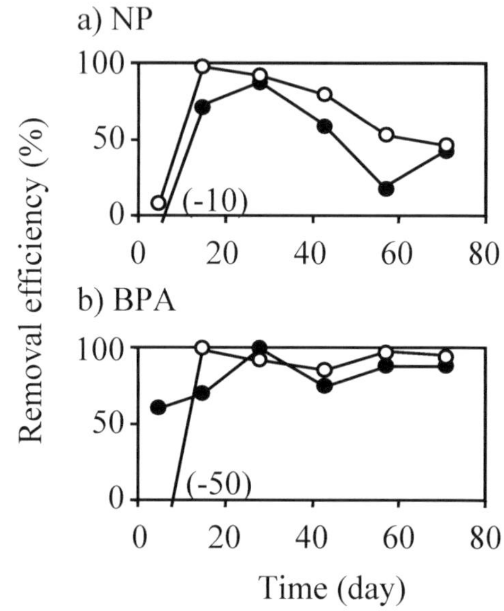

Figure 5. Removal efficiency of nonylphenol (NP:a) and bisphenol A (BPA:b) on the 20-cm column trial (closed symbol) and on the 50-cm column trial (open symbol)

For NP and BPA, the concentrations in the effluent from both columns exceeded those of the influent by the 10th day of feeding, which resulted in negative removal rate (Figure 5). Thereafter, the effluent concentrations of both the substances sharply decreased, which resulted in relatively high removal efficiency in the column. However, the effluent NP concentration made a gradual rise, resulting in gradual decrease of removal rate (Figure 5). This result suggests a reduction of the sorption capacity and low biodegradability of the columns. For BPA, a high removal rate was kept during the experiment except for the analytical data of the first sample (Figure 5). Effluent NP and BPA concentrations were unexpectedly high in the first sample, which could be caused due to desorption of these in possible contaminated soil.

For estrogens, high removal efficiency was stably observed during the experiment (95% or more). Because degradation of estrogens in soil based on a simple first-order kinetic process (Das *et al.*, 2004) was also suggested, soil aquifer treatment demonstrates to be effective in removal of estrogen from treated wastewater so far.

3.2 Results of Bioassay

No significant inhibition that exceeds EC50 was observed from the results of exposure of influent and effluent water in the test using luminescent marine bacteria.

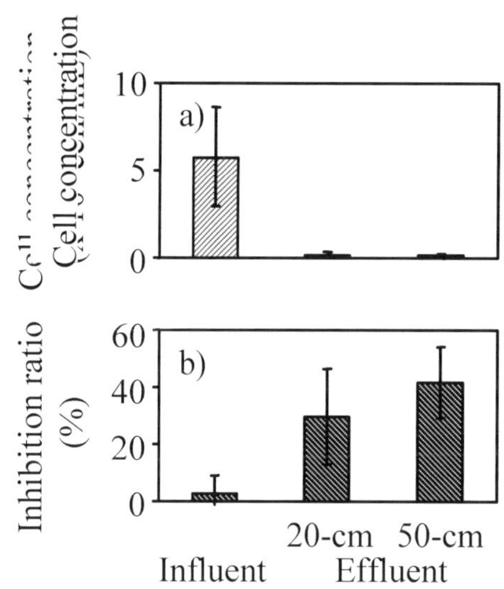

Figure 6. Results of ecological test of column influent and effluents with algal growth potential (a) and algal growth inhibition (b). Bars in this figure are mean of standard deviation (n=6).

In the AGP test, growth promotion by addition of the influent was observed. This phenomenon is presumed to be due to nutrients contained in the influent. In the exposure of the column effluent, no growth of algae was observed. In the AGI test, no sign of growth inhibition was seen in the influent, but a high level of inhibition was continuously observed in the column effluent (Figure 6). Judging from the fact that no sign of inhibition was observed in the column influent, it is suggested that eluted substances from the packed soil or chemical substances contained in the influent water could be denatured during the passage of the column to become inhibitors for algae. No reason for this inhibitor is clarified at this point.

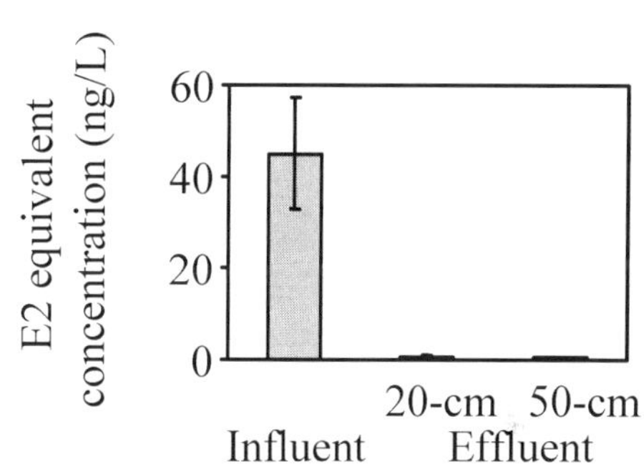

Figure 7. Estrogenic activities of the column influent and effluents with standard deviation (n=6).

In measurement of estrogenic activity using DNA recombinant yeast, a high level of activity was measured in the influent, while high level of activity was not detected in the effluent (Figure 7). According to the results of recent research (Nakada *et al.*, 2004), it is reported that estrogenic activity in the treated wastewater mostly comes from human-derived estrogen. These results are in good agreement with the results of instrumental analysis of estrogens. For activity of an extremely low concentration often observed in the effluent, no increase or decrease associated with the progress of the test was confirmed. In the results from bioassay, significant differences between the 20-cm and 50-cm column trials were not observed.

4. Conclusions

1) We conducted the SAT experiments using 20-cm and 50-cm columns packed with Ando sols collected from -90 to -120 cm in depth. The treated wastewater was chlorinated and dropped to each column at a rate of 5 mL/min for 80 days (240 mm/day). The experimental results revealed that high removal efficiencies were achieved for phosphorus, estrogens, BPA, estrogenic activity and, most of heavy metals as well as conventional COD_{Mn}, BOD, TOC and DOC on 60 days after operation. Penetration period from 30 to 50 days was necessary to achieve stable removal rates for COD_{Mn}, BOD, TOC and DOC.

2) The stable removal of NH_4^+-N was observed after the 40th day of the experiment, while low removal efficiency of total nitrogen was reconfirmed because of the increase in the

concentration of NO_3^--N in the column by nitrification. In addition, the removal of copper was not effective.
Decrease in removal efficiency was observed for NP with the increase in the amount of fed volume. For NP and copper, slight differences in the removal efficiency between the 20-cm and 50-cm column trials were observed. These results suggest the difference of sorption capacity of each column.

3) As a result of the bioassay study, no sign of inhibition for marine bacterial growth was confirmed either in the column influent or effluent. On the other hand, inhibition of algal growth was detected for the column effluent, suggesting that toxicant(s) for algae were eluted from the packed soil. It is necessary to identify substances that inhibited algal growth and to clarify condensation of chemicals in the soil and changes in microbial population in the column.

4) In this study, STP effluent was used as the column influent. Since the sample had lower concentration of suspended solids (Table 1) than that of raw sewage, clogging of the columns were not observed through the experiment. Furthermore, significant differences between the 20-cm and 50-cm column trial were not observed in the results both from instrumental analysis, except for NP and copper, and from bioassay. However, the change in the removal efficiency caused by differences in the length of packed soil layer, flow rate, and experiment period should be investigated in the future work because of limitation of experiment period and conditions. Although appropriate additional treatment, especially for the removal of nitrogen and copper, may be necessary for secondary effluent, the SAT system are expected to be an effective alternative to brush up the treated wastewater under appropriate design condition, which seems suitable to the developing countries seeking for both water resources and water quality managements through surface water and groundwater system.

5. Acknowledgement

The study was conducted as a part of a Core Research for Evolutional Science and Technology (CREST) entitled "Risk-based Management of Self-regulated Urban Water Recycle and Reuse System" in "Hydrological Modeling and Water Resources System" supported by Japan Science and Technology Agency.

6. References

Das B.S., Lee L.S., Rao P.S.C. and Hultgren R.P. (2004). Sorption and degradation of steroid hormones in soils during transport: Column studies and model evaluation. *Environ. Sci. Tech.*, **38** (5), 1460-1470.

Nakada N., Nyunoy H., Nakamura M., Hara A., Iguchi T. and Takada H. (2004). Identification of estrogenic compounds in wastewater effluent. *Environ. Toxicol. Chem.* **23** (12), 2807-2815, 2004.

Testing methods for industrial wastewater (1998). JIS 0102, Japanese Standards Association, Tokyo. (in Japanese).

Yakou Y., Takahashi A., Higashitani T. and Tanaka H. (1999). Measurement of environmental estrogen-like activity in wastewater using recombinant yeast. *Environ. Engineer. Res. Jap. Soc. Civil Engineer.*, **36**, 199-208.

Water-soluble Organic Micro-Pollutants in Municipal Wastewater and their Removal During Advanced Treatment

H. Shinohara[1)], T. Tanishima[1)], S. Kojima[1)], S. Managaki[1)], H. Takada[1)], N. Nakada[2)], H. Tanaka[3)], F. Nakajima[4)] and H. Furumai[5)]

1) Institute of Symbiotic Science and Technology, Tokyo University of Agriculture and Technology
Fuchu, Tokyo 183-8509, Japan e-mail: shige@cc.tuat.ac.jp
2) Water Environment Research Group, Public Works Research Institute
1-6, Minamihara, Tsukuba, Ibaraki, 305-8516, Japan e-mail: nakada55@pwri.go.jp
3) Research Center for Environmental Quality Control, Kyoto University
1-2, Yumihama, Otsu, Kyoto, 520-6222, Japan e-mail: htanaka@biwa.eqc.kyoto-u.ac.jp
4) Research Center for Advanced Science and Technology(RCAST), the University of Tokyo
4-6-1 Komaba, Meguro, Tokyo 153-8904, Japan e-mail: nakajima@env.t.u-tokyo.ac.jp
5) Department of Urban Engineering, the University of Tokyo
7-3-1 Hongo, Bunkyo, Tokyo 113-8656, Japan e-mail: furumai@env.t.u-tokyo.ac.jp

Abstract

To evaluate the potential of recycling and relocation of urban wastewater, broad-spectrum analysis of water-soluble organic micro-pollutants were conducted for secondary effluents from municipal sewage treatment plants in Tokyo, Japan. GC-MS and LC-MS/MS analysis demonstrated that the secondary effluents contained synthetic-detergent-derived pollutants (linear alkylbenzenesulfonates: LAS and fluorescent whiting agents: FWAs) with concentration range from hundreds ng/L to tens μg/L, phenolic endocrine disrupting chemicals (alkylphenols and bisphenol A) with tens to hundreds ng/L, natural estrogens (E1, E2 and E3) with sub ng/L to tens ng/L, and 16 species of pharmaceutical chemicals (e.g., crotamiton, triclosan, mefenamic acid) with sub ng/L to hundreds ng/L. Behaviors of the water-soluble pollutants during sand filtration and ozonation of the secondary effluent were studied under actual sewage treatment conditions. Ozonation proved to remove natural estrogens, FWAs and some pharmaceuticals with removal efficiency of more than 90 %. As a result of drastic decrease in E1 and E2 concentration by ozonation, calculated estrogenic activities in the effluents after ozonation were extremely low (<0.6 ng-E2 equivalent concentration/L), while those in the secondary effluents were 7.6 – 14.7 ng/L which were close to the threshold to cause endocrine disruption to aquatic organisms. The decreases in estrogenic activities were consistent with bioassay using DNA recombinant yeast.

Keywords: endocrine disrupting chemicals, LC-MS/MS, ozonation, pharmaceuticals, sand-filtration, synthetic detergent

1. Introduction

Growing in sizes and activities of cities increases demand for water resources. However, introduction of water from upper stream of rivers to the cities is sometimes associated with environmental problems (e.g., destruction of natural ecosystem through construction of dam). Facilitating recycling of water within the cities is one of the options to solve such problems. Recycling of the wastewater also could reduce the amounts of wastewater discharged to nearby rivers and coastal zones and contribute improvement of the water quality. For reuse and recycling of the wastewaters generated in the cities, understanding of the water quality is essential to relocate the reclaimed water to appropriate purposes. Among various kinds of wastewater generated in cities, the present study focused on effluents from sewage treatment plants (STPs) and the water quality was characterized to evaluate their potential for reuse and recycling.

To date, water quality of municipal wastewater has been evaluated in terms of conventional water quality parameters (i.e., suspended solid, bulk organic matter measured as COD and BOD or TOC, and nutrients). However, occurrence of water soluble organic micropollutants have been well documented (Alder *et al.*, 1997). Especially, recent papers demonstrated that secondary effluents from municipal STPs contain various man-made chemicals (e.g., nonylphenol and bisphenol A) which can mimic or modulate endogenous hormones (Jobling and Sumpter, 1993). These chemicals are collectively termed "endocrine disrupting chemical" (EDCs). Also significant concentrations of natural estrogens (i.e. estradiol: E2, estrone: E1 and estriol: E3) were detected in the secondary effluents (Desbrow *et al.*, 1998). Some of these compounds have been identified as contributors to estrogenic activity in environmental waters and to cause endocrine disruption in aquatic animals (Harries *et al.*, 1997; Desbrow *et al.*, 1998; Nakada *et al.*, 2004). The presence of the EDCs limits the reuse of the effluents for some purposes (e.g., water for artificial stream with keeping fishes). More recently, wide-spread distribution of pharmaceuticals and personal care products (PPCP) have been reported in sewage effluents and rivers (Ternes, 1998; Kolpin *et al.*, 2002). Although their effects on aquatic animals and human are still unknown, detection of the PPCP may raise concern on the recycling of the effluents. In these contexts, the present study conducted broad-spectrum analysis of secondary effluents for comprehensive understanding of the concentration range of water-soluble organic micro-pollutants in effluents.

Because of the presence of the anthropogenic compounds in the secondary effluents, the reuse and recycling of the effluents is limited. To improve water quality of secondary effluents, advanced treatments have been studied and implemented in some sewage treatment plants. There have been some studies on the removal of water-soluble organic micro-pollutants including estradiol (Alum *et al.*, 2004), bisphenol A (Lee *et al.*, 2003), NP (Petrovic *et al.*, 2003), and pharmaceuticals (Boyd *et al.*, 2003) during advanced treatments. However, most of them were conducted on laboratory scale or pilot plant scale, although a few reports were under actual sewage treatment conditions (Boyd *et al.*, 2003). Also coverage of the target compounds in the studies has been limited. Then, we studied the removal of the wide-range of water soluble contaminants through advance treatment of the secondary effluents (i.e., sand-filtration and ozonation) under actual wastewater treatment conditions.

2. Materials and methods

As samples for broad-spectrum analysis of secondary effluents, 24-hr-composite samples were taken from 4 municipal STPs in Tokyo in May, August, November 2002 and February 2003. Individual plants treat more than 1.9 x 10^5 m^3/day of sewage from more than 0.7 million. All the plants employ primary and secondary treatment using activated sludge. Hydraulic retention time (HRT) was ~ 8 hours. The composite samples of the secondary effluents were collected in solvent-rinsed amber 3L glass bottles, transported to laboratory in cool, and filtered with pre-baked glass fiber filter (GF/F, Whatman). Filtrates were stored at 5

°C until solid phase extraction (normally 24 hours after the filtration). To study the removal of organic micro-contaminants during advanced treatment, spot samples were collected from a municipal STP in Tokyo in July and November 2003 and June 2004. The plant serves 0.46 million people and treats 1.4 x 10^5 m^3 of sewage per day on average. It employs primary and secondary treatment using activated sludge with HRT of 12 hours. A part of the secondary effluent receives ozonation following sand filtration. Flow rate of sand-filtration was 110 m/day. Concentration of ozone and retention time in the reaction tank was 1.0 mg/L and 27 min., respectively. The secondary effluents and effluents after sand-filtration and ozonation were collected, transported, filtered and stored as described above.

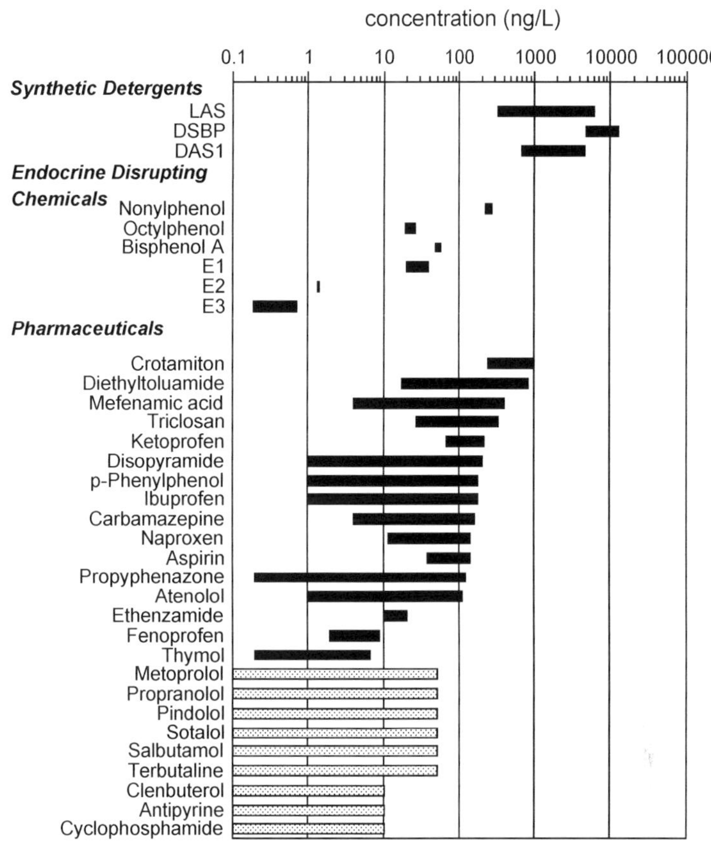

Figure 1. Concentration range of the water-soluble micro-pollutants in the secondary effluents. Gray bars: below detection limits (i.e., <50 ng/L or <10 ng/L)

Target compounds in the present study are illustrated in Figure 1. All target compounds were extracted by solid phase extraction. LAS and disopyramide, atenolol, metoprolol, propradolol, pindolol, sotalol, alfutaerol, terbutaline, clenbuterol, antipyrine, and cyclophosphaminde were quantified by LC-MS/MS (Theremoelectron TSQ Quantum mass spectrometer connected with Agilent 1100 series HPLC). Two FWAs species (i.e., DSBP and DAS1) were quantified by HPLC equipped with fluorescence detector. The other target compounds were quantified by GC-MS following appropriate purification and fractions. The details for analytical methods were described in Hayashi *et al.* (2002) and Nakada *et al.* (2004).

To directly determine estrogenic activity of the effluent samples, the extracts from the filtrates were subjected to bioassay using DNA recombinant yeast. The *Saccharomyces cerevisae*, carrying a human estrogen receptor α, was kindly provided from Dr. Sumpter, Brunel University in England and used for the estrogen receptor assay (Routledge and Sumpter, 1996). The procedure of the yeast estrogen screen was described in Yakou *et al.* (1999).

3. Results and discussion

GC-MS and LC-MS/MS analyses demonstrated the occurrence of various water-soluble organic micro pollutants including surfactants, FWAs, EDCs, natural estrogens, and pharmaceuticals, in the secondary effluents (Figure 2). This is reasonable because the municipal STPs employing activated sludge process are designed to remove suspended solid and biodegradable organic matter and, therefore, water-soluble and microbiologically-refractory chemicals escape adsorption to sludge and microbial attack and survive in the secondary effluents. Among the detected chemicals, synthetic-detergent-derived pollutants (i.e., LAS and FWAs) exhibited the highest concentrations. LAS concentrations ranged from 330 ng/L to 6100 ng/L. These are lower than 100000 ng/L above which chronic toxicity usually occurs with freshwater organisms (Lewis, 1991). However, effects on behavioral parameters could occur with freshwater fishes at much lower concentrations (i.e., 2000 ng/L; Hidaka and Tatsukawa, 1986) which is comparable to the concentration range observed in the secondary effluents. Concentrations of DSBP and DAS1 ranged from 4800 ng/L to 12500 ng/L and from 680 ng/L to 4560 ng/L, respectively. These concentration ranges are one to two orders of magnitude lower than the predicted-no-effect-concentration (PNEC) of FWAs (i.e., 100 μg/L, Richner *et al.*, 1997).

Among the EDCs determined in the present study, nonylphenol showed the highest concentrations (220 – 280 ng/L) followed by bisphenol A (49 – 55 ng/L) and octylphenol (19 – 25 ng/L). These concentrations were comparable to those reported for STP in Tokyo (Nakada *et al.*, 2004). Observed alkylphenol concentrations were one order of magnitude lower than the threshold concentrations for endocrine disruption with freshwater fish. Among natural estrogens, E1 (20 – 37 ng/L) was dominant over E2 (1.3 – 1.4 ng/L) and E3 (0.2 – 0.7 ng/L). Predominance of E1 over E2 is consistent with the report by Nakada *et al.* (2004) and

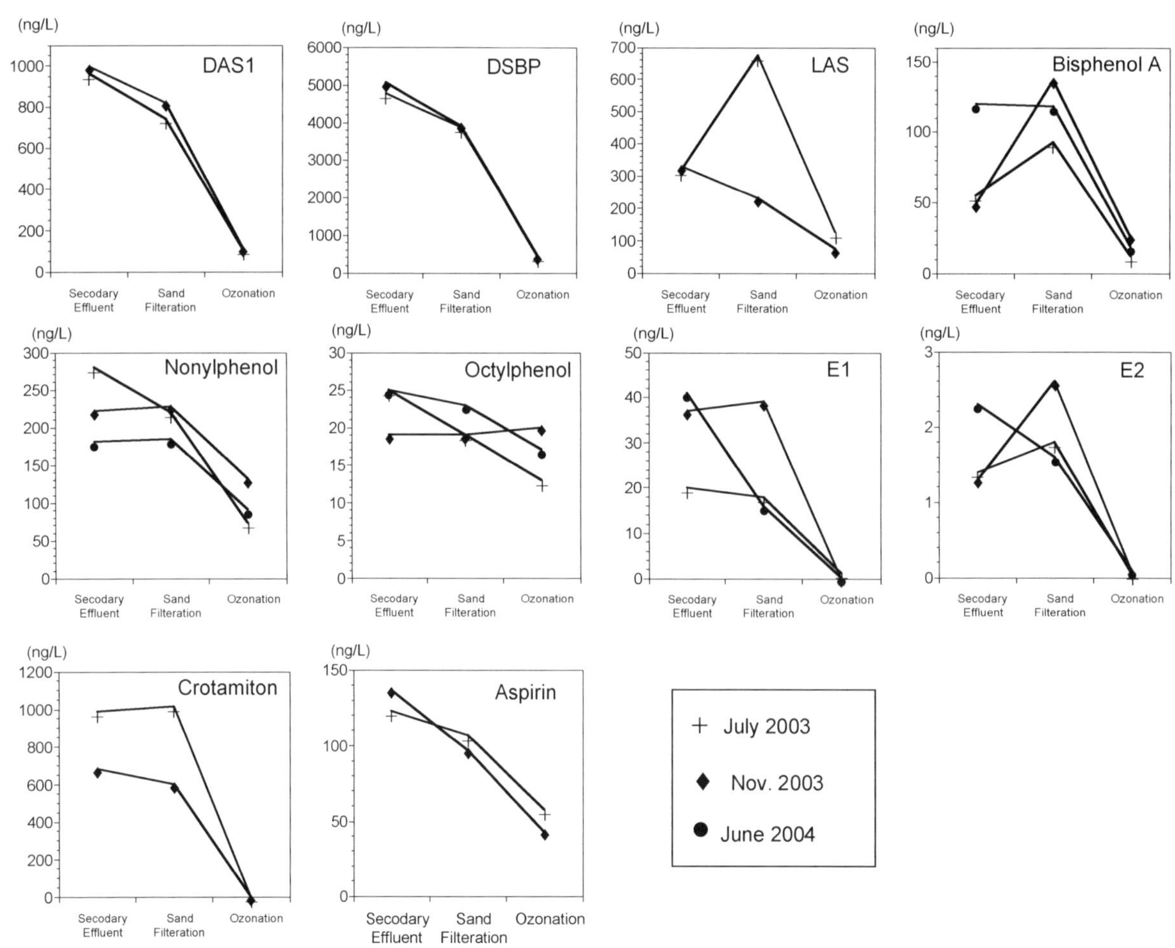

Figure 2. Removal of the water-soluble organic micro-contaminants during sand-filtration and ozonation

probably due to conversion from E2 to E1 during secondary treatment. Observed E1 concentrations were close to the threshold concentrations (50 ng/L), above which any endocrine disruption can be expected for wild fish (Routledge *et al.*, 1998). This may limit the reuse of the secondary effluents for some purposes (e.g., waters for streams and ponds with keeping fishes).

Twenty six species of pharmaceuticals were examined and 16 pharmaceuticals were significantly detected in the secondary effluents. The concentrations ranged from sub ng/L level to hundreds ng/L level. Crotamiton (245 – 987 ng/L), diethyltoluamide (17 – 815 ng/L), mefenamic acid (4 – 396 ng/L), triclosan (27 – 330 ng/L) and ketprofene (68 – 219 ng/L) frequently showed concentrations over 100 ng/L. These concentration ranges are comparable to those reported in European and U.S. sewage effluents (Boyd *et al.*, 2003; Ternes, 1998). Effects of these pharmaceuticals on aquatic organisms at these concentrations are still unknown because of the lack in the toxicological data.

Our study on advanced sewage treatment demonstrated that ozonation is effective to eliminate some of the water-soluble contaminants examined. Figure 3 shows concentrations of target compounds in the original secondary effluent and effluents after sand-filtration and ozonation. Approximately 90 % of FWAs originally present in the secondary effluents were removed by ozonation following sand-filtration, whereas only sand-filtration eliminated ~20% of the FWAs. However, ozonation is not so effective for LAS and alkylphenols and bisphenol A. Approximately 20 – 40 % of LAS, 26 – 60 % of NP, 50 – 105 % of OP, 16 – 53 % of BPA in the secondary effluents were still found after ozonation. Low removal efficiency of NP by ozonation is consistent with the reports by (Petrovic *et al.*, 2003) and can be partly attributed to formation of NP through the breakdown of ethoxylated nonylphenolic compounds during ozonation. On the other hand, less than 7 % of E1 and E2 in the secondary effluents were detected after ozonation, while no consistent removal was observed for sand-filtration. These observations ind

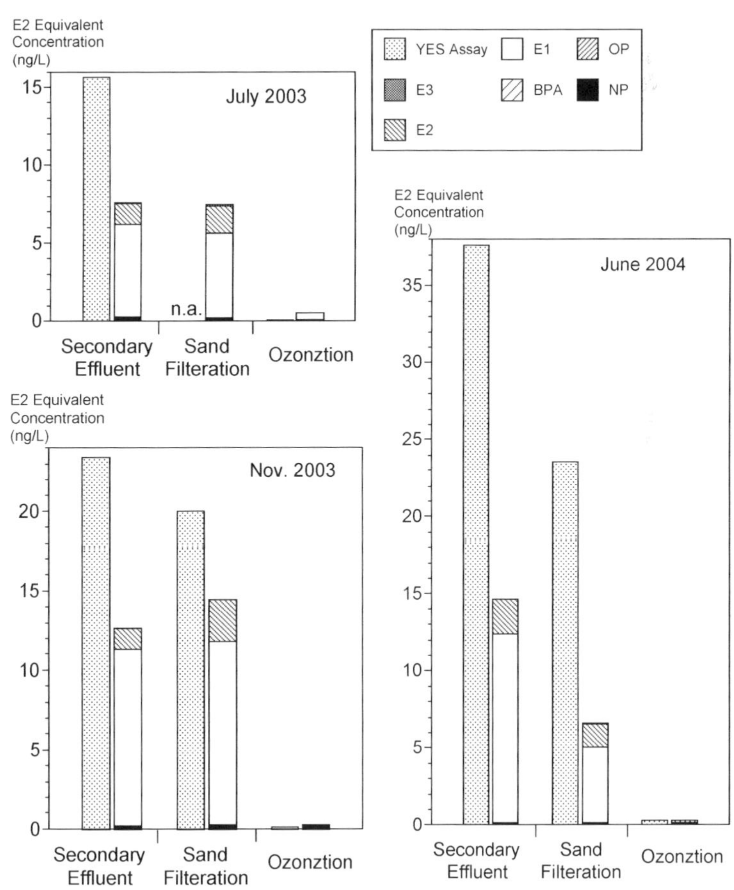

Figure3. Estradiol-equivalent concentrations in the sewage effluents. Left and right bars for individual effluents: results from bioassay (yeast) and from chemical analysis, respectively. n.a. : no data available.

icate that E1 and E2 are highly reactive toward ozone. The efficient removal of E1 and E2 by ozonation is consistent with the reports by (Alum *et al.*, 2004). Because of insignificant concentration of E3 in the secondary effluents, its removal through the advanced treatment could not be discussed. Efficient removal by ozonation was observed for some pharmaceuticals. Less than 0.5 % of crotamiton in the secondary effluents was detected in the effluents after ozonation while no significant removal was observed for sand-filtration. Similar efficient removal by ozonation was observed for naproxane and triclosan. Efficient removal of naproxane by ozonation is consistent with the report by Boyd *et al.* (2003). However, there observed some pharmaceuticals which showed no drastic decrease in their concentrations during ozonation. For example, approximately 50% of the aspirin originally present in the secondary effluents was detected even after ozonation. Our study implies that ozonation is not a perfect way to remove the pharmaceuticals in the sewage effluents, although it is effective for some species of pharmaceuticals.

To quantitatively discuss the change in the estrogenic activities during the advanced sewage treatments, 17 β-estradiol equivalent concentrations ($EEQ_{GC/MS}$) were calculated based on the instrumentally determined concentrations. $EEQ_{GC/MS}$ of individual compounds were calculated through multiplying their concentrations by relative potency (RP) of individual compounds. RP had been determined by bioassay using authentic standards (Yakou *et al.*, 1999) and the following values were applied in the present study, NP : 1.0×10^{-3}, OP : 6.0×10^{-4}, BPA : 6.0×10^{-5}, E1 : 0.3, E2 : 1.0, and E3 : 3.7×10^{-2}. Total $EEQ_{GC/MS}$ (sum of $EEQ_{GC/MS}$ of all the analytes) for the secondary effluent was calculated at 7.6 to 14.7 ng E2/L. These values were approximately 50% of $EEQ_{bioassay}$ directly determined for the crude extracts of the effluents by the bioassay. This difference can be explained by unidentified (non-quantified) EDCs in the effluent samples and/or enhancement of estrogenic activity by anionic surfactants in the effluent (Miyamoto *et al.*, 2002). These EEQ are comparable to the threshold concentration of E2 (10 ng/L), above which any endocrine disruption of wild fishes is expected. This implies that the secondary effluent may not be suitable for reuse as stream waters with wild fishes.

Figure 3 shows E1 and E2 contributed 78 – 88 % and 11 – 18 % of total $EEQ_{GC/MS}$ in the effluent, respectively. On the other hand, the contribution of synthetic chemicals, such as APs and BPA were not significant (< 4 %) in the secondary effluent. These results indicate that natural estrogens, especially E1, are the dominant environmental estrogens in the samples. The predominance of natural estrogens was consistent with Nakada *et al.* (2004). As a result of drastic decrease in natural estrogen concentrations by ozonation, total $EEQ_{GC/MS}$ was extremely reduced to 0.3 – 0.5 ng E2/L in the effluents after ozonation. Also direct determination of estrogenic activities in the effluents by the bioassay showed the consistent decrease and $EEQ_{bioassay}$ of 0.1 – 0.3 ng E2/L in the effluent after ozonation. These results suggest that the effluent after ozonation is free from endocrine disruption. This implies that ozonation increase the applicability of sewage effluents to reuse and recycling.

4. Conclusions

An array of water-soluble organic micro-pollutants was detected in the secondary effluents. These chemicals may limit the reuse and recycling of the effluents. Estrogenic activities in the secondary effluent were relevant to cause some endocrine disruption with wild fishes and may depress the potential of reuse of the effluent for some purposes as stream waters with keeping wild fishes. Ozonation efficiently removed some chemicals, including natural estrogens, in the secondary effluents. Consequently, estrogenic activity of the effluent after ozonation drastically decreased to the level much lower than the threshold of endocrine disruption with wild fishes. Ozonation increase the potential of the recycling of the sewage effluents. In the future study, risks associated with detected pharmaceutical and chemicals other than EDCs should be clarified and quantified. Also similar chemical and biological characterization of the other kinds of wastewater (e.g., street runoff) will be conducted in future studies. The present study was conducted for the effluents from STPs in Japan. Demand for recycling of urban wastewater will be elicited also in South East Asian cities in near future due to continuous population growth and drastic industrialization. Furthermore, less sufficient management of waste and different climatic conditions (frequent and heavy rain) could result in different combination of micro-pollutants in urban wastewaters. Thus, evaluation of urban wastewaters using the approach presented in the present study should be conducted in South East Asian cities in coming years.

5. Acknowledgement

The present research has been conducted as a part of a Core Research for Evolutional Science and Technology (CREST) entitled "Risk-based Management of Self-regulated Urban Water Recycle and Reuse System" in "Hydrological Modeling and Water Resources System" supported by Japan Science and Technology Agency. We thank the Bureau of Sewerage Tokyo Metropolitan Government for their cooperation in collecting sewage effluent samples. Several graduates and undergraduates in LOG (Laboratory of Organic Geochemistry, Tokyo University of Agriculture and Technology) provided welcome assistance with fieldwork.

6. References

Alder A.C., Siegrist H., Fent K., Egli T., Molnar E., Poiger T., Schaffner C. and Giger W. (1997). The fate of organic pollutants in wastewater and sludge treatment. Significant processes and impact of compound properties. *CHIMIA*, **51**(12), 922-928.

Alum A., Yoon Y., Westerhoff P. and Abbaszadegan M. (2004). Oxidation of bisphenol A, 17β-estradiol, and 17-ethynyl estradiol and byproduct estrogenicity. *Environ. Toxicol.*, **19**(3), 257-264.

Boyd G.R., Reemtsma H., Grimm D.A. and Mitra S. (2003). Pharmaceuticals and personal care products (PPCPs) in surface and treated waters of Louisiana, USA and Ontario, Canada. *Sci. Total Environ.*, **311**(1), 135-149.

Desbrow C., Routledge E.J., Brighty G.C., Sumpter J.P. and Waldock M. (1998). Identification of estrogenic chemicals in STW effluent. 1. Chemical fractionation and *in vitro* biological screening. *Environ. Sci. Technol.*, **32**(11), 1549-1558.

Harries J.E., Sheahan D.A., Jobling S., Matthiessen P., Neall P., Sumpter J.P., Taylor T. and Zaman N. (1997). Estrogenic activity in 5 United-kingdom rivers detected by measurement of vitellogenesis in caged male trout. *Environ. Toxicol. Chem.*, **16**(3), 534-542.

Hayashi Y., Managaki S. and Takada H. (2002). Fluorescent whitening agents (FWAs) in Tokyo Bay and adjacent rivers: their application as anthropogenic molecular markers in coastal environments. *Environ. Sci. Technol.*, **36**, 3556-3563.
Hidaka H. and Tatsukawa R. (1986). Variations by sex and body size in the avoidance tests of sodium linear laurylbenzene sulfonate and fenitrothion using a fish, medaka *oryzias latipes*. *Bull. Jap. Soc. Sci. Fisher.*, **52**(10), 1753-1757.

Jobling S. and Sumpter J.P. (1993). Detergent components in sewage effluent are weakly oestrogenic to fish -An *in vitro* study using rainbow trout (*Oncorhynchus mykiss*) hepatocytes. *Aqua. Toxicol.*, **27**(3-4), 361-372.

Kolpin D.W., Furlong E.T., Meyer M.T., Thurman E.M., Zaugg S.D., Barber L.B. and Buxton H.T. (2002). Pharmaceuticals, hormones, and other organic wastewater contaminants in U.S. streams, 1999-2000: A National Reconnaissance. *Environ. Sci. Technol.*, **36**(6), 1202-1211.

Lee J., Park H. and Yoon J. (2003). Ozonation characteristics of bisphenol A in water. *Environ. Technol.*, **24**(2), 241-248.

Lewis M.A. (1991). Chronic and sublethal toxicities of surfactants to aquatic animals: A review and risk assessment. *Wat. Res.*, **25**, 101-113.

Miyamoto N., Tanaka H., Tamamoto H. and Komori K. (2002). Fractionation method for estimating the cause of estrogen-like activity in sewage. *Proceedings of WEFTEC 2002.*

Nakada N., Nyunoy H., Nakamura M., Hara A., Iguchi T. and Takada H. (2004). Identification of estrogenic compounds in wastewater effluent. *Environ. Toxicol. Chem.* (in press)

Petrovic M., Diaz A., Ventura F. and Barcelo D. (2003). Occurrence and removal of estrogenic short-chain ethoxy nonylphenolic compounds and their halogenated derivatives during drinking water production. *Environ. Sci. Technol.*, **37**(19), 4442-4448.

Richner P., Kasching L. and Zellar M. (1997). Latest results from monitoring studies and environmental risk assessments (ERSs) of fluorescent whitening agents (FWAs). In: *the 7th annual meeting of SETAC-Europe "Prospects for the European Environment Beyond 2000", Amsterdam.*

Routledge E.J., Sheahan D., Desbrow C., Brighty G.C., Waldock M. and Sumpter J.P. (1998). Identification of estrogenic chemicals in STW effluent. 2. *In vivo* responses in trout and roach. *Environ. Sci. Technol.*, **32**(11), 1559-1565.

Routledge E.J. and Sumpter J.P. (1996). Estrogenic activity of surfactants and some of their degradation products assessed using a recombinant yeast screen. *Environ. Toxicol. Chem.*, **15**(3), 241-248.

Ternes T.A. (1998). Occurrence of drugs in German sewage treatment plants and rivers. *Wat. Res.*, **32**(11), 3245-3260.

Yakou Y., Takahashi A., Higashitani T. and Tanaka H. (1999). Measurement of environmental estrogen-like activity in wastewater using recombinant yeast. *Environ. Engineer. Res. Jap. Soc. Civil Engineer.*, **36**, 199-208.

Minimization and Stabilization of Leachate Through Internal Storage and Re-circulation Operation in Tropical Landfill

Chart Chiemchaisri[1)], Wilai Chiemchaisri[2)] and Joseph Traenkler[3)]

1) Department of Environmental Engineering, Kasetsart University
50 Phaholyothin Road, Bangkok 10900, Thailand
e-mail: fengccc@ku.ac.th
2) Department of Environmental Engineering, Kasetsart University
50 Phaholyothin Road, Bangkok 10900, Thailand
e-mail: fengwlc@ku.ac.th
3) ECO CITY Planning and Management Program
503 Jinling Zhong Road, Changzhou 213003, P.R. China
e-mail: j.trankler@trankler.de

Abstract

This research investigated the beneficial effect of using bioreactor landfill concept in accelerating the biodegradation of wastes while minimizing leachate problem in tropical landfill. Four pilot-scale landfill lysimeters were operated under different operational mode, i.e. 1) control (conventional landfill operation) 2) leachate re-circulation 3) internal storage without re-circulation and 4) leachate re-circulation with internal storage. The experiment was carried out to quantify the amount and characteristics of leachate produced from landfill lysimeters operated in tropical climatic rainfall event. Variation in leachate characteristics, methane content in landfill gas and water balance in the lysimeters was studied. From the experimental results, it was found that leachate re-circulation and storage operation yielded lower organic concentration in leachate whereas methane content in landfill gas was found highest among the lysimeters. Meanwhile leachate re-circulation enhanced degree of waste stabilization. Water balance evaluated by HELP model was in agreement with the experimental results. It was found that field capacity and porosity of waste was likely to be decreased along the experimental period. Rainfall intensity and leachate re-circulation practice affected leachate quantity as well as organic and nitrogen leaching from wastes. Meanwhile, leachate re-circulation techniques helped enhancing the rate of waste stabilization.

Keywords: bioreactor landfill, HELP model, leachate treatment, leachate recirculation, water balance, tropical climate

1. Introduction

Landfill has been widely used for municipal solid waste (MSW) disposal all over the world. Conventionally, landfill is designed to contain or store the wastes so that the exposure to human and environment could be minimized. Under the containment landfill concept, water infiltration into the landfill is minimized to reduce emission of leachate and gases into the environment. However, severe impacts are occasionally found in these landfills due to the failure of their containment system.

A bioreactor landfill is managed to accelerate decomposition of organic wastes by controlling moisture content of wastes, recycling of nutrients and seeding of microorganisms by leachate re-circulation. Several researches on landfill bioreactor in laboratory, pilot and full scale have been conducted during the past 30 years (Reinhart and Townsend 1998). Most of them have agreed that leachate circulation could enhance waste stabilization, leachate quality and landfill gas production (Pohland 1980; Titlebaum 1982; Barlaz et al. 1987; Stegmann and Ehrig 1989; Lay et al. 1998). However, some operational problems associated with leachate circulation were lack of appropriate circulation technique, leachate channeling and leachate ponding (Buivid et al. 1981).

Previous experiments on bioreactor landfill were mostly conducted under semi-arid climatic condition using wastes with relatively low moisture content. In contrary, municipal solid wastes from developing countries are mainly composed of easily biodegradable organic matter with high moisture content especially in tropical climate, therefore, providing different environmental conditions for microbial activities. This study focused on the application of bioreactor landfill concept in tropical landfill by using leachate re-circulation and internal storage techniques to enhance waste degradation while minimizing leachate discharge from landfill. The produced leachate during rainy season can be stored inside the waste cell for increasing the moisture content of wastes. It will be re-circulated back to maintain moisture content in top cover layer during dry period. As majority of leachate will be evaporated during irrigation practice, the quantity of leachate to be treated could be minimized.

2. Materials and methods

The experiment was conducted in pilot-scale landfill lysimeters. Four experiment units made of steel pipe of 0.90 m. diameter and 2.7-m height were used. The leachate drainage system was provided at the bottom of each lysimeter with a drainage pipe connected with 1-inch valve. Gravel layer (10-30 mm.) of 200 mm. thickness and geo-textile sheet were provided at the bottom of lysimeter to prevent clogging of suspended solids at the outlet. Leachate was circulated back to the lysimeter in the gravel layer placed below cover soil at the top of waste layer. Distribution of leachate was accomplished through perforated PVC pipe of 1-inch diameter installed in the gravel layer. Gas collection system was installed at the top of waste layer. It was made from perforated 1-inch diameter PVC pipe placed into a gravel pack to prevent clogging from solid particles in solid waste.

Municipal solid waste obtained from a waste transfer station in Bangkok, Thailand was used. It was mixed with digested sludge at a ratio of 4:1 (solid: digested sludge) before being filled into the lysimeters. Solid waste was placed in the lysimeter at 200 – 250 mm. thickness each time and compacted to a density of about 600 kg/m^3 until total height of 2 m was reached. The characteristics of solid wastes are shown in Table 1. Clay loam layer of 300 mm. was used to cover the top of each cell to prevent the release of gas through the cover soil layer.

The experimental conditions applied in each lysimeter were 1) Conventional landfill operation 2) Leachate re-circulation once a week. 3) Leachate re-circulation once a week and internal storage at 50 % of waste height and 4) Internal storage of leachate at 50% of waste height as shown in Figure1.

Table 1 Physical and chemical characteristics of solid wastes

Components	Percentage (wet wt. basis)	Parameters	Value
Food wastes	41.38	Moisture content	62.5% wet wt.
Plastic	22.33	Total solids	37.5% wet wt.
Paper	14.78	Volatile solids	70.1% dry wt.
Textile	6.86	Ash	29.9 % dry wt.
Wood	4.83	Carbon	38.9% dry wt.
Metals	0.76	Hydrogen	4.4% dry wt.
Foam & rubber	0.96	Oxygen	24.5% dry wt.
Bone & shell	1.11	Nitrogen	1.9% dry wt.
Glass	2.06	Phosphorus	0.21% dry wt.
Others	4.93	Sulfur	0.19% dry wt.
		Specific Weight	621 kg/m^3
		Field Capacity	430 mm./m
		Hydraulic Conductivity	$8.55*10^{-4}$ cm/s

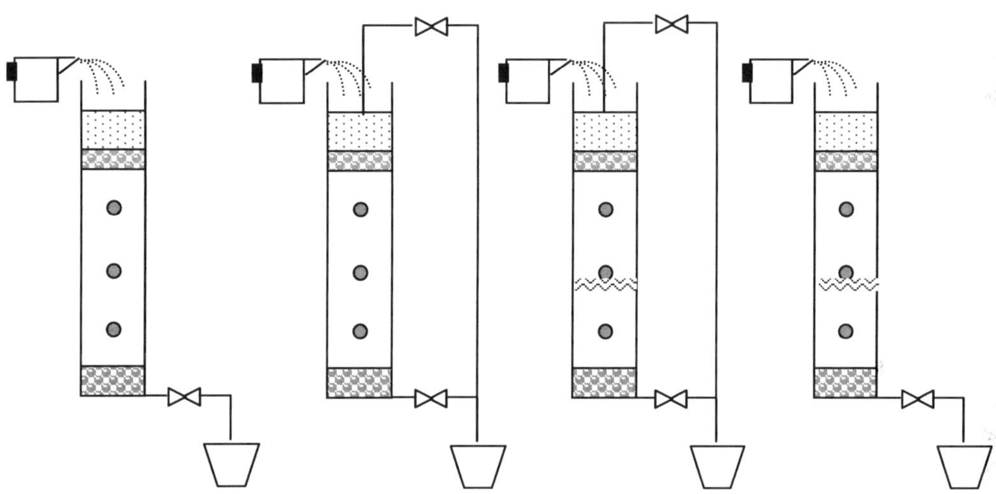

Figure 1 Lysimeter operation

In 1st year experiment, rainwater was added into the lysimeters to simulate wet period (rainy season) for 140 days. Afterwards, a dry period was simulated in which rainwater addition was terminated. Simulated rainfall (using actual rainwater) is applied daily on the top cover soil based on average at a rate of 70% that of average rainfall amount in Thailand. This is estimated to be equal to the water infiltrated through final cover soil determined by Hydrologic Evaluation of Landfill Performance (HELP) program. In 2nd year, the amount of rainwater added to each lysimeter was varied according to actual rainfall data of Bangkok, Thailand. The quantity of produced leachate was measured to verify the water balance in the lysimeters. Solid waste samples were analyzed for their temperature, density, moisture content, field capacity, total and volatile solids and chemical composition. Leachate samples from each lysimeter were collected and analyzed for pH, alkalinity, BOD, COD, VFA, alkalinity, TKN and monitoring of leachate quantities. All analyses were performed according to Standard Methods for the Examination of Water and Wastewater 18th Edition. Gas samples were collected once a week and analyzed for their composition (CO_2, O_2, N_2, CH_4) using a gas chromatography. (Agilent GC6890 with TCD, CRT1 Alltech stainless steel column).

3. Results and discussion

3.1 Lysimeter performance under constant rainwater input (1st year)

During this first year operation, the variation of physical characteristics of solid waste in each lysimeter was determined at different depths (30, 100 and 170 cm. from bottom of waste layer). It was found that the lysimeter temperature were mainly affected by ambient temperature and fluctuated within a range between 26 to 33^{0}C (ambient temperatures were between 30-35^{0}C). The temperature was found to be slightly lower at the middle and bottom part of lysimeter 3 and 4 as the wastes are immersed under stored leachate in the lysimeters.

The moisture content was found higher at the bottom part of the lysimeters. They were approximately 80% as compared to 50% in the upper part. However, there was not much difference in moisture content of wastes among the lysimeters operated during rainfall period. In the dry period, however, the lysimeters operated without re-circulation (lysimeter 1 and 4) had their moisture content dropped down whereas those of the others could be maintained relatively constant showing that the re-circulation help maintaining moisture content during the absence of rainfall.

Settlement rates of waste layer in the lysimeters are also determined. It was found that the settlement rates in lysimeter 1 and 2 were higher than that of lysimeter 3 and 4. Highest settlement was found in control lysimeter (7.5 cm. after 195 days) whereas the lysimeter with stored leachate had lowest settlement rate (2 cm. after 195 days).

For leachate characteristics, it was found that pH of leachate was maintained relatively constant at about 6 in all lysimeters while VFA/Alkalinity were maintaining in a range of 2-3. Leachate conductivity during rainfall period was slightly different among the lysimeter but maintained in the range of 30-40 mS/cm. In the latter dry period, a dropping trend in leachate conductivity was observed. BOD and COD concentration of leachate were found in a range of 40,000-60,000 mg/l and 40,000-70,000 mg/l respectively. High BOD/COD ratio suggested that the leachate was highly biodegradable which is normal for young landfill. TKN concentration in lysimeter 1 and 2 are slightly higher than that of lysimeter 3 and 4 but they were all in the range between 2,000-3,500 mg/l.

During the start-up period, they are gradually increased and were in the range of 5-20%. It was found that the lysimeter operated with leachate storage (lysimeter 3 and 4) has produced slightly higher methane content than the others. As the operation entered the dry period, the methane production in most of the lysimeters has been increased to about 30% but slightly higher in lysimeter 3. It was found that leachate re-circulation in combination with leachate storage could enhance waste biodegradation as it could help retaining high moisture content of wastes, especially during dry period, and improving the biodegradation of solid wastes.

3.2 Lysimeters performance under tropical rainfall condition (2nd year)

In the second year of experiment, the lysimeters were operated under the variation of rainwater input according to actual climatic data. During the experiment, it was found that the temperature of wastes was kept in range between 25-35°C. It was found that the lysimeter 1 (control) had average moisture content of 43.0%, 55.6% and 59.4% at the top, middle and bottom of waste layer. The lysimeter 2 with leachate re-circulation had average moisture content of 62.5%, 70.2% and 82.7%. In lysimeter 3 and 4, the middle and bottom parts were fully saturated whereas the top part had average moisture content of 71.7% and 51.7% respectively. The results suggested that higher moisture content could be maintained in the lysimeters with leachate re-circulation.

In term of waste settlement rate, it was found that the lysimeter 2 with leachate re-circulation had highest settlement of 10.8% followed by lysimeter 3 in the second year of experiment. The major part of settlement took place during rainy season as a result of higher waste degradation in that period. Leachate re-circulation helped promoting the waste settlement by allowing the leaching of hydrolyzed wastes into the bottom part of the lysimeters. The leaching of hydrolyzed wastes created higher porosity within the waste layer and the settlement took place once leachate had been drained out from the lysimeters. The lysimeter 3 and 4 with leachate storage had relatively low settlement because their bottom part was under water-saturated condition.

The comparison of initial and final chemical characteristics of solid wastes after two years of experiment shows that lysimeter 2 with leachate re-circulation had highest VS reduction from 70.1% to 39.7%, 19.1% and 9.9% at the top, middle and bottom part of the lysimeter respectively. Higher organic waste reduction was found at the bottom part of the lysimeter as the moisture content was maintained at higher level in wastes at such location. Higher moisture content promoted waste biodegradation by accelerating the hydrolysis step in waste stabilization process. These results clearly show that leachate re-circulation helped increasing waste biodegradation rate in the lysimeters.

BOD and COD concentration in leachate from lysimeter 1, 2 and 4 were much higher than that from lysimeter 3. The average BOD concentrations were 30150, 32140 and 32140 mg/l as compared to 1,120 mg/l of lysimeter 3. These results suggested that the application of leachate re-circulation and storage practice helped stabilizing organic substances in leachate and yielding low pollutant load if leachate were to be discharged from the landfill. BOD/COD ratios in leachate from the lysimeters were 0.51, 0.57, 0.49 and 0.58 respectively. TKN concentrations also had a similar decreasing trend as BOD and COD, but at much slower rate. They were reduced from 2000-2500 mg/l at the beginning to about 1500 mg/l at the end of the second year period. There was not much difference in term of TKN concentration in leachate among the lysimeters.

It was also found that methane content in gas from the lysimeter 1 was lowest at about 10% as compared to others (20-30%). As the operation entered raining period, methane content were raised to 24% in lysimeter 1 and 38% in lysimeter 3. Subsequent operation in dry period lowered methane content to about 13-17%. It is clearly show that rainwater affect the activities of methanogenic bacteria in the lysimeters. The lysimeter operation with re-circulation practice helped increased the moisture content in wastes especially during dry period and enhanced the microbial activities in such period.

The leaching of organic pollutants in terms of COD and TKN from the lysimeters was determined after two years experiment as shown in Table 2. It was found that the lysimeters with leachate re-circulation had lower COD leaching than the others. It was due to lower organic concentrations remaining in the drained leachate as they became more stabilized during re-circulation. The lysimeters with leachate re-circulation also gradually released organic substances at relatively constant rate over time whereas those without re-circulation flushed off major part of organic load at the beginning of rainy season. The leaching of TKN had similar pattern with those of COD. However, TKN leaching in the lysimeters with leachate re-circulation was found higher than the others in the second year. This was probably due to the fact that TKN was released at slower rate than COD. COD leaching in the second year reduced significantly from that of first year experiment while TKN leaching from lysimeters with re-circulation became higher. It suggests that TKN leaching from the landfill should be carefully taken into consideration in long term

Table 2 Comparison of COD and TKN leaching from solid wastes in 1st and 2nd year

Lysimeter No.	1	2	3	4
1st year				
Total COD Leaching (g)	12471	11914	9963	10116
Total TKN Leaching (g)	639	579	382	430
2nd year				
Total COD Leaching (g)	4140	1926	1067	3992
Total TKN Leaching (g)	341	974	711	299

3.3 Determination of water balance in the lysimeters

Water balance in the lysimeters was determined by the measurement of leachate amount drained from each lysimeters. The water balance equation can be described as follow.

$$V_I = V_L + V_S + V_E - V_R$$

Where V_I = Volume of rainwater added
V_L = Volume of drained leachate
V_E = Volume of water loss through evaporation
V_R = Volume of water produced during waste biodegradation

Total amount of rainwater input was 1,366 mm. or 869 liter to each lysimeter. At the beginning, the amount of leachate produced from lysimeter 2 and 3 were about 20 and 10 liter per week whereas those of lysimeter 1 and 4 were about 2 liter per week. This difference was due to the water stored in the lysimeter 2 and 3 from the first year experiment. As the operation entered rainy season period, the amount of leachate from all lysimeters increased. They were decreased again when dry period has been reached. The average volume of leachate produced from lysimeter 1 and 4 were found to be 18.80% of rainwater input whereas they were 59.41% and 36.41% in the lysimeter 2 and 3 respectively.

Hydrologic Evaluation of Landfill Performance (HELP) model was used to carry out the simulation of leachate production from lysimeters operated under different conditions. The input parameters to the program include climatic and rainfall data, geographical information of study site, waste characteristics (field capacity and porosity of wastes) and landfill components. It was found that the simulation results are fairly in agreement with the measured data depending on the input characteristic parameter of wastes, i.e. field capacity and porosity. The best-fitted simulation to the experimental results was obtained when using field capacity of 40% and porosity of 0.4. Comparing with initial field capacity of 50% at the beginning of the experiment, it would suggest that the field capacity of wastes probably decreased as the solid waste degradation and settlement took place.

Table 3 Comparison of leachate production in the lysimeters with HELP model

Items	Measured values				Calculated values			
	L1	L2	L3	L4	L1	L2	L3	L4
Rainfall (mm.)	1366.2	1366.2	1366.2	1366.2	1366.2	1366.2	1366.2	1366.2
Leachate production (%)	18.80	59.41	36.41	16.90	18.10	59.04	36.23	15.99
Evaporation (%)	80.00	32.89	63.52	81.20	80.71	33.28	62.95	81.81
Water storage (%)	1.20	7.70	0.70	1.90	1.19	7.68	0.82	2.20

4. Conclusions

From the experimental results obtained, the following conclusions can be drawn.

1) Leachate re-circulation in combination with leachate storage was found to be viable alternative operating techniques for tropical landfill. They could help retaining high moisture content of wastes during dry period and improving the biodegradation of solid wastes. The re-circulation practice enhanced degree of waste degradation in landfill whereas the incorporation of leachate storage in waste layer reduced organic and nitrogen concentrations and total loading in leachate.

2) Rainfall intensity and leachate re-circulation rate affected leaching behavior of organic substances from waste layer. Gradual leaching of pollutants from the lysimeters with re-circulation was observed whereas pulse leaching took place in non re-circulating lysimeters at the beginning of rainfall event. The leaching of organic carbon from the wastes took place much faster than organic nitrogen.

3) The determination of water balance in landfill by using HELP model was in agreement with the experimental results. It also suggested that field capacity and porosity of wastes decreased with time as the waste degradation took place. Average field capacity of 40% and porosity of 0.4 were suggested from the simulation.

4) The application of bioreactor landfill concept could be used to accelerate biodegradation of solid wastes in the lysimeter. Moreover, proper water management in the landfill can significantly reduce the discharge of pollutant load from landfill.

5. Acknowledgement

This research work is supported by Swedish International Development Cooperation Agency (SIDA) in Asian Regional Research Program on Environmental Technology (ARRPET).

6. References

Barlaz, M.A., M.W. Milke and R.K. Ham (1987), Gas production parameters in sanitary landfill simulators, *Waste Manage. Res.,* **5**, 27-39.

Buivid, M.G., D.L. Wise and M.J. Blanchet (1981), Fuel gas enhancement by controlled landfilling of municipal solid waste, *J. Resources Conserv.,* **6**, 3-20.

Lay, J.J., Y.Y. Li and T. Noike (1998), Developments of bacterial population and methanogenic activity in a laboratory scale landfill bioreactor, *Wat. Res.*, **32**(12), 3673-3679.

Pohland, F.G. (1980), Leachate recycle as landfill management option, *J. of Env. Eng. Div. ASCE,* **106**(EE6), 1057-1069.

Reinhart, D.R. and T.G. Townsend (1998), *Landfill Bioreactor Design and Operation*, Lewis Publishers, New York.

Stegmann, R. and H.J. Ehrig (1989), Enhancement of gas production in sanitary landfill sites-Experiences in West Germany, In: *Resource Recovery from Solid Waste*, pp.425-434.

Titlebaum, M.E. (1982), Organic carbon content stabilization through landfill leachate recirculation, *J. WPCF,* **54**(5), 428-433.

Application of Cleaner Technology and Photosynthetic Wastewater Treatment System in Noodle Processing Factory

Chart Chiemchaisri[1], Ryo Honda[2], Lumpoon Chaitrong[3], Kensuke Fukushi[4] and Kazuo Yamamoto[5]

1) Department of Environmental Engineering, Kasetsart University
50 Phaholyothin Road, Chatuchak, Bangkok 10900, Thailand
E-mail: fengccc@ku.ac.th
2) Department of Urban Engineering, University of Tokyo
7-3-1 Hongo, Bunkyo-Ku, Tokyo 113, Japan
E-mail: honda@env.t.u-tokyo.ac.jp
3) Department of Environmental Engineering, Kasetsart University
50 Phaholyothin Road, Chatuchak, Bangkok 10900, Thailand
E-mail: nuenglek@yahoo.com
4) Environmental Science Center, University of Tokyo
7-3-1 Hongo, Bunkyo-ku, Tokyo 113-0033, Japan
E-mail: fukushi@esc.u-tokyo.ac.jp
5) Environmental Science Center, University of Tokyo
7-3-1 Hongo, Bunkyo-ku, Tokyo 113-0033, Japan
E-mail: yamamoto@esc.u-tokyo.ac.jp

Abstract

This paper presents a case study on application of cleaner technology in small-scale noodle processing factory. From the results of waste audit in 10 small- and medium-scale noodle factories in Bangkok and surrounding provinces, it was found that wastewater discharged from these sources, with starch as the main constituent, varied in a range between 2.7-5.4 m^3/ton of product. The wastewater is majority produced during rice soaking and rinsing processes. Various technological options have been implemented in attempt to reduce raw material (starch) loss and water consumption. Some of the effective measures found included the modification of washing system from mechanical to pneumatic system, installation of a water reuse system for washing the rice. For carbon recovery purpose, wastewater was then further treated in pilot-scale photosynthetic wastewater treatment system where carbon could be transformed into protein in purple non-sulfur bacterial cells. The system has shown to be effective for the treatment of starch wastewater and cultivation of purple non-sulfur bacteria.

Keywords: cleaner technology, carbon recovery, noodle processing, purple non-sulfur bacteria, starch wastewater

1. Introduction

Noodle is one of typical and popular food for people in the Eastern part of Asia. In Thailand, rice noodle production in primarily serves domestic demand, through some export of rice noodles exist. In general, small- and medium-scale noodle industries sell their products in their local areas. Meanwhile, larger factories produce their products for sale in larger service areas or overseas.

Two typical rice noodles are produced in Thailand, unfermented noodles and fermented ones. Unfermented rice noodle production accounts for an estimated 66% of Thailand's total rice noodle production. Several types of unfermented rice noodles are produced in Thailand, including thick noodle (kwuay-teow), thin noodles (sen-lek), vermicelli noodles (sen-mee), and pre-cooked rolled noodles (kwuay-jahp). Of these, thick and vermicelli noodles account for the largest portion of unfermented noodle production. The Department of Industrial Works (DIW) recorded indicate that 476 thick noodle and 76 vermicelli noodle factories were registered nation-wide with DIW in 1994 (DANCED, 1999).

2. Typical noodle production and process wastewater

Typical production of rice noodle includes broken rice and water as two main raw materials for production of unfermented rice noodles. It involves washing and soaking of rice, pasting the rice and making the starch slurry. For thick noodle processing, the slurry is then used in sheeting prior to cooking and drying step. The final product has a moisture content of about 60%. For thin noodle processing, the sheeted slurry is cooked and dried in an oven to reduce its moisture content to about 20-30% before cutting process. The processing of vermicelli noodle is slightly different from thick and thin noodle. The starch slurry is passed through a filter press, yielding semi-dry starch solids. The starch solids are steamed before being extruded to form vermicelli noodles. The noodles are then cooked to produce fresh product with moisture content of 40%. Then it is dried to about 10% moisture content to produce dried product. Figure 1 shows the process flow diagram for noodle production.

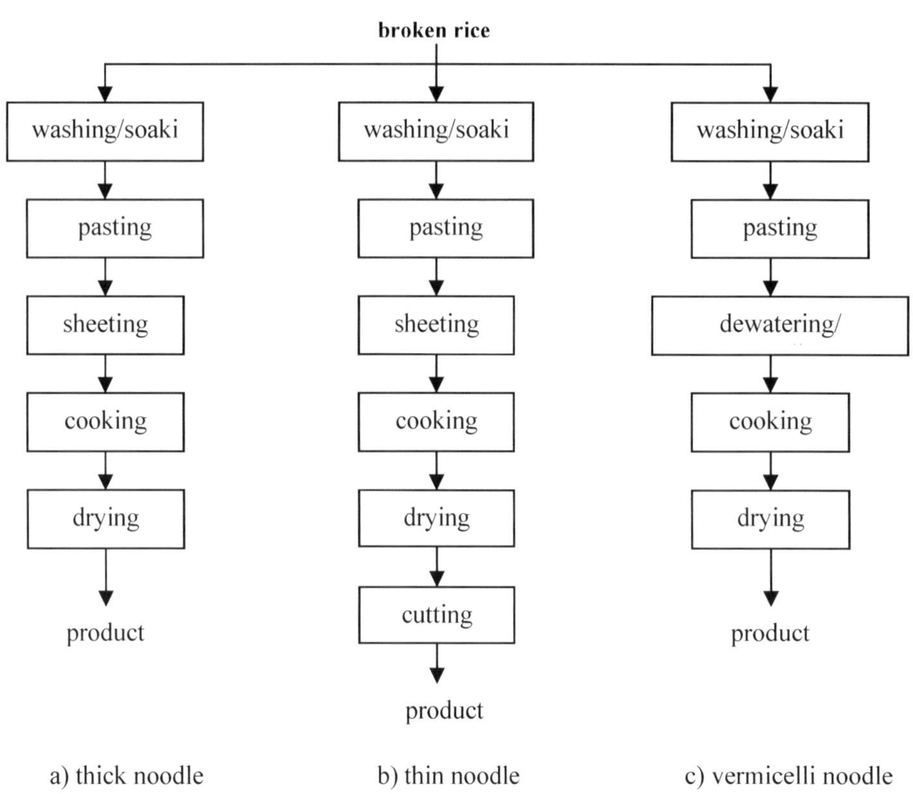

Figure 1. Process flow diagram for noodle production

The production of rice noodles produces wastes at various stage of process, i.e. raw material preparation, production, packaging and process line cleaning. The waste output can be categorized as wastewater, solid waste, exhausted gas and energy (heat) loss. Among those, wastewater mainly produced during rice soaking/rinsing and cleaning of pasting tank is of major concern. These process wastewater contains starch as its main constituents with high biochemical oxygen demand (BOD) and suspended solids (SS). In addition to the process wastewater, floor cleansing water also contribute significant amount of BOD and SS to the treatment system. Table 1 shows amount characteristics of wastewater from different processing stage in a typical noodle factory with production capacity of about 80 ton/day.

Table 1. Amount and characteristics of wastewater from a noodle processing factory

Activity	Wastewater amount (m^3/d)	Wastewater characteristics			Remark
		pH	BOD (mg/l)	SS (mg/l)	
Rice washing (soaking)	87	6.2	1,500	4,200	tank dia. 1.36 m. height 2.22 m.
Rice washing (rinsing)	87	6.1	800	170	
Pasting tank cleansing	1.5	4.7	6,500	3,060	tank dia. 1.56 m. height 1.50 m.
Floor cleansing	NA	11.7	6,000	2,220	practice once a day
Combined wastewater	approx. 210	4.7	2,450	1,750	

Raw material consumption and waste generation in the production of different categories of noodle was determined during the field survey of 10 rice noodle factories in Bangkok and surrounding area. Key figures in terms of broken rice, water and energy consumption as well as wastewater and solid waste generation were estimated as shown in Figure 2. It was found that there is much higher variation in water consumption per unit product as compared to broken rice. The variation of water consumption depends on processing equipment and operation (effectiveness of water use) which is much different from one factory to the others.

Table 2. Raw material consumption and waste generation in noodle production

Key figure	Thick noodle	Thin noodle	Fresh vermicelli noodle	Dried vermicelli noodle
Broken rice consumption (ton/ton product)	0.5	0.70-0.79	0.74-0.82	0.96-1.14
Water consumption (m^3/ton product)	1.9-2.7	2.4-4.2	2.8-4.0	3.7-8.6
Electricity consumption (kWh/ton product)	37	84-195	92-151	178-225
Wastewater (m^3/ton product)	-	2.7-5.4		
Solid waste (kg/ton product)	3-4	13-34	1-3	7-22

Source: DANCED (1999)

3. Application of cleaner technology in noodle production

The concept of cleaner technology has been adopted in noodle processing factory in attempt to reducing water consumption and wastewater discharge from the process. Moreover, prevention of starch loss from the process would also help reducing BOD concentration in discharged wastewater. As previously described, rice washing and rinsing has been identified as the main activity involving large amount of water. Therefore, the application of cleaner technology was focused around this activity.

The cleaner technology options that have been applied for reducing water consumption at the noodle factories were indicated as follows.

1. Changing rice washing system from mechanical (using impeller) to pneumatic system using compressed air. The introduction of compressed air helped increasing turbulence during washing while reducing water use. It had total investment of 1,720 US$ and 0.95 year payback period.

2. Installation of a water reuse system for rice washing process. A sand filter unit was used for the treatment and reuse water from washing process. This option had total investment of 2,070 US$ with 1.87 year payback period.

3. Installation of a rice classifier unit to reduce number of washings. Rice classifier helped removing the contaminants contained in raw materials thus reducing number of cycles in primary raw material washing step. This option had total investment of 37,050 US$ with 2.3 year payback period.

During one-year demonstration project when these cleaner technology options has been applied in 10 small- and medium-scale noodle processing factories in Thailand, it was found that the application of one or more of above options together with good housekeeping practices in the noodle processing factories could reduce water consumption by at least 10-30%. Furthermore, the operating cost of wastewater treatment system could be substantially reduced.

4. Application of photosynthetic wastewater treatment system to wastewater from a noodle processing factory

The wastewater produced from noodle production contains starch as its main constituent. Normally, starch particles are separated from wastewater in primary treatment step either by sedimentation or filtration using filter press. The clarified water is then further treated by biological wastewater treatment system such as stabilization pond, aerated lagoon or activated sludge system depending on availability of land. Whenever mechanical aeration is employed, substantial operation cost associated with high energy requirement in the treatment system is usually the main concern for the industries.

In this study, the feasibility of employing photosynthetic wastewater treatment system for the treatment and recovery of carbon source from wastewater under natural condition is investigated. Photosynthetic wastewater treatment system utilizing specific baterial groups has been successfully applied to food industrial wastes containing high BOD (Kobayashi and Tchan, 1973; Sasikala and Ramana, 1995). These bacteria were found to be very effective in the consumption of organic matter and treatment of organic wastewater (Nakadomi et al.,1999) while providing low environmental impact (Nakajima et al, 1997). Under natural condition, the growth of microalgae produce oxygen and it gives advantages to aerobic heterophs which compete with purple nonsulfur bacteria. Therefore, infrared-transmitting filter need to be used in order to promote selective growth of purple non-sulfur bacteria (Imhoff and Trueper, 1991). Ono-Izu (2002) showed that use of the filter is effective for inhibition of growth of microalgae, under constant light conditions illuminated with incandescent lamp. The main task of this study was to investigate the effect of infrared-transmitting filter on growth inhibition of microalgae under sunlight in open-air pilot-scale photobioreactor.

Pilot-scale experiment was conducted in a small factory in Bangkok, Thailand producing noodles with a capacity of 80 rice sacks/day. The factory production includes 2,880 ton/year of thick noodle, 1570 ton/year of thin noodle, 600 ton/year of vemicelli noodle and 370 ton/year of pre-cooked rolled noodle. Wastewater production from the process normally ranged from 60 to100 m^3/d.

Two of PVC tanks was used as photobioreactors (Figure 2). The size of tank is 1.1*1.1*0.9 m. One tank was covered with infrared-transmitting filter (TechnaLight IR, Tsutsunaka Plastic Industry, Japan), and the other was operated without the filter. The reactors were illuminated only with sunlight. Observed light intensity and radiation intensity at noon were approximately 100,000 lux and 790 w/m^2, respectively. Radiation intensity at noon below infrared – transmitting filter was approximately 370 w/m^2. The both tanks were covered with PVC sheet in order to prevent rain water from inflowing into the tanks. Four hundred liter of enriched culture of purple non-sulfur bacteria from existing wastewater treatment system (aerated pond) was filled in the both tanks. Raw wastewater was filtered through a filter press in order to recovery particulate starch. Forty liter of the filtrate was fed into the both reactors to maintain an HRT of about 10 days. The treatment tank was aerated intermittantly by mechanical agitation. The interval of agitation was controlled with timers. The value of pH was manually controlled between 6.5-9.0.

Figure 2. Photograph of pilot scale photobioreactor: (P1,left) with infrared-transmitting filter and (P2, right) without infrared-transmitting filter

Table 3 shows average characteristics of raw and filtered wastewater being fed into the photobioreactor during one month operation. Raw wastewater contained very high suspended solid (mainly strach particles) and organic matter (in terms of TOC and TCOD). Nevertheless, it could be effectively eliminated by an introduction of filter press for wastewater filtering. The wastewater also contain relatively high inorganic sulfur which could also upset the operation of photobioreactor if sulfur reducing bacteria became predominate in the system. Therefore, aeration provided by mechanical agitation was intermittently provided (1 min. on: 6 min. off) to prevent permanent drop of oxidation-reduction potential (ORP) and growth of sulfur reducing bacteria. Figure 3 shows that such operation could control ORP within a range of -200 to 400 mV in the photobioreactor.

Table 3. Wastewater characteristics from noodle processing factory

Wastewater	DOC [mg/l]	TOC [mg/l]	SCOD [mg/l]	TCOD [mg/l]]	SS [mg/l]	NH_3-N [mgN/l]	TKN [mgN/l]	Sulfite [mgS/l]	Sulfate [mgS/l]
Raw	712	5422	1,656	10,665	10,755	7.5	42	306	42
Filtered	452	529	1,308	2,233	208	8.5	32	199	137

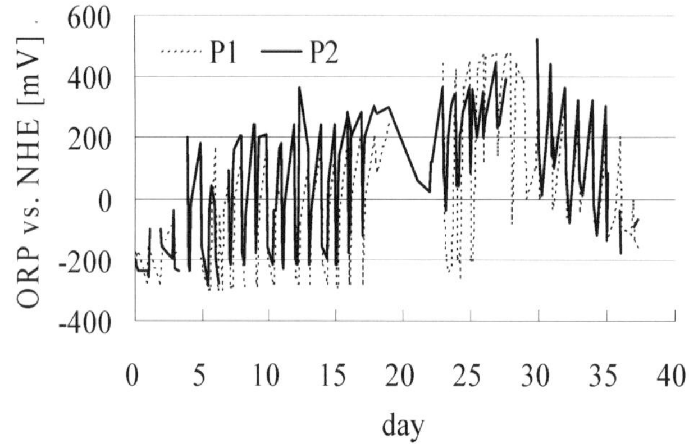

Figure 3. Variation of ORP in photobioreactor

During the operation of both photobioreactor (with and without infrared filter), it was found that organic matter in dissolved form could be effectively reduced. However, there is not much diference in term of treatment performance between both photobioreactor (Figure 3). Average organic removal efficiency of about 95% could be achieved. Table 4 shows the characteristics of water in both photobioreactor. It is also noted that OPR control could effectively prevent the growth sufur reducing bacteria as indicated by the presence of sulfate as the major part of inorganic sulfur.

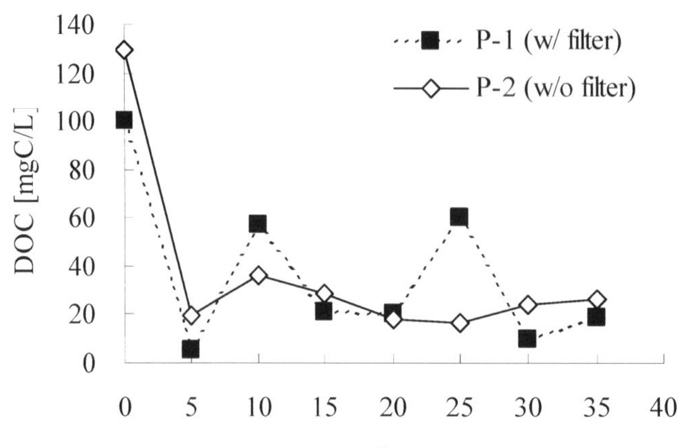

Figure 4. Variation of DOC concentration in photobioreactor

Table 4. Average water characteristics in photobioreactor

Reactor	DOC [mg/l]	TOC [mg/l]	COD [mg/l]	SS [mg/l]	TKN [mg/l]	Crude Protein [mg/l]	Sulfate [mgS/l]	TIS [a] [mgS/l]	Protein content [b]
P-1 (w/ filter)	28	215	620	452	25	176	205	216	0.34
P-2 (w/o filter)	24	188	590	432	33	204	251	263	0.48

a: Total inorganic sulfur
b: Crude protein / SS

Though there was not much difference in term of effluent qualities between both photobioreactor, neither growth of microalgae nor increase of Chl.a were observed in the photobioreactor with infrared filter (Figure 5). On the other hand, it was found that microalgae remarkably grew in photobioreactor without infrared filter starting from day 20-25 (Figure 6). Chlorophyll a (Chl.a) concentration also increased with growth of microalgae. These results shows that infrared-transmitting filter was effective for inhibition of algal growth. These preliminary results suggested that purple non-sulfur bacteria could be sucessfully grown under natural condition with high organic removal efficiency.

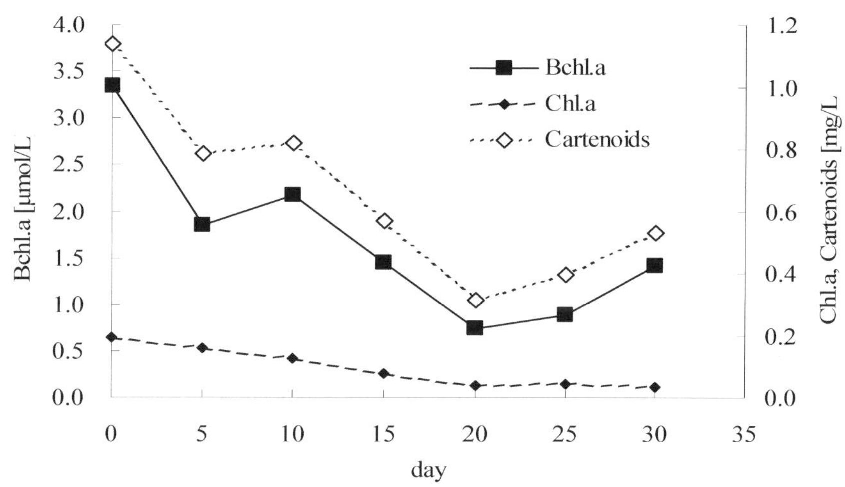

Figure 5. Photosynthetic pigments in photobioreactor with filter

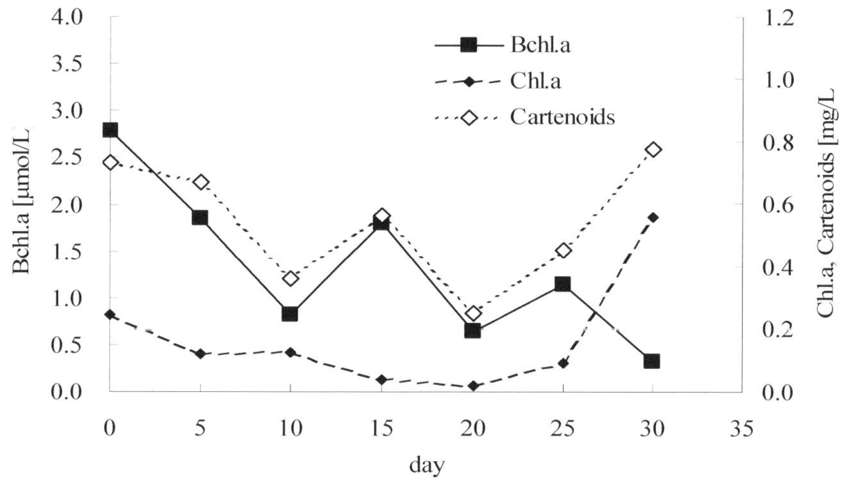

Figure 6. Photosynthetic pigments in photobioreactor w/o filter

5. Conclusions

Successful example on application of cleaner technology and photosynthetic wastewater treatment system in small-scale noodle processing factory is presented. Wastewater discharged from the production process, mainly from rice soaking and rinsing, contained starch as its main constituent and varied in a range between 2.7-5.4 m^3/ton of product. Modification of washing system from mechanical to pneumatic system, installation of water reuse system for rice washing process and rice classifier unit to reduce number of washing cycle were found to be viable options for reducing water consumption. The operation of pilot-scale photosynthetic wastewater treatment system at a noodle factory has shown promising results. In the system, carbon from starch discharged as wastewater could be transformed into purple non-sulfur bacterial cells. The system has shown to be effective for the reduction of organic matter. Infrared-transmitting filter installed at the top of photobioreactor was effective for inhibition of algal growth and promote the growth of purple non-sulfur bacteria under natural operating condition.

6. Acknowledgement

Cleaner technology audit in noodle processing factories was conducted as a part of "Project on Promotion of Cleaner Technology in Thai Industries" with financial support from Danish Cooperation for Environment and Development (DANCED). The research part on application of photosynthetic wastewater treatment system for carbon recovery in noodle processing factory is a collaborative research between Kasetsart University and University of Tokyo with financial support from University of Tokyo.

7. References

DANCED (1999). Rice noodle industry in Thailand. In *Case Study: Cleaner Technology, Project on Promotion of Cleaner Technology in Thai Industry*, Danish Cooperation for Environment and Development, Thailand Environment Institute and The Federation of Thai Industries, pp. 8-9.

Imhoff J. F. and Trueper H. G. 1991, The genus Rhodospirillum and related genera. In: *The Prokaryotes*, Balows A., Truper H.G., Dworkin M., Harder W., Schleifer K-H. (Eds), New York, Springer-Verlag. pp. 2141-2155.

Kobayashi M. and Tchan Y.T. (1973). Treatment of industrial waste solution and production of useful by-products using a photosynthetic bacterial method, *Wat.Res.*, **7**, 1219-1224.

Nakadomi H., Hiromatsu T., Takeno K. Watanabe M. and Sasaki K. (1999), Treatment of aquarium water by denitrifying photosynthetic bacteria using immobilized polyvinyl alcohol beads, *J. Biosci. Bioeng.*, **87**, 625-629.

Nakajima F., Kamiko N. and Yamamoto K. (1997), Organic wastewater treatment without greenhouse gas emission by photosynthetic bacteria, *Wat. Sci. Tech.*, **35**(8), 285-291.

Ono-Izu K. and Yamamoto K. (2002). Promotion of selective growth of purple nonsulfur bacteria on a wastewater treatment and simultaneous material production. J. Env. Systems Eng. JSCE, **699**, VII-22, 23-31 (in Japanese).

Sasikala C. and Ramana C.V. (1995), Biotechnological potentials of anoxygenic phototrophic bacteria. I. Production of single-cell protein, vitamins, ubiquinone, hormones, and enzymes and use in waste treatment, *Advance in Appli. Microbiol.*, **41**, 173-226.

Effect of Water Content and Loading on the Treatment of Waste Metal Working Fluids by Thermophilic Contact Oxidization Process

Xin Feng[1], Futoshi Kurisu[2] and Osami Yagi[3]

1) Environmental Engineering Course, Graduate School of Engineering, University of Tokyo,
7-3-1, Hongo, Bunkyo-ku, Tokyo 113-8656 Japan,
hyousin@env.t.u-tokyo.ac.jp
2) Research Center for Water Environment Technology, Graduate School of Engineering,
University of Tokyo, 7-3-1, Hongo, Bunkyo-ku, Tokyo 113-8656 Japan,
kurisu@env.t.u-tokyo.ac.jp
3) Research Center for Water Environment Technology, Graduate School of Engineering,
University of Tokyo, 7-3-1, Hongo, Bunkyo-ku, Tokyo 113-8656 Japan,
yagi@env.t.u-tokyo.ac.jp

Abstract

Waste metal working fluids (MWFs) contain complex and a large amount of organic chemicals. Some chemicals are resistant to the biodegradation. Conventionally, Waste MWFs is treated by the combination of treatment methods, such as deemulsification, incineration and activated sludge process, which are very complicated and costly. Therefore it is very important to develop effective treatment technologies for conserving the water environment. Thermophilic contact oxidation process (TCOP) is suitable for the treatment of wastewater containing high concentration of organic materials. In the previous research, it is demonstrated that TCOP is effective to treat waste MWFs. Therefore, in this study we examined the effect of water content and loading on the treatment of waste MWFs by semi continuous TCOP. Oil concentration was measured by infrared spectrophotometer and reductions of components were analyzed by GC/MS. The reductions of oil were around 75% for all water content ranging from 30% to 55%, and no significant difference of the reduction was observed among the components. In the case of effect of loading of waste MWFs, the highest loading of 15 ml/day inhibited the reduction. However, when the feeding interval was shortened from 48 hours to 12 hours, more than 60% of reduction ratio was kept. Therefore, the shortening of the feeding interval was effective to improve the loading of waste MWFs.

Keywords: loading; thermophilic contact oxidation process; water content; waste metal working fluids

1. Introduction

Metal working fluids (MWFs) are used as coolants and lubricants mainly in the metal processing industry (Gast *et al.*, 2002). Their compositions are very complicated. There are four main types of MWFs based on the amount of oil, namely straight oils, soluble oils, semisynthetic fluids, and synthetic fluids (Gordon, 2004). The latter three types are typically diluted with water to use. A large amount of waste MWFs are treated conventionally by deemulsification, incineration, coagulation, activated sludge process, or UF membrane process (Baker *et al.,* 1983 and Gast *et al.*, 2002). However, as the waste MWFs is oily wastewater, and some of organic substances in waste MWFs are resistant to biodegradation (Kim *et al.,* 1992, Kim *et al.,* 1994 and Gast *et al.,* 2004), the treatment process is very complicated and costly. In most Japanese factories, waste MWFs are treated by deemulsification first, then the separated oil is incinerated, and the remaining wastewater is treated by activated sludge process. It is necessary to develop simpler and cheaper process not only for developed countries but also for developing countries. Thermophilic contact oxidation process (TCOP) is suitable for the treatment of wastewater containing high concentration of organic materials (Kurisu *et al.,* 1996). Organic substances are degraded aerobically at high temperature by thermophilic bacteria in the reactor with absorbent media (Lim *et al.,* 1997). The degradation rate of organic materials and the indigenous decay of thermophilic bacteria are significantly high, so the process produces little sludge and can achieve high-speed treatment (Lapara and Alleman, 1999). The high-speed aerobic treatment is also supported by water absorbent media, as they hold wastewater and supply oxygen to the wastewater efficiently. Furthermore, there is no effluent discharge from this process, because water becomes to vapor at an aerobic and high temperature condition. As this process is conducted under high temperature above 50 degree C, it is considered to be more feasible in Southeast Asia region than in temperate region. The possibility of TCOP to treat waste MWFs has been proved by preliminary tests. In order to develop the treatment process, we examined the effect of water content and loading on the waste MWFs treatment in this study.

2. Materials and methods

2.1 Waste metal working fluids (Waste MWFs)

Two kinds of waste MWFs (waste MWFs A and waste MWFs B) from same industry were used in the study. The total organic carbon (TOC) value of waste MWF A was 20,000 mg/l, and oil concentration was 9,700 mg/l. pH was 8.0. The TOC value of waste MWF B was 13,000 mg/l, and oil concentration was 5,800 mg/l. pH was 8.0. The compositions of the waste MWFs were analyzed and shown in Table 2.

2.2 Reactors and their operations

We operated separate reactors to investigate the effect of water content (Run 1 – 4) and loading (Run 5 – 9) on the treatment of waste MWFs. The schematic diagram of reactors is shown in Figure 1. Cherry wood chips sized 2 - 6 mm diameter were used as the media. For Run 1 to Run 4, four wide-mouth bottles (500 ml) containing 75 g of wood chips were set in water bath. The temperature was kept at 60 degree C. At the beginning of the operation, 5 g of compost made from leaves, 1 g of yeast extract, and 30 ml of waste MWF A were added into the each four bottles. In order to adjust the water contents to 30%, 40%, 50%, and 55%, 2, 20, 45, and 62 ml of tap water was added respectively into the four bottles. The operation started after mixing. Air was sent into the bottles by air pumps at the flow rate of 100 ml/min. Waste MWF A was added every two days. Before feeding, the bottle was weighted and the weight reduction in two days was considered to be water evaporation. To keep the feeding of the four reactors same, waste MWF A equal to the

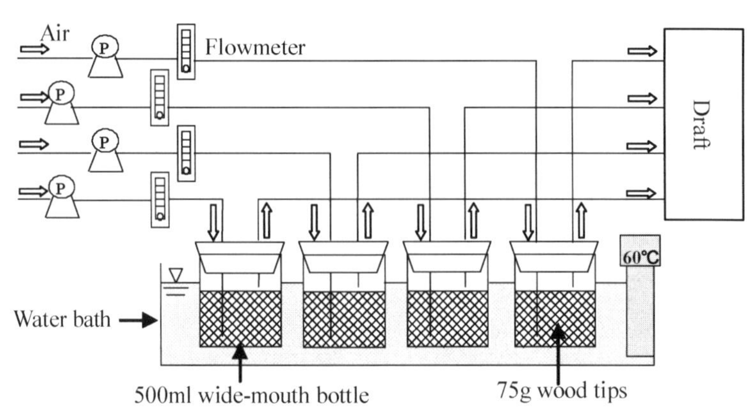

Figure 1. Schematic diagram of reactor

minimum water evaporation among the four reactors was added into the each four bottles. And water was added to the reactors to adjust the water content to the value set initially. The experiment was performed for 28 days. For Run 5 to Run 9, same reactor set-up was used with one additional bottle, five bottles in total. Five gram of compost made from leaves, 2 g of yeast extract, and 50 ml of waste MWF B were added into the each five bottles. The operation started after mixing. The feeding conditions and air flow rates were summarized in Table 1. The loading of waste MWFs B was different by two ways; by different amount of feeding per addition (Run 5 – 7) and by different interval of feeding (Run 7 – 9). The air flow rate was set in accordance with the loading, because the air flow rate was determined to assure the evaporation of water. The feeding per addition was fluctuated to keep the water content same among the reactors. The operations of Run 5, Run 6, and Run 7 were performed for 20 days, and the operations of Run 8 and Run 9 were performed for 16 days.

2.3 Oil concentration analysis

Oil concentration in the wood chips was determined by infrared spectrophotometer (wavelength range from 3.4 μm to 3.5 μm) after extraction. Half gram wood chips were freeze-dried, than extracted by 30 ml of S-316 (chlorotrifluoroethylene dimmer), with shaking at 150 rpm for 20 minutes. The extract was filtered with glass fiber filter (GS 25, 47mm diameter, Toyo Roshi Kaisha Ltd., Japan), and then measured by oil concentration meter (OCMA-350, Horiba, Japan). The oil concentration meter was calibrated by fuel oil. To check the recovery ratio of the oil extraction at different water content, known amount of waste MWF A were added to the media, and water contents were adjusted to 30%, 40%, 50%, and 55% respectively. Media samples were taken from each condition twice, and the oil content was analyzed. The measured value was not influenced by the water content and the recovery ratio of oil was 74% ± 12% (n = 8).

2.4 Composition analysis of waste MWFs by GC/MS

Compositions of waste MWFs were determined by GC/MS. Two milliliter of waste MWFs were evaporated to dryness, and methylated by 2.70 ml benzene:methanol (7:2) solution and 0.30 ml trimethylsilyldiazomethane. The methylated solution was evaporated to dryness again, and resuspended in 1 ml methanol. The methanol solution was analyzed by GC/MS. Composition analysis in reactors was performed as follows. One gram of wood chips was extracted by 30 ml methanol with shaking at 150 rpm for 20 minutes. After centrifugation at 3000 rpm for 5 minutes, 5 ml of supernatant was evaporated to dryness. The methylation was done as the same method of waste MWFs. The methylated solution was also evaporated to dryness again, and resuspended in 1 ml of methanol for

GC/MS analysis. Composition analysis was performed by GC/MS (GC-17A + QP5050,Shimazu, Japan) with a DB-5 capillary column (Polymer structures of 5% phenyl

Table 1. Feeding conditions and air flow rate for Run 5 – 9

	Feeding per addition		Feeding interval	Daily loading		Air flow rate
	(ml)	(mg-oil)	(hours)	(ml)	(mg-oil)	(ml/min)
Run 5	30	174	48	15	87	400
Run 6	15	87	48	7.5	44	200
Run 7	7.5	44	48	3.75	22	100
Run 8	7.5	44	24	7.5	44	200
Run 9	7.5	44	12	15	87	400

2003). The identification of the components was performed by comparison of their mass spectra with the database of CG-MS (Nist 62 lib).To confirm the components, standard methyl silicone, 30 m ×0.25mm, J&W). The temperature program for column oven was 40 degree C for 1 min, and then 7.5 degree C min^{-1} to 325 degree C held for 4 min (Gast *et al.*, samples were also analyzed by GC/MS. The components of waste MWFs were determined by the comparison of the mass spectra of waste MWFs with standard samples, and the concentration of each component was also calculated.

3. Results

3.1 Compositions of waste MWFs

Compositions of waste MWFs used in this study were analyzed by GC/MS. Nine and eleven components were identified in waste MWF A and B respectively (Table 2). The concentration of each component is also shown in Table 2.

3.2 Effect of water content

To investigate the effect of water content on the treatment of waste MWFs, four reactors were operated as Run 1 - 4. Water contents of the four reactors were 30%, 40%, 50%, and 55% respectively. Oil reduction in each Run is shown in Figure 2. The oil reduction ratios were about 50% after 2 days at the water contents of 30% and 40%. It was about 40% after 2 days at the water content of 50%, and was about 25% at the water content of 55%. Therefore the oil reduction decreased as the water content increased in the early period of the experiment. However the oil reduction ratios of all the reactors were around 75% after 28 days despite the difference of water content. The reductions of the nine components in waste MWF A after 2 days and 24 days are shown in Figure 3. The reductions of fatty acids (dodecanedioic acid, palmitic acid, linoleic acid, oleic acid, stearic acid) were distinctly different between the group of Run 1 and 2, and the group of Run 3 and 4. The reductions of the group of Run 3 and 4 were clearly higher than the group of Run 1 and 2 after 2 days. However the reductions became higher in both groups as time passed (data from days 2 to 24 was not shown). After 24 days, the reduction was above 77% for all conditions, and was the highest with Run3, was the lowest with Run 1. The reductions of alkene (1-tetradecene and 1-hexadecene) were about 100% after 24 days when the water contents were 30% - 50%. The reduction of dicyclohexylamine was the lowest in the nine

Table 2. Component concentration in waste MWFs

Group	Components	concentration of each component mg/l	
		Waste MWF A	Waste MWF B
Alkene	1-Tetradecene	72	116
	1-Hexadecene	348	116
Fatty acids	Nonanoic acid	-	150
	Decanoic acid	-	111
	Dodecanedioic acid	393	483
	Palmitic Acid	40	63
	Linoleic acid	110	60
	Oleic acid	237	571
	Stearic acid	72	52
Others	Dicyclohexylamine	300	1177
	Butoxyethoxyethanol	155	-
	p-t-Butylbenzoic acid	-	39

components. It was about 80% after 24 days, and remarkable differences due to the water content were not observed. The reduction of butoxyethoxyethanol was similar to that of fatty acids. The reductions of the group (Run 3 and 4) were clearly higher than the group (Run 1 and 2) after the 2 days. As the experiment period became longer, remarkable difference due to the water content was no longer observed. The reductions were around 95% after 24 days.

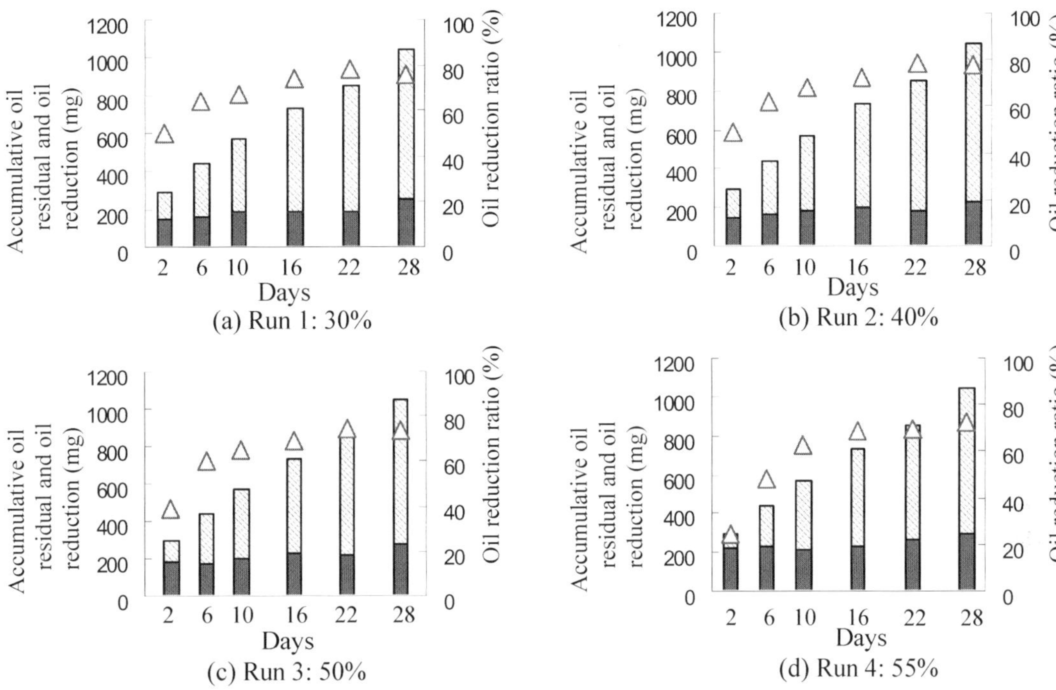

■Oil residual; ▨Oil reduction; △Oil reduction ratio.

Figure 2. Oil reductions at different water contents.

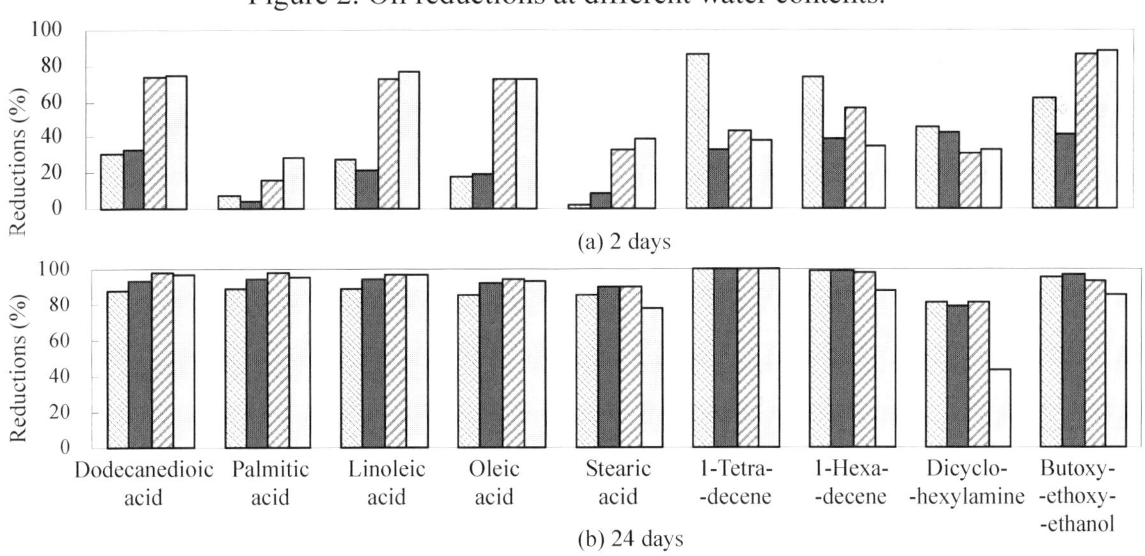

Figure 3. Reductions of individual components at different water contents after (a) 2 days and (b) 24 days. Run 1 (▧), Run 2 (■), Run 3 (▨), and Run 4 (□).

3.3 Effect of loading

Five reactors (Run 5, Run 6, Run 7, Run 8, and Run 9) were operated to examine the effect of loading of waste MWFs. Oil reduction in each reactor is shown in Fig.ure 4. The loading of Run 7 was the lowest in the five reactors. The oil reduction ratio of Run 7 was 57% after 20 days. The loading of Run 6 and Run 8 was the same and in the middle. The oil reduction ratio of Run 6 was 56% after 20 days. The oil reduction ratio of Run 8 was 49% after 16 days. Remarkable difference between Run 6 and Run 8 was not observed. They kept same oil reduction. The loading of Run 5 and Run 9 was the same and the highest in the five reactors, but the feeding interval of Run 9 was 12 hours, one quarter of 48 hours for Run 5. The oil reduction of Run 5 stopped after 10 days, and the oil reduction ratio was 25% after 20 days. In contrast, the oil reduction ratio of Run 9 was 60% after 16 days. Therefore, shortening of feeding interval could increase the loading of waste MWFs on the treatment.

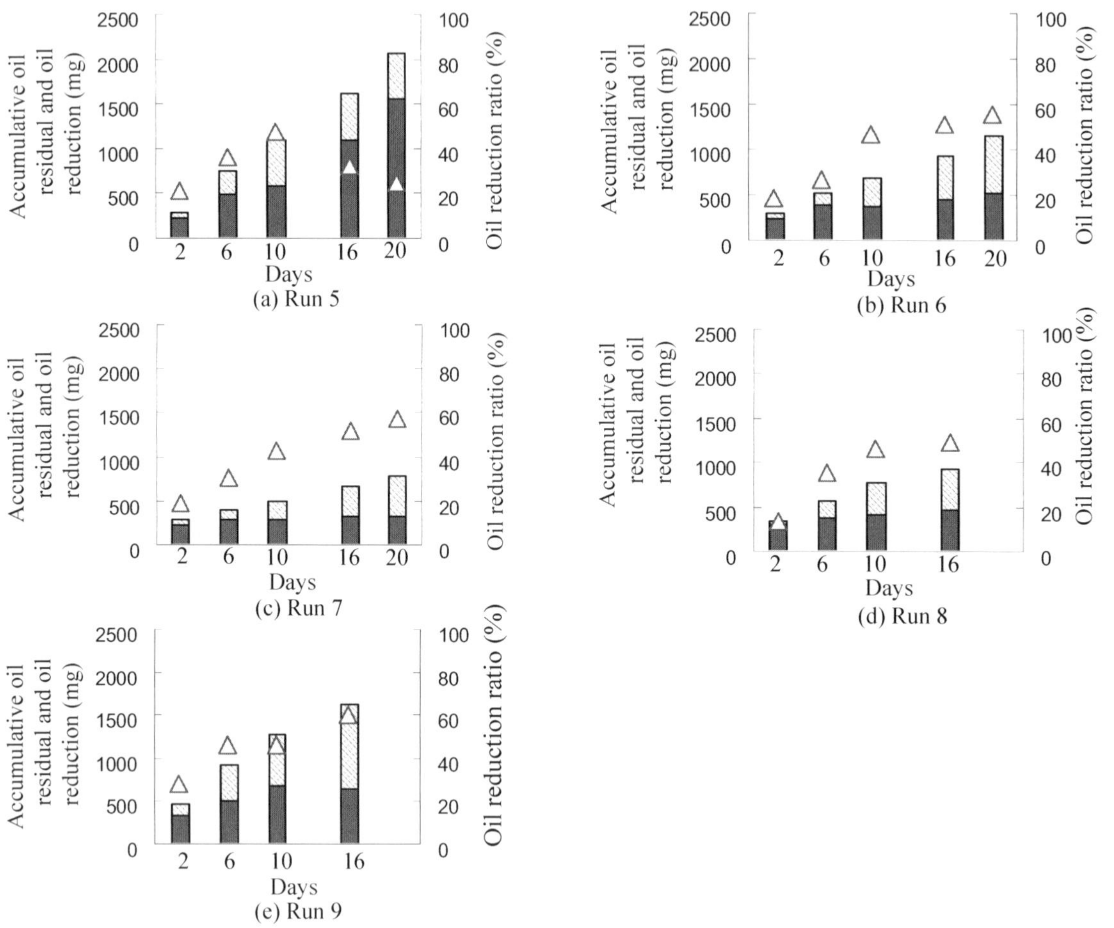

Figure 4. Oil reductions at different loading of waste MWFs. ■: Oil residual; ▧ : Oil reduction; △: Oil reduction rate.

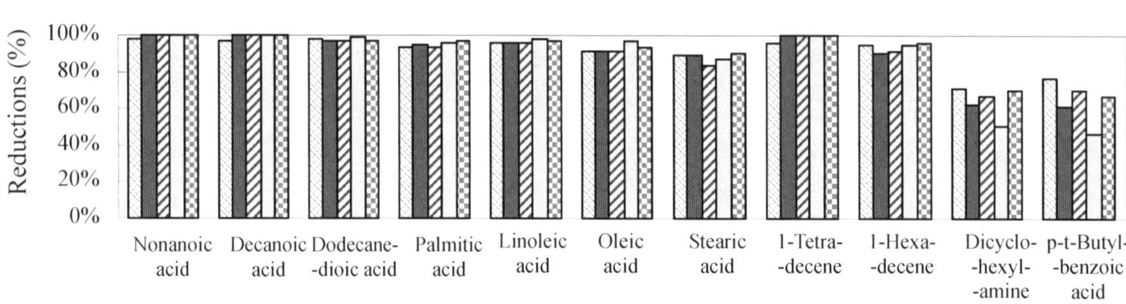

Figure 5. Reductions of individual components after 16 days at different loading of waste MWFs. Run 1 (▧), Run 2 (■), Run 3 (▨), Run 4 (□), Run 5 (▩).

Reductions of the eleven components in waste MWF B after 16 days are shown in Figure 5. Significant differences among the five reactors were not observed. However the reductions of dicyclohexylamine and p-t-butylbenzoic acid were obviously lower than the reductions of fatty acid (nonanoic acid, decanoic acid, dodecanedioic acid, palmitic acid, linoleic acid, oleic acid, and stearic acid) and alkene (1-tetradecene, 1-hexadecene).

4. Discussion

It is generally said that water content influence the reduction of organic loading by TCOP (Bach *et al.,* 1984 and Kurisu *et al.,* 1996). In this study, the effect of water contents on the treatment of waste MWFs by TCOP was examined. High reduction ratio was achieved regardless of water content ranging from 30% to 55% after 10 days in terms of both oil and individual components concentration. Only in the early period, the oil reduction was less under higher water content, and the favorable water content for the degradation was different between the groups of components. The reductions of components belonged to the same functional group were similar each other. These facts could be well explained by the shift of biological degradation activity. At first, the bacteria could exhibit higher activities under lower water content, possibly because the inocula for the reactors were compost, which had been acclimatized under lower water content. Under higher water content, bacterial consortia had higher activity to degrade fatty acids, and in contrast, they had lower activity for alkene degradation. However, the bacterial consortia could be acclimatized to wide range of water contents after 10 days. No needs to control the water content make the operation of TCOP simpler.

There are two ways to increase loading of waste MWFs for TCOP treatment. It can achieve either by increasing the amount of feeding, or by decreasing the interval between feeding. The comparison between Run 5 and Run 9 demonstrate that decreasing the interval of feeding was advantageous over increasing the amount of feeding. It is probably because under the same loading, frequent feeding can avoid large fluctuation of water content, and exceeding amount of inhibitory components.

Dicyclohexylamine was one of the important components in both waste MWF A and B. It is a toxic substance if it is released to the environment. The reduction of dicyclohexylamine was lower than fatty acids and alkene in all the tested runs. it is necessary to improve the treatment of dicyclohexylamine. Although P-t-Butylbenzoic acid only existed in waste MWF B, it was also taken into account as its reduction ratio was less than other components, and is also a toxic substance.

5. Conclusions

In this study we examined the effect of water content and loading on the treatment of waste MWFs by TCOP. We could achieve about 75% reduction of oil after 28 days. The reductions of oil as well as individual components were similar each other under all water contents ranging from 30% to 55%. Shorter interval of feeding was advantageous over more feeding at one time when increasing loading of this treatment.

6. Acknowledgement

The authors would like to acknowledge Mr. Tatsuo Sumino, Hitachi Plant Engineering & Construction Co., Ltd, for his contribution on the investigation of water-based MWF wastewater.

7. References

Bach P.D., Shoda M., and Kubota H. (1984) Composting of dewatered sewage sludge in continuously mixed isothermal reactor, *J. Ferment. Tech.*, 62 (3), 285-292.

Baker C.A., Claus G.W. and Taylor P.A. (1983) Predominant bacteria in an activated-sludge reactor for the degradation of cutting fluids, *Appl. Environ. Microbiol.*, 46(5), 1214-1223.

Gordon T., (2004) Metalworking fluid - The toxicity of a complex mixture, *J. Toxicol. Env. Health*, 67 (3), 209-219.

Kim B.R., Anderson S.G. and Zemla J.F. (1992) Aerobic treatment of metal-cutting-fluid waste-water, *Wat. Envir. Res.*, 64(3), 258-262.

Kim B.R., Rai D.N., Zemla J.F., Lipari F. and Harvath P.V. (1994) Biological removal of organic nitrogen and fatty-acids from metal-cutting-fluid waste-water, *Wat. Res.*, 28(6), 1453-1461.

Kurisu F., Mino T., and Matsuo T. (1996) Application of thermophilic contact oxidization process for treatment of excess sewage sludge, *Proc. of Environ. Eng. Res.*, 33, 39-46.

Kurisu F., Satoh H., and Mino T. (2002) Microbial community analysis of thermophilic contact oxidation process by using ribosomal RNA approaches and the quinone profile method, *Wat. Res.*, 36(2), 429-438.

Lapara T.M. and Alleman J.E. (1999) Thermophilic aerobic biological wastewater treatment, *Wat. Res.*, 33(4), 895-908.

Lim B.R., Huang X., Hu H.Y. and Fjuie K. (1997) Solid phase aerobic digestion of high strength organic wastewater using adsorbent polymer gel, *Wat.Sci.Tech.*, 35(7), 13-20.

Van der Gast C.J., Knowles C.J., Starkey M. and Thompson I.P. (2002) Selection of microbial consortia for treating metal-working fluids, *J. Indust. Microbiol. Biotechnol.*, 29(1), 20-27.

Van der Gast C.J., Whiteley A.S., Lilley A.K., Knowles C.J. and Thompson I.P. (2003) Bacterial community structure and function in a metal-working fluid, *Environ. Microbiol.*, 5(6), 453-461.

Van der Gast C.J., Whiteley A.S., Starkey M., Knowles C.J. and Thompson I.P. (2003) Bioaugmentation strategies for reting mixed chemical effluents, *Biotechnol. Progr.*, 19(4), 1156-1161.

Van der Gast C.J., Whiteley A.S. and Thompson I.P. (2004) Temporal dynamics and degradation activity of an bacterial inoculum for treating waste metal-working fluid, *Environ. Microbiol.*, 6(3), 254-263.

Keyword Index

Keywords are indexed by the first page number of the paper in which they appear.